# SCHAUM'S
## outlines

# Precalculus

## Third Edition

**Fred Safier**

*Professor of Mathematics*
*City College of San Francisco*

**Schaum's Outline Series**

New York   Chicago   San Francisco
Lisbon   London   Madrid   Mexico City
Milan   New Delhi   San Juan
Seoul   Singapore   Sydney   Toronto

The McGraw·Hill Companies

**FRED SAFIER** has an A.B. in physics from Harvard College and an M.S. in mathematics from Stanford University. Now retired, he was an instructor in mathematics at City College of San Francisco from 1967 to 2005 and is the author of numerous students' solution manuals in algebra, trigonometry, and precalculus.

4 5 6 7 8  9 10   QVS/QVS   1 9 8 7 6 5 4

ISBN  978-0-07-179559-3
MHID  0-07-179559-6

e-ISBN  978-0-07-179560-X (basic e-book)
e-MHID  0-07-179560-X

e-ISBN  978-0-07-181389-6 (enhanced e-book)
e-MHID  0-07-181389-1

Library of Congress Control Number: 2012948437

# *Preface to the Second Edition*

This edition has been expanded by material on average rate of change, price/demand, polar form of complex numbers, conic sections in polar coordinates, and the algebra of the dot product. An entire chapter (Chapter 45) is included as an introduction to differential calculus, which now appears in many precalculus texts. More than 30 solved and more than 110 supplementary problems have been added.

Thanks are due to Anya Kozorez and her staff at McGraw-Hill, and to Madhu Bhardwaj and her staff at International Typesetting and Composition. Also, the author would like to thank the users who sent him (mercifully few) corrections, in particular D. Mehaffey and B. DeRoes.

Most of all he owes thanks once again to his wife Gitta, whose careful checking eliminated numerous errors. Any further errors that users spot would be gratefully received at fsafier@ccsf.edu or fsafier@ccsf.cc.ca.us.

Fred Safier

# *Preface to the First Edition*

A course in precalculus is designed to prepare college students for the level of algebraic skills and knowledge that is expected in a calculus class. Such courses, standard at two-year and four-year colleges, review the material of algebra and trigonometry, emphasizing those topics with which familiarity is assumed in calculus. Key unifying concepts are those of functions and their graphs.

The present book is designed as a supplement to college courses in precalculus. The material is divided into forty-four chapters, and covers basic algebraic operations, equations, and inequalities, functions and graphs, and standard elementary functions including polynomial, rational, exponential, and logarithmic functions. Trigonometry is covered in Chapters 20 through 29, and the emphasis is on trigonometric functions as defined in terms of the unit circle. The course concludes with matrices, determinants, systems of equations, analytic geometry of conic sections, and discrete mathematics.

Each chapter starts with a summary of the basic definitions, principles, and theorems, accompanied by elementary examples. The heart of the chapter consists of solved problems, which present the material in logical order and take the student through the development of the subject. The chapter concludes with supplementary problems with answers. These provide drill on the material and develop some ideas further.

The author would like to thank his friends and colleagues, especially F. Cerrato, G. Ling, and J. Morell, for useful discussions. Thanks are also due to the staff of McGraw-Hill and to the reviewer of the text for their invaluable help. Most of all he owes thanks to his wife Gitta, whose careful line-by-line checking of the manuscript eliminated numerous errors. Any errors that remain are entirely his responsibility, and students and teachers who find errors are invited to send him email at fsafier@ccsf.cc.ca.us.

# Contents

# CHAPTER 1

# Preliminaries

## The Sets of Numbers Used in Algebra

The sets of numbers used in algebra are, in general, subsets of **R**, the set of real numbers.

### Natural Numbers *N*

The counting numbers, e.g., 1, 2, 3, 4, . . .

### Integers *Z*

The counting numbers, together with their opposites and 0, e.g., 0, 1, 2, 3, . . . $-1, -2, -3, \ldots$

### Rational Numbers *Q*

The set of all numbers that can be written as quotients $a/b$, $b \neq 0$, *a* and *b* integers, e.g., 3/17, 10/3, $-5.13, \ldots$

### Irrational Numbers *H*

All real numbers that are not rational numbers, e.g., $\pi, \sqrt{2}, \sqrt[3]{5}, -\pi/3, \ldots$

**EXAMPLE 1.1**   The number $-5$ is a member of the sets **Z, Q, R**. The number 156.73 is a member of the sets **Q, R**. The number $5\pi$ is a member of the sets **H, R**.

## Axioms for the Real Number System

There are two fundamental operations, addition and multiplication, that have the following properties (*a, b, c* arbitrary real numbers):

### Closure Laws

The sum $a + b$ and the product $a \cdot b$ or $ab$ are unique real numbers.

### Commutative Laws

$a + b = b + a$: order does not matter in addition.
$ab = ba$: order does not matter in multiplication.

### Associative Laws

$a + (b + c) = (a + b) + c$: grouping does not matter in repeated addition.
$a(bc) = (ab)c$: grouping does not matter in repeated multiplication.
*Note* (removing parentheses): Since $a + (b + c) = (a + b) + c$, $a + b + c$ can be written to mean either quantity
Also, since $a(bc) = (ab)c$, $abc$ can be written to mean either quantity.

### Distributive Laws

$a(b + c) = ab + ac$; also $(a + b)c = ac + bc$: multiplication is distributive over addition.

### Identity Laws

There is a unique number 0 with the property that $0 + a = a + 0 = a$.
There is a unique number 1 with the property that $1 \cdot a = a \cdot 1 = a$.

## Inverse Laws

For any real number $a$, there is a real number $-a$ such that $a + (-a) = (-a) + a = 0$.
For any nonzero real number $a$, there is a real number $a^{-1}$ such that $aa^{-1} = a^{-1}a = 1$.
$-a$ is called the additive inverse, or negative, of $a$.
$a^{-1}$ is called the multiplicative inverse, or reciprocal, of $a$.

**EXAMPLE 1.2**   Associative and commutative laws: Simplify $(3 + x) + 5$.

$$
\begin{aligned}
(3 + x) + 5 &= (x + 3) + 5 &&\text{Commutative law} \\
&= x + (3 + 5) &&\text{Associative law} \\
&= x + 8
\end{aligned}
$$

**EXAMPLE 1.3   FOIL (First Outer Inner Last).** Show that $(a + b)(c + d) = ac + ad + bc + bd$.

$$
\begin{aligned}
(a + b)(c + d) &= a(c + d) + b(c + d) &&\text{by the second form of the distributive law} \\
&= ac + ad + bc + bd &&\text{by the first form of the distributive law}
\end{aligned}
$$

## Zero Factor Laws

1. For every real number $a$, $a \cdot 0 = 0$.
2. If $ab = 0$, then either $a = 0$ or $b = 0$.

## Laws for Negatives

1. $-(-a) = a$
2. $(-a)(-b) = ab$
3. $-ab = (-a)b = a(-b) = -(-a)(-b)$
4. $(-1)a = -a$

## Subtraction and Division

**Definition of Subtraction:** $a - b = a + (-b)$

**Definition of Division:** $\dfrac{a}{b} = a \div b = a \cdot b^{-1}$. Thus, $b^{-1} = 1 \cdot b^{-1} = 1 \div b = \dfrac{1}{b}$.

*Note*: Since 0 has no multiplicative inverse, $a \div 0$ is not defined.

## Laws for Quotients

1. $-\dfrac{a}{b} = \dfrac{-a}{b} = \dfrac{a}{-b} = -\dfrac{-a}{-b}$

2. $\dfrac{-a}{-b} = \dfrac{a}{b}$

3. $\dfrac{a}{b} = \dfrac{c}{d}$ if and only if $ad = bc$.

4. $\dfrac{a}{b} = \dfrac{ka}{kb}$, for $k$ any nonzero real number. (Fundamental principle of fractions)

## Ordering Properties

The positive real numbers, designated by $R^+$, are a subset of the real numbers with the following properties:

1. If $a$ and $b$ are in $R^+$, then so are $a + b$ and $ab$.
2. For every real number $a$, either $a$ is in $R^+$, or $a$ is zero, or $-a$ is in $R^+$.

If $a$ is in $R^+$, $a$ is called positive; if $-a$ is in $R^+$, $a$ is called negative.

The number *a is less than b*, written $a < b$, if $b - a$ is positive. Then *b is greater than a*, written $b > a$. If *a* is either less than or equal to *b*, this is written $a \le b$. Then *b* is greater than or equal to *a*, written $b \ge a$.

**EXAMPLE 1.4** $3 < 5$ because $5 - 3 = 2$ is positive. $-5 < 3$ because $3 - (-5) = 8$ is positive.

The following may be deduced from these definitions:

1. $a > 0$ if and only if *a* is positive.
2. If $a \ne 0$, then $a^2 > 0$.
3. If $a < b$, then $a + c < b + c$.

4. If $a < b$, then $\begin{cases} ac < bc & \text{if } c > 0 \\ ac > bc & \text{if } c < 0 \end{cases}$

5. For any real number *a*, either $a > 0$, or $a = 0$, or $a < 0$.
6. If $a < b$ and $b < c$, then $a < c$.

## The Real Number Line

Real numbers may be represented by points on a line *l* such that to each real number *a* there corresponds exactly one point on *l*, and conversely.

**EXAMPLE 1.5** Indicate the set $\{3, -5, 0, 2/3, \sqrt{5}, -1.5, -\pi\}$ on a real number line.

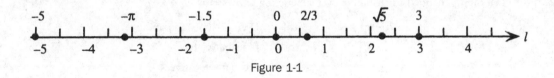

Figure 1-1

## Absolute Value of a Number

The absolute value of a real number *a*, written $|a|$, is defined as follows:

$$|a| = \begin{cases} a & \text{if } a \ge 0 \\ -a & \text{if } a < 0 \end{cases}$$

## Complex Numbers

Not all numbers are real numbers. The set *C* of numbers of the form $a + bi$, where *a* and *b* are real and $i^2 = -1$, is called the complex numbers. Since every real number *x* can be written as $x + 0i$, it follows that every real number is also a complex number.

**EXAMPLE 1.6** $3 + \sqrt{-4} = 3 + 2i$, $-5i$, $2\pi i$, $\dfrac{1}{2} + \dfrac{\sqrt{3}}{2} i$ are examples of nonreal complex numbers.

## Order of Operations

In expressions involving combinations of operations, the following order is observed:

1. Perform operations within grouping symbols first. If grouping symbols are nested inside other grouping symbols, proceed from the innermost outward.
2. Apply exponents before performing multiplications and divisions, unless grouping symbols indicate otherwise.
3. Perform multiplications and divisions, in order from left to right, before performing additions and subtractions (also from left to right), unless operation symbols indicate otherwise.

**EXAMPLE 1.7** Evaluate (a) $-5 - 3^2$, (b) $3 - 4[5 - 6(2 - 8)]$, (c) $[3 - 8 \cdot 5 - (-1 - 2 \cdot 3)] \cdot (3^2 - 5^2)^2$.

(a) $-5 - 3^2 = -5 - 9 = -14$

(b) $3 - 4[5 - 6(2 - 8)] = 3 - 4[5 - 6(-6)]$

$$= 3 - 4[5 + 36]$$

$$= 3 - 4[41] = 3 - 164 = -161$$

(c) $[3 - 8 \cdot 5 - (-1 - 2 \cdot 3)] \cdot (3^2 - 5^2)^2 = [3 - 8 \cdot 5 - (-1 - 6)] \cdot (9 - 25)^2$

$$= [3 - (8 \cdot 5) - (-7)] \cdot (-16)^2$$

$$= [3 - 40 + 7] \cdot 256$$

$$= -30 \cdot 256 = -7,680$$

## SOLVED PROBLEMS

**1.1.** Prove the extended distributive law $a(b + c + d) = ab + ac + ad$.

$$a(b + c + d) = a[(b + c) + d] \qquad \text{Associative law}$$

$$= a(b + c) + ad \qquad \text{Distributive law}$$

$$= ab + ac + ad \qquad \text{Distributive law}$$

**1.2.** Prove that multiplication is distributive over subtraction: $a(b - c) = ab - ac$.

$$a(b - c) = a[b + (-c)] \qquad \text{Definition of subtraction}$$

$$= ab + a(-c) \qquad \text{Distributive law}$$

$$= ab + (-ac) \qquad \text{Laws for negatives}$$

$$= ab - ac \qquad \text{Definition of subtraction}$$

**1.3.** Show that $-(a + b) = -a - b$.

$$-(a + b) = (-1)(a + b) \qquad \text{Laws for negatives}$$

$$= (-1)\,a + (-1)\,b \qquad \text{Distributive law}$$

$$= (-a) + (-b) \qquad \text{Laws for negatives}$$

$$= -a - b \qquad \text{Definition of subtraction}$$

**1.4.** Show that if $\dfrac{a}{b} = \dfrac{c}{d}$, then $ad = bc$.

Assume that $\dfrac{a}{b} = \dfrac{c}{d}$. By the definition of division, $\dfrac{a}{b} = \dfrac{c}{d}$ means $ab^{-1} = cd^{-1}$. Hence,

$$ad = ad \cdot 1 \qquad \text{Identity law}$$

$$= adbb^{-1} \qquad \text{Inverse law}$$

$$= ab^{-1}db \qquad \text{Associative and commutative laws}$$

$$= cd^{-1}db \qquad \text{By hypothesis}$$

$$= c \cdot 1 \cdot b \qquad \text{Inverse law}$$

$$= bc \qquad \text{Identity and commutative laws}$$

**1.5.** Prove that if $a < b$, then $a + c < b + c$.

Assume that $a < b$. Then $b - a$ is positive. But $b - a = b - a + 0 = b - a + c + (-c)$ by the identity and inverse laws. Since $b - a + c + (-c) = b - a + c - c = b + c - (a + c)$ by the definition of subtraction, the associative and commutative laws, and Problem 1.3, it follows that $b + c - (a + c)$ is positive. Hence $a + c < b + c$.

**1.6.** Identify as a member of the sets $N, Z, Q, H, R$, or $C$:

(a)  $-7$            (b)  $0.7$          (c)  $\sqrt{7}$;

(d)  $\dfrac{7}{0}$         (e)  $\sqrt{-7}$

(a)  $-7$ is a negative integer; hence it is also rational, real, and complex. $-7$ is in $Z, Q, R$, and $C$.

(b)  $0.7 = 7/10$; hence it is a rational number, hence real and complex. $0.7$ is in $Q, R$, and $C$.

(c)  $\sqrt{7}$; is an irrational number; hence it is also real and complex. $\sqrt{7}$; is in $H, R$, and $C$.

(d)  $\dfrac{7}{0}$ is not defined. This is not a member of any of these sets.

(e)  $\sqrt{-7}$ is not a real number, but it can be written as $i\sqrt{7}$; hence, it is a complex number. $\sqrt{-7}$ is in $C$.

**1.7.** Identify as true or false:

(a)  $-7 < -8$         (b)  $\pi = 22/7$        (c)  $x^2 \geq 0$ for all real $x$.

(a)  Since $(-8) - (-7) = -1$ is negative, $-8 < -7$, so the statement is false.

(b)  Since $\pi$ is an irrational number and $22/7$ is rational, the statement is false.

(c)  This follows from property 2 for inequalities; the statement is true.

**1.8.** Rewrite the following without using the absolute value symbol, and simplify:

(a)  $|3 - 5|$         (b)  $|3| - |5|$        (c)  $|2 - \pi|$

(d)  $|x - 5|$ if $x > 5$       (e)  $|x + 6|$ if $x < -6$

(a)  $|3 - 5| = |-2| = 2$        (b)  $|3| - |5| = 3 - 5 = -2$

(c)  Since $2 < \pi, 2 - \pi$ is negative. Hence $|2 - \pi| = -(2 - \pi) = \pi - 2$.

(d)  Given that $x > 5$, $x - 5$ is positive. Hence $|x - 5| = x - 5$.

(e)  Given that $x < -6$, $x - (-6) = x + 6$ is negative. Hence $|x + 6| = -(x + 6) = -x - 6$.

## SUPPLEMENTARY PROBLEMS

**1.9.** Identify the law that justifies each of the following statements:

(a)  $(2x + 3) + 5 = 2x + (3 + 5)$       (b)  $2x + (5 + 3x) = 2x + (3x + 5)$

(c)  $x^2(x + y) = x^2 \cdot x + x^2 \cdot y$       (d)  $100[0.01(50 - x)] = [100(0.01)](50 - x)$

(e)  If $a + b = 0$, then $b = -a$.       (f)  If $(x - 5)(x + 3) = 0$, then either $x - 5 = 0$ or $x + 3 = 0$.

*Ans.*  (a) Associative law for addition   (b) Commutative law for addition

       (c) Distributive law              (d) Associative law for multiplication

       (e) Inverse law for addition      (f) Zero factor law

**1.10.** Are the following statements true or false?

(a)  3 is a real number.          (b)  $\pi = 3.14$

(c)  $|x - 5| = x + 5$           (d)  Every rational number is also a complex number.

*Ans.*  (a) true; (b) false; (c) false; (d) true

**1.11.** Place the correct inequality sign between the following:

(a)  $9 \, ? -8$

(b)  $\pi \, ? \, 4$

(c)  $\dfrac{1}{3} \, ? \, 0.33$

(d)  $\dfrac{22}{7} \, ? \, \pi$

(e)  $-1.414 \, ? - \sqrt{2}$

*Ans.*  (a) $>$; (b) $<$; (c) $>$; (d) $>$; (e) $>$

**1.12.** Show that if $ad = bc$, then $\dfrac{a}{b} = \dfrac{c}{d}$. (*Hint:* Assume that $ad = bc$; then start with $ab^{-1}$ and transform it into $cd^{-1}$ in analogy with Problem 1.4.)

**1.13.** Show that $\dfrac{a}{b} = \dfrac{ak}{bk}$ follows from the law that $\dfrac{a}{b} = \dfrac{c}{d}$ if and only if $ad = bc$.

**1.14.** Rewrite the following without using the absolute value symbol, and simplify:

(a)  $|(-5) - [-(-9)]|$

(b)  $-|\sqrt{2} - 1.4|$

(c)  $|6 - x|$, if $x > 6$.

(d)  $-|-4 - x^2|$

*Ans.*  (a) $14$; (b) $1.4 - \sqrt{2}$; (c) $x - 6$; (d) $-4 - x^2$

**1.15.** Evaluate (a) $2 \cdot 3 - 4 \cdot 5^2$    (b) $7 + 3[2(5 - 8) - 4]$    (c) $\{4 \cdot 8 - 6[7 - (5 - 8)^2]\}^2$

*Ans.*  (a) $-94$; (b) $-23$; (c) $1936$

**1.16.** Consider the set $\left\{ -5, -\dfrac{5}{3}, 0, \sqrt{5}, \pi, \dfrac{50}{7}, \sqrt{625} \right\}$

(a)  Which members of this set are members of $N$?

(b)  Which members of this set are members of $Z$?

(c)  Which members of this set are members of $Q$?

(d)  Which members of this set are members of $H$?

*Ans.*  (a) $\sqrt{625}$; (b) $-5, 0, \sqrt{625}$; (c) $-5, -\dfrac{5}{3}, 0, \dfrac{50}{7}, \sqrt{625}$; (d) $\sqrt{5}, \pi$

**1.17.** A set is closed under an operation if the result of applying the operation to any members of the set is also a member of the set. Thus, the integers $Z$ are closed under $+$, while the irrational numbers $H$ are not, since, for example, $\pi + (-\pi) = 0$ which is not irrational. Identify as true or false:

(a)  $Z$ is closed under multiplication.

(b)  $H$ is closed under multiplication.

(c)  $N$ is closed under subtraction.

(d)  $Q$ is closed under addition.

(e)  $Q$ is closed under multiplication.

*Ans.*  (a) true; (b) false; (c) false; (d) true; (e) true

# Polynomials

## Definition of a Polynomial

A polynomial is an expression that can be written as a term or a sum of more than one term of the form $ax_1^{n_1}x_2^{n_2} \ldots x_m^{n_m}$, where the $a$ is a constant and the $x_1, \ldots, x_m$ are variables. A polynomial of one term is called a *monomial*. A polynomial of two terms is called a *binomial*. A polynomial of three terms is called a *trinomial*.

**EXAMPLE 2.1**  $5, -20, \pi, t, 3x^2, -15x^3y^2, \frac{2}{3}xy^4zw$ are monomials.

**EXAMPLE 2.2**  $x + 5, x^2 - y^2, 3x^5y^7 - \sqrt{3}x^3z$ are binomials.

**EXAMPLE 2.3**  $x + y + 4z, 5x^2 - 3x + 1, x^3 - y^3 + t^3, 8xyz - 5x^2y + 20t^3u$ are trinomials.

## The Degree of a Term

The degree of a term in a polynomial is the exponent of the variable, or, if more than one variable is present, the sum of the exponents of the variables. If no variables occur in a term, it is called a *constant term*. The degree of a constant term is 0.

**EXAMPLE 2.4**  (a) $3x^8$ has degree 8; (b) $12xy^2z^2$ has degree 5; (c) $\pi$ has degree 0.

## The Degree of a Polynomial

The degree of a polynomial with more than one term is the largest of the degrees of the individual terms.

**EXAMPLE 2.5**  (a) $x^4 + 3x^2 - 250$ has degree 4; (b) $x^3y^2 - 30x^4$ has degree 5; (c) $16 - x - x^{10}$ has degree 10; (d) $x^3 + 3x^2h + 3xh^2 + h^3$ has degree 3.

## Like and Unlike Terms

Two or more terms are called *like* terms if they are both constants, or if they contain the same variables raised to the same exponents, and differ only, if at all, in their constant coefficients. Terms that are not like terms are called unlike terms.

**EXAMPLE 2.6**  $3x$ and $5x$, $-16x^2y$ and $2x^2y$, $tu^5$ and $6tu^5$ are examples of like terms. 3 and $3x$, $x^2$ and $y^2$, $a^3b^2$ and $a^2b^3$ are examples of unlike terms.

## Addition

The sum of two or more polynomials is found by combining like terms. Order is unimportant, but polynomials in one variable are generally written in order of descending degree in their terms. A polynomial in one variable, $x$, can always be written in the form:

$$a_n x^n + a_{n-1} x^{n-1} + \cdots + a_1 x + a_0$$

This form is generally referred to as *standard* form. The degree of a polynomial written in standard form is immediately seen to be $n$.

**EXAMPLE 2.7**     $5x^3 + 6x^4 - 8x + 2x^2 = 6x^4 + 5x^3 + 2x^2 - 8x \text{ (degree 4)}$

**EXAMPLE 2.8**     $(x^3 - 3x^2 + 8x + 7) + (-5x^3 - 12x + 3) = x^3 - 3x^2 + 8x + 7 - 5x^3 - 12x + 3$
$$= -4x^3 - 3x^2 - 4x + 10$$

## Subtraction

The difference of two polynomials is found using the definition of subtraction: $A - B = A + (-B)$. Note that to subtract $B$ from $A$, write $A - B$.

**EXAMPLE 2.9**     $(y^2 - 5y + 7) - (3y^2 - 5y + 12) = (y^2 - 5y + 7) + (-3y^2 + 5y - 12)$
$$= y^2 - 5y + 7 - 3y^2 + 5y - 12$$
$$= -2y^2 - 5$$

## Multiplication

The product of two polynomials is found using various forms of the distributive property as well as the first law of exponents: $x^a x^b = x^{a+b}$

**EXAMPLE 2.10**     $x^3(3x^4 - 5x^2 + 7x + 2) = x^3 \cdot 3x^4 - x^3 \cdot 5x^2 + x^3 \cdot 7x + x^3 \cdot 2$
$$= 3x^7 - 5x^5 + 7x^4 + 2x^3$$

**EXAMPLE 2.11**   Multiply: $(x + 2y)(x^3 - 3x^2 y + xy^2)$
$$(x + 2y)(x^3 - 3x^2 y + xy^2) = (x + 2y)x^3 - (x + 2y)3x^2 y + (x + 2y)xy^2$$
$$= x^4 + 2x^3 y - 3x^3 y - 6x^2 y^2 + x^2 y^2 + 2xy^3$$
$$= x^4 - x^3 y - 5x^2 y^2 + 2xy^3$$

Often a vertical format is used for this situation:

$$
\begin{array}{r}
x^3 - 3x^2 y + xy^2 \\
x + 2y \\
\hline
x^4 - 3x^3 y + x^2 y^2 \\
2x^3 y - 6x^2 y^2 + 2xy^3 \\
\hline
x^4 - x^3 y - 5x^2 y^2 + 2xy^3
\end{array}
$$

## The FOIL (First Outer Inner Last) Method

The FOIL (First Outer Inner Last) method for multiplying two binomials:

$$(a + b)(c + d) = ac + ad + bc + bd$$

First Outer Inner Last

**EXAMPLE 2.12**     $(2x + 3)(4x + 5) = 8x^2 + 10x + 12x + 15 = 8x^2 + 22x + 15$

## Special Product Forms

| | |
|---|---|
| $(a + b)(a - b) = a^2 - b^2$ | Difference of two squares |
| $(a + b)^2 = (a + b)(a + b) = a^2 + 2ab + b^2$ | Square of a sum |
| $(a - b)^2 = (a - b)(a - b) = a^2 - 2ab + b^2$ | Square of a difference |

$$(a - b)(a^2 + ab + b^2) = a^3 - b^3 \qquad \text{Difference of two cubes}$$
$$(a + b)(a^2 - ab + b^2) = a^3 + b^3 \qquad \text{Sum of two cubes}$$
$$(a + b)^3 = (a + b)(a + b)^2 \qquad \text{Cube of a sum}$$
$$= (a + b)(a^2 + 2ab + b^2) = a^3 + 3a^2b + 3ab^2 + b^3$$
$$(a - b)^3 = (a - b)(a - b)^2 \qquad \text{Cube of a difference}$$
$$= (a - b)(a^2 - 2ab + b^2) = a^3 - 3a^2b + 3ab^2 - b^3$$

## Factoring

Factoring polynomials reverses the distributive operations of multiplication. A polynomial that cannot be factored is called *prime*. Common factoring techniques include: removing a common factor, factoring by grouping, reverse FOIL factoring, and special factoring forms.

**EXAMPLE 2.13**   Removing a monomial common factor: $3x^5 - 24x^4 + 12x^3 = 3x^3(x^2 - 8x + 4)$

**EXAMPLE 2.14**   Removing a nonmonomial common factor:

$$12(x^2 - 1)^4(3x + 1)^3 + 8x(x^2 - 1)^3(3x +1)^4 = 4(x^2 - 1)^3(3x + 1)^3[3(x^2 - 1) + 2x(3x + 1)]$$
$$= 4(x^2 - 1)^3(3x + 1)^3(9x^2 + 2x - 3)$$

It is important to note that the common factor in such problems consists of each base to the lowest exponent present in each term.

**EXAMPLE 2.15**   Factoring by grouping:

$$3x^2 + 4xy - 3xt - 4ty = (3x^2 + 4xy) - (3xt + 4ty) = x(3x + 4y) - t(3x + 4y) = (3x + 4y)(x - t)$$

Reverse FOIL factoring follows the patterns:

$$x^2 + (a + b)x + ab = (x + a)(x + b)$$
$$acx^2 + (bc + ad)xy + bdy^2 = (ax + by)(cx + dy)$$

**EXAMPLE 2.16**   Reverse FOIL factoring:

(a)   To factor $x^2 - 15x + 50$, find two factors of 50 that add to $-15$: $-5$ and $-10$.

$$x^2 - 15x + 50 = (x - 5)(x - 10)$$

(b)   To factor $4x^2 + 11xy + 6y^2$, find two factors of $4 \cdot 6 = 24$ that add to 11:8 and 3.

$$4x^2 + 11xy + 6y^2 = 4x^2 + 8xy + 3xy + 6y^2 = 4x(x + 2y) + 3y(x + 2y) = (x + 2y)(4x + 3y)$$

## Special Factoring Forms

$$a^2 - b^2 = (a + b)(a - b) \qquad \text{Difference of two squares}$$
$$a^2 + b^2 \text{ is prime.} \qquad \text{Sum of two squares}$$
$$a^2 + 2ab + b^2 = (a + b)^2 \qquad \text{Square of a sum}$$
$$a^2 - 2ab + b^2 = (a - b)^2 \qquad \text{Square of a difference}$$
$$a^3 + b^3 = (a + b)(a^2 - ab + b^2) \qquad \text{Sum of two cubes}$$
$$a^3 - b^3 = (a - b)(a^2 + ab + b^2) \qquad \text{Difference of two cubes}$$

## General Factoring Strategy

*Step* 1: Remove all factors common to all terms.
*Step* 2: Note the number of terms.

If the polynomial remaining after step 1 has two terms, look for a difference of two squares, or a sum or difference of two cubes.

If the polynomial remaining after step 1 has three terms, look for a perfect square or try reverse FOIL factoring.

If the polynomial remaining after step 1 has four or more terms, try factoring by grouping.

## SOLVED PROBLEMS

**2.1.** Find the degree of: (a) 12; (b) $35x^3$; (c) $3x^3 - 5x^4 + 3x^2 + 9$; (d) $x^8 - 64$

   (a)   This polynomial has one term and no variables. The degree is 0.

   (b)   This polynomial has one term. The exponent of the variable is 3. The degree is 3.

   (c)   This polynomial has four terms, of degrees 3,4,2,0, respectively. The largest of these is 4, hence the degree of the polynomial is 4.

   (d)   This polynomial has two terms, of degrees 8 and 0, respectively. The largest of these is 8, hence the degree of the polynomial is 8.

**2.2.** Find the degree of (a) $x^2y$    (b) $xy - y^3 + 7$    (c) $x^4 + 4x^3h + 6x^2h^2 + 4xh^3 + h^4$

   (a)   This polynomial has one term. The sum of the exponents of the variables is $2 + 1 = 3$, hence the degree of the polynomial is 3.

   (b)   This polynomial has three terms, of degrees 2,3,0, respectively. The largest of these is 3, hence the degree of the polynomial is 3.

   (c)   This polynomial has five terms, each of degree 4, hence the degree of the polynomial is 4.

**2.3.** If $A = x^2 - 6x + 10$ and $B = 3x^3 - 7x^2 + x + 1$, find (a) $A + B$ (b) $A - B$.

   (a)   $A + B = (x^2 - 6x + 10) + (3x^3 - 7x^2 + x + 1)$
   $$= x^2 - 6x + 10 + 3x^3 - 7x^2 + x + 1$$
   $$= 3x^3 - 6x^2 - 5x + 11$$

   (b)   $A - B = (x^2 - 6x + 10) - (3x^3 - 7x^2 + x + 1)$
   $$= x^2 - 6x + 10 - 3x^3 + 7x^2 - x - 1$$
   $$= -3x^3 + 8x^2 - 7x + 9$$

**2.4.** Add $8x^3 - y^3$ and $x^2 - 5xy^2 + y^3$.

$$(8x^3 - y^3) + (x^2 - 5xy^2 + y^3) = 8x^3 - y^3 + x^2 - 5xy^2 + y^3 = 8x^3 + x^2 - 5xy^2$$

**2.5.** Subtract $8x^3 - y^3$ from $x^2 - 5xy^2 + y^3$.

$$(x^2 - 5xy^2 + y^3) - (8x^3 - y^3) = x^2 - 5xy^2 + y^3 - 8x^3 + y^3 = -8x^3 + x^2 - 5xy^2 + 2y^3$$

**2.6.** Simplify: $3x^2 - 5x - (5x + 8 - (8 - 5x^2 + (3x^2 - x + 1)))$

$3x^2 - 5x - (5x + 8 - (8 - 5x^2 + (3x^2 - x + 1))) = 3x^2 - 5x - (5x + 8 - (8 - 5x^2 + 3x^2 - x + 1))$
$$= 3x^2 - 5x - (5x + 8 - (-2x^2 - x + 9))$$
$$= 3x^2 - 5x - (5x + 8 + 2x^2 + x - 9)$$
$$= 3x^2 - 5x - (2x^2 + 6x - 1)$$
$$= x^2 - 11x + 1$$

**2.7.** Multiply: (a) $12x^2(x^2 - xy + y^2)$; (b) $(a + b)(2a - 3)$; (c) $(3x - 1)(4x^2 - 8x + 3)$

  (a) $12x^2(x^2 - xy + y^2) = 12x^2 \cdot x^2 - 12x^2 \cdot xy + 12x^2 \cdot y^2 = 12x^4 - 12x^3y + 12x^2y^2$

  (b) $(a + b)(2a - 3) = a(2a - 3) + b(2a - 3)$
$$= 2a^2 - 3a + 2ab - 3b$$

  (c) $(3x - 1)(4x^2 - 8x + 3) = (3x - 1)4x^2 - (3x - 1)8x + (3x - 1)3$
$$= 12x^3 - 4x^2 - 24x^2 + 8x + 9x - 3$$
$$= 12x^3 - 28x^2 + 17x - 3$$

**2.8.** Multiply, using the vertical scheme: $(4p - 3q)(2p^3 - p^2q + pq^2 - 2q^3)$

$$
\begin{array}{r}
2p^3 - \phantom{1}p^2q + \phantom{1}pq^2 \phantom{+4p^2q^2} - 2q^3 \\
4p - \phantom{1}3q \\
\hline
8p^4 - \phantom{1}4p^3q + 4p^2q^2 \phantom{+3p^2q^2} - 8pq^3 \\
-6p^3q + 3p^2q^2 - 3pq^3 + 6q^4 \\
\hline
8p^4 - 10p^3q + 7p^2q^2 - 11pq^3 + 6q^4
\end{array}
$$

**2.9.** Multiply:

  (a) $(cx - d)(cx + d)$; (b) $(3x - 5)^2$; (c) $(2t - 5)(4t^2 + 10t + 25)$;

  (d) $4(-2x)(1 - x^2)^3$; (e) $[(r - s) + t][(r - s) - t]$

  (a) $(cx - d)(cx + d) = (cx)^2 - d^2 = c^2x^2 - d^2$

  (b) $(3x - 5)^2 = (3x)^2 - 2(3x) \cdot 5 + 5^2 = 9x^2 - 30x + 25$

  (c) $(2t - 5)(4t^2 + 10t + 25) = (2t)^3 - 5^3 = 8t^3 - 125$    using the difference of two cubes pattern.

  (d) $4(-2x)(1 - x^2)^3 = -8x(1 - x^2)^3$
$$= -8x(1 - 3x^2 + 3x^4 - x^6) \quad \text{using the cube of a difference pattern.}$$
$$= -8x + 24x^3 - 24x^5 + 8x^7$$

  (e) $[(r - s) + t][(r - s) - t] = (r - s)^2 - t^2 = r^2 - 2rs + s^2 - t^2$    using the difference of two squares pattern, followed by the square of a difference pattern.

**2.10.** Perform indicated operations: (a) $(x + h)^3 - (x - h)^3$; (b) $(1 + t)^4$.

  (a) $(x + h)^3 - (x - h)^3 = (x^3 + 3x^2h + 3xh^2 + h^3) - (x^3 - 3x^2h + 3xh^2 - h^3)$
$$= x^3 + 3x^2h + 3xh^2 + h^3 - x^3 + 3x^2h - 3xh^2 + h^3$$
$$= 6x^2h + 2h^3$$

  (b) $(1 + t)^4 = ((1 + t)^2)^2 = (1 + 2t + t^2)^2 = (1 + 2t)^2 + 2(1 + 2t)t^2 + t^4$
$$= 1 + 4t + 4t^2 + 2t^2 + 4t^3 + t^4 = 1 + 4t + 6t^2 + 4t^3 + t^4$$

**2.11.** Factor: (a) $15x^4 - 10x^3 + 25x^2$; (b) $x^2 + 12x + 20$; (c) $9x^2 - 25y^2$;

  (d) $6x^5 - 48x^4 - 54x^3$; (e) $5x^2 + 13xy + 6y^2$; (f) $P(1 + r) + P(1 + r)r$; (g) $x^3 - 64$;

  (h) $3(x + 3)^2(x - 8)^4 + 4(x + 3)^3(x - 8)^3$; (i) $x^4 - y^4 + x^3 - xy^2$; (j) $x^6 - 64y^6$

(a) $15x^4 - 10x^3 + 25x^2 = 5x^2(3x^2 - 2x + 5)$. After removing the common factor, the remaining polynomial is prime.

(b) $x^2 + 12x + 20 = (x + 10)(x + 2)$ using reverse FOIL factoring.

(c) $9x^2 - 25y^2 = (3x)^2 - (5y)^2 = (3x - 5y)(3x + 5y)$ using the difference of two squares pattern.

(d) $6x^5 - 48x^4 - 54x^3 = 6x^3 (x^2 - 8x - 9) = 6x^3 (x - 9)(x + 1)$ removing the common factor, then using reverse FOIL factoring.

(e) $5x^2 + 13xy + 6y^2 = (5x + 3y)(x + 2y)$ using reverse FOIL factoring.

(f) $P(1 + r) + P(1 + r)r = P(1 + r)(1 + r) = P(1 + r)^2$. Here, the common factor $P(1 + r)$ was removed from both terms.

(g) $x^3 - 64 = (x - 4)(x^2 + 4x + 16)$ using the difference of two cubes pattern.

(h) Removing the common factor from both terms and combining terms in the remaining factor yields:
$$3(x + 3)^2 (x - 8)^4 + 4(x + 3)^3 (x - 8)^3 = (x + 3)^2 (x - 8)^3 [3(x - 8) + 4(x + 3)]$$
$$= (x + 3)^2 (x - 8)^3 (7x - 12)$$

(i) $x^4 - y^4 + x^3 - xy^2 = (x^4 - y^4) + (x^3 - xy^2)$
$$= (x^2 - y^2)(x^2 + y^2) + x(x^2 - y^2)$$
$$= (x^2 - y^2)(x^2 + y^2 + x)$$
$$= (x - y)(x + y)(x^2 + y^2 + x)$$

(j) $x^6 - 64y^6 = (x^3 - 8y^3)(x^3 + 8y^3) = (x - 2y)(x^2 + 2xy + 4y^2)(x + 2y)(x^2 - 2xy + 4y^2)$

**2.12.** A special factoring technique that is occasionally of use involves adding a term to make a polynomial into a perfect square, then subtracting that term immediately. If the added term is itself a perfect square, then the original polynomial can be factored as the difference of two squares. Illustrate this technique for (a) $x^4 + 4y^4$; (b) $x^4 + 2x^2y^2 + 9y^4$.

(a) Since $x^4 + 4y^4 = (x^2)^2 + (2y^2)^2$, adding $2x^2(2y^2) = 4x^2y^2$ makes the polynomial into a perfect square. Then subtracting this quantity yields a difference of two squares, which can be factored:
$$x^4 + 4y^4 = x^4 + 4x^2y^2 + 4y^4 - 4x^2y^2$$
$$= (x^2 + 2y^2)^2 - (2xy)^2$$
$$= (x^2 + 2y^2 - 2xy)(x^2 + 2y^2 + 2xy)$$

(b) If the middle term of this polynomial were $6x^2y^2$ instead of $2x^2y^2$, the polynomial would be a perfect square. Therefore, adding and subtracting $4x^2y^2$ yields a difference of two squares, which can be factored:
$$x^4 + 2x^2y^2 + 9y^4 = x^4 + 6x^2y^2 + 9y^4 - 4x^2y^2$$
$$= (x^2 + 3y^2)^2 - (2xy)^2$$
$$= (x^2 + 3y^2 - 2xy)(x^2 + 3y^2 + 2xy)$$

## SUPPLEMENTARY PROBLEMS

**2.13.** Find the degree of (a) 8; (b) $8x^7$; (c) $5x^2 - 5x + 5$; (d) $5\pi^2 - 5\pi + 5$; (e) $x^2 + 2xy + y^2 - 6x + 8y + 25$

*Ans.* (a) 0;   (b) 7;   (c) 2;   (d) 0;   (e) 2

**2.14.** Let $P$ be a polynomial of degree $m$ and $Q$ be a polynomial of degree $n$. Show that (a) $PQ$ is a polynomial of degree $m + n$; (b) the degree of $P + Q$ is less than or equal to the larger of $m$, $n$.

**2.15.** Let $A = x^2 - xy + 2y^2$, $B = x^3 - y^3$, $C = 2x^2 - 5x + 4$, $D = 3x^2 - 2y^2$. Find

(a) $A + D$; (b) $BD$; (c) $B - Cx$; (d) $x^2A^2 - B^2$; (e) $AD - B^2$

*Ans.* (a) $4x^2 - xy$; (b) $3x^5 - 2x^3y^2 - 3x^2y^3 + 2y^5$; (c) $-x^3 - y^3 + 5x^2 - 4x$;

(d) $-2x^5y + 5x^4y^2 - 2x^3y^3 + 4x^2y^4 - y^6$;

(e) $3x^4 - 3x^3y + 4x^2y^2 + 2xy^3 - 4y^4 - x^6 + 2x^3y^3 - y^6$

**2.16.** Using the definitions of the previous problem, subtract $C$ from the sum of $A$ and $D$.

*Ans.* $2x^2 - xy + 5x - 4$

**2.17.** Perform indicated operations: (a) $-(x - 5)^2$; (b) $2x - (x - 3)^2$; (c) $5a(2a - 1)^2 - 3(a - 2)^3$;

(d) $-(4x + 1)^3 - 2(4x + 1)^2$

*Ans.* (a) $-x^2 + 10x - 25$; (b) $-x^2 + 8x - 9$; (c) $17a^3 - 2a^2 - 31a + 24$;

(d) $-64x^3 - 80x^2 - 28x - 3$

**2.18.** Perform indicated operations: (a) $-3(x - 2)^2$; (b) $-3 - 4(x + 4)^2$; (c) $4(x + 3)^2 - 3(x - 2)^2$;

(d) $(x + 3)(x + 4) - (x + 5)^2$; (e) $-(x + 2)^3 - (x + 2)^2 - 5(x + 2) + 10$

*Ans.* (a) $-3x^2 + 12x - 12$; (b) $-4x^2 - 32x - 67$; (c) $x^2 + 36x + 24$;

(d) $-3x - 13$; (e) $-x^3 - 7x^2 - 21x - 12$

**2.19.** Perform indicated operations: (a) $(x - h)^2 + (y - k)^2$; (b) $(x + h)^4 - x^4$;

(c) $R^2 - (R - x)^2$; (d) $(ax + by + c)^2$

*Ans.* (a) $x^2 - 2xh + h^2 + y^2 - 2yk + k^2$; (b) $4x^3h + 6x^2h^2 + 4xh^3 + h^4$;

(c) $2Rx - x^2$; (d) $a^2x^2 + b^2y^2 + c^2 + 2abxy + 2acx + 2bcy$

**2.20.** Factor: (a) $x^2 - 12x + 27$; (b) $x^2 + 10x + 25$; (c) $x^4 - 6x^2 + 9$; (d) $x^3 - 64$;

(e) $3x^2 - 7x - 10$; (f) $3x^3 + 15x^2 - 18x$; (g) $x^5 + x^2$; (h) $4x^4 - x^2 - 18$; (i) $x^4 - 11x^2y^2 + y^4$

*Ans.* (a) $(x - 3)(x - 9)$; (b) $(x + 5)^2$; (c) $(x^2 - 3)^2$; (d) $(x - 4)(x^2 + 4x + 16)$;

(e) $(3x - 10)(x + 1)$; (f) $3x(x + 6)(x - 1)$; (g) $x^2(x + 1)(x^2 - x + 1)$;

(h) $(x^2 + 2)(2x - 3)(2x + 3)$; (i) $(x^2 - 3xy - y^2)(x^2 + 3xy - y^2)$

**2.21.** Factor: (a) $t^2 + 6t - 27$; (b) $4x^3 - 20x^2 - 24x$; (c) $3x^2 - x - 14$; (d) $5x^2 - 3x - 14$; (e) $4x^6 - 37x^3 + 9$;

(f) $(x - 2)^3 - (x - 2)^2$; (g) $x^2 - 6x + 9 - y^2 - 2yz - z^2$; (h) $16x^4 - x^2y^2 + y^4$

*Ans.* (a) $(t + 9)(t - 3)$; (b) $4x(x + 1)(x - 6)$; (c) $(3x - 7)(x + 2)$; (d) $(5x + 7)(x - 2)$;

(e) $(4x^3 - 1)(x^3 - 9)$; (f) $(x - 2)^2(x - 3)$; (g) $(x - 3 - y - z)(x - 3 + y + z)$;

(h) $(4x^2 + y^2 - 3xy)(4x^2 + y^2 + 3xy)$

**2.22.** Factor: (a) $x^2 - 6xy + 9y^2$; (b) $x^4 - 5x^2 + 4$; (c) $x^4 - 3x^2 - 4$; (d) $x^3 + y^3 + x^2 - y^2$;

(e) $P + Pr + (P + Pr)r + [P + Pr + (P + Pr)r]r$; (f) $a^6x^6 - 64y^6$; (g) $a^6x^6 + 64y^6$

*Ans.*    (a)   $(x - 3y)^2$; (b)  $(x - 1)(x + 1)(x - 2)(x + 2)$; (c)  $(x - 2)(x + 2)(x^2 + 1)$;

(d)  $(x + y)(x^2 - xy + y^2 + x - y)$; (e)  $P(1 + r)^3$;

(f)  $(ax - 2y)(ax + 2y)(a^2x^2 + 2axy + 4y^2)(a^2x^2 - 2axy + 4y^2)$;

(g)  $(a^2x^2 + 4y^2)(a^4x^4 - 4a^2x^2y^2 + 16y^4)$

**2.23.**  Factor:  (a)  $x^5(x + 2)^3 + x^4(x + 2)^4$; (b)  $5x^4(3x - 5)^4 + 12x^5(3x - 5)^3$;

(c)  $2(x + 3)(x + 5)^4 + 4(x + 3)^2(x + 5)^3$; (d)  $3(5x + 2)^2(5)(3x - 4)^4 + (5x + 2)^3(4)(3x - 4)^3(3)$;

(e)  $5(x^2 + 4)^4(8x - 1)^2(2x) + 2(x^2 + 4)^5(8x - 1)(8)$

*Ans.*    (a)   $x^4(x + 2)^3(2x + 2)$; (b)  $x^4(3x - 5)^3(27x - 25)$; (c)  $2(x + 3)(x + 5)^3(3x + 11)$;

(d)  $3(5x + 2)^2(3x - 4)^3(35x - 12)$; (e)  $2(x^2 + 4)^4(8x - 1)(48x^2 - 5x + 32)$

# CHAPTER 3

# *Exponents*

## Natural Number Exponents

Natural number exponents are defined by:

$$x^n = xx \cdots x \qquad (n \text{ factors of } x)$$

**EXAMPLE 3.1**  (a) $x^5 = xxxxx$; (b) $5x^4yz^3 = 5xxxxyzzz$; (c) $5a^3b + 3(2ab)^3 = 5aaab + 3(2ab)(2ab)(2ab)$

## Zero as an Exponent

$x^0 = 1$ for $x$ any nonzero real number. $0^0$ is not defined.

## Negative Integer Exponents

Negative integer exponents are defined by:

$$x^{-n} = \frac{1}{x^n} \text{ for } x \text{ any nonzero real number.}$$

$0^{-n}$ is not defined for $n$ any positive integer.

**EXAMPLE 3.2**  (a) $x^{-5} = \frac{1}{x^5}$; (b) $4y^{-3} = 4 \cdot \frac{1}{y^3} = \frac{4}{y^3}$; (c) $5^{-3} = \frac{1}{5^3} = \frac{1}{125}$; (d) $-4^{-2} = -\frac{1}{4^2} = -\frac{1}{16}$;

(e) $3x^{-2}y^4 + 2(3x)^{-4}y^{-5}z^2 = 3 \cdot \frac{1}{x^2}y^4 + 2 \cdot \frac{1}{(3x)^4} \cdot \frac{1}{y^5}z^2 = \frac{3y^4}{x^2} + \frac{2z^2}{(3x)^4y^5}$

## Rational Number Exponents

$x^{1/n}$, the principal $n$th root of $x$, is defined, for $n$ an integer greater than 1, by:

If $n$ is odd, $x^{1/n}$ is the unique real number $y$ which, when raised to the $n$th power, gives $x$. If $n$ is even, then,

> if $x > 0$, $x^{1/n}$ is the positive real number $y$ which, when raised to the $n$th power, gives $x$.
> if $x = 0$, $x^{1/n} = 0$.
> if $x < 0$, $x^{1/n}$ is not a real number.

*Note:* The principal $n$th root of a positive number is positive.

**EXAMPLE 3.3**  (a) $8^{1/3} = 2$; (b) $(-8)^{1/3} = -2$; (c) $-8^{1/3} = -2$; (d) $16^{1/4} = 2$;
(e) $(-16)^{1/4}$ is not a real number; (f) $-16^{1/4} = -2$

$x^{m/n}$ is defined by: $x^{m/n} = (x^{1/n})^m$, provided $x^{1/n}$ is real.

$$x^{-m/n} = \frac{1}{x^{m/n}}$$

**EXAMPLE 3.4**  (a) $125^{2/3} = (125^{1/3})^2 = 5^2 = 25$; (b) $8^{-4/3} = \frac{1}{8^{4/3}} = \frac{1}{(8^{1/3})^4} = \frac{1}{2^4} = \frac{1}{16}$;

(c) $(-64)^{5/6}$ is not a real number.

## Laws of Exponents

For $a$ and $b$ rational numbers and $x$ and $y$ real numbers (avoiding even roots of negative numbers and division by 0):

$$x^a x^b = x^{a+b} \qquad (xy)^a = x^a y^a \qquad (x^a)^b = x^{ab}$$

$$\frac{x^a}{x^b} = x^{a-b} \qquad \frac{x^a}{x^b} = \frac{1}{x^{b-a}} \qquad \left(\frac{x}{y}\right)^a = \frac{x^a}{y^a}$$

$$\left(\frac{x}{y}\right)^{-m} = \left(\frac{y}{x}\right)^m \qquad \frac{x^{-n}}{y^{-m}} = \frac{y^m}{x^n}$$

In general, $x^{m/n} = (x^{1/n})^m = (x^m)^{1/n}$, provided $x^{1/n}$ is real.

Unless otherwise specified, it is generally assumed that variable bases represent positive numbers. With this assumption, then, write $(x^n)^{1/n} = x$. However, if this assumption does not hold, then:

$$(x^n)^{1/n} = x \qquad \text{if } n \text{ is odd, or if } n \text{ is even and } x \text{ is nonnegative}$$
$$(x^n)^{1/n} = |x| \qquad \text{if } n \text{ is even and } x \text{ is negative}$$

**EXAMPLE 3.5** If $x$ is known positive: (a) $(x^2)^{1/2} = x$; (b) $(x^3)^{1/3} = x$; (c) $(x^4)^{1/2} = x^2$; (d) $(x^6)^{1/2} = x^3$

**EXAMPLE 3.6** For general $x$: (a) $(x^2)^{1/2} = |x|$; (b) $(x^3)^{1/3} = x$; (c) $(x^4)^{1/2} = |x^2| = x^2$; (d) $(x^6)^{1/2} = |x^3|$

## Scientific Notation

In dealing with very large or very small numbers, scientific notation is often used. A number is written in scientific notation when it is expressed as a number between 1 and 10 multiplied by a power of 10.

**EXAMPLE 3.7** (a) $51,000,000 = 5.1 \times 10^7$; (b) $0.000\,000\,000\,035\,2 = 3.52 \times 10^{-11}$;

(c) $\dfrac{(50,000,000)(0.000\,000\,000\,6)}{(20,000)^3} = \dfrac{(5 \times 10^7)(6 \times 10^{-10})}{(2 \times 10^4)^3} = \dfrac{30 \times 10^{-3}}{8 \times 10^{12}} = 3.75 \times 10^{-15}$

## SOLVED PROBLEMS

In the following, bases are assumed to be positive unless otherwise specified:

**3.1.** Simplify (a) $2(3x^2 y)^3 (x^4 y^3)^2$; (b) $\dfrac{(4x^5 y^3)^2}{2(xy^4)^3}$

(a) $2(3x^2 y)^3 (x^4 y^3)^2 = 2 \cdot 3^3 x^6 y^3 \cdot x^8 y^6 = 54 x^{14} y^9$; (b) $\dfrac{(4x^5 y^3)^2}{2(xy^4)^3} = \dfrac{16 x^{10} y^6}{2x^3 y^{12}} = \dfrac{8x^7}{y^6}$

**3.2.** Simplify and write with positive exponents: (a) $\dfrac{x^2 y^{-3}}{x^3 y^3}$; (b) $\dfrac{(x^2 y^{-3})^{-2}}{(x^3 y^4)^{-4}}$; (c) $(x^2 + y^2)^{-2}$;

(d) $(3x^{-5})^{-2}(5y^{-4})^3$; (e) $(x^{-2} + y^{-2})^2$; (f) $\left(\dfrac{t^3 u^4}{4t^5 u^3}\right)^{-3}$

(a) $\dfrac{x^2 y^{-3}}{x^3 y^3} = x^{2-3} y^{-3-3} = x^{-1} y^{-6} = \dfrac{1}{xy^6}$; (b) $\dfrac{(x^2 y^{-3})^{-2}}{(x^3 y^4)^{-4}} = \dfrac{x^{-4} y^6}{x^{-12} y^{-16}} = x^{-4-(-12)} y^{6-(-16)} = x^8 y^{22}$;

(c) $(x^2 + y^2)^{-2} = \dfrac{1}{(x^2 + y^2)^2} = \dfrac{1}{x^4 + 2x^2 y^2 + y^4}$; (d) $(3x^{-5})^{-2}(5y^{-4})^3 = 3^{-2} x^{10} 5^3 y^{-12} = \dfrac{125 x^{10}}{9y^{12}}$;

(e) $(x^{-2} + y^{-2})^2 = x^{-4} + 2x^{-2} y^{-2} + y^{-4} = \dfrac{1}{x^4} + \dfrac{2}{x^2 y^2} + \dfrac{1}{y^4}$; (f) $\left(\dfrac{t^3 u^4}{4t^5 u^3}\right)^{-3} = \left(\dfrac{4t^5 u^3}{t^3 u^4}\right)^3 = \left(\dfrac{4t^2}{u}\right)^3 = \dfrac{64 t^6}{u^3}$

**3.3.** Simplify: (a) $x^{1/2}x^{1/3}$; (b) $x^{2/3}/x^{5/8}$; (c) $(x^4y^4)^{-1/2}$; (d) $(x^4 + y^4)^{-1/2}$

(a) $x^{1/2}x^{1/3} = x^{1/2 + 1/3} = x^{5/6}$; (b) $x^{2/3}/x^{5/8} = x^{2/3 - 5/8} = x^{1/24}$; (c) $(x^4y^4)^{-1/2} = x^{-2}y^{-2} = \dfrac{1}{x^2y^2}$;

(d) $(x^4 + y^4)^{-1/2} = \dfrac{1}{(x^4 + y^4)^{1/2}}$

**3.4.** Simplify: (a) $3x^{2/3}y^{3/4}(2x^{5/3}y^{1/2})^3$; (b) $\dfrac{(8x^2y^{2/3})^{2/3}}{2(x^{3/4}y)^3}$

(a) $3x^{2/3}y^{3/4}(2x^{5/3}y^{1/2})^3 = 3x^{2/3}y^{3/4} \cdot 8x^5y^{3/2} = 24x^{17/3}y^{9/4}$;

(b) $\dfrac{(8x^2y^{2/3})^{2/3}}{2(x^{3/4}y)^3} = \dfrac{8^{2/3}x^{4/3}y^{4/9}}{2x^{9/4}y^3} = \dfrac{4x^{4/3}y^{4/9}}{2x^{9/4}y^3} = \dfrac{2}{x^{9/4 - 4/3}y^{3 - 4/9}} = \dfrac{2}{x^{11/12}y^{23/9}}$

**3.5.** Simplify: (a) $x^{2/3}(x^2 + x + 3)$; (b) $(x^{1/2} + y^{1/2})^2$; (c) $(x^{1/3} - y^{1/3})^2$; (d) $(x^2 + y^2)^{1/2}$

(a) $x^{2/3}(x^2 + x + 3) = x^{2/3}x^2 + x^{2/3}x + 3x^{2/3} = x^{8/3} + x^{5/3} + 3x^{2/3}$

(b) $(x^{1/2} + y^{1/2})^2 = (x^{1/2})^2 + 2x^{1/2}y^{1/2} + (y^{1/2})^2 = x + 2x^{1/2}y^{1/2} + y$

(c) $(x^{1/3} - y^{1/3})^2 = (x^{1/3})^2 - 2x^{1/3}y^{1/3} + (y^{1/3})^2 = x^{2/3} - 2x^{1/3}y^{1/3} + y^{2/3}$

(d) This expression cannot be simplified.

**3.6.** Factor: (a) $x^{-4} + 3x^{-2} + 2$; (b) $x^{2/3} + x^{1/3} - 6$; (c) $x^{11/3} + 7x^{8/3} + 12x^{5/3}$

(a) $x^{-4} + 3x^{-2} + 2 = (x^{-2} + 1)(x^{-2} + 2)$ using reverse FOIL factoring.

(b) $x^{2/3} + x^{1/3} - 6 = (x^{1/3} + 3)(x^{1/3} - 2)$ using reverse FOIL factoring.

(c) $x^{11/3} + 7x^{8/3} + 12x^{5/3} = x^{5/3}(x^2 + 7x + 12) = x^{5/3}(x + 3)(x + 4)$ removing the monomial common factor, then using reverse FOIL factoring.

**3.7.** Remove common factors: (a) $(x + 2)^{-2} + (x + 2)^{-3}$; (b) $6x^5y^{-3} - 3y^{-4}x^6$;

(c) $4(3x + 2)^3 3(x + 5)^{-3} - 3(x + 5)^{-4}(3x + 2)^4$; (d) $5x^3(3x + 1)^{2/3} + 3x^2(3x + 1)^{5/3}$

The common factor in such problems, just as in the analogous polynomial problems, consists of each base raised to the smallest exponent present in each term.

(a) $(x + 2)^{-2} + (x + 2)^{-3} = (x + 2)^{-3}[(x + 2)^{-2 - (-3)} + 1] = (x + 2)^{-3}(x + 2 + 1) = (x + 2)^{-3}(x + 3)$

(b) $6x^5y^{-3} - 3y^{-4}x^6 = 3x^5y^{-4}(2y^{-3 - (-4)} - x^{6 - 5}) = 3x^5y^{-4}(2y - x)$

(c) $4(3x + 2)^3 3(x + 5)^{-3} - 3(x + 5)^{-4}(3x + 2)^4 = 3(3x + 2)^3(x + 5)^{-4}[4(x + 5) - (3x + 2)]$

$$= 3(3x + 2)^3(x + 5)^{-4}(x + 18)$$

(d) $5x^3(3x + 1)^{2/3} + 3x^2(3x + 1)^{5/3} = x^2(3x + 1)^{2/3}[5x + 3(3x + 1)^{5/3 - 2/3}]$

$$= x^2(3x + 1)^{2/3}[5x + 3(3x + 1)]$$

$$= x^2(3x + 1)^{2/3}(14x + 3)$$

**3.8.** Simplify: (a) $\dfrac{x^{p+q}}{x^{p-q}}$; (b) $(x^{p+1})^2(x^{p-1})^2$; (c) $\left(\dfrac{x^{mn}}{x^{n^2}}\right)^{1/n}$

(a) $\dfrac{x^{p+q}}{x^{p-q}} = x^{(p+q)-(p-q)} = x^{p+q-p+q} = x^{2q}$

(b) $(x^{p+1})^2(x^{p-1})^2 = x^{2(p+1)}x^{2(p-1)} = x^{(2p+2)+(2p-2)} = x^{4p}$

(c) $\left(\dfrac{x^{mn}}{x^{n^2}}\right)^{1/n} = \dfrac{x^{mn(1/n)}}{x^{n^2(1/n)}} = \dfrac{x^m}{x^n} = x^{m-n}$

**3.9.** Simplify, without assuming that variable bases are positive:

(a) $(x^4)^{1/4}$; (b) $(x^2y^4z^6)^{1/2}$; (c) $(x^3y^6z^9)^{1/3}$; (d) $[x(x+h)^2]^{1/2}$

(a) $(x^4)^{1/4} = |x|$; (b) $(x^2y^4z^6)^{1/2} = (x^2)^{1/2}(y^4)^{1/2}(z^6)^{1/2} = |x|\,|y^2|\,|z^3| = |x|y^2|z^3|$; (c) $(x^3y^6z^9)^{1/3} = (x^3)^{1/3}(y^6)^{1/3}(z^9)^{1/3} = xy^2z^3$;

(d) $[x(x+h)^2]^{1/2} = x^{1/2}[(x+h)^2]^{1/2} = x^{1/2}|x+h|$

**3.10.** (a) Write in scientific notation: The velocity of light is 186,000 mi/sec. (b) Find the number of seconds in a year and write the answer in scientific notation. (c) Express the distance light travels in 1 year in scientific notation.

(a) Moving the decimal point to the right of the first nonzero digit is a shift of 5 places: thus,
186,000 miles/sec = $1.86 \times 10^5$ mi/sec.

(b) 1 year = 365 days $\times$ 24 hours/day $\times$ 60 minutes/hour $\times$ 60 seconds/minute
= 31,536,000 seconds = $3.15 \times 10^7$ seconds.

(c) Since distance = velocity $\times$ time, the distance light travels in 1 year
= $(1.86 \times 10^5$ mi/sec$) \times (3.15 \times 10^7$ sec$) = 5.87 \times 10^{12}$ mi.

## SUPPLEMENTARY PROBLEMS

**3.11.** Simplify:(a) $(xy^3)^4(3x^2y)^3$; (b) $\dfrac{(x^2y^3)^3}{(2x^3y^4)^2}$

*Ans.*  (a) $27x^{10}y^{15}$; (b) $\dfrac{y}{4}$

**3.12.** Simplify: (a) $2(xy^{-3})^{-2}(4x^{-3}y^2)^{-1}$; (b) $\left(\dfrac{3x^3y^{-2}}{2xy^{-5}}\right)^{-2}$

*Ans.*  (a) $\dfrac{xy^4}{2}$; (b) $\dfrac{4}{9x^4y^6}$

**3.13.** Simplify, assuming all variable bases are positive: (a) $(8y^3z^4)^{2/3}$; (b) $(100x^8y^3)^{-1/2}$; (c) $\left(\dfrac{8x^4}{27y^6}\right)^{-2/3}$; (d) $\left(\dfrac{a^0z}{25x^6}\right)^{-3/2}$

*Ans.*  (a) $4y^2z^{8/3}$; (b) $\dfrac{1}{10x^4y^{3/2}}$; (c) $\dfrac{9y^4}{4x^{8/3}}$; (d) $\dfrac{125x^9}{z^{3/2}}$

**3.14.** Simplify, assuming all variable bases are positive:

(a) $(x^4y^3)^{1/2}(8x^6y)^{2/3}$; (b) $(9x^8y)^{-1/2}(16x^{-4}y^3)^{3/2}$

*Ans.*  (a) $4x^6y^{13/6}$; (b) $\dfrac{64y^4}{3x^{10}}$

**3.15.** Calculate: (a) $25^{-1/2} - 16^{-1/2}$; (b) $(25 - 16)^{-1/2}$; (c) $16^{3/4} + 16^{-3/4}$

*Ans.*  (a) $-\dfrac{1}{20}$; (b) $\dfrac{1}{3}$; (c) $\dfrac{65}{8}$

**3.16.** Simplify: (a) $x^0 + y^0 + (x + y)^0$; (b) $\left(\dfrac{8x^0y^5}{3x^5y^{-3}}\right)^{-2}$; (c) $\left(\dfrac{32x^2y^{-4}}{x^7y^6}\right)^{3/5}$; (d) $\dfrac{px^{p-1}}{q(x^{p/q})^{q-1}}$

*Ans.*  (a) $3$; (b) $\dfrac{9x^{10}}{64y^{16}}$; (c) $\dfrac{8}{x^3y^6}$; (d) $\dfrac{px^{p/q-1}}{q}$

**3.17.** Derive the laws $\dfrac{x^{-m}}{y^{-n}} = \dfrac{y^n}{x^m}$ and $\left(\dfrac{x}{y}\right)^{-m} = \left(\dfrac{y}{x}\right)^m$ from the definition of negative exponents and standard fraction operations.

**3.18.** Perform indicated operations: (a) $(x^{1/2} + y^{1/2})(x^{1/2} - y^{1/2})$; (b) $(x^{1/3} + y^{1/3})(x^{1/3} - y^{1/3})$;

(c) $(x^{1/3} + y^{1/3})(x^{2/3} - x^{1/3}y^{1/3} + y^{2/3})$; (d) $(x^{1/3} + y^{1/3})(x^{2/3} + x^{1/3}y^{1/3} + y^{2/3})$; (e) $(x^{2/3} - y^{2/3})^3$

*Ans.* (a) $x - y$; (b) $x^{2/3} - y^{2/3}$; (c) $x + y$; (d) $x + 2x^{2/3}y^{1/3} + 2x^{1/3}y^{2/3} + y$; (e) $x^2 - 3x^{4/3}y^{2/3} + 3x^{2/3}y^{4/3} - y^2$

**3.19.** Remove common factors: (a) $x^{-8}y^{-7} + x^{-7}y^{-8}$; (b) $x^{-5/3}y^3 - x^{-2/3}y^2$; (c) $x^{p+q} + x^p$;

(d) $4(x^2 + 4)^{3/2}(3x + 5)^{1/3} + (3x + 5)^{4/3}(x^2 + 4)^{1/2}3x$

*Ans.* (a) $x^{-8}y^{-8}(y + x)$; (b) $x^{-5/3}y^2(y - x)$; (c) $x^p(x^q + 1)$; (d) $(3x + 5)^{1/3}(x^2 + 4)^{1/2}(13x^2 + 15x + 16)$

**3.20.** Remove common factors: (a) $x^{-5} + 2x^{-4} + 2x^{-3}$; (b) $6x^2(x^2 - 1)^{3/2} + x^3(x^2 - 1)^{1/2}(6x)$;

(c) $-4x^{-5}(1 - x^2)^3 + x^{-4}(6x)(1 - x^2)^2$; (d) $x^{-4}(1 - 2x)^{-3/2} - 4x^{-5}(1 - 2x)^{-1/2}$

*Ans.* (a) $x^{-5}(1 + 2x + 2x^2)$; (b) $6x^2(x^2 - 1)^{1/2}(2x^2 - 1)$; (c) $2x^{-5}(1 - x^2)^2(5x^2 - 2)$; (d) $x^{-5}(1 - 2x)^{-3/2}(9x - 4)$

**3.21.** Remove common factors:

(a) $-(x - 2)^{-2}(3x - 7)^{-3} - 3(x - 2)^{-1}(3x - 7)^{-4}(3)$;

(b) $-4(x^2 - 4)^{-5}(2x)(x^2 + 4)^3 + 3(x^2 - 4)^{-4}(x^2 + 4)^2(2x)$

*Ans.* (a) $-(x - 2)^{-2}(3x - 7)^{-4}(12x - 25)$; (b) $(x^2 - 4)^{-5}(x^2 + 4)^2(2x)(-x^2 - 28)$

**3.22.** Remove common factors:

(a) $3(x + 3)^2(3x - 1)^{1/2} + (x + 3)^3\left(\dfrac{1}{2}\right)(3x - 1)^{-1/2}(3)$;

(b) $\dfrac{3}{2}(2x + 3)^{1/2}(3x + 4)^{4/3}(2) + (2x + 3)^{3/2}\left(\dfrac{4}{3}\right)(3x + 4)^{1/3}(3)$;

(c) $-\dfrac{3}{2}(4x^2 - 1)^{-5/2}(8x)(1 + x^2)^{2/3} + (4x^2 - 1)^{-3/2}\left(\dfrac{2}{3}\right)(1 + x^2)^{-1/3}(2x)$

*Ans.* (a) $\dfrac{3}{2}(x + 3)^2(3x - 1)^{-1/2}(7x + 1)$; (b) $(2x + 3)^{1/2}(3x + 4)^{1/3}(17x + 24)$;

(c) $\dfrac{4}{3}x(4x^2 - 1)^{-5/2}(1 + x^2)^{-1/3}(-5x^2 - 10)$

**3.23.** Simplify and write in scientific notation: (a) $(7.2 \times 10^{-3})(5 \times 10^{12})$;

(b) $(7.2 \times 10^{-3}) \div (5 \times 10^{12})$; (c) $\dfrac{(3 \times 10^{-5})(6 \times 10^{-3})^3}{(9 \times 10^{-12})^2}$

*Ans.* (a) $3.6 \times 10^{10}$; (b) $1.44 \times 10^{-15}$; (c) $8 \times 10^{10}$

**3.24.** There are approximately $6.01 \times 10^{23}$ atoms of hydrogen in one gram. Calculate the approximate mass in grams of one hydrogen atom.

*Ans.* $1.67 \times 10^{-24}$ grams

**3.25.** According to the United States Department of Commerce, the U.S. Gross Domestic Product (GDP) for 2006 was \$13,509,000,000,000. According to the United States Bureau of the Census, the U.S. population was 300,000,000 (October 2006). Write these figures in scientific notation and use the result to estimate the GDP per person as of 2006.

*Ans.* $1.3509 \times 10^{13}$, $3 \times 10^8$, $4.503 \times 10^4$ or \$45,030

**3.26.** In 2007, the federal debt limit was raised to \$8,965,000,000,000. Meanwhile, the U.S. population had increased to 301,000,000. Write these figures in scientific notation and use the result to estimate each U.S. inhabitant's share of the debt.

*Ans.* $8.965 \times 10^{12}$, $3.01 \times 10^8$, $2.9784 \times 10^4$ or \$29,784

# CHAPTER 4

# Rational and Radical Expressions

## Rational Expressions

A rational expression is one which can be written as the quotient of two polynomials. (Hence any polynomial is also a rational expression.) Rational expressions are defined for all real values of the variables except those that make the denominator equal to zero.

**EXAMPLE 4.1** $\dfrac{x^2}{y^3}$, $(y \neq 0)$; $\dfrac{x^2 - 5x + 6}{x^3 + 8}$, $(x \neq -2)$; $y^3 - 5y^2$; $x^3 - 3x^2 + 8x - \dfrac{2x}{x^2 - 1}$, $(x \neq \pm 1)$ are examples of rational expressions.

## Fundamental Principle of Fractions

For all real numbers $a$, $b$, $k$ ($b$, $k \neq 0$)

$$\frac{a}{b} = \frac{ak}{bk} \text{ (building to higher terms)} \qquad \frac{ak}{bk} = \frac{a}{b} \text{ (reducing to lower terms)}$$

**EXAMPLE 4.2**   Reducing to lowest terms: $\dfrac{x^2 - 2xy + y^2}{x^2 - y^2} = \dfrac{(x - y)^2}{(x - y)(x + y)} = \dfrac{x - y}{x + y}$

## Operations on Rational Expressions

Operations on rational expressions (all denominators assumed $\neq 0$):

$$\left(\frac{a}{b}\right)^{-1} = \frac{b}{a} \qquad \frac{a}{b} \cdot \frac{c}{d} = \frac{ac}{bd} \qquad \frac{a}{b} \div \frac{c}{d} = \frac{a}{b} \cdot \left(\frac{c}{d}\right)^{-1} = \frac{a}{b} \cdot \frac{d}{c} = \frac{ad}{bc}$$

$$\frac{a}{c} \pm \frac{b}{c} = \frac{a \pm b}{c} \qquad \frac{a}{b} \pm \frac{c}{d} = \frac{ad}{bd} \pm \frac{bc}{bd} = \frac{ad \pm bc}{bd}$$

*Note*: In addition of expressions with unequal denominators, the result is usually written in lowest terms, and the expressions are built to higher terms using the lowest common denominator (LCD).

**EXAMPLE 4.3**   Subtraction: $\dfrac{5}{x^3 y^2} - \dfrac{6}{x^2 y^4} = \dfrac{5y^2}{x^3 y^4} - \dfrac{6x}{x^3 y^4} = \dfrac{5y^2 - 6x}{x^3 y^4}$

## Complex Fractions

Complex fractions are expressions containing fractions in the numerator and/or denominator. They can be reduced to simple fractions by two methods:

*Method* 1: Combine numerator and denominator into single quotients, then divide.

**EXAMPLE 4.4** $\dfrac{\dfrac{x}{x - 1} - \dfrac{a}{a - 1}}{x - a} = \dfrac{\dfrac{x(a - 1) - a(x - 1)}{(x - 1)(a - 1)}}{x - a} = \dfrac{xa - x - ax + a}{(x - 1)(a - 1)} \div (x - a)$

$$= \frac{a - x}{(x - 1)(a - 1)} \cdot \frac{1}{x - a} = \frac{-1}{(x - 1)(a - 1)}$$

*Method* 2: Multiply numerator and denominator by the LCD of all internal fractions:

**EXAMPLE 4.5**
$$\frac{\dfrac{x}{y} - \dfrac{y}{x}}{\dfrac{x}{y^2} + \dfrac{y}{x^2}} = \frac{\dfrac{x}{y} - \dfrac{y}{x}}{\dfrac{x}{y^2} + \dfrac{y}{x^2}} \cdot \frac{x^2 y^2}{x^2 y^2} = \frac{x^3 y - xy^3}{x^3 + y^3} = \frac{xy(x-y)(x+y)}{(x+y)(x^2 - xy + y^2)} = \frac{xy(x-y)}{x^2 - xy + y^2}$$

Method 2 is more convenient when the fractions in numerator and denominator involve very similar expressions.

## Rational Expressions

Rational expressions are often written in terms of negative exponents.

**EXAMPLE 4.6**   Simplify: $x^{-3}y^5 - 3x^{-4}y^6$

This can be done in two ways, either by removing the common factor of $x^{-4}y^5$, as in the previous chapter, or by rewriting as the sum of two rational expressions:

$$x^{-3}y^5 - 3x^{-4}y^6 = \frac{y^5}{x^3} - \frac{3y^6}{x^4} = \frac{xy^5}{x^4} - \frac{3y^6}{x^4} = \frac{xy^5 - 3y^6}{x^4}.$$

## Radical Expressions

For $n$ a natural number greater than 1 and $x$ a real number, the $n$th root radical is defined to be the principal $n$th root of $x$:

$$\sqrt[n]{x} = x^{1/n}$$

If $n = 2$, write $\sqrt{x}$ in place of $\sqrt[2]{x}$.

The symbol $\sqrt{\phantom{x}}$ is called a radical, $n$ is called the index, and $x$ is called the radicand.

## Properties of Radicals

$$(\sqrt[n]{x})^n = x, \text{ if } \sqrt[n]{x} \text{ is defined} \qquad \sqrt[n]{x^n} = x, \text{ if } x \geq 0$$

$$\sqrt[n]{x^n} = x, \text{ if } x < 0, n \text{ odd} \qquad \sqrt[n]{x^n} = |x|, \text{ if } x < 0, n \text{ even}$$

$$\sqrt[n]{ab} = \sqrt[n]{a}\,\sqrt[n]{b} \qquad \sqrt[n]{\sqrt[m]{x}} = \sqrt[mn]{x}$$

$$\sqrt[n]{\frac{a}{b}} = \frac{\sqrt[n]{a}}{\sqrt[n]{b}}$$

Unless otherwise specified, it is normally assumed that variable bases represent nonnegative real numbers.

## Simplest Radical Form

Simplest radical form for radical expressions:

1. No radicand can contain a factor with an exponent greater than or equal to the index of the radical.
2. No power of the radicand and the index of the radical can have a common factor other than 1.
3. No radical appears in a denominator.
4. No fraction appears in a radical.

**EXAMPLE 4.7**

(a)  $\sqrt[3]{16x^3 y^5}$ violates condition 1. It is simplified as follows:

$$\sqrt[3]{16x^3 y^5} = \sqrt[3]{8x^3 y^3 \cdot 2y^2} = \sqrt[3]{8x^3 y^3} \cdot \sqrt[3]{2y^2} = 2xy \sqrt[3]{2y^2}$$

(b) $\sqrt[6]{t^3}$ violates condition 2. It is simplified as follows:

$$\sqrt[6]{t^3} = \sqrt[2\cdot3]{t^3} = \sqrt{\sqrt[3]{t^3}} = \sqrt{t}$$

(c) $\dfrac{12x^2}{\sqrt[4]{27xy^2}}$ violates condition 3. It is simplified as follows:

$$\frac{12x^2}{\sqrt[4]{27xy^2}} = \frac{12x^2}{\sqrt[4]{27xy^2}} \cdot \frac{\sqrt[4]{3x^3y^2}}{\sqrt[4]{3x^3y^2}} = \frac{12x^2\sqrt[4]{3x^3y^2}}{\sqrt[4]{81x^4y^4}} = \frac{12x^2\sqrt[4]{3x^3y^2}}{3xy} = \frac{4x\sqrt[4]{3x^3y^2}}{y}$$

(d) $\sqrt[4]{\dfrac{3x}{5y^3}}$ violates condition 4. It is simplified as follows:

$$\sqrt[4]{\frac{3x}{5y^3}} = \sqrt[4]{\frac{3x}{5y^3} \cdot \frac{5^3y}{5^3y}} = \sqrt[4]{\frac{375xy}{5^4y^4}} = \frac{\sqrt[4]{375xy}}{5y}$$

Satisfying condition 3 is often referred to as *rationalizing the denominator.*

The *conjugate* expression for a binomial of form $a + b$ is the expression $a - b$ and conversely.

**EXAMPLE 4.8**    Rationalize the denominator: $\dfrac{x-4}{\sqrt{x}-2}$

Multiply numerator and denominator by the conjugate expression for the denominator:

$$\frac{x-4}{\sqrt{x}-2} = \frac{x-4}{\sqrt{x}-2} \cdot \frac{\sqrt{x}+2}{\sqrt{x}+2} = \frac{(x-4)(\sqrt{x}+2)}{x-4} = \sqrt{x}+2$$

Expressions are not always written in simplest radical form. Often it is important to *rationalize the numerator.*

**EXAMPLE 4.9**    Rationalize the numerator: $\dfrac{\sqrt{x}-\sqrt{a}}{x-a}$

Multiply numerator and denominator by the conjugate expression for the *numerator.*

$$\frac{\sqrt{x}-\sqrt{a}}{x-a} = \frac{\sqrt{x}-\sqrt{a}}{x-a} \cdot \frac{\sqrt{x}+\sqrt{a}}{\sqrt{x}+\sqrt{a}} = \frac{x-a}{(x-a)(\sqrt{x}+\sqrt{a})} = \frac{1}{\sqrt{x}+\sqrt{a}}$$

## Conversion of Radical Expressions

Conversion of radical expressions to exponent form:

For $m, n$ positive integers ($n > 1$) and $x \geq 0$ when $n$ is even,

$$\sqrt[n]{x^m} = x^{m/n}$$

Conversely, $x^{m/n} = \sqrt[n]{x^m}$

Also, $x^{m/n} = (\sqrt[n]{x})^m$

**EXAMPLE 4.10**    (a) $\sqrt[3]{x} = x^{1/3}$;  (b) $\sqrt[4]{x^3} = x^{3/4}$;  (c) $\sqrt{x^5} = x^{5/2}$

## Operations with Complex Numbers

Complex numbers can be written in *standard form* $a + bi$. In this form, they can be combined using the operations defined for real numbers, together with the definition of the imaginary unit $i$: $i^2 = -1$. The conjugate of a complex number $z$ is denoted $\bar{z}$. If $z = a + bi$, then $\bar{z} = a - bi$

**EXAMPLE 4.11**    (a) Write $4 - \sqrt{-25}$ in standard form. (b) Find the conjugate of $3 - 7i$. (c) Simplify $(3 + 4i)^2$.

(a) $4 - \sqrt{-25} = 4 - \sqrt{25}\sqrt{-1} = 4 - 5i$

(b) The conjugate of $3 - 7i$ is $3 - (-7i)$ or $3 + 7i$.

(c) $(3 + 4i)^2 = 3^2 + 2 \cdot 3 \cdot 4i + (4i)^2 = 9 + 24i + 16i^2 = 9 + 24i - 16 = -7 + 24i$

## SOLVED PROBLEMS

**4.1.** Reduce to lowest terms: (a) $\dfrac{5x^2 - 8x + 3}{25x^2 - 9}$ (b) $\dfrac{x^3 - a^3}{x - a}$ (c) $\dfrac{(x + h)^2 - x^2}{h}$

(a) First factor numerator and denominator, then reduce by removing common factors:

$$\frac{5x^2 - 8x + 3}{25x^2 - 9} = \frac{(5x - 3)(x - 1)}{(5x - 3)(5x + 3)} = \frac{x - 1}{5x + 3}$$

(b) $\dfrac{x^3 - a^3}{x - a} = \dfrac{(x - a)(x^2 + ax + a^2)}{x - a} = x^2 + ax + a^2$

(c) $\dfrac{(x + h)^2 - x^2}{h} = \dfrac{x^2 + 2xh + h^2 - x^2}{h} = \dfrac{2xh + h^2}{h} = \dfrac{h(2x + h)}{h} = 2x + h$

**4.2.** Explain why every polynomial is also a rational expression.

A rational expression is one which can be written as the quotient of two polynomials. Every polynomial $P$ can be written as $P/1$, where numerator and denominator are polynomials; hence every polynomial is also a rational expression.

**4.3.** Perform indicated operations:

(a) $\dfrac{x^2 - 7x + 12}{x^2 - 9} \cdot \dfrac{x^3 - 6x^2 + 9x}{x^3 - 4x^2}$ (b) $\dfrac{x^2 - 4y^2}{xy + 2y^2} \div (x^2 - 3xy + 2y^2)$

(c) $\dfrac{1}{x + h} - \dfrac{1}{x}$ (d) $\dfrac{2}{x - 1} + \dfrac{3}{x + 1} - \dfrac{4x - 2}{x^2 - 1}$

(a) Factor all numerators and denominators, then reduce by removing any common factors.

$$\frac{x^2 - 7x + 12}{x^2 - 9} \cdot \frac{x^3 - 6x^2 + 9x}{x^3 - 4x^2} = \frac{(x - 3)(x - 4)}{(x - 3)(x + 3)} \cdot \frac{x(x - 3)^2}{x^2(x - 4)} = \frac{(x - 3)^2}{x(x + 3)}$$

(b) Change division to multiplication, then proceed as in (a).

$$\frac{x^2 - 4y^2}{xy + 2y^2} \div (x^2 - 3xy + 2y^2) = \frac{x^2 - 4y^2}{xy + 2y^2} \cdot \frac{1}{x^2 - 3xy + 2y^2} = \frac{(x - 2y)(x + 2y)}{y(x + 2y)} \cdot \frac{1}{(x - y)(x - 2y)}$$

$$= \frac{1}{y(x - y)}$$

(c) Find the lowest common denominator, then build to higher terms and perform the subtraction.

$$\frac{1}{x + h} - \frac{1}{x} = \frac{x}{x(x + h)} - \frac{(x + h)}{x(x + h)} = \frac{x - (x + h)}{x(x + h)} = \frac{-h}{x(x + h)}$$

(d) Proceed as in (c).

$$\frac{2}{x - 1} + \frac{3}{x + 1} - \frac{4x - 2}{x^2 - 1} = \frac{2(x + 1)}{(x - 1)(x + 1)} + \frac{3(x - 1)}{(x - 1)(x + 1)} - \frac{(4x - 2)}{(x - 1)(x + 1)}$$

$$= \frac{2(x + 1) + 3(x - 1) - (4x - 2)}{(x - 1)(x + 1)} = \frac{2x + 2 + 3x - 3 - 4x + 2}{(x - 1)(x + 1)}$$

$$= \frac{x + 1}{(x - 1)(x + 1)} = \frac{1}{x - 1}$$

**4.4.** Write each complex fraction as a simple fraction in lowest terms:

(a) $\dfrac{y + \dfrac{x^2}{y}}{y^2}$ (b) $\dfrac{\dfrac{x}{x - 1} - \dfrac{x}{x + 1}}{\dfrac{x}{x - 1} + \dfrac{x}{x + 1}}$ (c) $\dfrac{\dfrac{2}{x + 2} - 3}{\dfrac{4}{x} - x}$ (d) $\dfrac{\dfrac{1}{x} - \dfrac{1}{a}}{x - a}$

(a) Multiply numerator and denominator by $y$, the only internal denominator:

$$\frac{y + \frac{x^2}{y}}{y^2} = \frac{y + \frac{x^2}{y}}{y^2} \cdot \frac{y}{y} = \frac{y^2 + x^2}{y^3}$$

(b) Multiply numerator and denominator by $(x - 1)(x + 1)$, the LCD of the internal fractions:

$$\frac{\frac{x}{x-1} - \frac{x}{x+1}}{\frac{x}{x-1} + \frac{x}{x+1}} = \frac{\frac{x}{x-1} - \frac{x}{x+1}}{\frac{x}{x-1} + \frac{x}{x+1}} \cdot \frac{(x-1)(x+1)}{(x-1)(x+1)} = \frac{x(x+1) - x(x-1)}{x(x+1) + x(x-1)} = \frac{x^2 + x - x^2 + x}{x^2 + x + x^2 - x} = \frac{2x}{2x^2} = \frac{1}{x}$$

(c) Combine numerator and denominator into single quotients, then divide:

$$\frac{\frac{2}{x+2} - 3}{\frac{4}{x} - x} = \frac{\frac{2}{x+2} - \frac{3(x+2)}{x+2}}{\frac{4}{x} - \frac{x^2}{x}} = \frac{\frac{2 - 3x - 6}{x+2}}{\frac{4 - x^2}{x}} = \frac{-3x - 4}{x + 2} \div \frac{4 - x^2}{x}$$

$$= \frac{-3x - 4}{x + 2} \cdot \frac{x}{4 - x^2} = \frac{-3x^2 - 4x}{(x + 2)(4 - x^2)}$$

(d) Proceed as in (c):

$$\frac{\frac{1}{x} - \frac{1}{a}}{x - a} = \frac{\frac{a}{ax} - \frac{x}{ax}}{x - a} = \frac{\frac{a - x}{ax}}{x - a} = \frac{a - x}{ax} \div (x - a) = \frac{-(x - a)}{ax} \cdot \frac{1}{x - a} = -\frac{1}{ax}$$

**4.5.** Simplify: (a) $3(x + 3)^2(2x - 1)^{-4} - 8(x + 3)^3(2x - 1)^{-5}$    (b) $\dfrac{(x + h)^{-2} - x^{-2}}{h}$

(a) Remove the common factor $(x + 3)^2(2x - 1)^{-5}$ first:

$$3(x + 3)^2(2x - 1)^{-4} - 8(x + 3)^3(2x - 1)^{-5} = (x + 3)^2(2x - 1)^{-5}[3(2x - 1) - 8(x + 3)]$$

$$= (x + 3)^2(2x - 1)^{-5}[-2x - 27]$$

$$= -\frac{(x + 3)^2(2x + 27)}{(2x - 1)^5}$$

(b) Eliminate negative exponents, then multiply numerator and denominator by $x^2(x + h)^2$, the LCD of the internal denominators:

$$\frac{(x + h)^{-2} - x^{-2}}{h} = \frac{\frac{1}{(x + h)^2} - \frac{1}{x^2}}{h} \cdot \frac{x^2(x + h)^2}{x^2(x + h)^2} = \frac{x^2 - (x + h)^2}{hx^2(x + h)^2} = \frac{x^2 - x^2 - 2xh - h^2}{hx^2(x + h)^2}$$

$$= \frac{-2xh - h^2}{hx^2(x + h)^2} = \frac{h(-2x - h)}{hx^2(x + h)^2} = \frac{-2x - h}{x^2(x + h)^2}$$

**4.6.** Write in simplest radical form:

(a) $\sqrt{20x^3y^4z^5}$    (b) $\sqrt[3]{108x^5(x + y)^6}$    (c) $\sqrt{\dfrac{3x}{5y}}$    (d) $\sqrt[3]{\dfrac{2x^4}{9yz^2}}$

(a) Remove the largest possible perfect square factor, then apply the rule $\sqrt{ab} = \sqrt{a}\sqrt{b}$:

$$\sqrt{20x^3y^4z^5} = \sqrt{4x^2y^4z^4 \cdot 5xz} = \sqrt{4x^2y^4z^4} \cdot \sqrt{5xz} = 2xy^2z^2\sqrt{5xz}$$

(b) Remove the largest possible perfect cube factor, then apply the rule $\sqrt[3]{ab} = \sqrt[3]{a}\sqrt[3]{b}$:

$$\sqrt[3]{108x^5(x + y)^6} = \sqrt[3]{27x^3(x + y)^6}\sqrt[3]{4x^2} = 3x(x + y)^2 \cdot \sqrt[3]{4x^2}$$

(c) Build to higher terms so that the denominator is a perfect square, then apply $\sqrt{\dfrac{a}{b}} = \dfrac{\sqrt{a}}{\sqrt{b}}$:

$$\sqrt{\frac{3x}{5y}} = \sqrt{\frac{3x}{5y} \cdot \frac{5y}{5y}} = \sqrt{\frac{15xy}{25y^2}} = \frac{\sqrt{15xy}}{\sqrt{25y^2}} = \frac{\sqrt{15xy}}{5y}$$

(d) Build to higher terms so that the denominator is a perfect cube, then apply $\sqrt[3]{\dfrac{a}{b}} = \dfrac{\sqrt[3]{a}}{\sqrt[3]{b}}$:

$$\sqrt[3]{\frac{2x^4}{9yz^2}} = \sqrt[3]{\frac{2x^4}{9yz^2} \cdot \frac{3y^2z}{3y^2z}} = \sqrt[3]{\frac{6x^4y^2z}{27y^3z^3}} = \frac{\sqrt[3]{6x^4y^2z}}{\sqrt[3]{27y^3z^3}} = \frac{\sqrt[3]{x^3 \cdot 6xy^2z}}{3yz} = \frac{x\sqrt[3]{6xy^2z}}{3yz}$$

**4.7.** Rationalize the denominator:

(a) $\dfrac{x^3y^2}{\sqrt[3]{2xy^2}}$  (b) $\dfrac{\sqrt{x}}{\sqrt{x}+1}$  (c) $\dfrac{\sqrt{x}+\sqrt{h}}{\sqrt{x}-\sqrt{h}}$  (d) $\dfrac{x^2-16y^2}{\sqrt{x}-2\sqrt{y}}$

(a) Build to higher terms so that the denominator becomes the cube root of a perfect cube, then reduce:

$$\frac{x^3y^2}{\sqrt[3]{2xy^2}} = \frac{x^3y^2}{\sqrt[3]{2xy^2}} \cdot \frac{\sqrt[3]{4x^2y}}{\sqrt[3]{4x^2y}} = \frac{x^3y^2\sqrt[3]{4x^2y}}{\sqrt[3]{8x^3y^3}} = \frac{x^3y^2\sqrt[3]{4x^2y}}{2xy} = \frac{x^2y\sqrt[3]{4x^2y}}{2}$$

(b) Build to higher terms using $\sqrt{x}-1$, the conjugate expression for the denominator:

$$\frac{\sqrt{x}}{\sqrt{x}+1} = \frac{\sqrt{x}}{\sqrt{x}+1} \cdot \frac{\sqrt{x}-1}{\sqrt{x}-1} = \frac{x-\sqrt{x}}{x-1}$$

(c) Proceed as in (b):

$$\frac{\sqrt{x}+\sqrt{h}}{\sqrt{x}-\sqrt{h}} = \frac{\sqrt{x}+\sqrt{h}}{\sqrt{x}-\sqrt{h}} \cdot \frac{\sqrt{x}+\sqrt{h}}{\sqrt{x}+\sqrt{h}} = \frac{(\sqrt{x}+\sqrt{h})^2}{x-h} = \frac{x+2\sqrt{xh}+h}{x-h}$$

(d) Proceed as in (b):

$$\frac{x^2-16y^2}{\sqrt{x}-2\sqrt{y}} = \frac{x^2-16y^2}{\sqrt{x}-2\sqrt{y}} \cdot \frac{\sqrt{x}+2\sqrt{y}}{\sqrt{x}+2\sqrt{y}} = \frac{(x^2-16y^2)(\sqrt{x}+2\sqrt{y})}{x-4y} = (x+4y)(\sqrt{x}+2\sqrt{y})$$

**4.8.** Rationalize the numerator:

(a) $\dfrac{\sqrt{x}}{\sqrt{x}+1}$  (b) $\dfrac{\sqrt{x}+\sqrt{h}}{\sqrt{x}-\sqrt{h}}$  (c) $\dfrac{\sqrt{x+h}-\sqrt{x}}{h}$

(a) Build to higher terms using $\sqrt{x}$:

$$\frac{\sqrt{x}}{\sqrt{x}+1} = \frac{\sqrt{x}}{\sqrt{x}+1} \cdot \frac{\sqrt{x}}{\sqrt{x}} = \frac{x}{x+\sqrt{x}}$$

(b) Build to higher terms using $\sqrt{x}-\sqrt{h}$, the conjugate expression for the numerator:

$$\frac{\sqrt{x}+\sqrt{h}}{\sqrt{x}-\sqrt{h}} = \frac{\sqrt{x}+\sqrt{h}}{\sqrt{x}-\sqrt{h}} \cdot \frac{\sqrt{x}-\sqrt{h}}{\sqrt{x}-\sqrt{h}} = \frac{x-h}{(\sqrt{x}-\sqrt{h})^2} = \frac{x-h}{x-2\sqrt{xh}+h}$$

(c) Proceed as in (b):

$$\frac{\sqrt{x+h}-\sqrt{x}}{h} = \frac{\sqrt{x+h}-\sqrt{x}}{h} \cdot \frac{\sqrt{x+h}+\sqrt{x}}{\sqrt{x+h}+\sqrt{x}} = \frac{x+h-x}{h(\sqrt{x+h}+\sqrt{x})}$$

$$= \frac{h}{h(\sqrt{x+h}+\sqrt{x})} = \frac{1}{\sqrt{x+h}+\sqrt{x}}$$

**4.9.** Write in exponent notation: (a) $\sqrt{xy^3}$  (b) $\sqrt[3]{a^2b(x-y)^5}$

(a) $\sqrt{xy^3} = (xy^3)^{1/2} = x^{1/2}y^{3/2}$;  (b) $\sqrt[3]{a^2b(x-y)^5} = [a^2b(x-y)^5]^{1/3} = a^{2/3}b^{1/3}(x-y)^{5/3}$

**4.10.** Write as a sum or difference of terms in exponential notation:

(a) $\dfrac{x-1}{\sqrt{x}}$    (b) $\dfrac{x^3 - 6x^2 + 3x + 1}{6\sqrt[3]{x^5}}$

(a) $\dfrac{x-1}{\sqrt{x}} = \dfrac{x-1}{x^{1/2}} = \dfrac{x}{x^{1/2}} - \dfrac{1}{x^{1/2}} = x^{1/2} - x^{-1/2}$

(b) $\dfrac{x^3 - 6x^2 + 3x + 1}{6\sqrt[3]{x^5}} = \dfrac{x^3 - 6x^2 + 3x + 1}{6x^{5/3}} = \dfrac{x^3}{6x^{5/3}} - \dfrac{6x^2}{6x^{5/3}} + \dfrac{3x}{6x^{5/3}} + \dfrac{1}{6x^{5/3}}$

$$= \frac{1}{6}x^{4/3} - x^{1/3} + \frac{1}{2}x^{-2/3} + \frac{1}{6}x^{-5/3}$$

**4.11.** Write as a single fraction in lowest terms. Do not rationalize denominators.

(a) $\sqrt{x-2} + \dfrac{2}{\sqrt{x-2}}$    (b) $\dfrac{\sqrt{x^2-1} - \dfrac{x^2}{\sqrt{x^2-1}}}{x^2-1}$    (c) $\dfrac{x^2(x^2+9)^{-1/2} - \sqrt{x^2+9}}{x^2}$

(d) $\dfrac{(x^2-9)^{1/3}\, 3 - (4x)\left(\frac{1}{3}\right)(x^2-9)^{-2/3}(2x)}{[(x^2-9)^{1/3}]^2}$

(a) $\sqrt{x-2} + \dfrac{2}{\sqrt{x-2}} = \dfrac{\sqrt{x-2}}{1} + \dfrac{2}{\sqrt{x-2}} = \dfrac{\sqrt{x-2}\cdot\sqrt{x-2}}{\sqrt{x-2}} + \dfrac{2}{\sqrt{x-2}} = \dfrac{x-2+2}{\sqrt{x-2}}$

$$= \frac{x}{\sqrt{x-2}}$$

(b) Multiply numerator and denominator by $\sqrt{x^2-1}$, the only internal denominator:

$$\dfrac{\sqrt{x^2-1} - \dfrac{x^2}{\sqrt{x^2-1}}}{x^2-1} = \dfrac{\sqrt{x^2-1} - \dfrac{x^2}{\sqrt{x^2-1}}}{x^2-1} \cdot \dfrac{\sqrt{x^2-1}}{\sqrt{x^2-1}} = \dfrac{x^2-1-x^2}{(x^2-1)\sqrt{x^2-1}} = -\dfrac{1}{(x^2-1)^{3/2}}$$

(c) Rewrite in exponent notation, then remove the common factor $(x^2+9)^{-1/2}$ from the numerator:

$$\dfrac{x^2(x^2+9)^{-1/2} - \sqrt{x^2+9}}{x^2} = \dfrac{x^2(x^2+9)^{-1/2} - (x^2+9)^{1/2}}{x^2} = \dfrac{(x^2+9)^{-1/2}[x^2 - (x^2+9)^1]}{x^2}$$

$$= \dfrac{x^2 - x^2 - 9}{(x^2+9)^{1/2}x^2} = \dfrac{-9}{x^2(x^2+9)^{1/2}}$$

(d) Eliminate negative exponents, then multiply numerator and denominator by $3(x^2-9)^{2/3}$, the only internal denominator:

$$\dfrac{(x^2-9)^{1/3}\,3 - (4x)\left(\frac{1}{3}\right)(x^2-9)^{-2/3}(2x)}{[(x^2-9)^{1/3}]^2} = \dfrac{(x^2-9)^{1/3}\,3 - \dfrac{(4x)(2x)}{3(x^2-9)^{2/3}}}{(x^2-9)^{2/3}}$$

$$= \dfrac{(x^2-9)^{1/3}\,3 - \dfrac{(4x)(2x)}{3(x^2-9)^{2/3}}}{(x^2-9)^{2/3}} \cdot \dfrac{3(x^2-9)^{2/3}}{3(x^2-9)^{2/3}}$$

$$= \dfrac{9(x^2-9) - 8x^2}{3(x^2-9)^{4/3}}$$

$$= \dfrac{x^2 - 81}{3(x^2-9)^{4/3}}$$

**4.12.** Let $z = 4 - 7i$ and $w = -6 + 5i$ be two complex numbers. Find

(a) $z + w$    (b) $w - z$    (c) $wz$    (d) $\dfrac{w}{z}$    (e) $w^2 - i\bar{z}$

(a) $z + w = (4 - 7i) + (-6 + 5i) = 4 - 7i - 6 + 5i = -2 - 2i$

(b) $w - z = (-6 + 5i) - (4 - 7i) = -6 + 5i - 4 + 7i = -10 + 12i$

(c) Use FOIL: $wz = (-6 + 5i)(4 - 7i) = -24 + 42i + 20i - 35i^2 = -24 + 62i + 35 = 11 + 62i$

(d) To write the quotient of two complex numbers in standard form, multiply numerator and denominator of the quotient by the conjugate of the denominator:

$$\frac{w}{z} = \frac{-6 + 5i}{4 - 7i} = \frac{-6 + 5i}{4 - 7i} \cdot \frac{4 + 7i}{4 + 7i} = \frac{-59 - 22i}{16 - 49i^2} = \frac{-59 - 22i}{16 + 49} = \frac{-59 - 22i}{65} \text{ or } -\frac{59}{65} - \frac{22}{65}i$$

(e) $w^2 - \bar{z} = (-6 + 5i)^2 - i\overline{(4 - 7i)} = 36 - 60i + 25i^2 - i(4 + 7i) = 36 - 60i - 25 - 4i + 7 = 18 - 64i$

## SUPPLEMENTARY PROBLEMS

**4.13.** Reduce to lowest terms:

(a) $\dfrac{x^4 - y^4}{x^4 - 2x^2y^2 + y^4}$ 

(b) $\dfrac{x^3 + x^2 + x + 1}{x^3 + 3x^2 + 3x + 1}$ 

(c) $\dfrac{(x^2 + 1)^2 3x^2 - x^3(2x)(x^2 + 1)2}{(x^2 + 1)^4}$ 

(d) $\dfrac{(x + h)^3 - x^3}{h}$

Ans. (a) $\dfrac{x^2 + y^2}{x^2 - y^2}$; (b) $\dfrac{x^2 + 1}{x^2 + 2x + 1}$; (c) $\dfrac{3x^2 - x^4}{(x^2 + 1)^3}$; (d) $3x^2 + 3xh + h^2$

**4.14.** Perform indicated operations:

(a) $\dfrac{1}{(x + 1)(x + 2)} - \dfrac{3}{(x - 1)(x + 2)} + \dfrac{3}{(x - 1)(x + 1)}$ 

(b) $\dfrac{5}{x - 2} + \dfrac{3}{x + 2} - \dfrac{x - 1}{x^2 + 4}$

(c) $\dfrac{3x - 1}{(x^2 + 4)^2} + \dfrac{2x - 5}{x^2 + 4}$ 

(d) $(x^2 - 3x + 2) \cdot \dfrac{x^2 - 5x + 4}{x^3 - 6x^2 + 8x}$

Ans. (a) $\dfrac{1}{(x - 1)(x + 1)}$; (b) $\dfrac{7x^3 + 5x^2 + 36x + 12}{x^4 - 16}$; (c) $\dfrac{2x^3 - 5x^2 + 11x - 21}{(x^2 + 4)^2}$; (d) $\dfrac{x^2 - 2x + 1}{x}$

**4.15.** Write as a simple fraction in lowest terms:

(a) $\dfrac{\dfrac{1}{t - 1} + \dfrac{1}{t + 1}}{\dfrac{1}{t} - \dfrac{1}{t^2}}$ 

(b) $\dfrac{\dfrac{2x}{x + 1} - \dfrac{2a}{a + 1}}{x - a}$ 

(c) $\dfrac{(x^2 - 4)^3(2x) - x^2(3)(x^2 - 4)^2(2x)}{(x^2 - 4)^6}$

Ans. (a) $\dfrac{2t^3}{(t - 1)(t^2 - 1)}$; (b) $\dfrac{2}{(x + 1)(a + 1)}$; (c) $\dfrac{-4x^3 - 8x}{(x^2 - 4)^4}$

**4.16.** Write as a simple fraction in lowest terms:

(a) $\dfrac{(x + 5)^{-5} - (x + 5)^{-4}}{(x + 5)^{-3}}$; 

(b) $\dfrac{(x + h)^{-1} - x^{-1}}{h}$; 

(c) $\dfrac{x^{-2} - a^{-2}}{x - a}$

Ans. (a) $\dfrac{-4 - x}{(x + 5)^2}$; (b) $\dfrac{-1}{x(x + h)}$; (c) $\dfrac{-x - a}{a^2x^2}$

**4.17.** Write in simplest radical form: (a) $\sqrt[4]{48x^6y^7z^8}$ (b) $\sqrt[4]{\dfrac{15x^2}{8y^7z}}$ (c) $\dfrac{M_0}{\sqrt{1 - \dfrac{v^2}{c^2}}}$

Ans. (a) $2xyz^2\sqrt[4]{3x^2y^3}$; (b) $\dfrac{\sqrt[4]{30x^2yz^3}}{2y^2z}$; (c) $\dfrac{M_0c\sqrt{c^2 - v^2}}{c^2 - v^2}$

**4.18.** Rationalize the denominator: (a) $\dfrac{1}{\sqrt{a} - \sqrt{b}}$    (b) $\dfrac{\sqrt{x} + 2}{\sqrt{x} - 1}$

    *Ans.*  (a) $\dfrac{\sqrt{a} + \sqrt{b}}{a - b}$; (b) $\dfrac{x + 3\sqrt{x} + 2}{x - 1}$

**4.19.** Rationalize the numerator: (a) $\dfrac{\sqrt{x + 1} - \sqrt{a + 1}}{x - a}$    (b) $\dfrac{\dfrac{1}{\sqrt{x + h}} - \dfrac{1}{\sqrt{x}}}{h}$    (c) $\dfrac{\sqrt[3]{x} - \sqrt[3]{a}}{x - a}$

    *Ans.*  (a) $\dfrac{1}{\sqrt{x + 1} + \sqrt{a + 1}}$; (b) $\dfrac{-1}{\sqrt{x}\sqrt{x + h}(\sqrt{x} + \sqrt{x + h})}$; (c) $\dfrac{1}{\sqrt[3]{x^2} + \sqrt[3]{xa} + \sqrt[3]{a^2}}$

**4.20.** Write as a sum or difference of terms in exponential notation: (a) $\dfrac{3x^2 - 2x}{x\sqrt{x}}$;    (b) $\dfrac{4x^3 - 5x^2 - 8x + 1}{2x^{\frac{1}{3}}}$

    *Ans.*  (a) $3x^{1/2} - 2x^{-1/2}$; (b) $2x^{8/3} - \dfrac{5}{2}x^{5/3} - 4x^{2/3} + \dfrac{1}{2}x^{-1/3}$

**4.21.** Write as a simple fraction in lowest terms. Do not rationalize denominators.

  (a) $\dfrac{2x\sqrt{4 - x^2} + \dfrac{x^2 \cdot 2x}{\sqrt{4 - x^2}}}{4 - x^2}$    (b) $\dfrac{\dfrac{2}{3}x(x^2 + 4)^{1/2}(x^2 - 9)^{-2/3} - x(x^2 - 9)^{1/3}(x^2 + 4)^{-1/2}}{x^2 + 4}$

    *Ans.*  (a) $\dfrac{8x}{(4 - x^2)^{3/2}}$; (b) $\dfrac{-x^3 + 35x}{3(x^2 - 9)^{2/3}(x^2 + 4)^{3/2}}$

**4.22.** Write as a simple fraction in lowest terms. Do not rationalize denominators:

  (a) $\dfrac{x\left(\dfrac{1}{2}\right)(x^2 + 9)^{-1/2}(2x) - \sqrt{x^2 + 9}}{x^2}$    (b) $\dfrac{(x^2 - 1)^{3/2} - x\left(\dfrac{3}{2}\right)(x^2 - 1)^{1/2}(2x)}{(x^2 - 1)^3}$

  (c) $\dfrac{(x^2 - 1)^{4/3}(2x) - (x^2 + 4)\left(\dfrac{4}{3}\right)(x^2 - 1)^{1/3}(2x)}{(x^2 - 1)^{8/3}}$

    *Ans.*  (a) $\dfrac{-9}{x^2(x^2 + 9)^{1/2}}$; (b) $\dfrac{-1 - 2x^2}{(x^2 - 1)^{5/2}}$; (c) $\dfrac{-2x^3 - 38x}{3(x^2 - 1)^{7/3}}$

**4.23.** Let $z = 5 - 2i$, $w = -3 + i$. Write in standard form for complex numbers:

  (a) $z + w$;   (b) $z - w$;   (c) $zw$;   (d) $z/w$

    *Ans.*  (a) $2 - i$; (b) $8 - 3i$; (c) $-13 + 11i$; (d) $-\dfrac{17}{10} + \dfrac{1}{10}i$

**4.24.** Write in standard form for complex numbers:

  (a) $\sqrt{5^2 - 4 \cdot 1 \cdot 10}$    (b) $i^6$    (c) $(1 + 2i)^3$    (d) $(1 - i)/(2 + 3i) - (4 + 5i)/(6i^3)$

    *Ans.*  (a) $i\sqrt{15}$; (b) $-1$ or $-1 + 0i$; (c) $-11 - 2i$; (d) $\dfrac{59}{78} - \dfrac{41}{39}i$

**4.25.** For $z$, $w$ complex numbers, show:

  (a) $\overline{z + w} = \bar{z} + \bar{w}$    (b) $\overline{z - w} = \bar{z} - \bar{w}$    (c) $\overline{zw} = \bar{z}\bar{w}$    (d) $\overline{z/w} = \bar{z}/\bar{w}$

  (e) $\bar{z} = z$ if and only if $z$ is a real number.

# CHAPTER 5

# Linear and Nonlinear Equations

## Equations

An equation is a statement that two expressions are equal. An equation containing variables is in general neither true nor false; rather, its truth depends on the value(s) of the variable(s). For equations in one variable, a value of the variable which makes the equation true is called a *solution* of the equation. The set of all solutions is called the *solution set* of the equation. An equation which is true for all those values of the variable for which it is meaningful is called an *identity*.

## Equivalent Equations

Equations are equivalent if they have the same solution sets.

**EXAMPLE 5.1**   The equations $x = -5$ and $x + 5 = 0$ are equivalent. Each has the solution set $\{-5\}$.

**EXAMPLE 5.2**   The equations $x = 5$ and $x^2 = 25$ are not equivalent; the first has the solution set $\{5\}$, while the second has the solution set $\{-5, 5\}$.

The process of *solving* an equation consists of transforming it into an equivalent equation whose solution is obvious. Operations of transforming an equation into an equivalent equation include the following:

1. **ADDING** the same number to both sides. Thus, the equations $a = b$ and $a + c = b + c$ are equivalent.
2. **SUBTRACTING** the same number from both sides. Thus, the equations $a = b$ and $a - c = b - c$ are equivalent.
3. **MULTIPLYING** both sides by the same nonzero number. Thus, the equations $a = b$ and $ac = bc (c \neq 0)$ are equivalent.
4. **DIVIDING** both sides by the same nonzero number. Thus, the equations $a = b$ and $\frac{a}{c} = \frac{b}{c} (c \neq 0)$ are equivalent.
5. **SIMPLIFYING** expressions on either side of an equation.

## Linear Equations

A linear equation is one which is in the form $ax + b = 0$ or can be transformed into an equivalent equation in this form. If $a \neq 0$, a linear equation has exactly one solution. If $a = 0$ the equation has no solutions unless $b = 0$, in which case the equation is an identity. An equation which is not linear is called *nonlinear*.

**EXAMPLE 5.3**   $2x + 6 = 0$ is an example of a linear equation in one variable. It has one solution, $-3$. The solution set is $\{-3\}$.

**EXAMPLE 5.4**   $x^2 = 16$ is an example of a nonlinear equation in one variable. It has two solutions, 4 and $-4$. The solution set is $\{4, -4\}$.

Linear equations are solved by the process of *isolating the variable*. The equation is transformed into equivalent equations by simplification, combining all variable terms on one side, all constant terms on the other, then dividing both sides by the coefficient of the variable.

**EXAMPLE 5.5** Solve the equation $3x - 8 = 7x + 9$.

$$3x - 8 = 7x + 9 \qquad \text{Subtract } 7x \text{ from both sides.}$$
$$-4x - 8 = 9 \qquad \text{Add 8 to both sides.}$$
$$-4x = 17 \qquad \text{Divide both sides by } -4.$$
$$x = -\tfrac{17}{4} \qquad \text{Solution set: } \left\{ -\tfrac{17}{4} \right\}$$

## Quadratic Equations

A quadratic equation is one which is in the form $ax^2 + bx + c = 0$, $(a \neq 0)$ (*standard* form), or which can be transformed into this form. There are four methods for solving quadratic equations.

1. **FACTORING.** If the polynomial $ax^2 + bx + c$ has linear factors with rational coefficients, write it in factored form, then apply the zero-factor property that $AB = 0$ only if $A = 0$ or $B = 0$.
2. **SQUARE ROOT PROPERTY.** If the equation is in the form $A^2 = b$, where $b$ is a constant, then its solutions are found as $A = \sqrt{b}$ and $A = -\sqrt{b}$, generally written $A = \pm\sqrt{b}$.
3. **COMPLETING THE SQUARE.**
   a. Write the equation in the form $x^2 + px = q$.
   b. Add $p^2/4$ to both sides to form $x^2 + px + p^2/4 = q + p^2/4$.
   c. The left side is now a perfect square. Write $(x + p/2)^2 = q + p^2/4$ and apply the square root property.
4. **QUADRATIC FORMULA.** The solutions of $ax^2 + bx + c = 0$, $(a \neq 0)$ can always be written as:

$$x = \frac{-b \pm \sqrt{b^2 - 4ac}}{2a}$$

In general, a quadratic equation is solved by first checking whether it is easily factorable. If it is, then the factoring method is used; otherwise the quadratic formula is used.

**EXAMPLE 5.6** Solve $3x^2 + 5x + 2 = 0$
$$3x^2 + 5x + 2 = 0 \qquad \text{Polynomial is factorable using integers}$$
$$(3x + 2)(x + 1) = 0 \qquad \text{Apply the zero-factor property}$$
$$3x + 2 = 0 \qquad \text{or} \qquad x + 1 = 0$$
$$x = -\frac{2}{3} \qquad \text{or} \qquad x = -1$$

**EXAMPLE 5.7** Solve $x^2 + 5x + 2 = 0$
$$x^2 + 5x + 2 = 0 \qquad\qquad\qquad \text{Polynomial is not factorable, use formula}$$
$$x = \frac{-5 \pm \sqrt{5^2 - 4 \cdot 1 \cdot 2}}{2 \cdot 1} \qquad a = 1, b = 5, c = 2$$
$$x = \frac{-5 \pm \sqrt{17}}{2}$$

In the quadratic formula, the quantity $b^2 - 4ac$ is called the *discriminant*. The sign of this quantity determines the number of solutions of a quadratic equation:

| SIGN OF DISCRIMINANT | NUMBER OF REAL SOLUTIONS |
| --- | --- |
| positive | 2 |
| zero | 1 |
| negative | 0 |

Occasionally complex solutions are of interest. Then the discriminant determines the number and type of solutions:

| SIGN OF DISCRIMINANT | NUMBER AND TYPE OF SOLUTIONS |
| --- | --- |
| positive | 2 real solutions |
| zero | 1 real solution |
| negative | 2 imaginary solutions |

**EXAMPLE 5.8**   For $x^2 - 8x + 25 = 0$, find (a) all real solutions; (b) all complex solutions.

Use the quadratic formula with $a = 1$, $b = -8$, $c = 25$.

(a)
$$x = \frac{-(-8) \pm \sqrt{(-8)^2 - 4 \cdot 1 \cdot 25}}{2 \cdot 1}$$

$$x = \frac{8 \pm \sqrt{-36}}{2}$$

No real solution

(b)
$$x = \frac{-(-8) \pm \sqrt{(-8)^2 - 4 \cdot 1 \cdot 25}}{2 \cdot 1}$$

$$x = \frac{8 \pm \sqrt{-36}}{2}$$

$$x = 4 \pm 3i$$

Many equations which are not at first glance linear or quadratic can be reduced to linear or quadratic equations, or can be solved by a factoring method.

**EXAMPLE 5.9**   Solve $x^3 - 5x^2 - 4x + 20 = 0$

$$x^3 - 5x^2 - 4x + 20 = 0 \qquad \text{Factor by grouping}$$
$$x^2(x - 5) - 4(x - 5) = 0$$
$$(x - 5)(x^2 - 4) = 0$$
$$(x - 5)(x - 2)(x + 2) = 0$$
$$x = 5 \text{ or } x = 2 \text{ or } x = -2$$

## Equations Containing Radicals

Equations containing radicals require an additional operation: In general, the equation $a = b$ is not equivalent to the equation $a^n = b^n$; however, if $n$ is odd, they have the same real solutions. If $n$ is even, all solutions of $a = b$ are found among the solutions of $a^n = b^n$. Hence it is permissible to raise both sides of an equation to an odd power, and also permissible to raise both sides to an even power if all solutions of the resulting equation are checked to see if they are solutions of the original equation.

**EXAMPLE 5.10**   Solve $\sqrt{x + 2} = x - 4$

$$\sqrt{x + 2} = x - 4 \qquad \text{Square both sides.}$$
$$(\sqrt{x + 2})^2 = (x - 4)^2$$
$$x + 2 = x^2 - 8x + 16$$
$$0 = x^2 - 9x + 14$$
$$0 = (x - 2)(x - 7)$$
$$x = 2 \qquad \text{or} \qquad x = 7$$

Check: $x = 2$: $\sqrt{2 + 2} = 2 - 4$?  $\qquad x = 7$: $\sqrt{7 + 2} = 7 - 4$?

$\qquad\qquad 2 \neq -2$ $\qquad\qquad\qquad\qquad 3 = 3$

$\qquad\quad$ Not a solution $\qquad\qquad\qquad$ 7 is the only solution

## Applications: Formulas, Literal Equations, and Equations in More Than One Variable

In these situations, letters are used as coefficients rather than particular numbers. However, the procedures for solving for a specified variable are essentially the same; the other variables are simply treated as constants:

**EXAMPLE 5.11**   Solve $A = P + Prt$ for $P$.

This equation is linear in $P$, the specified variable. Factor out $P$, then divide by the coefficient of $P$.

$$A = P + Prt$$
$$A = P(1 + rt)$$
$$\frac{A}{1 + rt} = P$$
$$P = \frac{A}{1 + rt}$$

**EXAMPLE 5.12**　Solve $s = \frac{1}{2}gt^2$ for $t$.

This equation is quadratic in $t$, the specified variable. Isolate $t^2$, then apply the square root property.

$$s = \frac{1}{2}gt^2$$

$$\frac{2s}{g} = t^2$$

$$t = \pm\sqrt{\frac{2s}{g}}$$

Frequently, but not always, in applied situations, only the positive solutions are retained: $t = \sqrt{2s/g}$.

## Applications: Word Problems

Here, a situation is described and questions are posed in ordinary language. It is necessary to form a model of the situation using variables to stand for unknown quantities, construct an equation (later, an inequality or system of equations) that describes the relation among the quantities, solve the equation, then interpret the solution to answer the original questions.

**EXAMPLE 5.13**　A right triangle has sides whose lengths are three consecutive even integers. Find the lengths of the sides.

Sketch a figure as in Fig. 5-1:

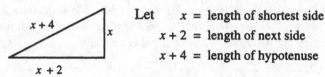

| Let | $x$ = | length of shortest side |
| | $x + 2$ = | length of next side |
| | $x + 4$ = | length of hypotenuse |

Figure 5-1

Now apply the Pythagorean theorem: In a right triangle with sides $a$, $b$, $c$, $a^2 + b^2 = c^2$. Hence,

$$x^2 + (x + 2)^2 = (x + 4)^2$$
$$x^2 + x^2 + 4x + 4 = x^2 + 8x + 16$$
$$2x^2 + 4x + 4 = x^2 + 8x + 16$$
$$x^2 - 4x - 12 = 0$$
$$(x - 6)(x + 2) = 0$$
$$x = 6 \quad \text{or} \quad x = -2$$

The negative answer is discarded. Hence, the lengths of the sides are: $x = 6$, $x + 2 = 8$, and $x + 4 = 10$.

## SOLVED PROBLEMS

**5.1.** Solve: $\dfrac{x}{5} - \dfrac{3x}{4} = 2 - \dfrac{x}{8}$

$$\frac{x}{5} - \frac{3x}{4} = 2 - \frac{x}{8} \qquad \text{Multiply both sides by 40, the LCD of all fractions.}$$

$$40 \cdot \frac{x}{5} - 40 \cdot \frac{3x}{4} = 80 - 40 \cdot \frac{x}{8}$$

$$8x - 30x = 80 - 5x$$

$$-17x = 80$$

$$x = -\frac{80}{17}$$

**5.2.** Solve: $2(3x + 4) + 5(6x - 7) = 7(5x - 4) + 1 + x$

Remove parentheses and combine like terms.

$$2(3x + 4) + 5(6x - 7) = 7(5x - 4) + 1 + x$$
$$6x + 8 + 30x - 35 = 35x - 28 + 1 + x$$
$$36x - 27 = 36x - 27$$

This statement is true for all (real) values of the variable; the equation is an identity.

**5.3.** Solve: $5x = 2x - (1 - 3x)$

Remove parentheses, combine like terms, and isolate the variable.

$$5x = 2x - 1 + 3x$$
$$5x = 5x - 1$$
$$0 = -1$$

The statement is true for no value of the variable; the equation has no solution.

**5.4.** Solve: $\dfrac{x + 5}{x - 3} = 7$

Multiply both sides by $x - 3$, the only denominator; then isolate $x$. *Note*: $x \neq 3$.

$$(x - 3)\frac{x + 5}{x - 3} = 7(x - 3)$$
$$x + 5 = 7x - 21$$
$$-6x = -26$$
$$x = \frac{13}{3}$$

**5.5.** Solve: $\dfrac{6}{x + 1} = 5 - \dfrac{6x}{x + 1}$

Multiply both sides by $x + 1$, the only denominator. *Note*: $x \neq -1$.

$$\frac{6}{x + 1} = 5 - \frac{6x}{x + 1}$$
$$(x + 1) \cdot \frac{6}{x + 1} = 5(x + 1) - (x + 1)\frac{6x}{x + 1}$$
$$6 = 5x + 5 - 6x$$
$$1 = -x$$
$$x = -1$$

In this case, since $x \neq -1$, there can be no solution.

**5.6.** Solve: $(x + 5)^2 + (2x - 7)^2 = 82$

Remove parentheses and combine like terms; the resulting quadratic equation is factorable.

$$(x + 5)^2 + (2x - 7)^2 = 82$$
$$x^2 + 10x + 25 + 4x^2 - 28x + 49 = 82$$
$$5x^2 - 18x - 8 = 0$$
$$(5x + 2)(x - 4) = 0$$
$$x = -\frac{2}{5} \quad \text{or} \quad x = 4$$

**5.7.** Solve: $5x^2 + 16x + 2 = 0$

This is not factorable in the integers; use the quadratic formula, with $a = 5$, $b = 16$, $c = 2$.

$$5x^2 + 16x + 2 = 0$$

$$x = \frac{-16 \pm \sqrt{16^2 - 4 \cdot 5 \cdot 2}}{2 \cdot 5}$$

$$x = \frac{-16 \pm \sqrt{216}}{10}$$

$$x = \frac{-16 \pm 6\sqrt{6}}{10}$$

$$x = \frac{-8 \pm 3\sqrt{6}}{5}$$

**5.8.** Solve $x^2 - 8x + 13 = 0$ by completing the square.

$$x^2 - 8x + 13 = 0$$
$$x^2 - 8x = -13 \qquad \left[\tfrac{1}{2}(-8)\right]^2 = (-4)^2 = 16$$
$$x^2 - 8x + 16 = 3 \qquad \text{Add 16 to both sides.}$$
$$(x - 4)^2 = 3$$
$$x - 4 = \pm\sqrt{3}$$
$$x = 4 \pm \sqrt{3}$$

**5.9.** Solve: $\dfrac{2}{x} + \dfrac{3}{x + 1} = 4$

$$\frac{2}{x} + \frac{3}{x + 1} = 4$$

$$x(x + 1)\frac{2}{x} + x(x + 1)\frac{3}{x + 1} = 4x(x + 1)$$

$$2(x + 1) + 3x = 4x^2 + 4x$$

$$5x + 2 = 4x^2 + 4x$$

$$0 = 4x^2 - x - 2$$

This is not factorable in the integers; use the quadratic formula, with $a = 4$, $b = -1$, $c = -2$.

$$x = \frac{-(-1) \pm \sqrt{(-1)^2 - 4(4)(-2)}}{2 \cdot 4}$$

$$x = \frac{1 \pm \sqrt{33}}{8}$$

**5.10.** Find all solutions, real and complex, for $x^3 - 64 = 0$.

First factor the polynomial as the difference of two cubes.

$$x^3 - 4^3 = 0$$
$$(x - 4)(x^2 + 4x + 16) = 0$$
$$x = 4 \text{ or } x^2 + 4x + 16 = 0$$

Now apply the quadratic formula to the quadratic factor, with $a = 1$, $b = 4$, $c = 16$.

$$x = \frac{-4 \pm \sqrt{4^2 - 4 \cdot 1 \cdot 16}}{2 \cdot 1}$$

$$x = \frac{-4 \pm \sqrt{-48}}{2}$$

$$x = \frac{-4 \pm 4i\sqrt{3}}{2}$$

$$x = -2 \pm 2i\sqrt{3}$$

Solutions: $4, -2 \pm 2i\sqrt{3}$.

**5.11.** Solve: $x^4 - 5x^2 - 36 = 0$

This is an example of an equation *in quadratic form*. It is convenient, although not necessary, to introduce the substitution $u = x^2$. Then $u^2 = x^4$ and the equation becomes:

$$u^2 - 5u - 36 = 0 \qquad \text{This is factorable in the integers.}$$
$$(u - 9)(u + 4) = 0$$
$$u = 9 \quad \text{or} \quad u = -4$$

Now undo the original substitution $x^2 = u$.

$$x^2 = 9 \quad \text{or} \quad x^2 = -4$$
$$x = \pm 3 \qquad \text{no real solution}$$

**5.12.** Solve: $x^{2/3} - x^{1/3} - 6 = 0$

This equation is in quadratic form. Introduce the substitution $u = x^{1/3}$. Then $u^2 = x^{2/3}$ and the equation becomes:

$$u^2 - u - 6 = 0$$
$$(u - 3)(u + 2) = 0$$
$$u = 3 \quad \text{or} \quad u = -2$$

Now undo the original substitution $x^{1/3} = u$.

$$x^{1/3} = 3 \quad \text{or} \quad x^{1/3} = -2$$
$$x = 3^3 \qquad x = (-2)^3$$
$$x = 27 \qquad x = -8$$

**5.13.** Solve: $\sqrt{2x} = \sqrt{x + 1} + 1$

Square both sides, noting that the right side is a binomial.

$$\sqrt{2x} = \sqrt{x + 1} + 1$$
$$(\sqrt{2x})^2 = (\sqrt{x + 1} + 1)^2$$
$$2x = x + 1 + 2\sqrt{x + 1} + 1$$

Now isolate the term containing the square root and square again.

$$x - 2 = 2\sqrt{x + 1}$$
$$(x - 2)^2 = (2\sqrt{x + 1})^2$$
$$x^2 - 4x + 4 = 4(x + 1)$$
$$x^2 - 4x + 4 = 4x + 4$$
$$x^2 - 8x = 0$$
$$x(x - 8) = 0$$

$x = 0$ or $x = 8$ Check: $x = 0: \sqrt{2 \cdot 0} = \sqrt{0 + 1} + 1?$ $x = 8: \sqrt{2 \cdot 8} = \sqrt{8 + 1} + 1?$

$$0 \neq 1 + 1 \qquad\qquad 4 = 3 + 1$$
$$\text{Not a solution} \qquad \text{8 is the only solution}$$

**5.14.** Solve the literal equation $S = 2xy + 2xz + 2yz$ for $y$.

This equation is linear in $y$, the specified variable. Since all terms involving $y$ are already on one side, get all terms not involving $y$ on the other side, then divide both sides by the coefficient of $y$.

$$S = 2xy + 2xz + 2yz$$
$$S - 2xz = 2xy + 2yz$$

$$S - 2xz = y(2x + 2z)$$

$$\frac{S - 2xz}{2x + 2z} = y$$

$$y = \frac{S - 2xz}{2x + 2z}$$

**5.15.** Solve $\frac{1}{p} + \frac{1}{q} = \frac{1}{f}$ for $f$.

This equation is linear in $f$, the specified variable. Multiply both sides by $pqf$, the LCD of all fractions, then divide both sides by the coefficient of $f$.

$$\frac{1}{p} + \frac{1}{q} = \frac{1}{f}$$

$$pqf \cdot \frac{1}{p} + pqf \cdot \frac{1}{q} = pqf \cdot \frac{1}{f}$$

$$qf + pf = pq$$

$$f(q + p) = pq$$

$$f = \frac{qp}{q + p}$$

**5.16.** Solve $s = \frac{1}{2}gt^2 - v_0 t + s_0$ for $t$.

This equation is quadratic in $t$, the specified variable. Get the equation into standard form for quadratic equations:

$$s = \frac{1}{2}gt^2 - v_0 t + s_0$$

$$\frac{1}{2}gt^2 - v_0 t + s_0 - s = 0$$

Now apply the quadratic formula with $a = \frac{1}{2}g$, $b = -v_0$, $c = s_0 - s$.

$$t = \frac{-(-v_0) \pm \sqrt{(-v_0)^2 - 4\left(\frac{1}{2}g\right)(s_0 - s)}}{2\left(\frac{1}{2}g\right)}$$

$$t = \frac{v_0 \pm \sqrt{v_0^2 - 2g(s_0 - s)}}{g}$$

**5.17.** $9000 is to be invested, part at 6% interest, and part at 10% interest. How much should be invested at each rate if a total return of 9% is desired?

Use the formula $I = Prt$ with $t$ understood to be one year. Let $x$ = amount invested at 6%; a tabular arrangement is helpful:

|  | *P*: AMOUNT INVESTED | *r*: RATE OF INTEREST | *I*: INTEREST EARNED |
|---|---|---|---|
| First account | $x$ | 0.06 | $0.06x$ |
| Second account | $9000 - x$ | 0.1 | $0.1(9000 - x)$ |
| Total investment | 9000 | 0.09 | $0.09(9000)$ |

Since the interest earned is the total of the interest on the two investments, write:

$$0.06x + 0.1(9000 - x) = 0.09(9000)$$

Solving yields:

$$0.06x + 900 - 0.1x = 810$$

$$-0.04x = -90$$

$$x = 2250$$

Therefore, $2250 should be invested at 6% and $9000 - x = \$6750$ should be invested at 10%.

**5.18.** A box with a square base and no top is to be made from a square piece of cardboard by cutting out a 3-inch square from each corner and folding up the sides. If the box is to hold 75 cubic inches, what size piece of cardboard should be used?

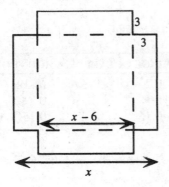 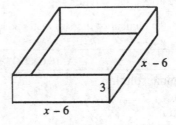

Figure 5-2

Sketch a figure (see Fig. 5-2).
Let $x$ = length of side of original piece. Then $x - 6$ = length of side of box.
Use volume = (length)(width)(height):

$$3(x - 6)^2 = 75$$

$$(x - 6)^2 = 25$$

$$x - 6 = \pm 5$$

$$x = 6 \pm 5$$

Thus, $x = 11$ in or $x = 1$ in. Clearly, the latter does not make sense; hence, the dimensions of the original cardboard must be 11 in square.

**5.19.** Two people have a walkie-talkie set with a range of $\frac{3}{4}$ mi. One of them starts walking at noon in an easterly direction, at a rate of 3 mph. Five minutes later the other person starts walking in a westerly direction, at a rate of 4 mph. At what time will they reach the range of the device?

Use distance = (rate)(time). Let $t$ = time since noon. A tabular arrangement is helpful.

| | TIME WALKED | RATE OF WALKING | DISTANCE |
|---|---|---|---|
| First person | $t$ | 3 | $3t$ |
| Second person | $t - \frac{5}{60}$ | 4 | $4\left(t - \frac{5}{60}\right)$ |

Since the distances add up to the total distance of $\frac{3}{4}$ mi, this yields:

$$3t + 4\left(t - \frac{5}{60}\right) = \frac{3}{4}$$

$$3t + 4t - \frac{1}{3} = \frac{3}{4}$$

$$7t = \frac{1}{3} + \frac{3}{4}$$

$$t = \frac{\frac{1}{3} + \frac{3}{4}}{7}$$

$$t = \frac{13}{84}$$

The time will be noon plus $\frac{13}{84}$ hours, or approximately 12:09 p.m.

**5.20.** A container is filled with 8 liters of a 20% salt solution. How many liters of pure water must be added to produce a 15% salt solution?

Let $x$ = the number of liters of water added. A tabular arrangement is helpful.

|  | AMOUNT OF SOLUTION | PERCENTAGE OF SALT | AMOUNT OF SALT |
|---|---|---|---|
| Original solution | 8 | 0.2 | (0.2)8 |
| Water | $x$ | 0 | 0 |
| Mixture | $8 + x$ | 0.15 | $0.15(8 + x)$ |

Since the amounts of salt in the original solutions and the added water must add up to the amount of salt in the mixture, this yields:

$$(0.2)8 + 0 = 0.15(8 + x)$$

$$1.6 = 1.2 + 0.15x$$

$$0.4 = 0.15x$$

$$x = \frac{0.4}{0.15} \quad \text{or} \quad 2\frac{2}{3} \text{ liters}$$

**5.21.** Machine A can perform a job in 6 hours, working alone. Machine B can complete the same job in 10 hours, working alone. How long would it take the two machines, working together, to complete the job?

Use quantity of work = (rate)(time). Note that if a machine can do a job in $x$ hours, it performs $1/x$ of the work in one hour; that is, its rate is $1/x$ job per hour. Let $t$ = the time worked by each machine. A tabular arrangement is helpful.

|  | RATE | TIME | QUANTITY OF WORK |
|---|---|---|---|
| Machine A | 1/6 | $t$ | $t/6$ |
| Machine B | 1/10 | $t$ | $t/10$ |

Since the quantity of work performed by the two machines totals to one entire job, this yields:

$$\frac{t}{6} + \frac{t}{10} = 1$$

$$30 \cdot \frac{t}{6} + 30 \cdot \frac{t}{10} = 30$$

$$5t + 3t = 30$$

$$8t = 30$$

$$t = \frac{15}{4}$$

The time would be $3\frac{3}{4}$ hours.

## SUPPLEMENTARY PROBLEMS

**5.22.** Solve: $3 - \dfrac{x}{8} = \dfrac{5x}{2} - \dfrac{2}{3}(x - 4) + 5$   *Ans.*   $-\dfrac{112}{47}$

**5.23.** Solve: $7(x - 6) - 6(x + 3) = 5(x - 6) - 2(3 + 2x)$   *Ans.*   No solution.

**5.24.** Solve: $\dfrac{5}{x} - \dfrac{4}{x(x - 2)} = \dfrac{x - 4}{x - 2}$   *Ans.*   7

**5.25.** Find all real solutions:

(a) $x^2 - 9x = 36$; (b) $3x^2 = 2x + 8$; (c) $4x^2 + 3x + 5 = 0$; (d) $x^2 - 5 = 2x + 3$;

(e) $(x - 8)(x + 6) = 32$; (f) $8x^2 - 3x + 4 = 3x^2 + 12$; (g) $(x - 5)^2 = 7$; (h) $4x^2 + 3x - 5 = 0$

*Ans.*   (a) $\{-3, 12\}$; (b) $\left\{-\dfrac{4}{3}, 2\right\}$; (c) no real solutions; (d) $\{-2, 4\}$;

(e) $\{-8, 10\}$; (f) $\left\{-1, \dfrac{8}{5}\right\}$; (g) $5 \pm \sqrt{7}$; (h) $\left\{\dfrac{-3 + \sqrt{89}}{8}, \dfrac{-3 - \sqrt{89}}{8}\right\}$

**5.26.** Solve:

(a) $\sqrt[3]{5x + 9} = -6$     (b) $\sqrt{5x + 9} = -6$        *Ans.*   (a) $-45$        (b) No solution.

**5.27.** Find all real solutions:

(a) $x^4 - x^2 - 6 = 0$; (b) $x^{2/3} - 3x^{1/3} - 4 = 0$; (c) $x^6 + 6x^3 - 16 = 0$

*Ans.*   (a) $\{-\sqrt{3}, \sqrt{3}\}$; (b) $\{-1, 64\}$; (c) $\{-2, \sqrt[3]{2}\}$

**5.28.** Solve: (a) $x - \sqrt{x} = 12$; (b) $\sqrt{2x + 1} + 1 = x$; (c) $\sqrt{4x + 1} - \sqrt{2x - 3} = 2$

*Ans.*   (a) $\{16\}$; (b) $\{4\}$; (c) $\{2, 6\}$

**5.29.** Find all complex solutions for $x^3 - 5x^2 + 4x - 20 = 0$   *Ans.*   $5, 2i, -2i$

**5.30.** Solve: $\dfrac{1}{p} + \dfrac{1}{q} = \dfrac{1}{f}$ for $q$.     *Ans.*   $q = \dfrac{pf}{p - f}$

**5.31.** Solve: $LI^2 + RI + \dfrac{1}{C} = 0$ for $I$.        *Ans.*   $I = \dfrac{-RC \pm \sqrt{R^2C^2 - 4LC}}{2LC}$

**5.32.** Solve: $(x - h)^2 + (y - k)^2 = r^2$ for $y$.        *Ans.*   $y = k \pm \sqrt{r^2 - (x - h)^2}$

**5.33.** Solve for $y$ in terms of $x$: (a) $3x - 5y = 8$; (b) $x^2 - 2xy + y^2 = 4$; (c) $\dfrac{x + y}{x - y} = 5$; (d) $x = \sqrt{y^2 - 2y}$

*Ans.*   (a) $y = \dfrac{3x - 8}{5}$; (b) $y = x + 2$ or $y = x - 2$; (c) $y = \frac{2}{3}x$; (d) $y = 1 \pm \sqrt{x^2 + 1}$

**5.34.** A rectangle has perimeter 44 cm. Find its dimensions if its length is 5 cm less than twice its width.

*Ans.*   Width = 9 cm, length = 13 cm

**5.35.** Solve the walkie-talkie problem (5.19) if the two people start walking at the same time, but the second person walks north.

*Ans.*   Exactly 12:09 p.m.

**5.36.** A shop wishes to blend coffee priced at $6.50 per pound with coffee priced at $9.00 per pound in order to yield 60 pounds of a blend to sell for $7.50 per pound. How much of each type of coffee should be used?

*Ans.* 36 pounds of the $6.50-per-pound coffee, 24 pounds of the $9.00-per-pound coffee.

**5.37.** A container is filled with 8 centiliters of a 30% acid solution. How many centiliters of pure acid must be added to produce a 50% acid solution?

*Ans.* 3.2 cl

**5.38.** A chemistry stockroom has two alcohol solutions, a 30% and a 75% solution. How many deciliters of each must be mixed to obtain 90 deciliters of a 65% solution?

*Ans.* 20 dl of the 30% solution, 70 dl of the 75% solution

**5.39.** A 6-gallon radiator is filled with a 40% solution of antifreeze in water. How much of the solution must be drained and replaced with pure antifreeze to obtain a 65% solution?

*Ans.* 2.5 gallons

**5.40.** Machine A can complete a job in 8 hours, working alone. Working together with machine B, the job can be completed in 5 hours. How long would it take machine B, working alone, to complete the job?

*Ans.* $13\frac{1}{3}$ hours

**5.41.** Machine A can do a job, working alone, in 4 hours less than machine B. Working together, they can complete the job in 5 hours. How long would it take each machine, working alone, to complete the job?

*Ans.* Machine A: 8.4 hours; machine B: 12.4 hours, approximately

# CHAPTER 6

# Linear and Nonlinear Inequalities

## Inequality Relations

The number *a is less than b*, written $a < b$, if $b - a$ is positive. Then *b is greater than a*, written $b > a$. If $a$ is either less than or equal to $b$, this is written $a \le b$. Then *b* is greater than or equal to *a*, written $b \ge a$. *Geometrical Interpretation*: If $a < b$, then $a$ is to the left of $b$ on a real number line (Fig. 6-1). If $a > b$, then $a$ is to the right of $b$.

**EXAMPLE 6.1**

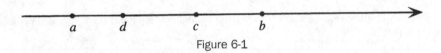

Figure 6-1

In Fig. 6-1, $a < d$ and $b > c$. Also, $a < c$ and $b > d$.

## Combined Inequalities and Intervals

If $a < x$ and $x < b$, the two statements are often combined to write: $a < x < b$. The set of all real numbers $x$ satisfying $a < x < b$ is called an *open interval* and is written $(a,b)$. Similarly, the set of all real numbers $x$ satisfying the combined inequality $a \le x \le b$ is called a closed interval and is written $[a,b]$. The following table shows various common inequalities and their interval representations.

| Inequality | Notation | Graph |
|---|---|---|
| $a < x < b$ | $(a,b)$ | ⟶ $x$, $(a$ ... $)b$ |
| $a \le x \le b$ | $[a,b]$ | ⟶ $x$, $[a$ ... $]b$ |
| $a < x \le b$ | $(a,b]$ | ⟶ $x$, $(a$ ... $]b$ |
| $a \le x < b$ | $[a,b)$ | ⟶ $x$, $[a$ ... $)b$ |
| $x > a$ | $(a,\infty)$ | ⟶ $x$, $(a$ |

| Inequality | Notation | Graph |
|:---:|:---:|:---:|
| $x \geq a$ | $[a, \infty)$ | |
| $x < b$ | $(-\infty, b)$ | |
| $x \leq b$ | $(-\infty, b]$ | |

## Inequality Statements Involving Variables

An inequality statement involving variables, like an equation, is in general neither true nor false; rather, its truth depends on the value(s) of the variable(s). For inequality statements in one variable, a value of the variable that makes the statement true is a solution to the inequality. The set of all solutions is called the *solution set* of the inequality.

## Equivalent Inequalities

Inequalities are equivalent if they have the same solution sets.

**EXAMPLE 6.2**   The inequalities $x < -5$ and $x + 5 < 0$ are equivalent. Each has the solution set consisting of all real numbers less than $-5$, that is, $(-\infty, -5)$.

The process of *solving* an inequality consists of transforming it into an equivalent inequality whose solution is obvious. Operations of transforming an inequality into an equivalent inequality include the following:

1. **ADDING OR SUBTRACTING:** The inequalities $a < b$, $a + c < b + c$, and $a - c < b - c$ are equivalent, for $c$ any real number.
2. **MULTIPLYING OR DIVIDING BY A POSITIVE NUMBER:** The inequalities $a < b$, $ac < bc$, and $a/c < b/c$ are equivalent, for $c$ any positive real number.
3. **MULTIPLYING OR DIVIDING BY A NEGATIVE NUMBER:** The inequalities $a < b$, $ac > bc$, and $a/c > b/c$ are equivalent, for $c$ any negative real number. Note that the sense of an inequality reverses upon multiplication or division by a negative number.
4. **SIMPLIFYING** expressions on either side of an inequality.

Similar rules apply for inequalities of the form $a > b$ and so on.

## Linear Inequalities

A linear inequality is one which is in the form $ax + b < 0$, $ax + b > 0$, $ax + b \leq 0$, or $ax + b \geq 0$, or can be transformed into an equivalent inequality in this form. In general, linear inequalities have infinite solution sets in one of the forms shown in the table above. Linear inequalities are solved by isolating the variable in a manner similar to solving equations.

**EXAMPLE 6.3**   Solve: $5 - 3x > 4$.

$$5 - 3x > 4$$
$$-3x > -1$$
$$x < \frac{1}{3}$$

Note that the sense of the inequality was reversed by dividing both sides by $-3$.
An inequality that is not linear is called nonlinear.

## Solving Nonlinear Inequalities

An inequality for which the left side can be written as a product or quotient of linear factors (or prime quadratic factors) can be solved through a *sign diagram*. If any such factor is not zero on an interval, then it is either positive on the whole interval or negative on the whole interval. Hence:

1. Determine the points where each factor is 0. These are called the *critical points*.
2. Draw a number line and show the critical points.
3. Determine the sign of each factor in each interval; then, using laws of multiplication or division, determine the sign of the entire quantity on the left side of the inequality.
4. Write the solution set.

**EXAMPLE 6.4**   Solve: $(x-1)(x+2) > 0$

The critical points are 1 and $-2$, where, respectively, $x-1$ and $x+2$ are zero. Draw a number line showing the critical points (Fig. 6-2). These points divide the real number line into the intervals $(-\infty,-2)$, $(-2,1)$, and $(1,\infty)$. In $(-\infty,-2)$, $x-1$ and $x+2$ are negative; hence the product is positive. In $(-2,1)$, $x-1$ is negative and $x+2$ is positive; hence the product is negative. In $(1,\infty)$, both factors are positive; hence the product is positive.

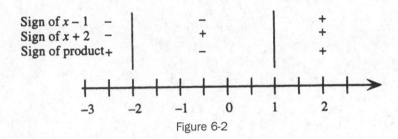

Figure 6-2

The inequality holds when $(x-1)(x+2)$ is *positive*. Hence the solution set consists of the intervals: $(-\infty,-2) \cup (1,\infty)$.

## SOLVED PROBLEMS

**6.1.** Solve: $3(y-5) -4(y+6) \le 7$

Eliminate parentheses, combine terms, and isolate the variable:

$$3(y-5) -4(y+6) \le 7$$

$$3y - 15 - 4y - 24 \le 7$$

$$-y - 39 \le 7$$

$$-y \le 46$$

$$y \ge -46$$

The solution set is $[-46,\infty)$.

**6.2.** Solve: $\dfrac{2x-3}{3} - \dfrac{5x+4}{6} > 5 - \dfrac{3x}{8}$

Multiply both sides by 24, the LCD of all fractions, then proceed as in the previous problem.

$$\frac{2x-3}{3} - \frac{5x+4}{6} > 5 - \frac{3x}{8}$$

$$24 \cdot \frac{(2x-3)}{3} - 24 \cdot \frac{(5x+4)}{6} > 120 - 24 \cdot \frac{3x}{8}$$

$$16x - 24 - 20x - 16 > 120 - 9x$$

$$-4x - 40 > 120 - 9x$$
$$5x > 160$$
$$x > 32$$

The solution set is $(32, \infty)$.

**6.3.** Solve: $-8 < 2x - 7 \le 5$

A combined inequality of this type can be solved by isolating the variable in the middle.

$$-8 < 2x - 7 \le 5$$
$$-1 < 2x \le 12$$
$$-\frac{1}{2} < x \le 6$$

The solution set is $(-\frac{1}{2}, 6]$.

**6.4.** Solve: $0 < 3 - 5x \le 10$

$$0 < 3 - 5x \le 10$$
$$-3 < -5x \le 7$$
$$\frac{3}{5} > x \ge -\frac{7}{5}$$
$$-\frac{7}{5} \le x < \frac{3}{5}$$

The solution set is $[-\frac{7}{5}, \frac{3}{5})$.

**6.5.** A chemical solution is to be kept between $-30$ and $-22.5°C$. To what range in Fahrenheit degrees does this correspond?

Write $-30 < C < -22.5$ and use $C = \frac{5}{9}(F - 32)$.

$$-30 < C < -22.5$$
$$-30 < \frac{5}{9}(F - 32) < -22.5$$
$$-54 < F - 32 < -40.5$$
$$-22 < F < -8.5$$

The range is between $-22$ and $-8.5°F$.

**6.6.** Solve: $x^2 - 8x \le 20$

Get 0 on the right side, put the left side into factored form, then form a sign diagram.

$$x^2 - 8x - 20 \le 0$$
$$(x - 10)(x + 2) \le 0$$

The critical points are 10 and $-2$, where, respectively, $x - 10$ and $x + 2$ are zero. Draw a number line showing the critical points (Fig. 6-3).

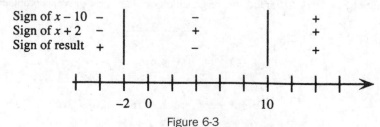

Figure 6-3

The critical points divide the real number line into the intervals $(-\infty, -2)$, $(-2, 10)$, and $(10, \infty)$. In $(-\infty, -2)$, $x - 10$ and $x + 2$ are negative, hence the product is positive. In $(-2, 10)$, $x - 10$ is negative and $x + 2$ is positive; hence the product is negative. In $(10, \infty)$, both factors are positive; hence the product is positive. The equation part of the inequality is satisfied at both critical points, and the inequality holds when $(x + 2)(x - 10)$ is negative; hence the solution set is $[-2, 10]$.

**6.7.** Solve: $2x^2 + 2 \geq 5x$

Get 0 on the right side, put the left side into factored form, then form a sign diagram

$$2x^2 - 5x + 2 \geq 0$$

$$(x - 2)(2x - 1) \geq 0$$

Draw a number line showing the critical points $\frac{1}{2}$ and 2 (Fig. 6-4).

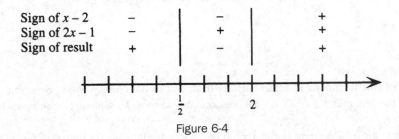

Figure 6-4

The critical points divide the real number line into the intervals $(-\infty, \frac{1}{2})$, $(\frac{1}{2}, 2)$, and $(2, \infty)$. The product has sign, respectively, positive, negative, positive in these intervals. The equation part of the inequality is satisfied at both critical points, and the inequality holds when $(2x - 1)(x - 2)$ is positive, hence the solution set is $(-\infty, \frac{1}{2}] \cup [2, \infty)$.

**6.8.** Solve: $x^3 < x^2 + 6x$

Get 0 on the right side, put the left side into factored form, then form a sign diagram.

$$x^3 - x^2 - 6x < 0$$

$$x(x - 3)(x + 2) < 0$$

Draw a number line showing the critical points $-2$, 0, and 3 (Fig. 6-5).

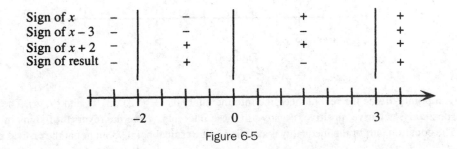

Figure 6-5

The critical points divide the real number line (Fig. 6-5) into the intervals $(-\infty, -2)$, $(-2, 0)$, $(0, 3)$, and $(3, \infty)$. The product has sign, respectively, negative, positive, negative, positive in these intervals. The inequality holds when $x(x - 3)(x + 2)$ is negative, hence the solution set is $(-\infty, -2) \cup (0, 3)$.

**6.9.** Solve: $\dfrac{x + 5}{x - 3} \leq 0$

Draw a number line showing the critical points $-5$ and 3 (Fig. 6-6).

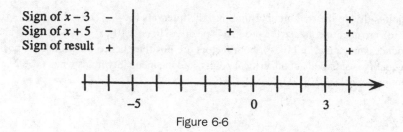

Figure 6-6

The critical points divide the real number line into the intervals $(-\infty, -5)$, $(-5, 3)$, and $(3, \infty)$. The quotient has sign, respectively, positive, negative, positive in these intervals. The equation part of the inequality is satisfied at the critical point $-5$, but not at the critical point 3, since the expression $\frac{x+5}{x-3}$ is not defined there. The inequality holds when $\frac{x+5}{x-3}$ is negative; hence the solution set is $[-5, 3)$.

**6.10.** Solve: $\frac{2x}{x-3} \geq 3$

The solution of this inequality statement differs from that of the corresponding equation. If both sides were multiplied by the denominator $x-3$, it would be necessary to consider separately the cases where this is positive, zero, or negative.

It is preferable to get 0 on the right side and combine the left side into one fraction, then form a sign diagram.

$$\frac{2x}{x-3} - 3 \geq 0$$

$$\frac{2x}{x-3} - \frac{3(x-3)}{x-3} \geq 0$$

$$\frac{9-x}{x-3} \geq 0$$

Draw a number line showing the critical points 3 and 9 (Fig. 6-7).

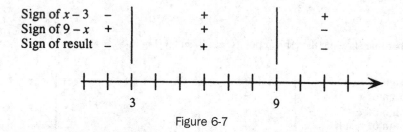

Figure 6-7

The critical points divide the real number line into the intervals $(-\infty, 3)$, $(3, 9)$, and $(9, \infty)$. The quotient has sign, respectively, negative, positive, negative in these intervals. (Note the reversal of signs in the chart for $9-x$.) The equation part of the inequality is satisfied at the critical point 9, but not at the critical point 3, since the expression $\frac{9-x}{x-3}$ is not defined there. The inequality holds when $\frac{9-x}{x-3}$ is positive, hence the solution set is $(3, 9]$.

**6.11.** Solve: $\dfrac{(x-2)^{1/3}(2x+3)^2}{(x+5)^3(x^2+4)} \geq 0$

Draw a number line showing the critical points $-5, -\frac{3}{2}$, and 2 (Fig. 6-8). Note that the factor $x^2+4$ has no critical point; its sign is positive for all real $x$; hence it has no effect on the sign of the result.

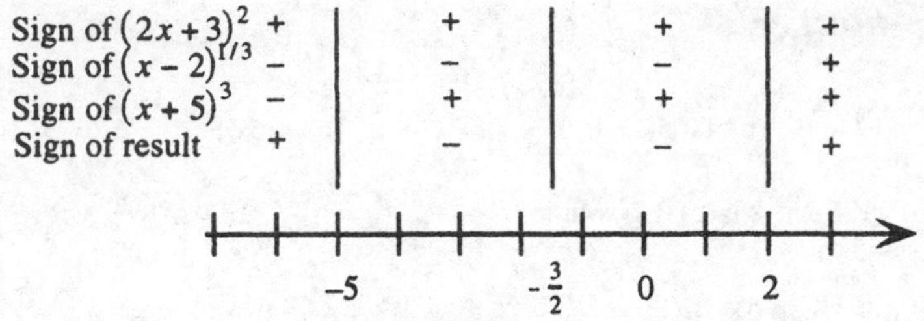

Sign of $(2x + 3)^2$    +
Sign of $(x - 2)^{1/3}$    –
Sign of $(x + 5)^3$    –
Sign of result    +

Figure 6-8

The critical points divide the real number line into the intervals $(-\infty, -5), (-5, -\frac{3}{2}), (-\frac{3}{2}, 2)$, and $(2, \infty)$. The quotient has sign, respectively, positive, negative, negative, positive in these intervals. (Note that the factor $(2x + 3)^2$ is positive except at its critical point.) The equation part of the inequality is satisfied at the critical points $-\frac{3}{2}$ and 2, but not at the critical point $-5$. The inequality holds when the expression under consideration is positive; hence the solution set is $(-\infty, -5) \cup \{-\frac{3}{2}\} \cup [2, \infty)$.

**6.12.** For what values of $x$ does the expression $\sqrt{9 - x^2}$ represent a real number?

The expression represents a real number when the quantity $9 - x^2$ is nonnegative. Solve the inequality statement $9 - x^2 \geq 0$, or $(3 - x)(3 + x) \geq 0$, by drawing a number line showing the critical points 3 and $-3$ (Fig. 6-9).

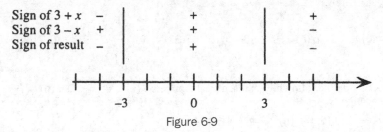

Sign of $3 + x$    –
Sign of $3 - x$    +
Sign of result    –

Figure 6-9

The critical points divide the real number line into the intervals $(-\infty, -3), (-3, 3)$, and $(3, \infty)$. The product has sign, respectively, negative, positive, negative in these intervals. The equation part of the inequality is satisfied at the critical points, and the inequality holds when $9 - x^2$ is positive, hence the expression $\sqrt{9 - x^2}$ represents a real number when $x$ is in $[-3, 3]$.

**6.13.** For what values of $x$ does the expression $\sqrt{\dfrac{x}{(2 - x)(5 + x)}}$ represent a real number?

The expression represents a real number when the quantity under the radical is nonnegative. Solve the inequality $\dfrac{x}{(2 - x)(5 + x)} \geq 0$ by drawing a number line showing the critical points $-5$, 0, and 2 (Fig. 6-10).

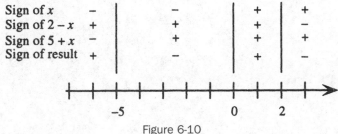

Sign of $x$    –
Sign of $2 - x$    +
Sign of $5 + x$    –
Sign of result    +

Figure 6-10

The critical points divide the real number line into the intervals $(-\infty, -5), (-5, 0), (0, 2)$, and $(2, \infty)$. The quotient has sign, respectively, positive, negative, positive, negative in these intervals. The equation part of the inequality is satisfied only at the critical point 0, and the inequality holds when the quantity under the radical is positive; hence the entire expression represents a real number when $x$ is in $(-\infty, -5) \cup [0, 2)$.

## SUPPLEMENTARY PROBLEMS

**6.14.** Solve: (a) $\dfrac{2x + 7}{5} < \dfrac{5x - 3}{2}$; (b) $0.05(2x - 3) + 0.02x > 15$; (c) $4(5x - 6) - 3(6x - 3) > 2x + 1$

　　*Ans.*　(a) $(\frac{29}{21}, \infty)$; (b) $(126.25, \infty)$; (c) No solution

**6.15.** Solve: (a) $-0.01 < x - 5 < 0.01$; (b) $\dfrac{1}{2} \le \dfrac{5x - 6}{4} < 7$; (c) $-6 < 3 - 7x \le 8$

　　*Ans.*　(a) $(4.99, 5.01)$; (b) $[\frac{8}{5}, \frac{34}{5})$ (c) $[-\frac{5}{7}, \frac{9}{7})$

**6.16.** Solve: (a) $5x - x^2 < 6$; (b) $(x + 6)^2 \ge (2x - 1)^2$; (c) $t^2 + (t + 1)^2 > (t + 2)^2$

　　*Ans.*　(a) $(-\infty, 2) \cup (3, \infty)$; (b) $[-\frac{5}{3}, 7]$; (c) $(-\infty, -1) \cup (3, \infty)$

**6.17.** Solve: (a) $x^2 \le 1$; (b) $x^2 + 1 < 1$; (c) $\dfrac{1}{x} < 1$; (d) $\dfrac{1}{x^2} \le 1$; (e) $\dfrac{1}{1 - x^2} \le 1$

　　*Ans.*　(a) $[-1, 1]$; (b) no solution; (c) $(-\infty, 0) \cup (1, \infty)$; (d) $(-\infty, -1] \cup [1, \infty)$; (e) $(-\infty, -1) \cup \{0\} \cup (1, \infty)$

**6.18.** Solve: (a) $5 > \dfrac{x + 3}{x}$; (b) $\dfrac{-9x^2}{x^2 - 9} \le 0$; (c) $\dfrac{x^2 - 4x}{3x^2 - 12} \ge 0$

　　*Ans.*　(a) $(-\infty, 0) \cup (\frac{3}{4}, \infty)$; (b) $(-\infty, -3) \cup \{0\} \cup (3, \infty)$; (c) $(-\infty, -2) \cup [0, 2) \cup [4, \infty)$

**6.19.** For what values of $x$ do the following represent real numbers ? (a) $\sqrt{x^2 - 25}$; (b) $\sqrt{\dfrac{x - 4}{x + 4}}$

　　*Ans.*　(a) $(-\infty, -5] \cup [5, \infty)$; (b) $(-\infty, -4) \cup [4, \infty)$

**6.20.** For what values of $x$ do the following represent real numbers?

(a) $\dfrac{1}{\sqrt{x^2 - 16}}$; (b) $\dfrac{1}{\sqrt{36 - x^2}}$

　　*Ans.*　(a) $(-\infty, -4) \cup (4, \infty)$; (b) $(-6, 6)$

# CHAPTER 7

# Absolute Value in Equations and Inequalities

## Absolute Value of a Number

The absolute value of a real number $a$, written $|a|$, was defined (Chapter 1) as follows:

$$|a| = \begin{cases} a & \text{if } a \geq 0 \\ -a & \text{if } a < 0 \end{cases}$$

## Absolute Value, Interpreted Geometrically

Geometrically, the absolute value of a real number is the distance of that number from the origin (see Fig. 7-1).

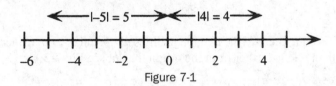

Figure 7-1

Similarly, the distance between two real numbers $a$ and $b$ is the absolute value of their difference: $|a - b|$ or $|b - a|$.

## Properties of Absolute Values

$$|-a| = |a| \qquad |a| = \sqrt{a^2}$$

$$|ab| = |a||b| \qquad |a + b| \leq |a| + |b| \text{ (Triangle inequality)}$$

**EXAMPLE 7.1** (a) $|-5| = |5| = 5$; (b) $|-6| = 6$; $\sqrt{(-6)^2} = \sqrt{36} = 6$, thus, $|-6| = \sqrt{(-6)^2}$.

**EXAMPLE 7.2** (a) $|-5x^2| = |-5||x^2| = 5x^2$; (b) $|3y| = |3||y| = 3|y|$

**EXAMPLE 7.3** Triangle inequality: $|5 + (-7)| = 2 \leq |5| + |-7| = 5 + 7 = 12$

## Absolute Value in Equations

Since $|a|$ is the distance of $a$ from the origin,

1. The equation $|a| = b$ is equivalent to the two equations $a = b$ and $a = -b$, for $b > 0$. (The distance of $a$ from the origin will equal $b$ precisely when $a$ equals $b$ or $-b$.)
2. The equation $|a| = |b|$ is equivalent to the two equations $a = b$ and $a = -b$.

**EXAMPLE 7.4**   Solve: $|x + 3| = 5$

Transform into equivalent equations that do not contain the absolute value symbol and solve:

$$x + 3 = 5 \quad \text{or} \quad x + 3 = -5$$
$$x = 2 \qquad\qquad x = -8$$

**EXAMPLE 7.5**   Solve: $|x - 4| = |3x + 1|$

Transform into equivalent equations that do not contain the absolute value symbol and solve:

$$x - 4 = 3x + 1 \quad \text{or} \quad x - 4 = -(3x + 1)$$
$$-2x = 5 \qquad\qquad x - 4 = -3x - 1$$
$$x = -\frac{5}{2} \qquad\qquad 4x = 3$$
$$x = \frac{3}{4}$$

## Absolute Value in Inequalities

For $b > 0$,

1. The inequality $|a| < b$ is equivalent to the double inequality $-b < a < b$. (Since the distance of $a$ from the origin is *less* than $b$, $a$ is closer to the origin than $b$; see Fig. 7-2.)

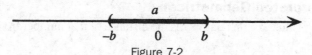

Figure 7-2

2. The inequality $|a| > b$ is equivalent to the two inequalities $a > b$ and $a < -b$. (Since the distance of $a$ from the origin is *greater* than $b$, $a$ is farther from the origin than $b$; see Fig 7-3.)

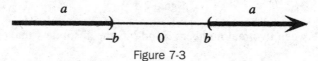

Figure 7-3

**EXAMPLE 7.6**   Solve: $|x - 5| > 3$

Transform into equivalent inequalities that do not contain the absolute value symbol and solve:

$$x - 5 > 3 \quad \text{or} \quad x - 5 < -3$$
$$x > 8 \qquad\qquad x < 2$$

## SOLVED PROBLEMS

**7.1.** Solve: $|x - 7| = 2$

Transform into equivalent equations that do not contain the absolute value symbol and solve:

$$x - 7 = 2 \quad \text{or} \quad x - 7 = -2$$
$$x = 9 \qquad\qquad x = 5$$

**7.2.** Solve: $|x + 5| = 0.01$

$$x + 5 = 0.01 \quad \text{or} \quad x + 5 = -0.01$$
$$x = -4.99 \qquad\qquad x = -5.01$$

**7.3.** Solve: $|6x + 7| = 10$

$$6x + 7 = 10 \quad \text{or} \quad 6x + 7 = -10$$

$$6x = 3 \qquad\qquad 6x = -17$$

$$x = \frac{1}{2} \qquad\qquad x = -\frac{17}{6}$$

**7.4.** Solve: $5|x| - 3 = 6$

First isolate the absolute value expression, then write the two equivalent equations that do not contain the absolute value symbol.

$$5\left|x\right| = 9$$

$$\left|x\right| = \frac{9}{5}$$

$$x = \frac{9}{5} \quad \text{or} \quad x = -\frac{9}{5}$$

**7.5.** Solve: $3|5 - 2x| + 4 = 9$

First isolate the absolute value expression.

$$3\left|5 - 2x\right| = 5$$

$$\left|5 - 2x\right| = \frac{5}{3}$$

Now write and solve the two equivalent equations that do not contain the absolute value symbol.

$$5 - 2x = \frac{5}{3} \quad \text{or} \quad 5 - 2x = -\frac{5}{3}$$

$$-2x = -\frac{10}{3} \qquad\qquad -2x = -\frac{20}{3}$$

$$x = \frac{5}{3} \qquad\qquad x = \frac{10}{3}$$

**7.6.** Solve: $|5x - 3| = -8$

Since the absolute value of a number is never negative, this equation has no solution.

**7.7.** Solve: $|2x - 5| = |8x + 3|$

Transform into equivalent equations that do not contain the absolute value symbol and solve:

$$2x - 5 = 8x + 3 \quad \text{or} \quad 2x - 5 = -(8x + 3)$$

$$-6x = 8 \qquad\qquad 2x - 5 = -8x - 3$$

$$x = -\frac{4}{3} \qquad\qquad 10x = 2$$

$$x = \frac{1}{5}$$

**7.8.** Solve: $|x + 5| > 3$

Transform into equivalent inequalities that do not contain the absolute value symbol and solve:

$$x + 5 > 3 \quad \text{or} \quad x + 5 < -3$$
$$x > -2 \qquad\qquad x < -8$$

Solution: $(-\infty, -8) \cup (-2, \infty)$

**7.9.** Solve: $|x - 3| \leq 10$

Transform into an equivalent double inequality and solve:

$$-10 \leq x - 3 \leq 10$$
$$-7 \leq x \leq 13$$

Solution: $[-7, 13]$

**7.10.** Solve: $4|2x - 7| + 5 < 19$

Isolate the absolute value symbol, then transform into an equivalent double inequality and solve:

$$4\left|2x - 7\right| < 14$$
$$\left|2x - 7\right| < \frac{7}{2}$$
$$-\frac{7}{2} < 2x - 7 < \frac{7}{2}$$
$$\frac{7}{2} < 2x < \frac{21}{2}$$
$$\frac{7}{4} < x < \frac{21}{4}$$

Solution: $\left(\frac{7}{4}, \frac{21}{4}\right)$

**7.11.** Solve: $|5x - 3| > -1$

Since the absolute value of a real number is always positive or zero—hence, always greater than any negative number—all real numbers are solutions.

**7.12.** Write as an inequality statement with and without the absolute value symbol and graph the solutions on a number line: The distance between $x$ and $a$ is less than $\delta$.

In terms of the absolute value symbol, this statement becomes $|x - a| < \delta$. Rewrite as a double linequality and solve:

$$-\delta < x - a < \delta$$
$$a - \delta < x < a + \delta$$

The graph is shown in Fig. 7-4:

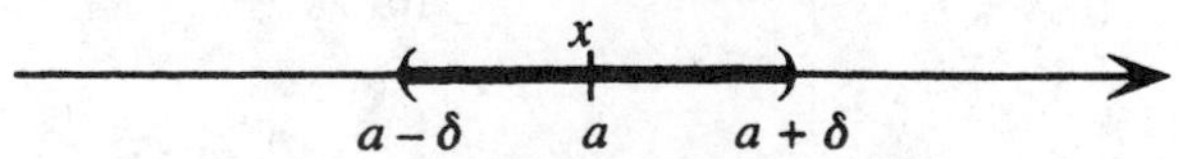

Figure 7-4

## SUPPLEMENTARY PROBLEMS

**7.13.** Prove: $|ab| = |a||b|$. (*Hint:* Consider the cases separately for various signs of $a$ and $b$.)

**7.14.** (a) Prove: for any real number $x$, $-|x| \leq x \leq |x|$. (b) Use part (a) to prove the triangle inequality.

**7.15.** Write as an equation or inequality and solve:

(a) The distance between $x$ and 3 is equal to 7. (b) 5 is twice the distance between $x$ and 6.

(c) The distance between $x$ and $-3$ is more than 2.

*Ans.*   (a) $|x - 3| = 7$; $\{-4, 10\}$; (b) $5 = 2|x - 6|$; $\left\{\frac{17}{2}, \frac{7}{2}\right\}$; (c) $|x + 3| > 2$; $(-\infty, -5) \cup (-1, \infty)$

**7.16.** Solve: (a) $|x + 8| = 5$; (b) $|x + 5| < 8$; (c) $|x - 3| \geq 4$

*Ans.*   (a) $\{-13, -3\}$; (b) $(-13, 3)$; (c) $(-\infty, -1] \cup [7, \infty)$

**7.17.** Solve: (a) $|x| + 8 = 5$; (b) $2|x| + 5 \leq 8$; (c) $|x + 8| - 5 > 1$

*Ans.*   (a) no solution; (b) $\left[-\frac{3}{2}, \frac{3}{2}\right]$; (c) $(-\infty, -14) \cup (-2, \infty)$

**7.18.** Solve: $|5 - 2x| = 3|x + 1|$     *Ans.*   $\left\{-8, \frac{2}{5}\right\}$

**7.19.** Solve: $|3 - 5x| \geq 9$     *Ans.*   $\left(-\infty, -\frac{6}{5}\right] \cup \left[\frac{12}{5}, \infty\right)$

**7.20.** Solve $|3x + 4| + 5 < 1$     *Ans.*   No solution

**7.21.** Solve: (a) $0 < |x - 5| < 8$; (b) $0 < |2x + 3| \leq 7$; (c) $0 < |x - c| \leq \delta$

*Ans.*   (a) $(-3, 5) \cup (5, 13)$; (b) $[-5, -3/2) \cup (-3/2, 2]$; (c) $[c - \delta, c) \cup (c, c + \delta]$

# CHAPTER 8

# Analytic Geometry

## Cartesian Coordinate System

A Cartesian coordinate system consists of two perpendicular real number lines, called *coordinate axes,* that intersect at their origins. Generally one line is horizontal and called the *x*-axis, and the other is vertical and called the *y*-axis. The axes divide the coordinate plane, or *xy*-plane, into four parts, called *quadrants,* and numbered first, second, third, and fourth, or I, II, III, and IV. Points on the axes are not in any quadrant.

## One-to-One Correspondence

A one-to-one correspondence exists between ordered pairs of numbers $(a,b)$ and points in the coordinate plane (Fig. 8-1). Thus,

1. To each point $P$ there corresponds an ordered pair of numbers $(a,b)$ called the coordinates of $P$. $a$ is called the *x-coordinate* or *abscissa*; $b$ is called the *y-coordinate* or *ordinate*.
2. To each ordered pair of numbers there corresponds a point, called the graph of the ordered pair. The graph can be indicated by a dot.

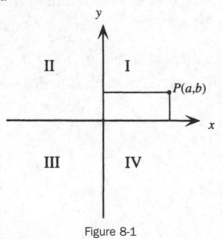

Figure 8-1

## Distance between Two Points

The distance between two points $P_1(x_1,y_1)$ and $P_2(x_2,y_2)$ in a Cartesian coordinate system is given by the *distance formula*:

$$d(P_1,P_2) = \sqrt{(x_2 - x_1)^2 + (y_2 - y_1)^2}$$

**EXAMPLE 8.1**   Find the distance between $(-3,5)$ and $(4,-1)$.

Label $P_1(x_1,y_1) = (-3,5)$ and $P_2(x_2,y_2) = (4,-1)$. Then substitute into the distance formula.

$$\begin{aligned}
d(P_1,P_2) &= \sqrt{(x_2 - x_1)^2 + (y_2 - y_1)^2} \\
&= \sqrt{[4 - (-3)]^2 + [(-1) - 5]^2} \\
&= \sqrt{7^2 + (-6)^2} = \sqrt{85}
\end{aligned}$$

## Graph of an Equation

The graph of an equation in two variables is the graph of its solution set, that is, of all ordered pairs $(a,b)$ that satisfy the equation. Since there are ordinarily an infinite number of solutions, a *sketch* of the graph is generally sufficient. A simple approach to finding a sketch of a graph is to find several solutions, plot them, then connect the dots with a smooth curve or line.

**EXAMPLE 8.2**   Sketch the graph of the equation $x - 2y = 10$.

Form a table of values; then plot the points and connect them. The graph is a straight line, as shown in Fig. 8-2.

| $x$ | −2 | 0 | 2 | 4 | 6 | 8 | 10 |
|-----|----|----|----|----|----|----|----|
| $y$ | −6 | −5 | −4 | −3 | −2 | −1 | 0 |

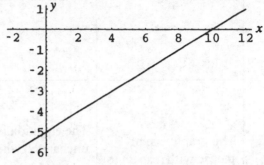

Figure 8-2

## Intercepts

The coordinates of the points where the graph of an equation crosses the $x$-axis and $y$-axis have special names:

1. The $x$-coordinate of a point where the graph crosses the $x$-axis is called the $x$-intercept of the graph. To find it, set $y = 0$ and solve for $x$.
2. The $y$-coordinate of a point where the graph crosses the $y$-axis is called the $y$-intercept of the graph. To find it, set $x = 0$ and solve for $y$.

**EXAMPLE 8.3**   In the previous example, the $x$-intercept of the graph is 10, since the graph crosses the $x$-axis at $(10,0)$; and the $y$-intercept is $-5$, since the graph crosses the $y$-axis at $(0,-5)$.

**EXAMPLE 8.4**   Find the intercepts of the graph of the equation $y = 4 - x^2$.

Set $x = 0$; then $y = 4 - 0^2 = 4$. Hence the $y$-intercept is 4.
Set $y = 0$. If $0 = 4 - x^2$, then $x^2 = 4$; thus $x = \pm 2$. Hence 2 and $-2$ are the $x$-intercepts.

## Symmetry

Symmetry is an important aid to graphing more complicated equations: A graph is

1. Symmetric with respect to the $y$-axis if $(-a,b)$ is on the graph whenever $(a,b)$ is on the graph. ($y$-axis symmetry)
2. Symmetric with respect to the $x$-axis if $(a,-b)$ is on the graph whenever $(a,b)$ is on the graph. ($x$-axis symmetry)
3. Symmetric with respect to the origin if $(-a,-b)$ is on the graph whenever $(a,b)$ is on the graph. (origin symmetry)
4. Symmetric with respect to the line $y = x$ if $(b,a)$ is on the graph whenever $(a,b)$ is on the graph.

## Tests for Symmetry

Tests for symmetry (Fig. 8-3):

1. If substituting $-x$ for $x$ leads to the same equation, the graph has symmetry with respect to the $y$-axis.
2. If substituting $-y$ for $y$ leads to the same equation, the graph has symmetry with respect to the $x$-axis.
3. If simultaneously substituting $-x$ for $x$ and $-y$ for $y$ leads to the same equation, the graph has symmetry with respect to the origin.

CHAPTER 8   Analytic Geometry

| Terminology | Test | Illustration |
|---|---|---|
| The graph is symmetric with respect to the $y$-axis | The equation is unchanged when $x$ is replaced by $-x$ | 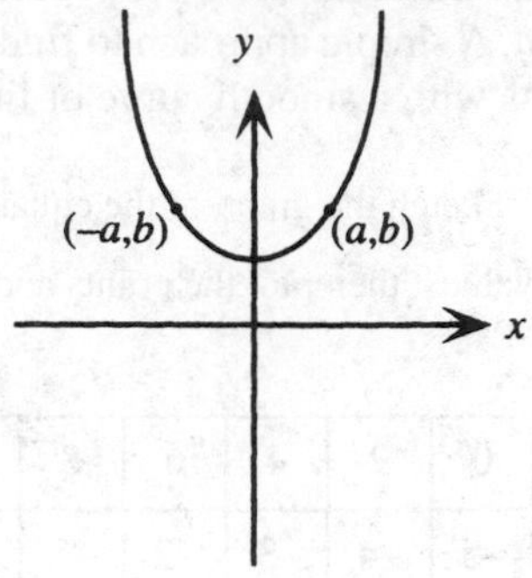 |
| The graph is symmetric with respect to the $x$-axis | The equation is unchanged when $y$ is replaced by $-y$ | 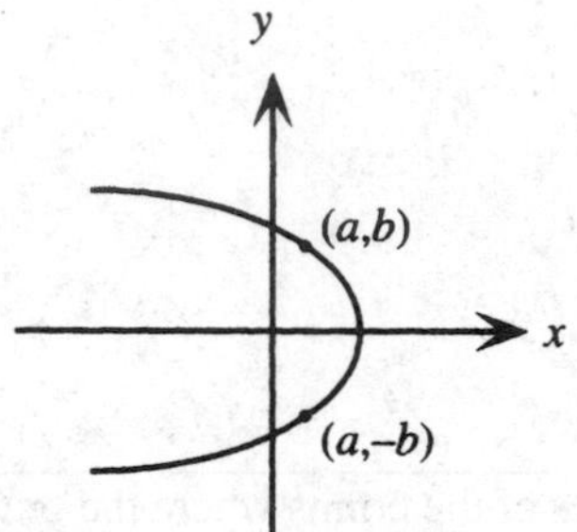 |
| The graph is symmetric with respect to the origin | The equation is unchanged when $x$ is replaced by $-x$ and $y$ is replaced by $-y$ | 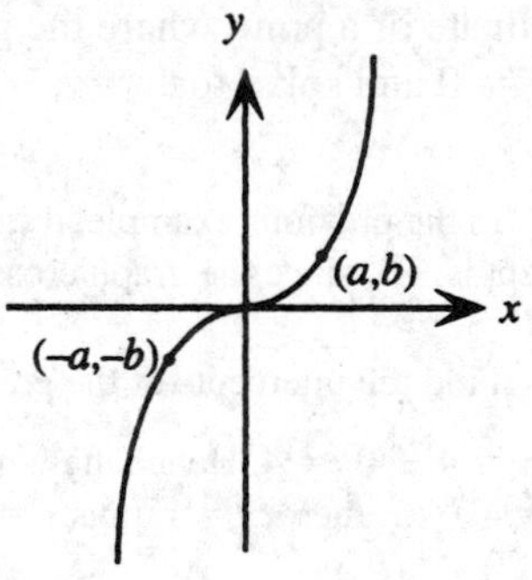 |
| The graph is symmetric with respect to the line $y = x$ | The equation is unchanged when $x$ and $y$ are interchanged | 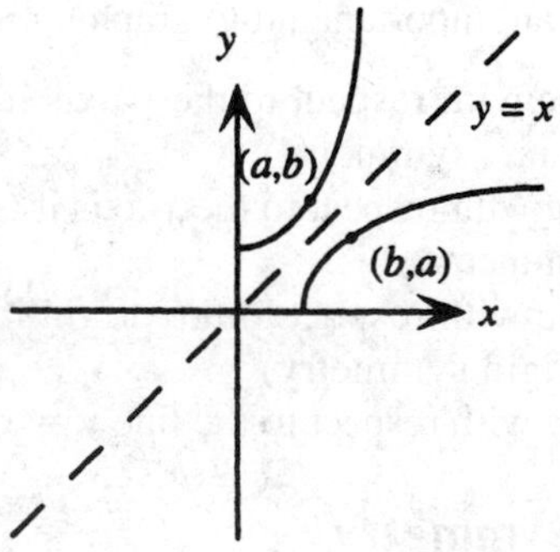 |

Figure 8-3

*Note:* a graph may have none of these three symmetries, one, or all three. It is not possible for a graph to have exactly two of these three symmetries.

The fourth symmetry is less commonly tested:

4. If interchanging the letters $x$ and $y$ leads to the same equation, the graph has symmetry with respect to the line $y = x$.

**EXAMPLE 8.5**  Test the equation $y = 4 - x^2$ for symmetry and draw the graph.

Substitute $-x$ for $x$: $y = 4 - (-x)^2 = 4 - x^2$. Since the equation is unchanged, the graph has $y$-axis symmetry (see Fig. 8-4).
Substitute $-y$ for $y$: $-y = 4 - x^2$; $y = -4 + x^2$. Since the equation is changed, the graph does not have $x$-axis symmetry. It is not possible for the graph to have origin symmetry; see the previous note. Since the graph has $y$-axis symmetry, it is only necessary to find points with nonnegative values of $x$, and then reflect the graph through the $y$-axis.

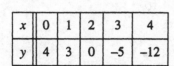

| $x$ | 0 | 1 | 2 | 3 | 4 |
|---|---|---|---|---|---|
| $y$ | 4 | 3 | 0 | -5 | -12 |

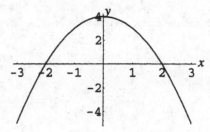

Figure 8-4

## Circle

A circle with center $C(h,k)$ and radius $r > 0$ is the set of all points in the plane that are $r$ units from $C$ (Fig. 8-5).

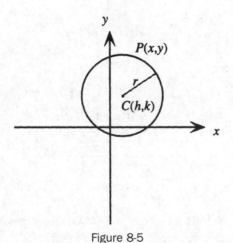

Figure 8-5

## Equation of a Circle

The equation of a circle with center $C(h,k)$ and radius $r > 0$ can be written as (standard form)

$$(x - h)^2 + (y - k)^2 = r^2$$

If the center of the circle is the origin $(0,0)$, this reduces to

$$x^2 + y^2 = r^2$$

If $r = 1$ the circle is called a *unit circle*.

## Midpoint of a Line Segment

The midpoint of a line segment with endpoints $P_1(x_1,y_1)$ and $P_2(x_2,y_2)$ is given by the midpoint formula:

$$\text{Midpoint of } P_1P_2 = \left(\frac{x_1 + x_2}{2}, \frac{y_1 + y_2}{2}\right)$$

**SOLVED PROBLEMS**

**8.1.** Prove the distance formula.

In Fig. 8-6, $P_1$ and $P_2$ are shown. Introduce $Q(x_2, y_1)$ as shown. Then the distance between $P_1$ and $Q$ is the difference in their $x$-coordinates, $|x_2 - x_1|$; similarly, the distance between $Q$ and $P_2$ is the difference in their $y$-coordinates, $|y_2 - y_1|$. In the right triangle $P_1P_2Q$, apply the Pythagorean theorem: $d^2 = |x_2 - x_1|^2 + |y_2 - y_1|^2 = (x_2 - x_1)^2 + (y_2 - y_1)^2$, since $|a|^2 = a^2$ by the properties of absolute values. Hence, taking the square root and noting that $d$, the distance, is always positive, $d(P_1,P_2) = \sqrt{(x_2 - x_1)^2 + (y_2 - y_1)^2}$:

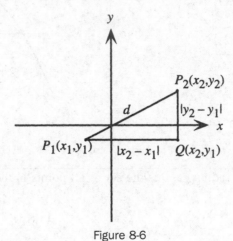

Figure 8-6

**8.2.** Find the distance $d(P_1,P_2)$ given

(a) $P_1(-5,-4), P_2(-8,0)$; (b) $P_1(2\sqrt{2},2\sqrt{2}), P_2(0,5\sqrt{2})$; (c) $P_1(x,x^2), P_2(x + h,(x + h)^2)$

(a) Substitute $x_1 = -5, y_1 = -4, x_2 = -8, y_2 = 0$ into the distance formula:

$$d = \sqrt{(x_2 - x_1)^2 + (y_2 - y_1)^2}$$
$$= \sqrt{[(-8) - (-5)]^2 + [0 - (-4)]^2}$$
$$= \sqrt{9 + 16} = \sqrt{25} = 5$$

(b) Substitute $x_1 = 2\sqrt{2}, y_1 = 2\sqrt{2}, x_2 = 0, y_2 = 5\sqrt{2}$ into the distance formula:

$$d = \sqrt{(x_2 - x_1)^2 + (y_2 - y_1)^2}$$
$$= \sqrt{(0 - 2\sqrt{2})^2 + (5\sqrt{2} - 2\sqrt{2})^2}$$
$$= \sqrt{(-2\sqrt{2})^2 + (3\sqrt{2})^2}$$
$$= \sqrt{8 + 18} = \sqrt{26}$$

(c) Substitute $x_1 = x, y_1 = x^2, x_2 = x + h, y_2 = (x + h)^2$ into the distance formula and simplify.

$$d = \sqrt{(x_2 - x_1)^2 + (y_2 - y_1)^2}$$
$$= \sqrt{(x + h - x)^2 + [(x + h)^2 - x^2]^2}$$
$$= \sqrt{h^2 + (2xh + h^2)^2}$$
$$= \sqrt{h^2 + 4x^2h^2 + 4xh^3 + h^4}$$

**8.3.** Analyze intercepts and symmetry, then sketch the graph:

(a) $y = 12 - 4x$; (b) $y = x^2 + 3$; (c) $y^2 + x = 5$; (d) $2y = x^3$.

(a) Set $x = 0$, then $y = 12 - 4 \cdot 0 = 12$. Hence 12 is the $y$-intercept.

Set $y = 0$, then $0 = 12 - 4x$; thus $x = 3$. Hence 3 is the $x$-intercept.

Substitute $-x$ for $x$: $y = 12 - 4(-x)$; $y = 12 + 4x$. Since the equation is changed, the graph (see Fig. 8-7) does not have $y$-axis symmetry.

Substitute $-y$ for $y$: $-y = 12 - 4x$; $y = -12 + 4x$. Since the equation is changed, the graph does not have $x$-axis symmetry.

Substitute $-x$ for $x$ and $-y$ for $y$: $-y = 12 - 4(-x)$; $y = -12 - 4x$. Since the equation is changed, the graph does not have origin symmetry.

Form a table of values; then plot the points and connect them. The graph is a straight line.

| $x$ | −1 | 0 | 1 | 2 | 3 | 4 | 5 |
|---|---|---|---|---|---|---|---|
| $y$ | 16 | 12 | 8 | 4 | 0 | −4 | −8 |

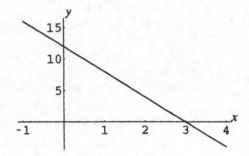

Figure 8-7

(b) Set $x = 0$, then $y = 0^2 + 3 = 3$. Hence 3 is the $y$-intercept.

Set $y = 0$, then $0 = x^2 + 3$. This has no real solution; hence there is no $x$-intercept.

Substitute $-x$ for $x$: $y = (-x)^2 + 3 = x^2 + 3$. Since the equation is unchanged, the graph (Fig. 8-8) has $y$-axis symmetry.

Substitute $-y$ for $y$: $-y = x^2 + 3$; $y = -x^2 - 3$. Since the equation is changed, the graph does not have $x$-axis symmetry.

It is not possible for the graph to have origin symmetry. Since the graph has $y$-axis symmetry, it is only necessary to find points with nonnegative values of $x$, and then reflect the graph through the $y$-axis.

| $x$ | 0 | 1 | 2 | 3 | 4 |
|---|---|---|---|---|---|
| $y$ | 3 | 4 | 7 | 12 | 19 |

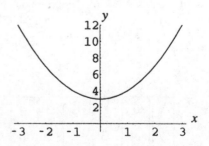

Figure 8-8

(c) Set $x = 0$, then $y^2 + 0 = 5$; thus $y = \pm\sqrt{5}$. Hence $\pm\sqrt{5}$ are the $y$-intercepts.

Set $y = 0$, then $x = 5$; hence 5 is the $x$-intercept.

Substitute $-x$ for $x$: $y^2 - x = 5$. Since the equation is changed, the graph (see Fig. 8-9) does not have $y$-axis symmetry.

Substitute $-y$ for $y$: $(-y)^2 + x = 5$; $y^2 + x = 5$. Since the equation is unchanged, the graph has $x$-axis symmetry.

It is not possible for the graph to have origin symmetry. Since the graph has $x$-axis symmetry, it is only necessary to find points with nonnegative values of $y$, and then reflect the graph through the $x$-axis.

| x | 5 | 4 | 1 | -4 | -11 |
|---|---|---|---|----|-----|
| y | 0 | 1 | 2 | 3  | 4   |

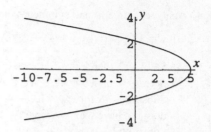

Figure 8-9

(d) Set $x = 0$, then $2y = 0^3$; thus $y = 0$. Hence 0 is the $y$-intercept.

Set $y = 0$, then $2 \cdot 0 = x^3$; thus $x = 0$. Hence 0 is the $x$-intercept.

Substitute $-x$ for $x$: $2y = (-x)^3$; $2y = -x^3$. Since the equation is changed, the graph (Fig. 8-10) does not have $y$-axis symmetry.

Substitute $-y$ for $y$: $2(-y) = x^3$; $2y = -x^3$. Since the equation is changed, the graph does not have $x$-axis symmetry.

Substitute $-x$ for $x$ and $-y$ for $y$: $-2y = (-x)^3$; $2y = x^3$. Since the equation is unchanged, the graph has origin symmetry.

From a table of values for positive $x$, plot the points and connect them, then reflect the graph through the origin.

| x | 0 | 1             | 2 | 3              | 4  |
|---|---|---------------|---|----------------|----|
| y | 0 | $\frac{1}{2}$ | 4 | $\frac{27}{2}$ | 32 |

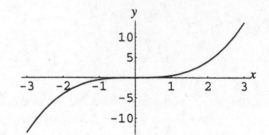

Figure 8-10

**8.4.** Analyze intercepts and symmetry, then sketch the graph:

(a) $y = |x| - 4$; (b) $4x^2 + y^2 = 36$; (c) $|x| + |y| = 3$; (d) $x^2y = 12$.

(a) Proceeding as in the previous problem, the $x$-intercepts are $\pm 4$ and the $y$-intercept is $-4$. The graph has $y$-axis symmetry. Form a table of values for positive $x$, plot the points and connect them, then reflect the graph (Fig. 8-11) through the $y$-axis.

| x | 0  | 1  | 2  | 3  | 4 | 5 | 6 |
|---|----|----|----|----|---|---|---|
| y | -4 | -3 | -2 | -1 | 0 | 1 | 2 |

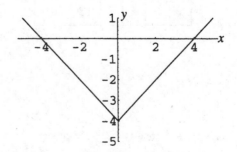

Figure 8.11

(b) The $x$-intercepts are $\pm 3$ and the $y$-intercepts are $\pm 6$.

The graph has $x$-axis, $y$-axis, and origin symmetry. Form a table of values for positive $x$ and $y$, plot the points and connect them, then reflect the graph (Fig. 8-12), first through the $y$-axis, then through the $x$-axis.

| $x$ | 0 | 1 | 2 | 3 |
|---|---|---|---|---|
| $y$ | 6 | $\sqrt{32} \approx 5.6$ | $\sqrt{20} \approx 4.4$ | 0 |

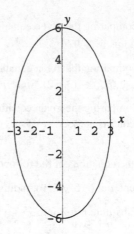

Figure 8-12

(c)  The $x$-intercepts are $\pm 3$ and the $y$-intercepts are $\pm 3$.
The graph has $x$-axis, $y$-axis, and origin symmetry. Form a table of values for positive $x$ and $y$, plot the points and connect them, then reflect the graph (Fig. 8-13), first through the $y$-axis, then through the $x$-axis.

| $x$ | 0 | 1 | 2 | 3 |
|---|---|---|---|---|
| $y$ | 3 | 2 | 1 | 0 |

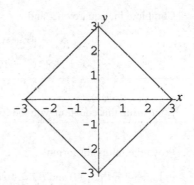

Figure 8-13

(d)  There are no $x$- or $y$-intercepts.
The graph has $y$-axis symmetry. Form a table of values for positive $x$, plot the points and connect them, then reflect the graph (Fig. 8-14) through the $y$-axis.

| $x$ | 0 | 1 | 2 | 3 | 4 |
|---|---|---|---|---|---|
| $y$ | undefined | 12 | 3 | 4/3 | 3/4 |

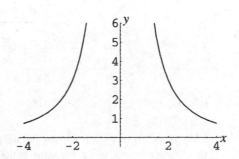

Figure 8-14

**8.5.**  Find the center and radius for the circles with the following equations:

(a) $x^2 + y^2 = 9$; (b) $(x - 3)^2 + (y + 2)^2 = 25$; (c) $(x + 5)^2 + \left(y + \frac{1}{2}\right)^2 = 21$

(a) Comparing the given equation with the form $x^2 + y^2 = r^2$, the center is at the origin. Since $r^2 = 9$, the radius is $\sqrt{9} = 3$.

(b) Comparing the given equation with the form $(x - h)^2 + (y - k)^2 = r^2$, $h = 3$ and $-k = 2$; hence the center is at $(h,k) = (3,-2)$. Since $r^2 = 25$, the radius is $\sqrt{25} = 5$.

(c) Comparing the given equation with the form $(x - h)^2 + (y - k)^2 = r^2$, $-h = 5$ and $-k = \frac{1}{2}$; hence the center is at $(h,k) = (-5, -\frac{1}{2})$. Since $r^2 = 21$, the radius is $\sqrt{21}$.

**8.6.** Find the equations of the following circles: (a) center at origin, radius 7; (b) center at $(2,-3)$, radius $\sqrt{14}$; (c) center at $(-5\sqrt{2}, 0)$, radius $5\sqrt{2}$.

(a) Substitute $r = 7$ into $x^2 + y^2 = r^2$. The equation is $x^2 + y^2 = 49$.

(b) Substitute $h = 2, k = -3, r = \sqrt{14}$ into $(x - h)^2 + (y - k)^2 = r^2$.

The equation is $(x - 2)^2 + [y - (-3)]^2 = (\sqrt{14})^2$ or $(x - 2)^2 + (y + 3)^2 = 14$.

(c) Substitute $h = -5\sqrt{2}, k = 0, r = 5\sqrt{2}$ into $(x - h)^2 + (y - k)^2 = r^2$.

The equation is $[x - (-5\sqrt{2})]^2 + (y - 0)^2 = (5\sqrt{2})^2$ or $(x + 5\sqrt{2})^2 + y^2 = 50$.

**8.7.** Find the center and radius of the circle with equation $x^2 + y^2 - 4x - 12y = 9$.

Complete the square on $x$ and $y$.

$$x^2 - 4x + y^2 - 12y = 9 \qquad\qquad \left[\tfrac{1}{2}(-4)\right]^2 = 4; \quad \left[\tfrac{1}{2}(-12)\right]^2 = 36$$

$$x^2 - 4x + 4 + y^2 - 12y + 36 = 4 + 36 + 9 \qquad \text{Add } 4 + 36 \text{ to both sides}$$

$$(x - 2)^2 + (y - 6)^2 = 49$$

Comparing this equation with the form $(x - h)^2 + (y - k)^2 = r^2$, the center is at $(h,k) = (2,6)$ and the radius is 7.

**8.8.** Prove the midpoint formula.

In Fig. 8-15, $P_1(x_1,y_1)$ and $P_2(x_2,y_2)$ are given. Let $(x,y)$ be the unknown coordinates of the midpoint $M$. Project the points $M, P_1, P_2$ to the $x$-axis as shown.

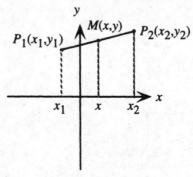

Figure 8-15

From plane geometry it is known that the projected segments are in the same ratio as the original segments. Hence the distance from $x_1$ to $x$ is the same as the distance from $x$ to $x_2$. Thus, $x_2 - x = x - x_1$. Solving for $x$ yields

$$-2x = -x_1 - x_2$$

$$x = \frac{x_1 + x_2}{2}$$

Similarly, it can be shown by projecting onto the $y$-axis that $y = \dfrac{y_1 + y_2}{2}$.

**8.9.** Find the midpoint $M$ of the segment $P_1P_2$ given $P_1(3,-8)$, $P_2(-6,6)$.

Substitute $x_1 = 3$, $y_1 = -8$, $x_2 = -6$, $y_2 = 6$ into the midpoint formula. Then

$$\left(\frac{x_1 + x_2}{2}, \frac{y_1 + y_2}{2}\right) = \left(\frac{3 + (-6)}{2}, \frac{(-8) + 6}{2}\right) = \left(-\frac{3}{2}, -1\right)$$

are the coordinates of $M$.

**8.10.** Find the equation of a circle given that $(0,6)$ and $(8,-8)$ are the endpoints of a diameter.

*Step* 1. The center is the midpoint of the diameter. Find the coordinates of the center from the midpoint formula.

$$\left(\frac{x_1 + x_2}{2}, \frac{y_1 + y_2}{2}\right) = \left(\frac{0 + 8}{2}, \frac{6 + (-8)}{2}\right) = (4, -1)$$

*Step* 2. The radius is the distance from the center to either of the given endpoints. Find the radius from the distance formula.

$$\sqrt{(x_2 - x_1)^2 + (y_2 - y_1)^2} = \sqrt{(4 - 0)^2 + [(-1) - 6]^2} = \sqrt{16 + 49} = \sqrt{65}$$

*Step* 3. Substitute the calculated radius and coordinates of the center into the standard form for the equation of a circle. $r = \sqrt{65}$, $(h, k) = (4,-1)$.

$$(x + h)^2 + (y - k)^2 = r^2$$

$$(x - 4)^2 + [y - (-1)]^2 = (\sqrt{65})^2$$

$$(x - 4)^2 + (y + 1)^2 = 65$$

**8.11.** Show that the triangle with vertices $A(1,3)$, $B(-1,2)$, $C(5, -5)$ is a right triangle.

*Step* 1. First find the lengths of the sides from the distance formula

$$d(A,B) = \sqrt{(x_2 - x_1)^2 + (y_2 - y_1)^2} = \sqrt{[(-1) - 1]^2 + (2 - 3)^2} = \sqrt{5} = c$$

$$d(B,C) = \sqrt{(x_2 - x_1)^2 + (y_2 - y_1)^2} = \sqrt{[5 - (-1)]^2 + [(-5) - 2]^2} = \sqrt{85} = a$$

$$d(A,C) = \sqrt{(x_2 - x_1)^2 + (y_2 - y_1)^2} = \sqrt{(5 - 1)^2 + [(-5) - 3]^2} = \sqrt{80} = b$$

*Step* 2. Apply the converse of the Pythagorean theorem.

Since $a^2 = (\sqrt{85})^2 = 85$ and $b^2 + c^2 = (\sqrt{80})^2 + (\sqrt{5})^2 = 80 + 5 = 85$, the relation $a^2 = b^2 + c^2$ is satisfied; hence the triangle is a right triangle.

**8.12.** Show that $P(-12,11)$ lies on the perpendicular bisector of the line segment joining $A(0,-3)$ and $B(6,15)$.

The perpendicular bisector of a segment consists of all points that are equidistant from its endpoints. Thus if $PA = PB$, then $P$ lies on the perpendicular bisector of $AB$. From the distance formula,

$$d(A,P) = \sqrt{(x_2 - x_1)^2 + (y_2 - y_1)^2} = \sqrt{[(-12) - 0]^2 + [11 - (-3)]^2} = \sqrt{340} = PA$$

$$d(P,B) = \sqrt{(x_2 - x_1)^2 + (y_2 - y_1)^2} = \sqrt{[6 - (-12)]^2 + (15 - 11)^2} = \sqrt{340} = PB$$

Hence $PA = PB$ and $P$ lies on the perpendicular bisector of $AB$.

**8.13.** Find an equation for the perpendicular bisector of the line segment joining $A(7,-8)$ and $B(-2, 5)$.

The perpendicular bisector of a segment consists of all points that are equidistant from its endpoints. Thus if $PA = PB$, then $P$ lies on the perpendicular bisector of $AB$. Let $P$ have the unknown coordinates $(x,y)$. Then, from the distance formula, $PA = PB$ if

$$PA = \sqrt{(x - 7)^2 + [y - (-8)]^2} = \sqrt{[x - (-2)]^2 + (y - 5)^2} = PB$$

Squaring both sides and simplifying yields

$$(x - 7)^2 + [y - (-8)]^2 = [x - (-2)]^2 + (y - 5)^2$$

$$x^2 - 14x + 49 + y^2 + 16y + 64 = x^2 + 4x + 4 + y^2 - 10y + 25$$

$$18x - 26y = 84$$

$$9x - 13y = 42$$

This is the equation satisfied by all points equidistant from $A$ and $B$. Hence, it is the equation of the perpendicular bisector of $AB$.

## SUPPLEMENTARY PROBLEMS

**8.14.** Describe the set of points that satisfy the relations: (a) $x = 0$; (b) $x > 0$; (c) $xy < 0$; (d) $y > 1$.

   *Ans.*   (a) All points on the $y$-axis; (b) all points to the right of the $y$-axis;

   (c) all points in the second and fourth quadrants; (d) all points above the line $y = 1$.

**8.15.** Find the distance between the following pairs of points: (a) $(0,-7)$ and $(7,0)$; (b) $(-3\sqrt{3},-3)$ and $(3\sqrt{3}, 3)$,

   *Ans.*   (a) $7\sqrt{2}$; (b) 12

**8.16.** Find the length and the midpoint of the line segment with the given endpoints:

   (a) $A(1,8)$, $B(-3,4)$; (b) $A(3,-7)$, $B(0,8)$; (c) $A(1,\sqrt{2})$, $B(-1,5\sqrt{2})$

   *Ans.*   (a) length $4\sqrt{2}$, midpoint $(-1,6)$; (b) length $3\sqrt{26}$, midpoint $\left(\frac{3}{2},\frac{1}{2}\right)$; (c) length 6, midpoint $(0,3\sqrt{2})$

**8.17.** Analyze the following for symmetry. Do not sketch graphs:

   (a) $xy^2 = 4$; (b) $x^3y = 4$; (c) $|xy| = 4$; (d) $x^2 + xy = 4$;

   (e) $x^2 + y + y^2 = 4$; (f) $x^2 + xy + y^2 = 4$

   *Ans.*   (a) $x$-axis symmetry; (b) origin symmetry; (c) $x$-axis, $y$-axis, origin symmetry;

   (d) origin symmetry; (e) $y$-axis symmetry; (f) origin symmetry

**8.18.** Analyze symmetry and intercepts, then sketch graphs of the following:

   (a) $3x + 4y + 12 = 0$      (b) $y^2 = 10 + x$

   (c) $y^2 - x^2 = 9$      (d) $|y| - |x| = 3$

   *Ans.*   (a) Fig. 8-16: $x$-intercept $-4$, $y$-intercept $-3$, no symmetry

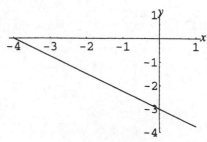

Figure 8-16

(b) Fig. 8-17: $x$-intercept $-10$, $y$-intercepts $\pm\sqrt{10}$, $x$-axis symmetry

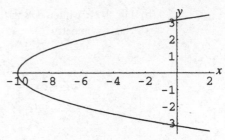

Figure 8-17

(c) Fig. 8-18: no $x$-intercept, $y$-intercept $\pm3$, $x$-axis, $y$-axis, origin symmetry

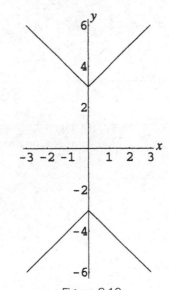

Figure 8-18

(d) Fig. 8-19: no $x$-intercept, $y$-intercepts $\pm3$, $x$-axis, $y$-axis, origin symmetry

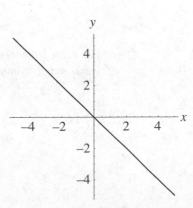

Figure 8-19

**8.19.** Analyze symmetry and intercepts, then sketch graphs of the following:

(a) $x + y = 0$; (b) $y + |x| = 4$; (c) $x^2 = 4|y|$;

(d) $|y| + x^2 = 4$; (e) $|x| = 4y^2$; (f) $-xy^2 = 4$

*Ans.* (a) Fig. 8-20: $x$-intercepts 0, $y$-intercept 0, origin symmetry

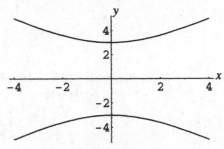

Figure 8-20

(b) Fig. 8-21: $x$-intercepts $\pm 4$, $y$-intercept 4, $y$-axis symmetry

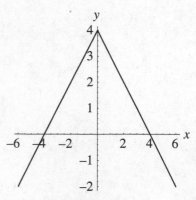

Figure 8-21

(c) Fig. 8-22: $x$-intercept 0, $y$-intercept 0, $x$-axis, $y$-axis, origin symmetry

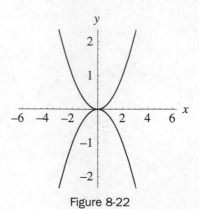

Figure 8-22

(d) Fig. 8-23: $x$-intercepts $\pm 2$, $y$-intercepts $\pm 4$, $x$-axis, $y$-axis, origin symmetry

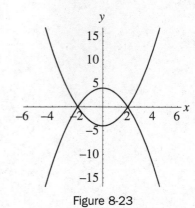

Figure 8-23

(e) Fig. 8-24: $x$-intercept 0, $y$-intercept 0, $x$-axis, $y$-axis, origin symmetry

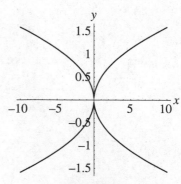

Figure 8-24

(f) Fig. 8-25: no intercepts, *x*-axis symmetry

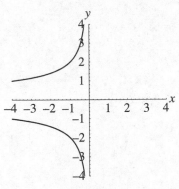

Figure 8-25

**8.20.** Find the equations of the following circles: (a) center $(5,-2)$, radius $\sqrt[4]{10}$; (b) center $\left(\frac{1}{2},-\frac{5}{2}\right)$, diameter 3;

(c) center $(3,8)$, passing through the origin; (d) center $(-3,-4)$, tangent to the *y*-axis.

*Ans.* (a) $(x-5)^2 + (y+2)^2 = \sqrt{10}$; (b) $\left(x-\frac{1}{2}\right)^2 + \left(y+\frac{5}{2}\right)^2 = \frac{9}{4}$;

(c) $(x-3)^2 + (y-8)^2 = 73$; (d) $(x+3)^2 + (y+4)^2 = 9$

**8.21.** Find the equations of the following circles: (a) center $(5,2)$, $(3,-1)$ is a point on the circle;

(b) $(5,-5)$ and $(-3,-9)$ are end points of a diameter.

*Ans.* (a) $(x-5)^2 + (y-2)^2 = 13$; (b) $(x-1)^2 + (y+7)^2 = 20$

**8.22.** For the following equations, determine whether they represent circles, and if so, find the center and radius:

(a) $x^2 + y^2 + 8x + 2y = 5$; (b) $x^2 + y^2 - 4x - 8y + 20 = 0$; (c) $2x^2 + 2y^2 - 6x + 14y = 3$;

(d) $x^2 + y^2 + 12x + 20y + 200 = 0$

*Ans.* (a) circle; center $(-4,-1)$, radius $\sqrt{22}$; (b) this is not a circle; the graph consists only of the point $(2,4)$;

(c) circle, center $\left(\frac{3}{2},-\frac{7}{2}\right)$, radius 4; (d) this is not a circle; there are no points on the graph.

**8.23.** Show that the triangle with vertices $(-10,7)$, $(-6,-2)$, and $(3,2)$ is isosceles.

**8.24.** Show that the triangle with vertices $(4,\sqrt{3})$, $(5,0)$, and $(6,\sqrt{3})$ is equilateral.

**8.25.** Show that the triangle with vertices $(6,9)$, $(1,1)$, and $(9,-4)$ is an isosceles right triangle.

**8.26.** Show that the quadrilateral with vertices $(-3,-3)$, $(5,-1)$, $(7,7)$, and $(-1,5)$ is a rhombus.

**8.27.** Show that the quadrilateral with vertices $(7,2)$, $(10,0)$, $(8,-3)$, and $(5,-1)$ is a square.

**8.28.** (a) Find the equation of the perpendicular bisector of the line segment with endpoints $(-2,-5)$ and $(7,-1)$.

(b) Show that the equation of the perpendicular bisector of the line segment with endpoints $(x_1,y_1)$ and $(x_2,y_2)$

can be written $\dfrac{x-\bar{x}}{y_2-y_1} + \dfrac{y-\bar{y}}{x_2-x_1} = 0$, where $(\bar{x},\bar{y})$ are the coordinates of the midpoint of the segment.

*Ans.* (a) $18x + 8y = 21$

# CHAPTER 9

# Functions

## Definition of Function

A function $f$ from set $D$ to set $E$ is a rule or correspondence that assigns to each element $x$ of set $D$ exactly one element $y$ of set $E$. The set $D$ is called the *domain* of the function. The element $y$ of $E$ is called the *image* of $x$ under $f$, or the value of $f$ at $x$, and is written $f(x)$. The subset $R$ of $E$ consisting of all images of elements of $D$ is called the *range* of the function. The members of the domain $D$ and range $R$ are referred to as the input and output values, respectively.

**EXAMPLE 9.1** Let $D$ be the set of all words in English having fewer than 20 letters. Let $f$ be the rule that assigns to each word the number of letters in the word. Then $E$ can be the set of all integers (or some larger set); $R$ is the set $\{x \in N | 1 \leq x < 20\}$. $f$ assigns to the word "truth" the number 5; this would be written $f(\text{truth}) = 5$. Moreover, $f(a) = 1$ and $f(\text{president}) = 9$.

Note that a function assigns a unique function value to each element in its domain; however, more than one element may be assigned the same function value.

**EXAMPLE 9.2** Let $D$ be the set of real numbers and $g$ be the rule given by $g(x) = x^2 + 3$. Find: $g(4)$, $g(-4)$, $g(a) + g(b)$, $g(a + b)$. What is the range of $g$?

Find values of $g$ by substituting for $x$ in the rule $g(x) = x^2 + 3$:

$$g(4) = 4^2 + 3 = 19 \qquad g(-4) = (-4)^2 + 3 = 19$$

$$g(a) + g(b) = a^2 + 3 + b^2 + 3 = a^2 + b^2 + 6$$

$$g(a + b) = (a + b)^2 + 3 = a^2 + 2ab + b^2 + 3$$

The range of $g$ is found by noting that the square of a number, $x^2$, is always greater than or equal to zero. Hence $g(x) = x^2 + 3 \geq 3$. Thus, the range of $g$ is $\{y \in R | y \geq 3\}$.

## Function Notation

A function is indicated by the notation $f : D \to E$. The effect of a function on an element of $D$ is then written $f : x \to f(x)$. A picture of the type shown in Fig. 9-1 is often used to visualize the function relationship.

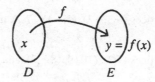

Figure 9-1

## Domain and Range

The domain and range of a function are normally sets of real numbers. If a function is defined by an expression and the domain is not stated, the domain is assumed to be the set of all real numbers for which the expression is defined. This set is called the *implied domain*, or the *largest possible domain*, of the function.

**EXAMPLE 9.3**   Find the (largest possible) domain for (a) $f(x) = \dfrac{x-3}{x+6}$; (b) $g(x) = \sqrt{x-5}$; (c) $h(x) = x^2 - 4$

(a)   The expression $\dfrac{x-3}{x+6}$ is defined for all real numbers $x$ except when $x + 6 = 0$, that is, when $x = -6$. Thus the domain of $f$ is $\{x \in \boldsymbol{R} | x \neq -6\}$.

(b)   The expression $\sqrt{x-5}$ is defined when $x - 5 \geq 0$, that is, when $x \geq 5$. Thus the domain of $g$ is $\{x \in \boldsymbol{R} | x \geq 5\}$.

(c)   The expression $x^2 - 4$ is defined for all real numbers. Thus the domain of $h$ is $\boldsymbol{R}$.

## Graph of a Function

The graph of a function $f$ is the graph of all points $(x,y)$ such that $x$ is in the domain of $f$, and $y = f(x)$.

## Vertical Line Test

Since for each value of $x$ in the domain of $f$ there is exactly one value of $y$ such that $y = f(x)$, a vertical line $x = c$ can cross the graph of a function at most once. Thus, if a vertical line crosses a graph more than once, the graph is not the graph of a function.

## Increasing, Decreasing, and Constant Functions

1. If, for all $x$ in an interval, as $x$ increases, the value of $f(x)$ increases, thus, the graph of the function rises from left to right, then the function $f$ is called an *increasing function on the interval*. A function that is increasing throughout its domain is referred to as an *increasing function*. Algebraically, then, $f$ is increasing on $(a, b)$ if for all $x_1, x_2$ in $(a, b)$, when $x_1 < x_2, f(x_1) < f(x_2)$.
2. If, for all $x$ in an interval, as $x$ increases, the value of $f(x)$ decreases, thus, the graph of the function falls from left to right, then the function $f$ is called a *decreasing function on the interval*. A function that is decreasing throughout its domain is referred to as a *decreasing function*. Algebraically, then, $f$ is decreasing on $(a, b)$ if for all $x_1, x_2$ in $(a, b)$, when $x_1 < x_2, f(x_1) > f(x_2)$.
3. If the value of a function does not change on an interval, thus, the graph of the function is a horizontal line segment, then the function is called a *constant function on the interval*. A function that is constant throughout its domain is referred to as a *constant function*. Algebraically, then, $f$ is constant on $(a, b)$ if for all $x_1, x_2$ in $(a, b), f(x_1) = f(x_2)$.

**EXAMPLE 9.4**   Given the graph of $f(x)$ shown in Fig. 9-2, assuming the domain of $f$ is $\boldsymbol{R}$, identify the intervals on which $f$ is increasing or decreasing:

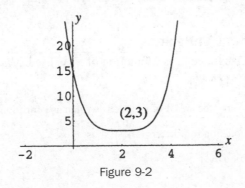

Figure 9-2

As $x$ increases through the domain of $f$, $y$ decreases until $x = 2$, then increases. Thus the function is decreasing on $(-\infty, 2)$ and increasing on $(2, \infty)$.

## Even and Odd Functions

1. If, for all $x$ in the domain of a function $f$, $f(-x) = f(x)$, the function is called an *even* function. Since, for an even function, the equation $y = f(x)$ is not changed when $-x$ is substituted for $x$, the graph of an even function has $y$-axis symmetry.

2. If, for all $x$ in the domain of a function $f$, $f(-x) = -f(x)$, the function is called an *odd* function. Since, for an odd function, the equation $y = f(x)$ is not changed when $-x$ is substituted for $x$ and $-y$ is substituted for $y$, the graph of an odd function has origin symmetry.
3. Most functions are neither even nor odd.

**EXAMPLE 9.5**  Determine whether the following functions are even, odd, or neither:

(a) $f(x) = 7x^2$  (b) $g(x) = 4x + 6$  (c) $h(x) = 6x - \sqrt[3]{x}$  (d) $F(x) = \dfrac{4}{x - 6}$

(a) Consider $f(-x)$. $f(-x) = 7(-x)^2 = 7x^2$. Since $f(-x) = f(x)$, $f$ is an even function.

(b) Consider $g(-x)$. $g(-x) = 4(-x) + 6 = -4x + 6$. Also, $-g(x) = -(4x + 6) = -4x - 6$. Since neither $g(-x) = g(x)$ nor $g(-x) = -g(x)$ is the case, the function $g$ is neither even nor odd.

(c) Consider $h(-x)$. $h(-x) = 6(-x) - \sqrt[3]{-x} = -6x + \sqrt[3]{x}$. Thus, $h(-x) = -h(x)$ and $h$ is an odd function.

(d) Consider $F(-x)$. $F(-x) = \dfrac{4}{-x - 6} = -\dfrac{4}{x + 6}$. Since neither $F(-x) = F(x)$ nor $F(-x) = -F(x)$ is the case, the function $F$ is neither even nor odd.

## Average Rate of Change of a Function

Let $f$ be a function. The *average rate of change of $f(x)$* with respect to $x$ over the interval $[a,b]$ is defined as

$$\frac{\text{Change in } f(x)}{\text{Change in } x} = \frac{f(b) - f(a)}{b - a}$$

Over an interval from $x$ to $x + h$ this quantity becomes

$$\frac{f(x + h) - f(x)}{h}$$

which is referred to as the *difference quotient*.

**EXAMPLE 9.6**  Find the average rate of change of $f(x) = x^2$ on the interval $[1,4]$.

Calculate: $\dfrac{f(4) - f(1)}{4 - 1} = \dfrac{4^2 - 1^2}{3} = 5$.

**EXAMPLE 9.7**  Find the difference quotient for $f(x) = x^2$.

$$\frac{f(x + h) - f(x)}{h} = \frac{(x + h)^2 - x^2}{h} = \frac{x^2 + 2xh + h^2 - x^2}{h} = \frac{2xh + h^2}{h} = 2x + h, \text{ for } h \neq 0.$$

## Independent and Dependent Variables

In applications, if $y = f(x)$, the language "$y$ is a function of $x$" is used. $x$ is referred to as the *independent variable*, and $y$ as the *dependent variable*.

**EXAMPLE 9.8**  In the formula $A = \pi r^2$, the area $A$ of a circle is written as a function of the radius $r$. To write the radius as a function of the area, solve this equation for $r$ in terms of $A$, thus: $r^2 = \dfrac{A}{\pi}$, $r = \pm\sqrt{\dfrac{A}{\pi}}$. Since the radius is a positive quantity, $r = \sqrt{\dfrac{A}{\pi}}$ gives $r$ as a function of $A$.

## SOLVED PROBLEMS

**9.1.** Which of the following equations defines $y$ as a function of $x$?

(a) $y = x^2 + 4$; (b) $x = y^2 + 5$; (c) $y = \sqrt{x - 5}$; (d) $y = 5$; (e) $x^2 - y^2 = 36$.

(a) Since for each value of $x$ there is exactly one corresponding value of $y$, this defines $y$ as a function of $x$.

(b) Let $x = 6$. Then $6 = y^2 + 5$; thus $y^2 = 1$ and $y = \pm 1$. Since for at least one value of $x$ there correspond two values of $y$, this equation does not define $y$ as a function of $x$.

(c) Since for each value of $x$ there is exactly one corresponding value of $y$, this defines $y$ as a function of $x$. Note that the radical symbol defines $y$ as the *positive* square root only.

(d) Since for each value of $x$ there is exactly one corresponding value of $y$, namely 5, this defines $y$ as a function of $x$.

(e) Let $x = 10$. Then $10^2 - y^2 = 36$, thus $y^2 = 64$ and $y = \pm 8$. Since for at least one value of $x$ there correspond two values of $y$, this equation does not define $y$ as a function of $x$.

**9.2.** Given $f(x) = x^2 - 4x + 2$, find (a) $f(5)$; (b) $f(-3)$; (c) $f(a)$; (d) $f(a + b)$; (e) $f(a) + f(b)$.

Replace $x$ by the various input values provided:

(a) $f(5) = 5^2 - 4 \cdot 5 + 2 = 7$; (b) $f(-3) = (-3)^2 - 4(-3) + 2 = 23$; (c) $f(a) = a^2 - 4a + 2$

(d) Here $x$ is replaced by the entire quantity $a + b$.

$$f(a + b) = (a + b)^2 - 4(a + b) + 2 = a^2 + 2ab + b^2 - 4a - 4b + 2$$

(e) Here $x$ is replaced by $a$ and by $b$, then the results are added. $f(a) = a^2 - 4a + 2$; $f(b) = b^2 - 4b + 2$; hence $f(a) + f(b) = a^2 - 4a + 2 + b^2 - 4b + 2 = a^2 + b^2 - 4a - 4b + 4$

**9.3.** Given $g(x) = -2x^2 + 3x$, find and simplify (a) $g(h)$; (b) $g(x + h)$; (c) $\dfrac{g(x + h) - g(x)}{h}$

(a) Replace $x$ by $h$. $g(h) = -2h^2 + 3h$.

(b) Replace $x$ by the entire quantity $x + h$.

$$g(x + h) = -2(x + h)^2 + 3(x + h) = -2x^2 - 4xh - 2h^2 + 3x + 3h$$

(c) Use the result of part (b).

$$\frac{g(x + h) - g(x)}{h} = \frac{[-2(x + h)^2 + 3(x + h)] - (-2x^2 + 3x)}{h}$$

$$= \frac{-2x^2 - 4xh - 2h^2 + 3x + 3h + 2x^2 - 3x}{h}$$

$$= \frac{-4xh - 2h^2 + 3h}{h} = -4x - 2h + 3$$

**9.4.** Given $f(x) = \dfrac{1}{x^2}$ and $g(x) = 4 - x^2$, find (a) $f(a)g(b)$; (b) $f(g(a))$; (c) $g(f(b))$.

(a) To find $f(a)g(b)$, substitute, then multiply: $f(a) = \dfrac{1}{a^2}$; $g(b) = 4 - b^2$; hence $f(a)g(b) = \left(\dfrac{1}{a^2}\right)(4 - b^2) = \dfrac{4 - b^2}{a^2}$.

(b) To find $f(g(a))$, first substitute $a$ into the rule for $g$ to obtain $g(a) = 4 - a^2$, then substitute this into the rule for $f$ to obtain $f(g(a)) = f(4 - a^2) = \dfrac{1}{(4 - a^2)^2}$.

(c) To find $g(f(b))$, first substitute $b$ into the rule for $f$ to obtain $f(b) = \dfrac{1}{b^2}$, then substitute this into the rule for $g$ to obtain $g(f(b)) = g\left(\dfrac{1}{b^2}\right) = 4 - \left(\dfrac{1}{b^2}\right)^2$.

**9.5.** Find the domain for each of the following functions: (a) $f(x) = 3x - x^3$; (b) $f(x) = \dfrac{5}{x^2 - 9}$;

(c) $f(x) = \dfrac{x^2 - 3x + 2}{x^3 + 2x^2 - 24x}$; (d) $f(x) = \sqrt{x + 5}$; (e) $f(x) = \sqrt{x^2 - 8x + 12}$; (f) $f(x) = \sqrt[3]{\dfrac{x + 1}{x^3 - 8}}$.

(a) This is an example of a polynomial function. Since the polynomial is defined for all real $x$, the domain of the function is all real numbers, $\mathbf{R}$.

(b) The expression $\dfrac{5}{x^2 - 9}$ is defined for all real numbers except if the denominator is 0. This occurs when $x^2 - 9 = 0$; thus $x = \pm 3$. The domain is therefore $\{x \in \mathbf{R} | x \neq \pm 3\}$.

(c) The expression on the right is defined for all real numbers except if the denominator is 0. This occurs when $x^3 + 2x^2 - 24x = 0$, or $x(x - 4)(x + 6) = 0$, thus $x = 0, 4, -6$. The domain is therefore $\{x \in \mathbf{R} | x \neq 0, 4, -6\}$.

(d) The expression $\sqrt{x + 5}$ is defined as long as the expression under the radical is nonnegative. This occurs when $x + 5 \geq 0$ or $x \geq -5$. The domain is therefore $\{x \in \mathbf{R} | x \geq -5\}$, or the interval $[-5, \infty)$.

(e) The expression on the right is defined as long as the expression under the radical is nonnegative. Solving $x^2 - 8x + 12 \geq 0$ by the methods of Chapter 6, $x \leq 2$ or $x \geq 6$ is obtained. The domain is therefore $\{x \in \mathbf{R} | x \leq 2 \text{ or } x \geq 6\}$.

(f) The cube root is defined for all real numbers. Thus the expression on the right is defined for all real numbers except if the denominator is 0. This occurs when $x^3 - 8 = 0$ or $(x - 2)(x^2 + 2x + 4) = 0$, thus only when $x = 2$. The domain is therefore $\{x \in \mathbf{R} | x \neq 2\}$.

**9.6.** Write the circumference $C$ of a circle as a function of its area $A$.

In Example 9.5 the radius $r$ of a circle was expressed as a function of its area $A$: $r = \sqrt{\dfrac{A}{\pi}}$. Since $C = 2\pi r$, it follows that $C = 2\pi\sqrt{\dfrac{A}{\pi}}$ expresses $C$ as a function of $A$.

**9.7.** A theater operator estimates that 500 tickets can be sold if they are priced at \$7 per ticket, and that for each \$.25 increase in the price of a seat, two fewer seats will be sold. Express the revenue $R$ as a function of the number $n$ of \$.25 price increases of a ticket.

The price of a ticket is $7 + 0.25n$ and the number of tickets sold is $500 - 2n$. Since revenue = (number of seats sold) $\times$ (price per seat), $R = (7 + 0.25n)(500 - 2n)$.

**9.8.** A field is to be marked off in the shape of a rectangle, with one side formed by a straight river. If 100 feet is available for fencing, express the area $A$ of the rectangle as a function of the length of one of the two equal sides $x$:

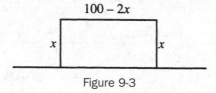

Figure 9-3

Since there are two sides of length $x$, the remaining side has length $100 - 2x$.
Since Area = length $\times$ width for a rectangle,
$A = x(100 - 2x)$.

**9.9.** A rectangle is inscribed in a circle of radius $r$ (see Fig. 9-4). Express the area $A$ of the rectangle as a function of one side $x$ of the rectangle.

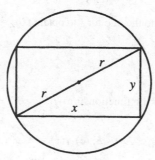

Figure 9-4

From the Pythagorean theorem, it is clear that the sides of the rectangle are related by $x^2 + y^2 = (2r)^2$. Thus $y = \sqrt{4r^2 - x^2}$ and $A = x\sqrt{4r^2 - x^2}$.

**9.10.** A right circular cylinder is inscribed in a right circular cone of height $H$ and base radius $R$ (Fig. 9-5). Express the volume $V$ of the cylinder as a function of its base radius $r$.

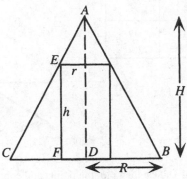

Figure 9-5

In the figure, a cross-section through the axis of the cone and cylinder is shown. Triangle $ADC$ is similar to triangle $EFC$, hence ratios of corresponding sides are equal. In particular, $\dfrac{EF}{FC} = \dfrac{AD}{DC}$ thus, $\dfrac{h}{R-r} = \dfrac{H}{R}$. Solving for $h$, $h = \dfrac{H}{R}(R - r)$. Since for a right circular cylinder, $V = \pi r^2 h$, the volume of this cylinder is $V = \pi \dfrac{H}{R}(R - r)r^2$.

**9.11.** Let $F(x) = mx$, $G(x) = x^2$.

(a) Show that $F(kx) = kF(x)$. (b) Show that $F(a + b) = F(a) + F(b)$. (c) Show that neither of these relations holds in general for the function $G$.

(a) $F(kx) = m(kx) = mkx = kmx = kF(x)$

(b) $F(a + b) = m(a + b) = ma + mb = F(a) + F(b)$

(c) For the function $G$, compare $G(kx) = (kx)^2 = k^2x^2$ with $kG(x) = kx^2$. These are only equal for the special cases $k = 0$ or $k = 1$. Similarly, compare $G(a + b) = (a + b)^2 = a^2 + 2ab + b^2$ with $G(a) + G(b) = a^2 + b^2$. These are only equal in case $a = 0$ or $b = 0$.

**9.12.** Make a table of values and draw graphs of the following functions: (a) $f(x) = 4$;

(b) $f(x) = \dfrac{4x + 3}{5}$; (c) $f(x) = 4x - x^2$; (d) $f(x) = \begin{cases} 4 & \text{if } x \geq 0 \\ -4 & \text{if } x < 0 \end{cases}$; (e) $f(x) = \begin{cases} 4 & \text{if } x \geq 2 \\ -x & \text{if } -1 < x < 2 \\ x + 2 & \text{if } x \leq -1 \end{cases}$

(a) Form a table of values; then plot the points and connect them. The graph (Fig. 9-6) is a horizontal straight line.

| $x$ | $-2$ | $0$ | $2$ | $4$ |
|---|---|---|---|---|
| $y$ | $4$ | $4$ | $4$ | $4$ |

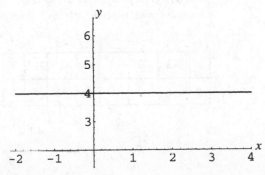

Figure 9-6

(b) Form a table of values; then plot the points and connect them. The graph (Fig. 9-7) is a straight line.

| x | −2 | 0 | 2 | 4 |
|---|---|---|---|---|
| y | −1 | $\frac{3}{5}$ | $\frac{11}{5}$ | $\frac{19}{5}$ |

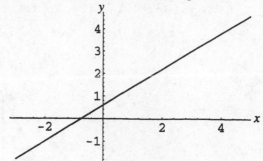

Figure 9-7

(c) Form a more extensive table of values; then plot the points and connect them. The graph (Fig. 9-8) is a smooth curve.

| x | −4 | −2 | 0 | 2 | 4 | 6 | 8 |
|---|---|---|---|---|---|---|---|
| y | −32 | −12 | 0 | 4 | 0 | −12 | −32 |

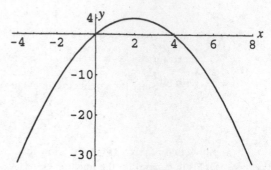

Figure 9-8

(d) Form a table of values. The graph (Fig. 9-9) is discontinuous at the point where $x = 0$.

| x | −4 | −2 | 0 | 2 | 4 |
|---|---|---|---|---|---|
| y | −4 | −4 | 4 | 4 | 4 |

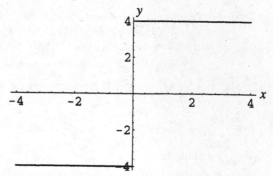

Figure 9-9

(e) Form a table of values. The graph (Fig. 9-10) is discontinuous at the point where $x = 2$. Note that the graph consists of three separate "pieces," since the rule defining the function does so also.

| x | −3 | −2 | −1 | 0 | 1 | 2 | 3 |
|---|---|---|---|---|---|---|---|
| y | −1 | −2 | 1 | 0 | −1 | 4 | 4 |

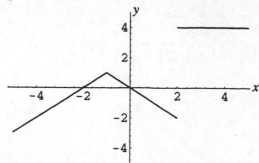

Figure 9-10

**9.13.** Find the range for each of the functions defined in the previous problem.

(a) $f(x) = 4$. The only possible function value is 4, hence the range is $\{4\}$.

(b) $f(x) = \dfrac{4x + 3}{5}$. Set $k = \dfrac{4x + 3}{5}$ and solve for $x$ in terms of $k$ to obtain $x = \dfrac{5k - 3}{4}$. There are no restrictions on $k$, hence the range is $R$.

(c) $f(x) = 4x - x^2$. Set $k = 4x - x^2$ and solve for $x$ in terms of $k$ to obtain $x = 2 \pm \sqrt{4 - k}$. This expression represents a real number only if $k \leq 4$; hence the range is $(-\infty, 4]$.

(d) $f(x) = \begin{cases} 4 & \text{if } x \geq 0 \\ -4 & \text{if } x < 0 \end{cases}$. The only possible function values are 4 and $-4$; hence the range is the set $\{4, -4\}$.

(e) $f(x) = \begin{cases} 4 & \text{if } x \geq 2 \\ -x & \text{if } -1 < x < 2. \\ x + 2 & \text{if } x \leq -1 \end{cases}$

If $x \leq -1, x + 2 \leq 1$; thus $f$ can take on any value in $(-\infty, 1]$. If $-1 < x < 2, -2 < -x < 1$; this adds nothing to the range. If $x \geq 2, f(x) = 4$; hence the range consists of the set union $(-\infty, 1] \cup \{4\}$.

**9.14.** Find the average rate of change for (a) $f(x) = 7x + 12$ on $[2,8]$; (b) $f(x) = \dfrac{3 - 5x}{9}$ on $[-5, 0]$

(a) $\dfrac{f(8) - f(2)}{8 - 2} = \dfrac{(7 \cdot 8 + 12) - (7 \cdot 2 + 12)}{6} = \dfrac{68 - 26}{6} = 7$

(b) $\dfrac{f(0) - f(-5)}{0 - (-5)} = \dfrac{\left(\dfrac{3 - 5 \cdot 0}{9}\right) - \left(\dfrac{3 - 5(-5)}{9}\right)}{5} = \dfrac{\dfrac{3}{9} - \dfrac{28}{9}}{5} = \dfrac{-25/9}{5} = -\dfrac{5}{9}$

**9.15.** Find the difference quotient for (a) $f(x) = x^3$; (b) $f(x) = \dfrac{1}{x^2}$

(a) $\dfrac{f(x + h) - f(x)}{h} = \dfrac{(x + h)^3 - x^3}{h} = \dfrac{x^3 + 3x^2h + 3xh^2 + h^3 - x^3}{h}$

$= \dfrac{3x^2h + 3xh^2 + h^3}{h} = 3x^2 + 3xh + h^2$

(b) $\dfrac{f(x + h) - f(x)}{h} = \dfrac{\dfrac{1}{(x + h)^2} - \dfrac{1}{x^2}}{h} = \dfrac{x^2 - (x + h)^2}{hx^2(x + h)^2} = \dfrac{x^2 - x^2 - 2xh - h^2}{hx^2(x + h)^2}$

$= \dfrac{-2xh - h^2}{hx^2(x + h)^2} = \dfrac{-2x - h}{x^2(x + h)^2}$

## SUPPLEMENTARY PROBLEMS

**9.16.** Let $F$ be any function whose domain contains $-x$ whenever it contains $x$. Define:

$$g(x) = \dfrac{F(x) + F(-x)}{2} \text{ and } h(x) = \dfrac{F(x) - F(-x)}{2}.$$

(a) Show that $g$ is an even function and $h$ is an odd function.

(b) Show that $F(x) = g(x) + h(x)$. Thus, any function can be written as the sum of an odd function and an even function.

(c) Show that the only function which is both even and odd is $f(x) = 0$.

**9.17.** Are the following functions even, odd, or neither?

(a) $f(x) = \dfrac{x^3}{x^4 + 1}$; (b) $f(x) = \dfrac{x^4}{x^3 + 1}$; (c) $f(x) = |x| - \dfrac{1}{x^2}$; (d) $f(x) = (x - 1)^3 + (x + 1)^3$

*Ans.* (a) odd; (b) neither; (c) even; (d) odd

**9.18.** Find the domain for the following functions:

(a) $f(x) = \sqrt{x-3}$; (b) $f(x) = \sqrt{3-x}$; (c) $f(x) = \dfrac{1}{\sqrt{x-3}}$; (d) $f(x) = \dfrac{1}{\sqrt{3-x}}$

*Ans.*    (a) $[3,\infty)$; (b) $(-\infty,3]$; (c) $(3,\infty)$; (d) $(-\infty,3)$

**9.19.** Find the domain for the following functions:

(a) $g(x) = |x-3|$; (b) $g(x) = \dfrac{x^2+9}{x-3}$; (c) $g(x) = \sqrt{\dfrac{x-3}{x^2-3x+2}}$; (d) $g(x) = \sqrt{x^3-9x^2}$

*Ans.*    (a) $R$; (b) $\{x \in R | x \neq 3\}$; (c) $(1,2) \cup [3, \infty)$; (d) $[9, \infty)$

**9.20.** The income tax rate in a certain state is 4% on taxable income up to \$30,000, 5% on taxable income between \$30,000 and \$50,000, and 6% on taxable income over \$50,000. Express the income tax $T(x)$ as a function of taxable income $x$.

*Ans.*    $T(x) = \begin{cases} 0.04x & \text{if } 0 < x \leq 30{,}000 \\ 1200 + 0.05(x - 30{,}000) & \text{if } 30{,}000 < x \leq 50{,}000 \\ 2200 + 0.06(x - 50{,}000) & \text{if } 50{,}000 < x \end{cases}$

**9.21.** (a) Express the length of a diagonal $d$ of a square as a function of the length of one side $s$. (b) Express $d$ as a function of the area $A$ of the square. (c) Express $d$ as a function of the perimeter $P$ of the square.

*Ans.*    (a) $d(s) = s\sqrt{2}$; (b) $d(A) = \sqrt{2A}$; (c) $d(P) = \dfrac{P\sqrt{2}}{4}$

**9.22.** (a) Express the area $A$ of an equilateral triangle as a function of one side $s$. (b) Express the perimeter of the triangle $P$ as a function of the area $A$.

*Ans.*    (a) $A(s) = s^2\sqrt{3}/4$; (b) $P(A) = (6\sqrt{A})/\sqrt[4]{3}$

**9.23.** An equilateral triangle of side $s$ is inscribed in a circle of radius $r$.

(a) Express $s$ as a function of $r$. (b) Express the area $A$ of the triangle as a function of $r$. (c) Express the area $A$ of the triangle as a function of $a$, the area of the circle.

*Ans.*    (a) $s(r) = r\sqrt{3}$; (b) $A(r) = \dfrac{3r^2\sqrt{3}}{4}$; (c) $A(a) = \dfrac{3a\sqrt{3}}{4\pi}$

**9.24.** (a) Express the volume $V$ of a sphere as a function of its radius $r$. (b) Express the surface area $S$ of the sphere as a function of $r$. (c) Express $r$ as a function of $S$. (d) Express $V$ as a function of $S$.

*Ans.*    (a) $V(r) = \dfrac{4}{3}\pi r^3$; (b) $S(r) = 4\pi r^2$; (c) $r(S) = \sqrt{\dfrac{S}{4\pi}}$; (d) $V(S) = \dfrac{1}{3}\sqrt{\dfrac{S^3}{4\pi}}$

**9.25.** A right circular cylinder is inscribed in a sphere of radius $R$. ($R$ is a constant.)

(a) Express the height $h$ of the cylinder as a function of the radius $r$ of the cylinder.

(b) Express the total surface area $S$ of the cylinder as a function of $r$.

(c) Express the volume $V$ of the cylinder as a function of $r$.

*Ans.*    (a) $h(r) = 2\sqrt{R^2 - r^2}$; (b) $S(r) = 4\pi r\sqrt{R^2 - r^2} + 2\pi r^2$; (c) $V(r) = 2\pi r^2\sqrt{R^2 - r^2}$

**9.26.** Which of Figs. 9-11 to 9-14 are graphs of functions?

(a) Figure 9-11

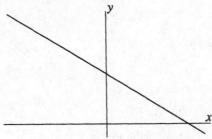

Figure 9-11

(b) Figure 9-12

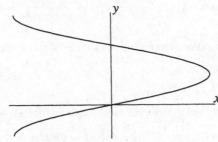

Figure 9-12

(c) Figure 9-13

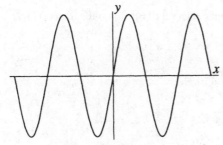

Figure 9-13

(d) Figure 9-14

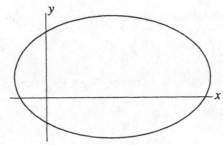

Figure 9-14

*Ans.* (a) and (c) are graphs of functions; (b) and (d) fail the vertical line test and are not graphs of functions.

**9.27.** Given $f(x) = x^2 - 3x + 1$, find (a) $f(2)$; (b) $f(-3)$; (c) $\dfrac{f(2 + h) - f(2)}{h}$.

*Ans.* (a) $-1$; (b) $19$; (c) $1 + h$

**9.28.** Given $f(x) = \dfrac{1}{x} - x$, find (a) $f(2)$; (b) $f(-3)$; (c) $\dfrac{f(3 + h) - f(3)}{h}$.

*Ans.* (a) $-\dfrac{3}{2}$; (b) $\dfrac{8}{3}$; (c) $\dfrac{-10 - 3h}{3(3 + h)}$

**9.29.** The distance $s$ an object falls from rest in time $t$ seconds is given in feet by $s(t) = 16t^2$. Find (a) $s(2)$; (b) $s(3)$;

(c) $\dfrac{s(3.01) - s(3)}{.01}$

*Ans.* (a) $64$ feet; (b) $144$ feet; (c) $96.16$ feet

**9.30.** Given $f(x) = \dfrac{3x + 1}{x - 3}$, find and write in simplest form: (a) $f(f(b))$; (b) $\dfrac{f(x) - f(a)}{x - a}$

*Ans.* (a) $b$; (b) $\dfrac{-10}{(x - 3)(a - 3)}$

**9.31.** Given $f(x) = x^2$, find and write in simplest form: (a) $f(f(b))$; (b) $\dfrac{f(x) - f(a)}{x - a}$; (c) $\dfrac{f(x + h) - f(x)}{h}$

*Ans.* (a) $b^4$; (b) $x + a$; (c) $2x + h$

**9.32.** Given $f(x) = \dfrac{1}{x}$, find and write in simplest form: (a) $f(f(b))$; (b) $\dfrac{f(x) - f(a)}{x - a}$; (c) $\dfrac{f(x + h) - f(x)}{h}$

*Ans.* (a) $b$; (b) $\dfrac{-1}{ax}$; (c) $\dfrac{-1}{x(x + h)}$

**9.33.** Given $f(x) = \dfrac{x}{1 + x^2}$, find and write in simplest form: (a) $f(f(b))$; (b) $\dfrac{f(x) - f(a)}{x - a}$

Ans.   (a) $\dfrac{b + b^3}{1 + 3b^2 + b^4}$; (b) $\dfrac{1 - ax}{(1 + x^2)(1 + a^2)}$.

**9.34.** Find the average rate of change for $f(x) = 9x - 7$ on the interval $[0,5]$.

Ans.   9

**9.35.** (a) Find the average rate of change for $f(x) = \sqrt{x}$ on the interval $[4,9]$.

(b) Find the difference quotient for $f(x) = \sqrt{x}$. Rationalize the numerator in the answer.

Ans.   (a) $\dfrac{1}{5}$; (b) $\dfrac{\sqrt{x + h} - \sqrt{x}}{h} = \dfrac{1}{\sqrt{x + h} + \sqrt{x}}$

**9.36.** Find the average rate of change for $f(x) = x^2 - 6x + 9$ (a) on the interval $[0,6]$; (b) on the interval $[1,7]$.

Ans.   (a) 0; (b) 2

**9.37.** Find the average rate of change for $f(x) = \dfrac{1}{x + 6}$ on the interval $[0,5]$.

Ans.   $-\dfrac{1}{66}$

**9.38.** Find the difference quotient for (a) $f(x) = \dfrac{x}{x + 1}$; (b) $f(x) = \sqrt{2x - 1}$. Rationalize the numerator in the answer.

(a) $\dfrac{1}{(x + 1)(x + h + 1)}$; (b) $\dfrac{\sqrt{2(x + h) - 1} - \sqrt{2x - 1}}{h} = \dfrac{2}{\sqrt{2(x + h) - 1} + \sqrt{2x - 1}}$

# CHAPTER 10

# Linear Functions

## Definition of Linear Function

A linear function is any function specified by a rule of form $f\colon x \to mx + b$, where $m \neq 0$. If $m = 0$, the function is not considered to be a linear function; a function $f(x) = b$ is called a *constant function*. The graph of a linear function is always a straight line. The graph of a constant function is a horizontal straight line.

## Slope of a Line

The slope of a line that is not parallel to the $y$-axis is defined as follows (see Figs. 10-1 and 10-2): Let $(x_1, y_1)$ and $(x_2, y_2)$ be distinct points on the line. Then the slope of the line is given by

$$m = \frac{y_2 - y_1}{x_2 - x_1} = \frac{\text{change in } y}{\text{change in } x} = \frac{\text{rise}}{\text{run}}$$

(a) Positive slope (line rises) (Fig. 10-1)  (b) Negative slope (line falls) (Fig. 10-2)

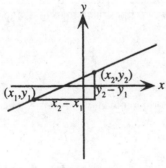

Figure 10-1

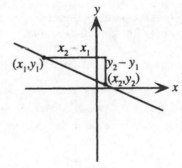

Figure 10-2

**EXAMPLE 10.1**  Find the slope of the lines through (a) (5,3) and (8,12); (b) (3,−4) and (−5,6).

(a) Identify $(x_1, y_1) = (5,3)$ and $(x_2, y_2) = (8,12)$. Then $m = \dfrac{y_2 - y_1}{x_2 - x_1} = \dfrac{12 - 3}{8 - 5} = 3$.

(b) Identify $(x_1, y_1) = (3,−4)$ and $(x_2, y_2) = (−5,6)$. Then $m = \dfrac{y_2 - y_1}{x_2 - x_1} = \dfrac{6 - (−4)}{−5 - 3} = -\dfrac{5}{4}$.

## Horizontal and Vertical Lines

1. A horizontal line (a line parallel to the $x$-axis) has slope 0, since any two points on the line have the same $y$ coordinates. A horizontal line has an equation of the form $y = k$. (See Fig. 10-3.)
2. A vertical line (a line parallel to the $y$-axis) has undefined slope, since any two points on the line have the same $x$ coordinates. A vertical line has an equation of the form $x = h$. (See Fig. 10-4.)

(*a*) Horizontal line                  (*b*) Vertical line

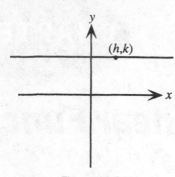

Figure 10-3

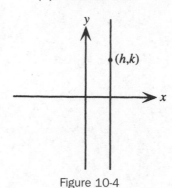

Figure 10-4

## Equation of a Line

The equation of a line can be written in several forms. Among the most useful are:

1. **SLOPE-INTERCEPT FORM:** The equation of a line with slope $m$ and $y$ intercept $b$ is given by $y = mx + b$.
2. **POINT-SLOPE FORM:** The equation of a line passing through $(x_0, y_0)$ with slope $m$ is given by $y - y_0 = m(x - x_0)$.
3. **STANDARD FORM:** The equation of a line can be written as $Ax + By = C$, where $A, B, C$ are integers with no common factors; $A$ and $B$ are not both zero.

**EXAMPLE 10.2**   Find the equation of the line passing through $(-6, 4)$ with slope $\frac{2}{3}$.

Use the point-slope form of the equation of a line: $y - 4 = \frac{2}{3}[x - (-6)]$. This can then be simplified to slope-intercept form: $y = \frac{2}{3}x + 8$. In standard form, this would become $2x - 3y = -24$.

## Parallel Lines

If two nonvertical lines are parallel, their slopes are equal. Conversely, if two lines have the same slope, they are parallel; two vertical lines are also parallel.

**EXAMPLE 10.3**   Find the equation of a line through $(3, -8)$ parallel to $5x + 2y = 7$.

First find the slope of the given line by isolating the variable $y$: $y = -\frac{5}{2}x + \frac{7}{2}$. Thus the given line has slope $-\frac{5}{2}$. Hence the desired line has slope $-\frac{5}{2}$ and passes through $(3, -8)$. Use the point-slope form to obtain $y - (-8) = -\frac{5}{2}(x - 3)$, which is written in standard form as $5x + 2y = -1$.

## Perpendicular Lines

If a line is horizontal, any line perpendicular to it is vertical, and conversely. If two nonvertical lines, with slopes $m_1$ and $m_2$, are perpendicular, then their slopes satisfy $m_1 m_2 = -1$ or $m_2 = -1/m_1$.

**EXAMPLE 10.4**   Find the equation of a line through $(3, -8)$ perpendicular to $5x + 2y = 7$.

The given line was found in the previous example to have slope $-\frac{5}{2}$. Hence the desired line has slope $\frac{2}{5}$ and passes through $(3, -8)$. Use the point-slope form to obtain $y - (-8) = \frac{2}{5}(x - 3)$, which is written in standard form as $2x - 5y = 46$.

## SOLVED PROBLEMS

**10.1.** For any linear function of form $f(x) = mx + b$ show that $\dfrac{f(x + h) - f(x)}{h} = m$.

     Given $f(x) = mx + b$, it follows that $f(x + h) = m(x + h) + b$, hence

$$\frac{f(x + h) - f(x)}{h} = \frac{[m(x + h) + b] - [mx + b]}{h} = \frac{mx + mh + b - mx - b}{h} = \frac{mh}{h} = m$$

**10.2.** Which of the following rules represent linear functions?

(a)  $f(x) = \frac{2}{3}$         (b)  $f(x) = \frac{2}{3}x + 7$         (c)  $f(x) = \frac{2}{3x} + 7$

Only (b) represents a linear function. The rule in (a) represents a constant function, while the rule in (c) is referred to as a nonlinear function.

**10.3.** Find the equation of the horizontal line through $(5,-3)$.

A horizontal line has an equation of the form $y = k$. In this case, the constant $k$ must be $-3$. Hence the required equation is $y = -3$.

**10.4.** Find the equation of the vertical line through $(5,-3)$.

A vertical line has an equation of the form $x = h$. In this case, the constant $h$ must be 5. Hence the required equation is $x = 5$.

**10.5.** Find the equation of the line through $(-6,8)$ with slope $\frac{3}{4}$. Write the answer in slope-intercept form and also in standard form.

Use the point-slope form of the equation of a line, with $m = \frac{3}{4}$ and $(x_0,y_0) = (-6,8)$. Then the equation of the line can be written:

$$y - 8 = \frac{3}{4}[x - (-6)]$$

Simplifying, this becomes $y = \frac{3}{4}x + \frac{25}{2}$ in slope-intercept form, and $-3x + 4y = 50$ in standard form.

**10.6.** Find the equation of the line through the points $(3,-4)$ and $(-7,2)$. Write the answer in slope-intercept form and also in standard form.

First, find the slope of the line: Identify $(x_1,y_1) = (3,-4)$ and $(x_2,y_2) = (-7,2)$. Then

$$m = \frac{y_2 - y_1}{x_2 - x_1} = \frac{2 - (-4)}{-7 - 3} = -\frac{3}{5}$$

Now, use the point-slope form of the equation of a line, with $m = -\frac{3}{5}$. Choose either of the given points, say, $(3,-4) = (x_0,y_0)$. Then the equation of the line can be written:

$$y - (-4) = -\frac{3}{5}(x - 3)$$

Simplifying, this becomes $y = -\frac{3}{5}x - \frac{11}{5}$ in slope-intercept form and $3x + 5y = -11$ in standard form.

**10.7.** (a) Show that the equation of a line with $x$-intercept $a$ and $y$-intercept $b$, where neither $a$ nor $b$ is 0, can be written as $\frac{x}{a} + \frac{y}{b} = 1$. (This is known as the *two-intercept* form of the equation of a line.) (b) Write the equation of the line with $x$-intercept 5 and $y$-intercept $-6$ in standard form.

(a) The line passes through the points $(a,0)$ and $(0,b)$. Hence its slope can be found from the definition of slope as $m = \frac{y_2 - y_1}{x_2 - x_1} = \frac{b - 0}{0 - a} = -\frac{b}{a}$. Using the slope-intercept form, the equation of the line can be written as $y = -\frac{b}{a}x + b$, or $\frac{b}{a}x + y = b$. Dividing by $b$ yields $\frac{x}{a} + \frac{y}{b} = 1$ as required.

(b) Using the result of part (a) gives $\frac{x}{5} + \frac{y}{-6} = 1$. Clearing of fractions yields $6x - 5y = 30$ in standard form as required.

**10.8.** It is shown in calculus that the slope of the line drawn tangent to the parabola $y = x^2$ at the point $(a,a^2)$ has slope $2a$. Find the equation of the line tangent to $y = x^2$ (a) at $(3, 9)$; (b) at $(a, a^2)$.

(a) Since the line has slope $2 \cdot 3 = 6$, use the point-slope form to find the equation of a line through $(3,9)$ with slope 6: $y - 9 = 6(x - 3)$ or $y = 6x - 9$.

(b) Since the line has slope $2a$, use the point-slope form to find the equation of a line through $(a,a^2)$ with slope $2a$: $y - a^2 = 2a(x - a)$ or $y = 2ax - a^2$.

**10.9.** Prove that two nonvertical lines are parallel if and only if they have the same slope. (See Fig. 10-5.)

Let $l_1$ and $l_2$ be two different lines with slopes, respectively, $m_1$ and $m_2$, and $y$-intercepts, respectively, $b_1$ and $b_2$.

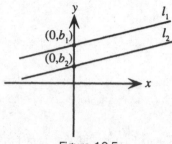

Figure 10-5

Then the lines have equations $y = m_1 x + b_1$ and $y = m_2 x + b_2$. The lines will intersect at some point $(x,y)$ if and only if for some $x$ the values of $y$ are equal, that is,

$$m_1 x + b_1 = m_2 x + b_2; \text{ thus, } (m_1 - m_2)x = b_2 - b_1$$

This is possible, that is, the lines intersect, if and only if $m_1 \neq m_2$. Hence the lines are parallel if and only if $m_1 = m_2$.

**10.10.** Find the equation of the line through $(5,-3)$ parallel to (a) $y = 3x - 5$; (b) $2x + 7y = 4$; (c) $x = -1$.

(a) Any line parallel to the given line will have the same slope as the given line. Since the given line is written in slope-intercept form, its slope is clearly seen to be 3. The equation of a line through $(5,-3)$ with slope 3 is found from the point-slope form to be $y - (-3) = 3(x - 5)$. Simplifying yields $y = 3x - 18$.

(b) It is possible to proceed as in (a); however, the equation of the given line must be analyzed to find its slope. An alternative method is to note that any line parallel to the given line can be written as $2x + 7y = C$. Then, since $(5,-3)$ must satisfy the equation, $2 \cdot 5 + 7(-3) = C$; hence $C = -11$ and $2x + 7y = -11$ is the required equation.

(c) Proceeding as in (b), note that any line parallel to the given line must be vertical, hence must have an equation of the form $x = h$. In this case, $h = 5$; hence $x = 5$ is the required equation.

**10.11.** Prove that if two lines with slopes $m_1$ and $m_2$ are perpendicular, then $m_1 m_2 = -1$. (Fig. 10-6.)

The slopes of the lines must have opposite signs. In the figure, $m_1$ is chosen (arbitrarily) positive and $m_2$ is chosen negative.

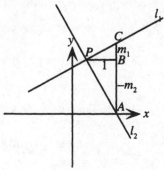

Figure 10-6

Since $l_1$ has slope $m_1$, a run of 1 (segment $PB$) yields a rise of $m_1$ along $l_1$ (segment $CB$). Similarly, since $l_2$ has slope $m_2$, a run of 1 yields a (negative) rise of $m_2$ along $l_2$, thus segment $AB$ has length $-m_2$. Since the lines are perpendicular, the triangles $PCB$ and $APB$ are similar. Hence ratios of corresponding sides are equal; it follows that

$$\frac{CB}{PB} = \frac{PB}{AB}$$

$$\frac{m_1}{1} = \frac{1}{-m_2}$$

$$m_1 m_2 = -1$$

**10.12.** Find the equation of the line through $(8,-2)$ perpendicular to (a) $y = \frac{4}{5}x + 2$; (b) $x + 3y = 6$; (c) $x = 7$.

(a) Any line perpendicular to the given line will have slope $m$ satisfying $\frac{4}{5}m = -1$; thus, $m = -\frac{5}{4}$. The equation of a line through $(8,-2)$ with slope $-\frac{5}{4}$ is found from the point-slope form to be $y - (-2) = -\frac{5}{4}(x - 8)$. Simplifying yields $5x + 4y = 32$.

(b) First, determine the slope of the given line. Isolating the variable $y$, the equation is seen to be equivalent to $y = -\frac{1}{3}x + 2$; hence the slope is $-\frac{1}{3}$. Any line perpendicular to the given line will have slope $m$ satisfying $-\frac{1}{3}m = -1$; thus, $m = 3$. The equation of a line through $(8,-2)$ with slope 3 is found from the point-slope form to be $y - (-2) = 3(x - 8)$. Simplifying yields $y = 3x - 26$.

(c) Since the given line is vertical, any line perpendicular to the given line must be horizontal, hence must have an equation of the form $y = k$. In this case, $k = -2$; hence $y = -2$ is the required equation.

**10.13.** Find the rule for a linear function, given $f(0) = 5$ and $f(10) = 12$.

Since the graph of a linear function is a straight line, this is equivalent to finding the equation of a line in slope-intercept form, given a $y$-intercept of 5. Since the line passes through $(0,5)$ and $(10,12)$, the slope is determined; $m = \dfrac{12 - 5}{10 - 0} = \dfrac{7}{10}$. Hence the equation of the line is $y = \frac{7}{10}x + 5$ and the rule for the function is $f(x) = \frac{7}{10}x + 5$.

**10.14.** Find a general expression for the rule for a linear function, given $f(a)$ and $f(b)$.

Since the graph of a linear function is a straight line, this is equivalent to finding the equation, in slope-intercept form, of a line passing through $(a, f(a))$ and $(b, f(b))$. Clearly, a line passing through these two points will have slope $m = \dfrac{f(b) - f(a)}{b - a}$. From the point-slope form, the equation of the line can be written as $y - f(a) = \dfrac{f(b) - f(a)}{b - a}(x - a)$ or $y = \dfrac{f(b) - f(a)}{b - a}(x - a) + f(a)$. Thus the rule for the function is

$$f(x) = \frac{f(b) - f(a)}{b - a}(x - a) + f(a)$$

**10.15.** Find the rule for a linear function, given $f(10) = 25,000$ and $f(25) = 10,000$.

Apply the formula from the previous problem with $a = 10$ and $b = 25$. Then

$$f(x) = \frac{10,000 - 25,000}{25 - 10}(x - 10) + 25,000$$

$$= -1000x + 35,000$$

**10.16.** Suppose the cost of producing 50 units of a given commodity is \$27,000, while the cost of producing 100 units of the same commodity is \$38,000. If the cost function $C(x)$ is assumed to be linear, find a rule for $C(x)$. Use the rule to estimate the cost of producing 80 units of the commodity.

This is equivalent to finding the equation of a straight line passing through $(50,27000)$ and $(100,38000)$. The slope of this line is $m = \dfrac{38,000 - 27,000}{100 - 50} = 220$; hence from the point-slope form the equation of the line is

$y - 27,000 = 220(x - 50)$. Simplifying yields $y = 220x + 16,000$; thus, the rule for the function is $C(x) = 220x + 16,000$.

The cost of producing 80 units of the commodity is given by $C(80) = 220 \cdot 80 + 16,000$ or $\$33,600$.

**10.17.** If the value of a piece of equipment is *depreciated linearly* over a 20-year period, the value $V(t)$ can be described as a linear function of time $t$.

(a) Find a rule for $V(t)$ assuming that the value at time $t = 0$ is $V_0$ and that the value after 20 years is zero.

(b) Use the rule to find the value after 12 years of a piece of equipment originally valued at $\$7500$.

(a) This is equivalent to finding the equation of a line of form $V = mt + b$, where the $V$-intercept is $b = V_0$, and the line passes through $(0,V_0)$ and $(20,0)$. The slope is given by $m = \dfrac{0 - V_0}{20 - 0} = -\dfrac{V_0}{20}$; hence the equation is $V = -\dfrac{V_0}{20}t + V_0$ and the rule for the function is $V(t) = -\dfrac{V_0}{20}t + V_0$.

(b) In this case, $V_0 = 7500$ and the value of $V(12)$ is required. Since the rule for the function is now
$V(t) = -\frac{7500}{20}t + 7500 = 7500 - 375t$, $V(12) = 7500 - 375 \cdot 12 = 3000$ and the value is $\$3000$.

## SUPPLEMENTARY PROBLEMS

**10.18.** Write the following equations in standard form:

(a) $y = 3x - 2$; (b) $y = -\frac{1}{2}x + 8$; (c) $y = \frac{2}{3}x - \frac{3}{5}$

*Ans.* (a) $3x - y = 2$; (b) $x + 2y = 16$; (c) $10x - 15y = 9$

**10.19.** Write the following equations in slope-intercept form:

(a) $2x + 6y = 7$; (b) $3x - 5y = 15$; (c) $\frac{1}{2}x + \frac{2}{3}y = \frac{3}{4}$

*Ans.* (a) $y = -\frac{1}{3}x + \frac{7}{6}$; (b) $y = \frac{3}{5}x - 3$; (c) $y = -\frac{3}{4}x + \frac{9}{8}$

**10.20.** Find the equation of a line in standard form given:

(a) The line is horizontal and passes through $\left(\frac{2}{3}, \frac{3}{4}\right)$.

(b) The line has slope $-0.3$ and passes through $(1.3, -5.6)$.

(c) The line has $x$-intercept 7 and slope $-4$.

(d) The line is parallel to $y = 3 - 2x$ and passes through the origin.

(e) The line is perpendicular to $3x - 5y = 7$ and passes through $\left(-\frac{5}{2}, \frac{8}{3}\right)$.

(f) The line passes through $(a,b)$ and $(c,d)$.

*Ans.* (a) $4y = 3$      (b) $30x + 100y = -521$      (c) $4x + y = 28$

        (d) $2x + y = 0$      (e) $10x + 6y = -9$      (f) $(b - d)x + (c - a)y = (b - d)a + (c - a)b$

**10.21.** Find the equation of a line in slope-intercept form given:

(a) The line is horizontal and passes through $(-3,8)$.

(b) The line has slope $-\frac{2}{3}$ and passes through $(-5,1)$.

(c) The line has $x$-intercept $-2$ and slope $\frac{1}{2}$.

(d) The line is parallel to $2x + 5y = 1$ and passes through $(2,-8)$.

(e) The line is perpendicular to $y = \frac{3}{8}x - 1$ and passes through $(6,0)$.

*Ans.*   (a) $y = 8$;          (b) $y = -\frac{2}{3}x - \frac{7}{3}$;          (c) $y = \frac{1}{2}x + 1$;

   (d) $y = -\frac{2}{5}x - \frac{36}{5}$;   (e) $y = -\frac{8}{3}x + 16$

**10.22.** Find the slope and the $y$-intercept for (a) $y = 5 - 3x$; (b) $2x + 6y = 9$; (c) $x + 5 = 0$.

*Ans.*   (a) slope $-3$, $y$-intercept 5; (b) slope $-\frac{1}{3}$, $y$-intercept $\frac{3}{2}$; (c) slope undefined, no $y$-intercept

**10.23.** Find the possible slopes of a line that passes through $(4,3)$ so that the portion of the line in the first quadrant forms a triangle of area 27 with the positive coordinate axes.

*Ans.*   $-\frac{3}{2}$ or $-\frac{3}{8}$

**10.24.** Repeat problem 10.23 except with the triangle having area 24.

*Ans.*   $-\frac{3}{4}$ is the only possible slope.

**10.25.** Recall from geometry that the line drawn tangent to a circle is perpendicular to the radius line drawn to the point of tangency. Use this fact to find the equation of the line tangent to

(a) the circle $x^2 + y^2 = 25$ at $(-3,4)$

(b) the circle $(x - 2)^2 + (y + 4)^2 = 4$ at $(2,-2)$

*Ans.*   (a) $3x - 4y = -25$; (b) $y = -2$

**10.26.** It is shown in calculus that the slope of the line drawn tangent to the curve $y = x^3$ at the point $(a,a^3)$ has slope $3a^2$. Find the equation of the line tangent to $y = x^3$ at (a) $(2,8)$; (b) $(a,a^3)$.

*Ans.*   (a) $y = 12x - 16$; (b) $y = 3a^2x - 2a^3$

**10.27.** The line drawn perpendicular to the tangent line to a curve at the point of tangency is called the *normal* line. Find the equation of the normal line to $y = x^3$ at $(2,8)$. (See the previous problem.)

*Ans.*   $y = -\frac{1}{12}x + \frac{49}{6}$

**10.28.** An *altitude* of a triangle is a line drawn from a vertex of the triangle perpendicular to the opposite side of the triangle. Find the equation of the altitude drawn from $A(0,0)$ to the side formed by $B(3,4)$ and $C(5,-2)$.

*Ans.*   $x - 3y = 0$

**10.29.** A *median* of a triangle is a line drawn from a vertex of the triangle to the midpoint of the opposite side of the triangle. Find the equation of the median drawn from $A(5,-2)$ to the side formed by $B(-3,9)$ and $C(4,-7)$.

*Ans.*   $2x + 3y = 4$

**10.30.** Find a rule for a linear function given $f(5) = -7$ and $f(-5) = 10$.

*Ans.*   $f(x) = -\frac{17}{10}x + \frac{3}{2}$

**10.31.** Find a rule for a linear function given $f(0) = a$ and $f(c) = b$.

*Ans.* $f(x) = \dfrac{b - a}{c} x + a$

**10.32.** In depreciation situations (see Problem 10.17), it is common that a piece of equipment has a residual value after it has been linearly depreciated over its entire lifetime.

(a) Find a rule for $V(t)$, the value of a piece of equipment, assuming that the value at time $t = 0$ is $V_0$ and that the value after 20 years is $R$.

(b) Use the rule to find the value after 12 years of a piece of equipment originally valued at $7500, assuming that it has a residual value after 20 years of $500.

*Ans.* (a) $V(t) = \dfrac{R - V_0}{20} t + V_0$; (b) $3300

# Transformations and Graphs

## Elementary Transformations

The graphs of many functions can be regarded as arising from more basic graphs as a result of one or more elementary transformations. The elementary transformations considered here are shifting, stretching and compression, and reflection with respect to a coordinate axis.

## Basic Function

Given a basic function $y = f(x)$ with the graph shown in Fig. 11-1, the following transformations have easily identified effects on the graph.

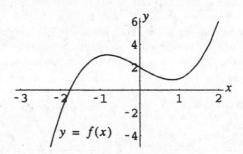

Figure 11-1

## Vertical Shifting

The graph of $y = f(x) + k$, for $k > 0$, is the same as the graph of $y = f(x)$ *shifted up* $k$ units. The graph of $y = f(x) + k$, for $k < 0$, is the same as the graph of $y = f(x)$ *shifted down* $k$ units.

**EXAMPLE 11.1**  For the basic function shown in Fig. 11-1, graph $y = f(x)$ and $y = f(x) + 2$ on the same coordinate system (Fig. 11-2), $y = f(x)$ and $y = f(x) - 2.5$ on the same coordinate system (Fig. 11-3).

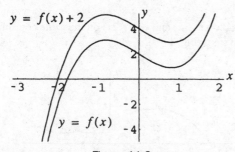

Figure 11-2

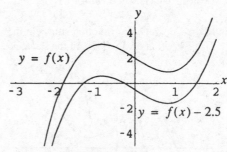

Figure 11-3

## Vertical Stretching and Compression

The graph of $y = af(x)$, for $a > 1$, is the same as the graph of $y = f(x)$ *stretched,* with respect to the $y$-axis, by a factor of $a$. The graph of $y = af(x)$, for $0 < a < 1$, is the same as the graph of $y = f(x)$ *compressed*, with respect to the $y$-axis, by a factor of $1/a$.

**EXAMPLE 11.2**   For the basic function shown in Fig. 11-1, graph $y = f(x)$ and $y = 2f(x)$ on the same coordinate system (Fig. 11-4); $y = f(x)$ and $y = \frac{1}{3}f(x)$ on the same coordinate system (Fig. 11-5).

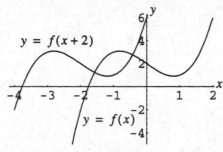

Figure 11-4

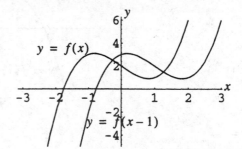

Figure 11-5

## Horizontal Shifting

The graph of $y = f(x + h)$, for $h > 0$, is the same as the graph of $y = f(x)$ *shifted left h* units. The graph of $y = f(x - h)$, for $h > 0$, is the same as the graph of $y = f(x)$ *shifted right h* units.

**EXAMPLE 11.3**   For the basic function shown in Fig. 11-1, graph $y = f(x)$ and $y = f(x + 2)$ on the same coordinate system (Fig. 11-6); $y = f(x)$ and $y = f(x - 1)$ on the same coordinate system (Fig. 11-7).

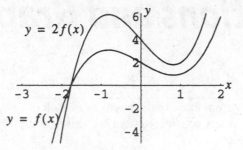

Figure 11-6

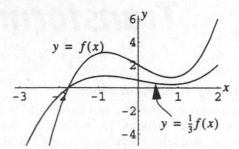

Figure 11-7

## Horizontal Stretching and Compression

The graph of $y = f(ax)$, for $a > 1$, is the same as the graph of $y = f(x)$ *compressed,* with respect to the $x$-axis, by a factor of $a$. The graph of $y = f(ax)$, for $0 < a < 1$, is the same as the graph of $y = f(x)$ *stretched,* with respect to the $x$-axis, by a factor of $1/a$.

**EXAMPLE 11.4**   For the basic function shown in Fig. 11-1, graph $y = f(x)$ and $y = f(2x)$ on the same coordinate system (Fig 11-8); $y = f(x)$ and $y = f\left(\frac{1}{2}x\right)$ on the same coordinate system (Fig. 11-9).

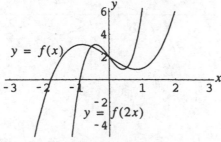

Figure 11-8

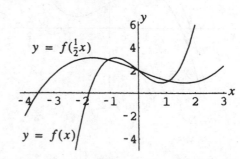

Figure 11-9

## Reflection with Respect to a Coordinate Axis

The graph of $y = -f(x)$ is the same as the graph of $y = f(x)$ reflected across the $x$-axis. The graph of $y = f(-x)$ is the same as the graph of $y = f(x)$ reflected across the $y$-axis.

**EXAMPLE 11.5**  For the basic function shown in Fig. 11-1, graph $y = f(x)$ and $y = -f(x)$ on the same coordinate system (Fig. 11-10); $y = f(x)$ and $y = f(-x)$ on the same coordinate system (Fig. 11-11).

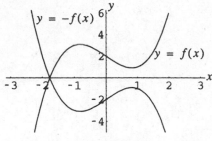

Figure 11-10

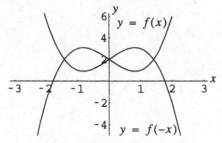

Figure 11-11

## SOLVED PROBLEMS

**11.1.** Explain why, for positive $h$, the graph of $y = f(x) + h$ is displaced *up* by $h$ units from the graph of $y = f(x)$, while the graph of $y = f(x + h)$ is displaced *left* by $h$ units.

Consider the point $(a, f(a))$ on the graph of $y = f(x)$. The point $(a, f(a) + h)$ on the graph of $y = f(x) + h$ can be regarded as the corresponding point. This point has $y$-coordinate $h$ units more than that of the original point $(a, f(a))$, and thus has been displaced *up* $h$ units.

It is not helpful to regard the point $(a, f(a + h))$ as the corresponding point on the graph of $y = f(x + h)$. Rather, consider the point with $x$-coordinate $a - h$; thus $y$-coordinate $f(a - h + h) = f(a)$. Then the point $(a - h, f(a))$ is easily seen to have $x$-coordinate $h$ units less than that of the original point $(a, f(a))$; thus, it has been displaced *left h* units.

**11.2.** Explain why the graph of an even function is unchanged by a reflection with respect to the $y$-axis.

A reflection with respect to the $y$-axis replaces the graph of $y = f(x)$ with the graph of $y = f(-x)$. Since for an even function $f(-x) = f(x)$, the graph of an even function is unchanged.

**11.3.** Explain why the graph of an odd function is altered in exactly the same way by reflection with respect to the $x$-axis or the $y$-axis.

A reflection with respect to the $x$-axis replaces the graph of $y = f(x)$ with the graph of $y = -f(x)$, while a reflection with respect to the $y$-axis replaces the graph of $y = f(x)$ with the graph of $y = f(-x)$. Since for an odd function $f(-x) = -f(x)$, the two reflections have exactly the same effect.

**11.4.** Given the graph of $y = |x|$ as shown in Fig. 11-12, sketch the graphs of (a) $y = |x| - 1$; (b) $y = |x - 2|$; (c) $y = |x + 2| - 1$; (d) $y = -2|x| + 3$.

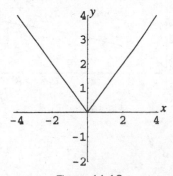

Figure 11-12

(a) The graph of $y = |x| - 1$ (Fig. 11-13) is the same as the graph of $y = |x|$ shifted down 1 unit.

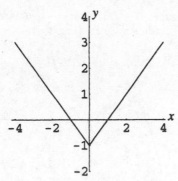

Figure 11-13

(b) The graph of $y = |x - 2|$ (Fig. 11-14) is the same as the graph of $y = |x|$ shifted right 2 units.

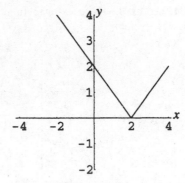

Figure 11-14

(c) The graph of $y = |x + 2| - 1$ (Fig. 11-15) is the same as the graph of $y = |x|$ shifted left 2 units and then down 1 unit.

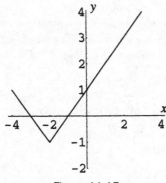

Figure 11-15

(d) The graph of $y = -2|x| + 3$ (Fig. 11-16) is the same as the graph of $y = |x|$ stretched by a factor of 2, reflected with respect to the $x$-axis, and shifted up 3 units.

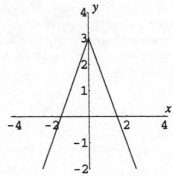

Figure 11-16

**11.5.** Given the graph of $y = \sqrt{x}$ as shown in Fig. 11-17, sketch the graphs of (a) $y = \sqrt{-x}$; (b) $y = -3\sqrt{x}$; (c) $y = \frac{1}{2}\sqrt{x} + 3$; (d) $y = -1.5\sqrt{x - 1} + 2$.

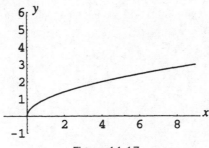

Figure 11-17

(a) The graph of $y = \sqrt{-x}$ (Fig. 11-18) is the same as the graph of $y = \sqrt{x}$ reflected with respect to the $y$-axis.

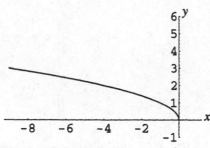

Figure 11-18

(b) The graph of $y = -3\sqrt{x}$ (Fig. 11-19) is the same as the graph of $y = \sqrt{x}$ stretched by a factor of 3 with respect to the $y$-axis and reflected with respect to the $x$-axis.

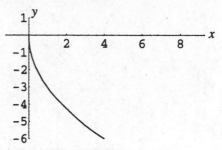

Figure 11-19

(c) The graph of $y = \frac{1}{2}\sqrt{x + 3}$ (Fig. 11-20) is the same as the graph of $y = \sqrt{x}$ shifted left 3 units and compressed by a factor of 2 with respect to the $y$-axis.

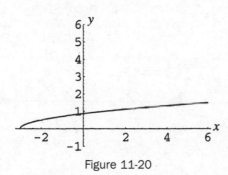

Figure 11-20

(d) The graph of $y = -1.5\sqrt{x - 1} + 2$ (Fig. 11-21) is the same as the graph of $y = \sqrt{x}$ shifted right 1 unit, stretched by a factor of 1.5 and reflected with respect to the $x$-axis, and shifted up 2 units.

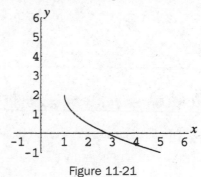

Figure 11-21

**11.6.** Given the graph of $y = x^3$ as shown in Fig. 11-22, sketch the graphs of (a) $y = 4 - x^3$; (b) $y = (\frac{1}{2}x)^3 - \frac{1}{2}$.

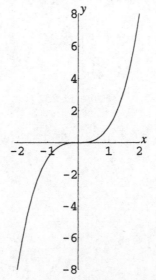

Figure 11-22

(a)  The graph of $y = 4 - x^3$ (Fig. 11-23) is the same as the graph of $y = x^3$ reflected with respect to the $x$-axis and shifted up 4 units.

(b)  The graph of $y = \left(\frac{1}{2}x\right)^3 - \frac{1}{2}$ (Fig. 11-24) is the same as the graph of $y = x^3$ stretched by a factor of 2 with respect to the $x$-axis and shifted down $\frac{1}{2}$ unit.

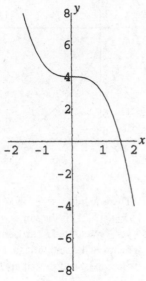

Figure 11-23

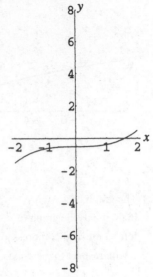

Figure 11-24

## SUPPLEMENTARY PROBLEMS

**11.7.**  Given the graph of $y = x^{1/3}$ as shown in Fig. 11-25, sketch the graphs of:

(a) $y = 2x^{1/3} + 1$; (b) $y = 2(x + 1)^{1/3}$; (c) $y = 2 - x^{1/3}$; (d) $y = (-2x)^{1/3} - 1$.

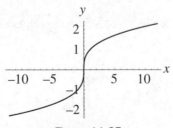

Figure 11-25

*Ans.*   (a)  See Fig. 11-26; (b)  see Fig. 11-27; (c)  see Fig. 11-28; (d)  see Fig. 11-29.

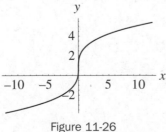

Figure 11-26

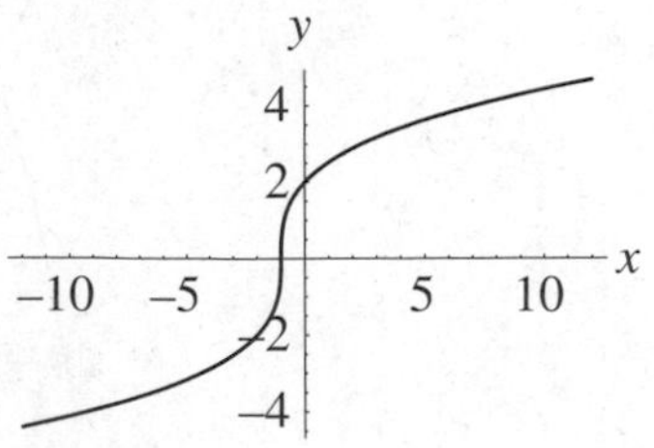

Figure 11-27

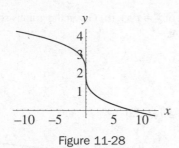

Figure 11-28

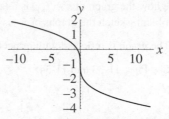

Figure 11-29

**11.8.** (a) Describe how the graph of $y = |f(x)|$ is related to the graph of $y = f(x)$. (b) Given the graph of $y = x^2$ as shown in Fig. 11-30, sketch first the graph of $y = x^2 - 4$, then the graph of $y = |x^2 - 4|$.

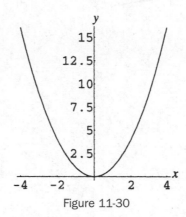

Figure 11-30

*Ans.* (a) The portions of the graph above the *x*-axis are identical to the original, while the portions of the graph below the *x*-axis are reflected with respect to the *x*-axis.

(b) See Figs. 11-31 and 11-32.

$y = x^2 - 4$

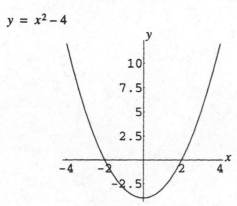

Figure 11-31

$y = |x^2 - 4|$

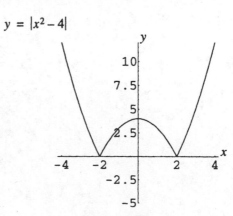

Figure 11-32

**11.9.** (a) Describe how the graph of $x = f(y)$ is related to the graph of $y = f(x)$. (b) Given the graphs shown in the previous problem, sketch the graphs of $x = y^2$ and $x = |y^2 - 4|$.

    *Ans.*   (a)  The graph is reflected with respect to the line $y = x$.

            (b)  See Figs. 11-33 and 11-34.

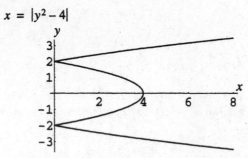

Figure 11-33                     Figure 11-34

# CHAPTER 12

# Quadratic Functions

## Definition of Quadratic Function

A quadratic function is any function specified by a rule that can be written as $f: x \to ax^2 + bx + c$, where $a \neq 0$. The form $ax^2 + bx + c$ is called *standard* form.

**EXAMPLE 12.1** $f(x) = x^2, f(x) = 3x^2 - 2x + 15, f(x) = -3x^2 + 5$, and $f(x) = -2(x + 5)^2$ are examples of quadratic functions. $f(x) = 3x + 5$ and $f(x) = x^3$ are examples of nonquadratic functions.

## Basic Quadratic Functions

The basic quadratic functions are the functions $f(x) = x^2$ and $f(x) = -x^2$. The graph of each is a *parabola* with vertex at the origin $(0,0)$ and axis of symmetry the $y$-axis (Figs. 12-1 and 12-2).

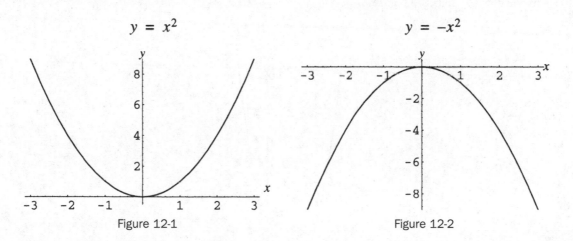

$$y = x^2$$

Figure 12-1

$$y = -x^2$$

Figure 12-2

## Graph of a General Quadratic Function

Any quadratic function can be written in the form $f(x) = a(x - h)^2 + k$ by completing the square. Therefore, any quadratic function has a graph that can be regarded as the result of performing simple transformations on the graph of one of the two basic functions, $f(x) = x^2$ and $f(x) = -x^2$. Thus the graph of any quadratic function is a parabola.

**EXAMPLE 12.2** The quadratic function $f(x) = 2x^2 - 12x + 4$ can be rewritten as follows:

$$
\begin{aligned}
f(x) &= 2x^2 - 12x + 4 \\
&= 2(x^2 - 6x) + 4 \\
&= 2(x^2 - 6x + 9) - 18 + 4 \\
&= 2(x - 3)^2 - 14
\end{aligned}
$$

## Parabola Opening Up

The graph of the function $f(x) = a(x - h)^2 + k$, for positive $a$, is the same as the graph of the basic quadratic function $f(x) = x^2$ stretched by a factor of $a$ (if $a > 1$) or compressed by a factor of $1/a$ (if $0 < a < 1$), and shifted left, right, up, or down so that the point $(0,0)$ becomes the vertex $(h,k)$ of the new graph. The graph of $f(x) = a(x - h)^2 + k$ is symmetric with respect to the line $x = h$. The graph is referred to as a parabola opening *up*.

## Parabola Opening Down

The graph of the function $f(x) = a(x - h)^2 + k$, for negative $a$, is the same as the graph of the basic quadratic function $f(x) = -x^2$ stretched by a factor of $|a|$ (if $|a| > 1$) or compressed by a factor of $1/|a|$ (if $0 < |a| < 1$), and shifted left, right, up, or down so that the point $(0,0)$ becomes the vertex $(h,k)$ of the new graph. The graph of $f(x) = a(x - h)^2 + k$ is symmetric with respect to the line $x = h$. The graph is referred to as a parabola opening *down*.

## Maximum and Minimum Values

For positive $a$, the quadratic function $f(x) = a(x - h)^2 + k$ has a *minimum* value of $k$. This value is attained when $x = h$. For negative $a$, the quadratic function $f(x) = a(x - h)^2 + k$ has a *maximum* value of $k$. This value, also, is attained when $x = h$.

**EXAMPLE 12.3**    Consider the function $f(x) = x^2 + 4x - 7$. By completing the square, this can be written as $f(x) = x^2 + 4x + 4 - 4 - 7 = (x + 2)^2 - 11$. Thus the graph of the function is the same as the graph of $f(x) = x^2$ shifted left 2 units and down 11 units; see Fig. 12-3.

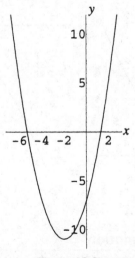

Figure 12-3

The graph is a parabola with vertex $(-2, -11)$, opening up. The function has a minimum value of $-11$. This minimum value is attained when $x = -2$.

**EXAMPLE 12.4**    Consider the function $f(x) = 6x - x^2$. By completing the square, this can be written as $f(x) = -x^2 + 6x = -(x^2 - 6x) = -(x^2 - 6x + 9) + 9 = -(x - 3)^2 + 9$. Thus the graph of the function is the same as the graph of $f(x) = -x^2$ shifted right 3 units and up 9 units. The graph is shown in Fig. 12-4.

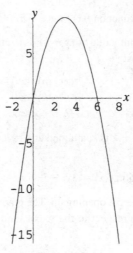

Figure 12-4

The graph is a parabola with vertex (3,9), opening down. The function has a maximum value of 9. This value is attained when $x = 3$.

## Domain and Range

The domain of any quadratic function is $R$, since $ax^2 + bx + c$ or $a(x - h)^2 + k$ is always defined for any real number $x$. For positive $a$, since the quadratic function has a minimum value of $k$, the range is $[k, \infty)$. For negative $a$, since the quadratic function has a maximum value of $k$, the range is $(-\infty, k]$.

**SOLVED PROBLEMS**

**12.1.** Show that the vertex of the parabola $y = ax^2 + bx + c$ is located at $\left(-\dfrac{b}{2a}, \dfrac{4ac - b^2}{4a}\right)$.

Completing the square on $y = ax^2 + bx + c$ gives, in turn,

$$y = a\left(x^2 + \frac{b}{a}x\right) + c$$

$$= a\left(x^2 + \frac{b}{a}x + \frac{b^2}{4a^2}\right) - \frac{b^2}{4a} + c$$

$$= a\left(x + \frac{b}{2a}\right)^2 + \frac{4ac - b^2}{4a}$$

Thus, the parabola $y = ax^2 + bx + c$ is obtained from the parabola $y = ax^2$ by shifting an amount $-b/2a$ with respect to the $x$-axis and an amount $(4ac - b^2)/(4a)$ with respect to the $y$-axis. Since the vertex of $y = ax^2$ is at $(0,0)$, the vertex of $y = ax^2 + bx + c$ is as specified.

**12.2.** Analyze the intercepts of the graph of $y = ax^2 + bc + c$.

For $x = 0$, $y = c$. Hence, the graph always has one $y$-intercept, at $(0,c)$.
For $y = 0$, the equation becomes $0 = ax^2 + bx + c$. The number of solutions of this equation depends on the value of the discriminant $b^2 - 4ac$ (Chapter 5). Thus if $b^2 - 4ac$ is negative, the equation has no solutions and the graph has no $x$-intercepts. If $b^2 - 4ac$ is zero, the equation has one solution, $x = -b/2a$, and the graph has one $x$-intercept.

If $b^2 - 4ac$ is positive, the equation has two solutions, $x = \dfrac{-b + \sqrt{b^2 - 4ac}}{2a}$ and $x = \dfrac{-b - \sqrt{b^2 - 4ac}}{2a}$,

and the graph has two $x$-intercepts.

Note that the $x$-intercepts are symmetrically placed with respect to the line $x = -b/2a$.

**12.3.** Show that, for positive $a$, the quadratic function $f(x) = a(x - h)^2 + k$ has a minimum value of $k$, attained at $x = h$.

For all real $x$, $x^2 \geq 0$. Thus, the minimum value of $(x - h)^2$ is 0, and this minimum value is attained at $x = h$. For positive $a$ and arbitrary $k$, it follows that:

$$(x - h)^2 \geq 0$$
$$a(x - h)^2 \geq 0$$
$$a(x-h)^2 + k \geq k$$

Thus, the minimum value of $a(x - h)^2 + k$ is $k$, attained at $x = h$.

**12.4.** Analyze and graph the quadratic function $f(x) = 3x^2 - 5$.

The graph is a parabola with vertex $(0, -5)$, opening up. The graph is the same as the graph of the basic parabola $y = x^2$ stretched by a factor of 3 with respect to the $y$-axis, and shifted down 5 units. The graph is shown in Fig. 12-5.

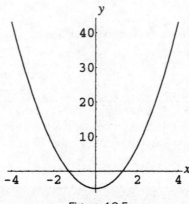

Figure 12-5

**12.5.** Analyze and graph the quadratic function $f(x) = -1 - \frac{1}{3}x^2$.

The function can be rewritten as $f(x) = -\frac{1}{3}x^2 - 1$. The graph is a parabola with vertex $(0, -1)$, opening down. The graph is the same as the graph of the basic parabola $y = -x^2$ compressed by a factor of 3 with respect to the $y$-axis, and shifted down 1 unit. The graph is shown in Fig. 12-6.

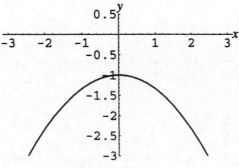

Figure 12-6

**12.6.** Analyze and graph the quadratic function $f(x) = 2x^2 - 6x$.

Completing the square, this can be rewritten:

$$f(x) = 2(x^2 - 3x) = 2\left(x^2 - 3x + \frac{9}{4}\right) - 2 \cdot \frac{9}{4} = 2\left(x - \frac{3}{2}\right)^2 - \frac{9}{2}$$

Hence the graph is a parabola with vertex $\left(\frac{3}{2}, -\frac{9}{2}\right)$, opening up. The parabola is the same as the graph of the basic parabola $y = x^2$ stretched by a factor of 2 with respect to the $y$-axis, and shifted right $\frac{3}{2}$ units and down $\frac{9}{2}$ units. The graph is shown in Fig. 12-7.

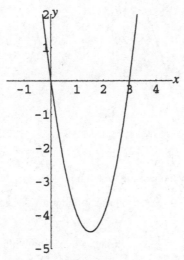

Figure 12-7

**12.7.** Analyze and graph the quadratic function $f(x) = \frac{1}{2}x^2 + 2x + 3$.

Completing the square, this can be rewritten:

$$f(x) = \frac{1}{2}(x^2 + 4x) + 3 = \frac{1}{2}(x^2 + 4x + 4) - 2 + 3 = \frac{1}{2}(x + 2)^2 + 1$$

Hence the graph is a parabola with vertex $(-2, 1)$, opening up. The parabola is the same as the graph of the basic parabola $y = x^2$ compressed by a factor of 2 with respect to the $y$-axis, and shifted left 2 units and up 1 unit. The graph is shown in Fig. 12-8.

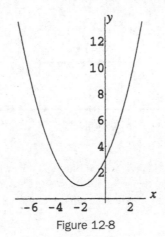

Figure 12-8

**12.8.** Analyze and graph the quadratic function $f(x) = -2x^2 + 4x + 5$.

Completing the square, this can be rewritten as $f(x) = -2(x - 1)^2 + 7$. Hence the graph is a parabola with vertex $(1, 7)$, opening down. The parabola is the same as the graph of the basic parabola $y = -x^2$ stretched by a factor of 2 with respect to the $y$-axis and shifted right 1 unit and up 7 units. The graph is shown in Fig. 12-9.

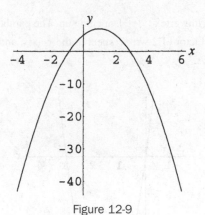

Figure 12-9

**12.9.** State the domain and range for each quadratic function in Problems 12.4−12.8.

In Problem 12.4, the function $f(x) = 3x^2 - 5$ has a minimum value of $-5$. Therefore, the domain is $\boldsymbol{R}$ and the range is $[-5, \infty)$.

In Problem 12.5, the function $f(x) = -1 - \frac{1}{3}x^2$ has a maximum value of $-1$. Therefore, the domain is $\boldsymbol{R}$ and the range is $(-\infty, -1]$.

In Problem 12.6, the function $f(x) = 2x^2 - 6x$ has a minimum value of $-\frac{9}{2}$. Therefore, the domain is $\boldsymbol{R}$ and the range is $\left[-\frac{9}{2}, \infty\right)$.

In Problem 12.7, the function $f(x) = \frac{1}{2}x^2 + 2x + 3$ has a minimum value of 1. Therefore, the domain is $\boldsymbol{R}$ and the range is $[1, \infty)$.

In Problem 12.8, the function $f(x) = -2x^2 + 4x + 5$ has a maximum value of 7. Therefore, the domain is $\boldsymbol{R}$ and the range is $(-\infty, 7]$.

**12.10.** A field is to be marked off in the shape of a rectangle, with one side formed by a straight river. If 100 feet is available for fencing, find the dimensions of the rectangle of maximum possible area. (See Problem 9.8.)

Let $x =$ length of one of the two equal sides (Fig. 12-10).

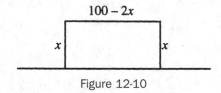

Figure 12-10

In Problem 9.8, it was shown that the area $A = x(100 - 2x)$. Rewriting this in standard form, this becomes $A = -2x^2 + 100x$. Completing the square gives $A = -2(x - 25)^2 + 1250$. Thus the maximum area of 1250 square feet is attained when $x = 25$. Thus the dimensions are 25 feet by 50 feet for maximum area.

**12.11.** In the previous problem, what is the domain of the area function $A(x)$? Graph the function on this domain.

The domain of an abstract quadratic function is $\boldsymbol{R}$, since $ax^2 + bx + c$ is defined and real for all real $x$. In a practical application, this domain may be restricted by physical considerations. Here the area must be positive; hence both $x$ and $100 - 2x$ must be positive. Thus $\{x \in \boldsymbol{R} \mid 0 < x < 50\}$ is the domain of $A(x)$. The graph of $A = -2(x - 25)^2 + 1250$ is the same as the graph of the basic parabola $y = -x^2$ stretched by a factor of 2 with respect to the $y$-axis and shifted right 25 units and up 1250 units. The graph is shown in Fig. 12-11.

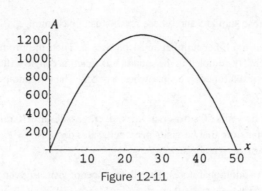

Figure 12-11

**12.12.** A projectile is thrown up from the ground with an initial velocity of 144 ft/sec². Its height $h(t)$ at time $t$ is given by $h(t) = -16t^2 + 144t$. Find its maximum height and the time when the projectile hits the ground.

The quadratic function $h(t) = -16t^2 + 144t$ can be written as $h(t) = -16\left(t - \frac{9}{2}\right)^2 + 324$ by completing the square. Thus the function attains a maximum value of 324 $\left(\text{when } t = \frac{9}{2}\right)$, that is, the maximum height of the projectile is 324 feet.

The projectile hits the ground when the function value is 0. Solving $-16t^2 + 144t = 0$ or $-16t(t - 9) = 0$ yields $t = 0$ (the starting time) or $t = 9$. Thus the projectile hits the ground after 9 seconds.

**12.13.** A suspension bridge is built with its cable hanging between two vertical towers in the form of a parabola. The towers are 400 feet apart and rise 100 feet above the horizontal roadway, while the center point of the cable is 10 feet above the roadway. Introduce a coordinate system as shown.

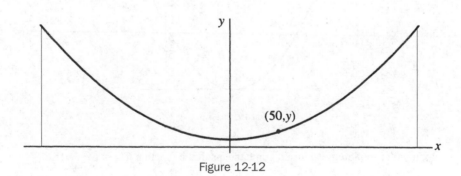

Figure 12-12

(a)  Find the equation of the parabola in the given coordinate system.

(b)  Find the height above the roadway of a point 50 feet from the center of the span.

(a)  Since the vertex of the parabola is at $(0,10)$, the equation of the parabola can be written as $y = ax^2 + 10$. At the right-hand tower, 200 feet from the center, the cable is 100 feet high; thus, the point $(200,100)$ is on the parabola. Substituting yields $100 = a(200)^2 + 10$; hence, $a = 90/40,000$ or $9/4000$. The equation of the parabola is

$$y = \frac{9x^2}{4000} + 10$$

(b)  Here the $x$-coordinate of the point is given as 50. Substituting in the equation yields

$$y = \frac{9(50)^2}{4000} + 10 = 15.625 \text{ feet}$$

**12.14.** Find two real numbers whose sum is $S$ and whose product is a maximum.

Let one number be $x$; then the other number must be $S - x$. Then the product is a quadratic function of $x$: $P(x) = x(S - x) = -x^2 + Sx$. By completing the square, this function can be written as $P(x) = -(x - S/2)^2 + S^2/4$. Thus the maximum value of the function occurs when $x = S/2$. The two numbers are both $S/2$.

**12.15.** A salesperson finds that if he visits 20 stores per week, average sales are 30 units per store each week; however, for each additional store that he visits per week, sales decrease by 1 unit. How many stores should he visit each week to maximize overall sales?

Let $x$ represent the number of additional stores. Then the number of visits is given by $20 + x$ and the corresponding sales are $30 - x$ per store. Total sales are then given by $S(x) = (30 - x)(20 + x) = 600 + 10x - x^2$. This is a quadratic function. Completing the square gives $S(x) = -(x - 5)^2 + 625$. This has a maximum value when $x = 5$; thus, the salesperson should visit 5 additional stores, a total of 25 stores, to maximize overall sales.

## SUPPLEMENTARY PROBLEMS

**12.16.** Show that, for negative $a$, the quadratic function $f(x) = a(x - h)^2 + k$ has a maximum value of $k$, attained at $x = h$.

**12.17.** Find the maximum or minimum value and graph the quadratic function $f(x) = x^2 + 6x + 9$.

*Ans.* Minimum value: 0 when $x = -3$. (See Fig. 12-13.)

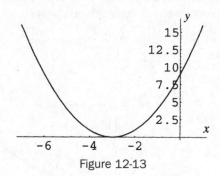

Figure 12-13

**12.18.** Find the maximum or minimum value and graph the quadratic function $f(x) = 6x^2 - 15x$.

*Ans.* Minimum value: $-\dfrac{75}{8}$ when $x = \dfrac{5}{4}$. (See Fig. 12-14.)

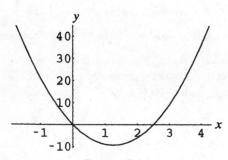

Figure 12-14

**12.19.** Find the maximum or minimum value and graph the quadratic function $f(x) = -\dfrac{3}{2}x^2 - \dfrac{4}{3}x + 6$.

*Ans.* Maximum value: $\dfrac{170}{27}$ when $x = -\dfrac{4}{9}$. (See Fig. 12-15.)

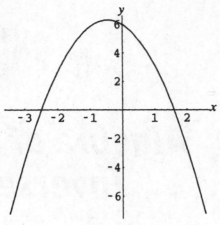

Figure 12-15

**12.20.** State the domain and range for each quadratic function:

(a) $f(x) = 3(x - 2)^2 + 5$; (b) $f(x) = -\frac{1}{2}(x + 3)^2 - 7$; (c) $f(x) = 6 - x^2$; (d) $f(x) = x^2 - 8x$

*Ans.* (a) domain: $R$, range $[5,\infty)$; (b) domain: $R$, range $(-\infty,-7]$;

(c) domain: $R$, range $(-\infty,6]$; (d) domain: $R$, range $[-16,\infty)$

**12.21.** A projectile is thrown up from an initial height of 72 feet with an initial velocity of 160 ft/sec$^2$. Its height $h(t)$ at time $t$ is given by $h(t) = -16t^2 + 160t + 72$. Find its maximum height, the time when this maximum height is reached, and the time when the projectile hits the ground.

*Ans.* Maximum height: 472 feet. Time of maximum height: 5 seconds.

Projectile hits ground: $5 + \sqrt{118}/2 \approx 10.4$ seconds.

**12.22.** 1500 feet of chain link fence are to be used to construct six animal cages as in Fig. 12-16.

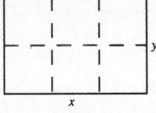

Figure 12-16

Express the total enclosed area as a function of the width $x$. Find the maximum value of this area and the dimensions that yield this area.

*Ans.* Area: $A(x) = \frac{1}{4}x(1500 - 3x)$. Maximum value: 46,875 square feet. Dimensions: 250 feet by 187.5 feet.

**12.23.** Find two real numbers whose difference is $S$ and whose product is a minimum.

*Ans.* $S/2$ and $-S/2$

**12.24.** A basketball team finds that if it charges \$25 per ticket, the average attendance per game is 400. For each \$.50 decrease in the price per ticket, attendance increases by 10. What ticket price yields the maximum revenue?

*Ans.* \$22.50

# CHAPTER 13

# Algebra of Functions; Inverse Functions

## Algebraic Combinations of Functions

Algebraic combinations of functions can be obtained in several ways: Given two functions $f$ and $g$, the sum, difference, product, and quotient functions can be defined as follows:

| NAME | DEFINITION | DOMAIN |
|---|---|---|
| Sum | $(f + g)(x) = f(x) + g(x)$ | The set of all $x$ that are in the domain of both $f$ and $g$ |
| Difference | $(f - g)(x) = f(x) - g(x)$ | The set of all $x$ that are in the domain of both $f$ and $g$ |
| Difference | $(g - f)(x) = g(x) - f(x)$ | The set of all $x$ that are in the domain of both $f$ and $g$ |
| Product | $(fg)(x) = f(x)g(x)$ | The set of all $x$ that are in the domain of both $f$ and $g$ |
| Quotient | $\left(\dfrac{f}{g}\right)(x) = \dfrac{f(x)}{g(x)}$ | The set of all $x$ that are in the domain of both $f$ and $g$, with $g(x) \neq 0$ |
| Quotient | $\left(\dfrac{g}{f}\right)(x) = \dfrac{g(x)}{f(x)}$ | The set of all $x$ that are in the domain of both $f$ and $g$, with $f(x) \neq 0$ |

**EXAMPLE 13.1**  Given $f(x) = x^2$ and $g(x) = \sqrt{x - 2}$, find $(f + g)x$ and $(f/g)(x)$ and state the domains of the functions.

$(f + g)(x) = f(x) + g(x) = x^2 + \sqrt{x - 2}$. Since the domain of $f$ is $R$ and the domain of $g$ is $\{x \in R \,|\, x \geq 2\}$ the domain of this function is also $\{x \in R \,|\, x \geq 2\}$.

$\left(\dfrac{f}{g}\right)(x) = \dfrac{x^2}{\sqrt{x - 2}}$. The domain of this function is the same as the domain of $f + g$, with the further restriction that $g(x) \neq 0$, that is, $\{x \in R \,|\, x > 2\}$.

## Definition of Composite Function

The composite function $f \circ g$ of two functions $f$ and $g$ is defined by:

$$f \circ g(x) = f(g(x))$$

The domain of $f \circ g$ is the set of all $x$ in the domain of $g$ such that $g(x)$ is in the domain of $f$.

**EXAMPLE 13.2**  Given $f(x) = 3x - 8$ and $g(x) = 1 - x^2$, find $f \circ g$ and state its domain.

$f \circ g(x) = f(g(x)) = f(1 - x^2) = 3(1 - x^2) - 8 = -5 - 3x^2$. Since the domains of $f$ and $g$ are both $R$, the domain of $f \circ g$ is also $R$.

**EXAMPLE 13.3** Given $f(x) = x^2$ and $g(x) = \sqrt{x - 5}$, find $f \circ g$ and state its domain.

$f \circ g(x) = f(g(x)) = f(\sqrt{x - 5}) = (\sqrt{x - 5})^2 = x - 5$. The domain of $f \circ g$ is not all of $\mathbf{R}$. Since the domain of $g$ is $\{x \in \mathbf{R} \mid x \geq 5\}$, the domain of $f \circ g$ is the set of all $x \geq 5$ in the domain of $f$, that is, all of $\{x \in \mathbf{R} \mid x \geq 5\}$.

Fig. 13-1 shows the relationships among $f$, $g$, and $f \circ g$.

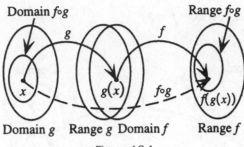

Figure 13-1

## One-to-One Functions

A function with domain $D$ and range $R$ is called a *one-to-one* function if exactly one element of set $D$ corresponds to each element of set $R$.

**EXAMPLE 13.4** Let $f(x) = x^2$ and $g(x) = 2x$. Show that $f$ is not a one-to-one function and that $g$ is a one-to-one function.

The domain of $f$ is $\mathbf{R}$. Since $f(3) = f(-3) = 9$, that is, the elements 3 and $-3$ in the domain of $f$ correspond to 9 in the range, $f$ is not one-to-one.

The domain and range of $g$ are both $\mathbf{R}$. Let $k$ be an arbitrary real number. If $2x = k$, then the only $x$ that corresponds to $k$ is $x = k/2$. Thus $g$ is one-to-one.

A function with domain $D$ and range $R$ is one-to-one if either of the following equivalent conditions is satisfied.

1. Whenever $f(u) = f(v)$ in $R$, then $u = v$ in $D$.
2. Whenever $u \neq v$ in $D$, then $f(u) \neq f(v)$ in $R$.

## Horizontal Line Test

Since for each value of $y$ in the domain of a one-to-one function $f$ there is exactly one $x$ such that $y = f(x)$, a horizontal line $y = c$ can cross the graph of a one-to-one function at most once. Thus, if a horizontal line crosses a graph more than once, the graph is not the graph of a one-to-one function.

## Definition of Inverse Function

Let $f$ be a one-to-one function with domain $D$ and range $R$. Since for each $y$ in $R$ there is exactly one $x$ in $D$ such that $y = f(x)$, define a function $g$ with domain $R$ and range $D$ such that $g(y) = x$. Then $g$ reverses the correspondence defined by $f$. The function $g$ is called the *inverse function* of $f$.

## Function-Inverse Function Relationship

If $g$ is the inverse function of $f$, then, by the above definition,

1. $g(f(x)) = x$ for every $x$ in $D$.
2. $f(g(y)) = y$ for every $y$ in $R$.

## Notation for Inverse Functions

If $f$ is a one-to-one function with domain $D$ and range $R$, then the inverse function of $f$ with domain $R$ and range $D$ is often denoted by $f^{-1}$. Then $f^{-1}$ is also a one-to-one function and $x = f^{-1}(y)$ if and only if $y = f(x)$. With this notation, the function-inverse function relationship becomes:

1. $f^{-1}(f(x)) = x$ for every $x$ in $D$.
2. $f(f^{-1}(y)) = y$ for every $y$ in $R$.

Figure 13-2 shows the relationship between $f$ and $f^{-1}$.

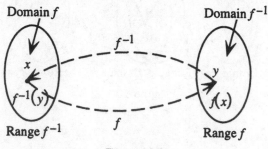

Figure 13-2

## To Find the Inverse Function for a Given Function $f$:

1. Verify that $f$ is one-to-one.
2. Solve the equation $y = f(x)$ for $x$ in terms of $y$, if possible. This gives an equation of form $x = f^{-1}(y)$.
3. Interchange $x$ and $y$ in the equation found in step 2. This gives an equation of the form $y = f^{-1}(x)$.

**EXAMPLE 13.5**　Find the inverse function for $f(x) = 3x - 1$.

First, show that $f$ is one-to-one. Assume $f(u) = f(v)$. Then it follows that

$$3u - 1 = 3v - 1$$
$$3u = 3v$$
$$u = v$$

Thus, $f$ is one-to-one. Now solve $y = 3x - 1$ for $x$ to obtain

$$y = 3x - 1$$
$$y + 1 = 3x$$
$$x = \frac{y + 1}{3}$$

Now interchange $x$ and $y$ to obtain $y = f^{-1}(x) = \dfrac{x + 1}{3}$.

## Graph of an Inverse Function

The graphs of $y = f(x)$ and $y = f^{-1}(x)$ are symmetric with respect to the line $y = x$.

**SOLVED PROBLEMS**

**13.1.** Given $f(x) = ax + b$ and $g(x) = cx + d$, $a, c \neq 0$, find $f + g, f - g, fg, f/g$ and state their domains.

$$(f + g)(x) = f(x) + g(x) = ax + b + cx + d = (a + c)x + (b + d)$$
$$(f - g)(x) = f(x) - g(x) = (ax + b) - (cx + d) = ax + b - cx - d = (a - c)x + (b - d)$$
$$(fg)(x) = f(x)g(x) = (ax + b)(cx + d) = acx^2 + (ad + bc)x + bd$$

Since $R$ is the domain of both $f$ and $g$, the domain of each of these function is $R$.

$$(f/g)(x) = f(x)/g(x) = (ax + b)/(cx + d)$$

The domain of this function is $\{x \in R \mid x \neq -d/c\}$.

**13.2.** Given $f(x) = \dfrac{x + 1}{x^2 - 4}$ and $g(x) = \dfrac{2}{x}$, find $f + g, f - g, fg, f/g$ and state their domains.

$$(f + g)(x) = f(x) + g(x) = \frac{x + 1}{x^2 - 4} + \frac{2}{x} = \frac{x(x + 1) + 2(x^2 - 4)}{x(x^2 - 4)} = \frac{3x^2 + x - 8}{x(x^2 - 4)}$$

$$(f - g)(x) = f(x) - g(x) = \frac{x + 1}{x^2 - 4} - \frac{2}{x} = \frac{x(x + 1) - 2(x^2 - 4)}{x(x^2 - 4)} = \frac{-x^2 + x + 8}{x(x^2 - 4)}$$

$$(fg)(x) = f(x)g(x) = \frac{x + 1}{x^2 - 4} \cdot \frac{2}{x} = \frac{2x + 2}{x(x^2 - 4)}$$

Since the domain of $f$ is $\{x \in R \mid x \neq -2, 2\}$ and the domain of $g$ is $\{x \in R \mid x \neq 0\}$, the domain of each of these functions is $\{x \in R \mid x \neq -2, 2, 0\}$.

$$\left(\frac{f}{g}\right)(x) = \frac{f(x)}{g(x)} = \frac{x + 1}{x^2 - 4} \div \frac{2}{x} = \frac{x + 1}{x^2 - 4} \cdot \frac{x}{2} = \frac{x^2 + x}{2x^2 - 8}$$

The domain of this function may not be apparent from its final form. From the definition of the quotient function, the domain of this function must be those elements of $\{x \in R \mid x \neq -2, 2, 0\}$ for which $g(x)$ is not 0. Since $g(x)$ is never 0, the domain of this quotient function is $\{x \in R \mid x \neq -2, 2, 0\}$.

**13.3.** If $f$ and $g$ are even functions, show that $f + g$ and $fg$ are even functions.

$(f + g)(-x) = f(-x) + g(-x)$ and $(fg)(-x) = f(-x)g(-x)$ by definition. Since $f$ and $g$ are even functions, $f(-x) + g(-x) = f(x) + g(x)$ and $f(-x)g(-x) = f(x)g(x)$. Therefore,

$$(f + g)(-x) = f(-x) + g(-x) = f(x) + g(x) = (f + g)(x) \text{ and}$$
$$(fg)(-x) = f(-x)g(-x) = f(x)g(x) = (fg)(x),$$

that is, $f + g$ and $fg$ are even functions.

**13.4.** Given $f(x) = \sqrt{1 - x}$ and $g(x) = \sqrt{x^2 - 4}$, find $g + f, g - f, gf, f/g$ and state their domains.

$$(g + f)(x) = g(x) + f(x) = \sqrt{x^2 - 4} + \sqrt{1 - x}$$
$$(g - f)(x) = g(x) - f(x) = \sqrt{x^2 - 4} - \sqrt{1 - x}$$
$$(gf)(x) = g(x)f(x) = \sqrt{x^2 - 4} \cdot \sqrt{1 - x}$$

Since the domain of $f$ is $\{x \in R \mid x \leq 1\}$ and the domain of $g$ is $\{x \in R \mid x \leq -2 \text{ or } x \geq 2\}$, the domain of each of these functions is the intersection of these two sets, that is, $\{x \in R \mid x \leq -2\}$.

$$\left(\frac{f}{g}\right)(x) = \frac{f(x)}{g(x)} = \frac{\sqrt{1 - x}}{\sqrt{x^2 - 4}}$$

The domain of this function is those elements of $\{x \in R \mid x \leq -2\}$ for which $g(x) \neq 0$, that is, $\{x \in R \mid x < -2\}$.

**13.5.** Given $f(x) = x^4$ and $g(x) = 3x + 5$, find $f \circ g$ and $g \circ f$ and state their domains.

$$f \circ g(x) = f(g(x)) = f(3x + 5) = (3x + 5)^4$$
$$g \circ f(x) = g(f(x)) = g(x^4) = 3x^4 + 5$$

Since the domains of $f$ and $g$ are both $R$, the domains of $f \circ g$ and $g \circ f$ are also $R$.

**13.6** Given $f(x) = |x|$ and $g(x) = -5$, find $f \circ g$ and $g \circ f$ and state their domains.

$$f \circ g(x) = f(g(x)) = f(-5) = |-5| = 5$$
$$g \circ f(x) = g(f(x)) = g(|x|) = -5$$

Since the domains of $f$ and $g$ are both $R$, the domains of $f \circ g$ and $g \circ f$ are also $R$.

**13.7.** Given $f(x) = \sqrt{x - 6}$ and $g(x) = x^2 + 5x$, find $f \circ g$ and $g \circ f$ and state their domains.

$$f \circ g(x) = f(g(x)) = f(x^2 + 5x) = \sqrt{x^2 + 5x - 6}$$
$$g \circ f(x) = g(f(x)) = g(\sqrt{x - 6}) = (\sqrt{x - 6})^2 + 5\sqrt{x - 6} = x - 6 + 5\sqrt{x - 6}$$

Since the domain of $g$ is $R$, the domain of $f \circ g$ is the set of all real numbers with $g(x)$ in the domain of $f$, that is, $g(x) \geq 6$, or $x^2 + 5x \geq 6$, or $\{x \in R | x \geq 1 \text{ or } x \leq -6\}$.
Since the domain of $f$ is $\{x \in R | x \geq 6\}$, the domain of $g \circ f$ is the set of all numbers in this set with $f(x)$ in the domain of $g$; this is all of $\{x \in R | x \geq 6\}$.

**13.8.** Given $f(x) = |x - 1|$ and $g(x) = 1/x$, find $f \circ g$ and $g \circ f$ and state their domains.

$$f \circ g(x) = f(g(x)) = f\left(\frac{1}{x}\right) = \left|\frac{1}{x} - 1\right| \text{ or } \left|\frac{1 - x}{x}\right|$$

$$g \circ f(x) = g(f(x)) = g(|x - 1|) = \frac{1}{|x - 1|}$$

Since the domain of $g$ is $\{x \in R | x \neq 0\}$, the domain of $f \circ g$ is the set of all nonzero real numbers with $g(x)$ in the domain of $f$, that is, $\{x \in R | x \neq 0\}$.
Since the domain of $f$ is $R$, the domain of $g \circ f$ is the set of all real numbers with $f(x)$ in the domain of $g$, that is, $\{x \in R | x \neq 1\}$.

**13.9.** Given $f(x) = \sqrt{x^2 + 5}$ and $g(x) = \sqrt{4 - x^2}$, find $f \circ g$ and $g \circ f$ and state their domains.

$$f \circ g(x) = f(g(x)) = f(\sqrt{4 - x^2}) = \sqrt{(\sqrt{4 - x^2})^2 + 5} = \sqrt{4 - x^2 + 5} = \sqrt{9 - x^2}$$

$$g \circ f(x) = g(f(x)) = g(\sqrt{x^2 + 5}) = \sqrt{4 - (\sqrt{x^2 + 5})^2} = \sqrt{4 - (x^2 + 5)} = \sqrt{-1 - x^2}$$

Since the domain of $g$ is $\{x \in R | -2 \leq x \leq 2\}$, the domain of $f \circ g$ is the set of all numbers in this set with $f(x)$ in the domain of $g$; this is all of $\{x \in R | -2 \leq x \leq 2\}$. Note that the domain of $f \circ g$ cannot be determined from its final form.
Since $-1 - x^2$ is negative for all real $x$, the domain of $g \circ f$ is empty.

**13.10.** Find a composite function form for each of the following:

(a) $y = (5x - 3)^4$    (b) $y = \sqrt{1 - x^2}$    (c) $y = \dfrac{1}{(x^2 - 5x + 6)^{2/3}}$

(a)   Let $y = u^4$ and $u = 5x - 3$. Then $y = f(u)$ and $u = g(x)$; hence $y = f(g(x))$.

(b)   Let $y = \sqrt{u}$ and $u = 1 - x^2$. Then $y = f(u)$ and $u = g(x)$; hence $y = f(g(x))$.

(c)   Let $y = u^{-2/3}$ and $u = x^2 - 5x + 6$. Then $y = f(u)$ and $u = g(x)$; hence $y = f(g(x))$.

**13.11.**   A spherical balloon is being inflated at the constant rate of $6\pi$ ft³/min. Express its radius $r$ as a function of time $t$ (in minutes), assuming that $r = 0$ when $t = 0$.

Express the radius $r$ as a function of the volume $V$ and $V$ as a function of the time $t$.

Since $V = \dfrac{4}{3}\pi r^3$ for a sphere, solve for $r$ to obtain $r = f(V) = \sqrt[3]{\dfrac{3V}{4\pi}}$. $V$ is a linear function of $t$ with slope $6\pi$;

since $V = 0$ when $t = 0$, $V = g(t) = 6\pi t$. Hence, $r = f(g(t)) = \sqrt[3]{\dfrac{3(6\pi t)}{4\pi}} = \sqrt[3]{\dfrac{9t}{2}}$ feet.

**13.12.**   The revenue (in dollars) from the sale of $x$ units of a certain product is given by the function $R(x) = 20x - x^2/200$. The cost (in dollars) of producing $x$ units is given by the function $C(x) = 4x + 8000$. Find the profit on sales of $x$ units.

The profit function $P(x)$ is given by $P(x) = (R - C)(x)$. Hence

$$
\begin{aligned}
P(x) &= (R - C)(x) \\
&= R(x) - C(x) \\
&= (20x - x^2/200) - (4x + 8000) \\
&= 20x - x^2/200 - 4x - 8000 \\
&= -x^2/200 + 16x - 8000
\end{aligned}
$$

**13.13.**   In the previous problem, if the demand $x$ and the price $p$ (in dollars) for the product are related by the function $x = f(p) = 4000 - 200p$, $0 \le p \le 20$, write the profit as a function of the demand $p$.

$$
\begin{aligned}
F(p) &= P \circ f(p) = P(f(p)) = P(4000 - 200p) \\
&= -(4000 - 200p)^2/200 + 16(4000 - 200p) - 8000
\end{aligned}
$$

**13.14.**   In the previous problem, find the price which would yield the maximum profit and also find this maximum profit.

Simplifying, obtain

$$
\begin{aligned}
F(p) &= -(4000 - 200p)^2/200 + 16(4000 - 200p) - 8000 \\
&= -(16{,}000{,}000 - 1{,}600{,}000p + 40{,}000p^2)/200 + 64{,}000 - 3200p - 8000 \\
&= -80{,}000 + 8000p - 200p^2 + 64{,}000 - 3200p - 8000 \\
&= -200p^2 + 4800p - 24{,}000
\end{aligned}
$$

This is a quadratic function. Completing the square yields $F(p) = -200(p - 12)^2 + 4800$. The function attains a maximum value (maximum profit) of \$4800 when the price $p = \$12$.

**13.15.**   Show that every increasing function is one-to-one on its domain.

Let $f$ be an increasing function, that is, for every $a,b$ in the domain of $f$, if $a < b$, then $f(a) < f(b)$. Now, if $u \ne v$, then either $u < v$ or $u > v$. Thus either $f(u) < f(v)$ or $f(u) > f(v)$; in either case, $f(u) \ne f(v)$ and $f$ is a one-to-one function.

**13.16.** Determine whether or not each of the following functions is one-to-one.

(a) $f(x) = 5$; (b) $f(x) = 5x$; (c) $f(x) = x^2 + 5$; (d) $f(x) = \sqrt{x - 5}$

(a) Since $f(2) = 5$ and $f(3) = 5$, this function is not one-to-one.

(b) Assume $f(u) = f(v)$. Then it follows that $5u = 5v$; hence $u = v$. Therefore, $f$ is one-to-one.

(c) Since $f(2) = 9$ and $f(-2) = 9$, this function is not one-to-one.

(d) Assume $f(u) = f(v)$. Then it follows that

$$\sqrt{u - 5} = \sqrt{v - 5}$$
$$u - 5 = v - 5$$
$$u = v$$

Therefore, $f$ is one-to-one.

**13.17.** Use the function-inverse function relationship to show that $f$ and $g$ are inverses of each other, and sketch the graphs of $f$ and $g$ and the line $y = x$ on the same Cartesian coordinate system.

(a) $f(x) = 2x - 3$          $g(x) = \dfrac{x + 3}{2}$

(b) $f(x) = x^2 + 3, x \geq 0$      $g(x) = \sqrt{x - 3}, x \geq 3$

(c) $f(x) = -\sqrt{4 - x}, x \leq 4$     $g(x) = 4 - x^2, x \leq 0$

(a) Note first that Dom $f$ = Range $g = \mathbf{R}$.
  Also Dom $g$ = Range $f = \mathbf{R}$.

$$g(f(x)) = g(2x - 3) = \frac{2x - 3 + 3}{2}$$

$$= x$$

$$f(g(y)) = f\left(\frac{y + 3}{2}\right) = 2\left(\frac{y + 3}{2}\right) - 3$$

$$= y$$

The line $y = x$ is shown dashed in Fig. 13-3.

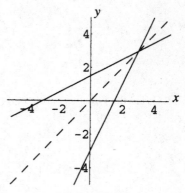

Figure 13-3

(b) Note first that Dom $f$ = Range $g = [0, \infty)$.
  Also Dom $g$ = Range $f = [3, \infty)$.

$$g(f(x)) = g(x^2 + 3) = \sqrt{x^2 + 3 - 3}$$

$$= \sqrt{x^2} = x \quad \text{on } [0, \infty).$$

$$f(g(y)) = f(\sqrt{y - 3}) = ((\sqrt{y - 3})^2 + 3)$$

$$= y - 3 + 3 = y$$

The line $y = x$ is shown dashed in Fig. 13-4.

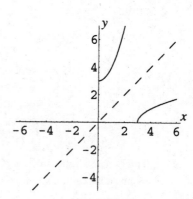

Figure 13-4

(c) Note first that Dom $f$ = Range $g$ = $(-\infty, 4]$.
Also Dom $g$ = Range $f$ = $(-\infty, 0]$.

$$g(f(x)) = g(-\sqrt{4-x}) = 4 - (-\sqrt{4-x})^2$$

$$= 4 - (4 - x) = x$$

$$f(g(y)) = f(4 - y^2) = -\sqrt{4 - (4 - y^2)}$$

$$= -\sqrt{y^2} = y \quad \text{on } (-\infty, 0]$$

The line $y = x$ is shown dashed in Fig. 13-5.

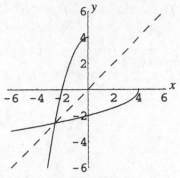

Figure 13-5

**13.18.** The following functions are one-to-one. Find the inverse functions for each.

(a) $f(x) = 4x - 1$

(b) $f(x) = \dfrac{2}{x+3}$

(c) $f(x) = x^2 - 9, x \geq 0$

(d) $f(x) = 4 + (x+3)^2, x \leq -3$

(a) Set $y = 4x - 1$.
Solve for $x$ in terms of $y$.

$$4x - 1 = y$$
$$4x = y + 1$$
$$x = \frac{y+1}{4}$$

Interchange $x$ and $y$.

$$y = f^{-1}(x) = \frac{x+1}{4}$$

*Note:*

Dom $f$ = Range $f^{-1}$ = $\mathbf{R}$

Dom $f^{-1}$ = Range $f$ = $\mathbf{R}$

(b) Set $y = 2/(x+3)$.
Solve for $x$ in terms of $y$.

$$y = \frac{2}{x+3}$$

$$x + 3 = \frac{2}{y}$$

$$x = \frac{2}{y} - 3$$

Interchange $x$ and $y$.

$$y = f^{-1}(x) = \frac{2}{x} - 3$$

*Note:*

Dom $f$ = Range $f^{-1}$ = $(-\infty, -3) \cup (-3, \infty)$

Dom $f^{-1}$ = Range $f$ = $(-\infty, 0) \cup (0, \infty)$

(c) Set $y = x^2 - 9, x \geq 0$.

Solve for $x$ in terms of $y$.

$$x^2 - 9 = y$$
$$x^2 = y + 9$$
$$x = \sqrt{y + 9}$$

(since $x$ must be nonnegative)

Interchange $x$ and $y$.

$$y = f^{-1}(x) = \sqrt{x + 9}$$

*Note:*

Dom $f$ = Range $f^{-1}$ = $[0, \infty)$

Dom $f^{-1}$ = Range $f$ = $[-9, \infty)$

(d) Set $y = 4 + (x+3)^2, x \leq -3$.

Solve for $x$ in terms of $y$.

$$4 + (x+3)^2 = y$$
$$(x+3)^2 = y - 4$$
$$x + 3 = -\sqrt{y - 4}$$

(since $x + 3$ must be nonpositive)

$$x = -3 - \sqrt{y - 4}$$

Interchange $x$ and $y$.

$$y = f^{-1}(x) = -3 - \sqrt{x - 4}$$

*Note:*

Dom $f$ = Range $f^{-1}$ = $(-\infty, -3]$

Dom $f^{-1}$ = Range $f$ = $[4, \infty)$

**13.19.** The function $F(x) = (x - 4)^2$ is not one-to-one. Find the inverse of the function defined by restricting the domain of $F$ to (a) $x \geq 4$; (b) $x \leq 4$.

(a) First, show that the function $f(x) = (x - 4)^2$, $x \geq 4$, is one-to-one.

Assume $f(u) = f(v)$. Then it follows that

$$(u - 4)^2 = (v - 4)^2, \qquad u, v \geq 4$$
$$u - 4 = \pm \sqrt{(v - 4)^2}$$
$$u = 4 \pm (v - 4) \qquad \text{since } v \geq 4$$

Now, since $u$ must be greater than or equal to 4, the positive sign must be chosen.

$$u = 4 + v - 4 = v$$

Therefore, $f$ is one-to-one.

Now set $y = f(x) = (x - 4)^2$, $x \geq 4$, and solve for $x$ in terms of $y$.

$$(x - 4)^2 = y$$
$$x - 4 = \sqrt{y} \quad \text{since } x \geq 4$$
$$x = 4 + \sqrt{y}$$

Interchange $x$ and $y$ to obtain $y = f^{-1}(x) = 4 + \sqrt{x}$.

*Note:* $\text{Dom} f = \text{Range} f^{-1} = [4, \infty)$. $\text{Dom} f^{-1} = \text{Range} f = [0, \infty)$

(b) First, show that the function $f(x) = (x - 4)^2$, $x \leq 4$, is one-to-one.

Assume $f(u) = f(v)$. Then it follows that

$$(u - 4)^2 = (v - 4)^2, \qquad u, v \leq 4$$
$$u - 4 = \pm \sqrt{(v - 4)^2}$$
$$u = 4 \pm (4 - v) \qquad \text{since } v \leq 4$$

Now, since $u$ must be less than or equal to 4, the negative sign must be chosen.

$$u = 4 - (4 - v) = v$$

Therefore, $f$ is one-to-one.

Now set $y = f(x) = (x - 4)^2$, $x \leq 4$, and solve for $x$ in terms of $y$.

$$(x - 4)^2 = y$$
$$x - 4 = -\sqrt{y} \quad \text{since } x \leq 4$$
$$x = 4 - \sqrt{y}$$

Interchange $x$ and $y$ to obtain $y = f^{-1}(x) = 4 - \sqrt{x}$.

*Note:* $\text{Dom} f = \text{Range} f^{-1} = (-\infty, 4]$. $\text{Dom} f^{-1} = \text{Range} f = [0, \infty)$.

## SUPPLEMENTARY PROBLEMS

**13.20.** Show that if $f$ and $g$ are odd functions, then $f + g$ and $f - g$ are odd functions, but $fg$ and $f/g$ are even functions.

**13.21.** Given $f(x) = \dfrac{3x - 1}{5}$ and $g(x) = \dfrac{5x + 1}{3}$, find $f \circ g$ and $g \circ f$ and state their domains.

*Ans.* $f \circ g(x) = g \circ f(x) = x$ for all $x \in \mathbf{R}$.

**13.22.** The revenue (in dollars) from the sale of $x$ units of a certain product is given by the function $R(x) = 60x - x^2/100$. The cost (in dollars) of producing $x$ units is given by the function $C(x) = 15x + 40{,}000$. Find the profit on sales of $x$ units.

Ans. $P(x) = -x^2/100 + 45x - 40{,}000$

**13.23.** In the previous problem, suppose that the demand $x$ and the price $p$ (in dollars) for the product are related by the function $x = f(p) = 5000 - 50p$ $\quad 0 \le p \le 100$. Write the profit as a function of the demand $p$.

Ans. $F(p) = -(5000 - 50p)^2/100 + 45(5000 - 50p) - 40{,}000$

**13.24.** In the previous problem, find the price which would yield the maximum profit and also find this maximum profit.

Ans. Price of $55 yields a maximum profit of $10,625.

**13.25.** A 300-foot-long cable, originally of diameter 5 inches, is submerged in seawater. Because of corrosion, the surface area of the cable diminishes at the rate of 1250 in$^2$/year. Express the diameter $d$ of the cable as a function of time $t$ (in years).

Ans. $d = 5 - \dfrac{25t}{72\pi}$ inches

**13.26.** Show that every decreasing function is one-to-one on its domain.

**13.27.** A function is *periodic* if there exists some nonzero real number $p$, called a period, such that $f(x + p) = f(x)$ for all $x$ in the domain of the function. Show that no periodic function is one-to-one.

**13.28.** Show that the graphs of $f^{-1}$ and $f$ are reflections of each other in the line $y = x$ by verifying the following: (a) If $P(u,v)$ is on the graph of $f$, then $Q(v,u)$ is on the graph of $f^{-1}$. (b) The midpoint of line segment $PQ$ is on the line $y = x$. (c). The line $PQ$ is perpendicular to the line $y = x$.

**13.29.** The following functions are one-to-one. Find the inverse functions for each.

(a) $f(x) = 5 - 10x$ 　　　 (b) $f(x) = \dfrac{4x}{x - 2}$ 　　　 (c) $f(x) = \dfrac{x + 5}{3x - 1}$

(d) $f(x) = 2 - x^3$ 　　　 (e) $f(x) = \sqrt{9 - x^2},\ 0 \le x \le 3$ 　　　 (f) $f(x) = 3 - \sqrt{x - 2}$

Ans. (a) $f^{-1}(x) = \dfrac{5 - x}{10}$; (b) $f^{-1}(x) = \dfrac{2x}{x - 4}$; (c) $f^{-1}(x) = \dfrac{x + 5}{3x - 1}$; (d) $f^{-1}(x) = \sqrt[3]{2 - x}$;

(e) $f^{-1}(x) = \sqrt{9 - x^2},\ 0 \le x \le 3$; (f) $f^{-1}(x) = (3 - x)^2 + 2,\ x \le 3$

**13.30.** The following functions are one-to-one. Find the inverse function for each.

(a) $f(x) = 2 + \sqrt{4 - x^2}$ 　　 $0 \le x \le 2$ 　　　 (b) $f(x) = 2 + \sqrt{4 - x^2}$ 　　 $-2 \le x \le 0$

(c) $f(x) = 2 - \sqrt{4 - x^2}$ 　　 $0 \le x \le 2$ 　　　 (d) $f(x) = 2 - \sqrt{4 - x^2}$ 　　 $-2 \le x \le 0$

Ans. (a) $f^{-1}(x) = \sqrt{4x - x^2}$ 　　 $2 \le x \le 4$; (b) $f^{-1}(x) = -\sqrt{4x - x^2}$ 　　 $2 \le x \le 4$;

(c) $f^{-1}(x) = \sqrt{4x - x^2}$ 　　 $0 \le x \le 2$; (d) $f^{-1}(x) = -\sqrt{4x - x^2}$ 　　 $0 \le x \le 2$

# CHAPTER 14

# Polynomial Functions

## Definition of Polynomial Function

A polynomial function is any function specified by a rule that can be written as $f: x \rightarrow a_n x^n + a_{n-1} x^{n-1} + \cdots + a_1 x + a_0$, where $a_n \neq 0$. $n$ is the degree of the polynomial function. The domain of a polynomial function, unless otherwise specified, is $R$.

## Special Polynomial Functions

Special polynomial function types have already been discussed:

| DEGREE | EQUATION | NAME | GRAPH |
|---|---|---|---|
| $n = 0$ | $f(x) = a_0$ | Constant function | Horizontal straight line |
| $n = 1$ | $f(x) = a_1 x + a_0$ | Linear function | Straight line with slope $a_1$ |
| $n = 2$ | $f(x) = a_2 x^2 + a_1 x + a_0$ | Quadratic function | Parabola |

## Integer Power Functions

If $f$ has degree $n$ and all coefficients except $a_n$ are zero, then $f(x) = ax^n$, where $a = a_n \neq 0$. Then if $n = 1$, the graph of the function is a straight line through the origin. If $n = 2$, the graph of the function is a parabola with vertex at the origin. If $n$ is an odd integer, the function is an odd function. If $n$ is an even integer, the function is an even function.

**EXAMPLE 14.1** Draw graphs of (a) $f(x) = x^3$; (b) $f(x) = x^5$; (c) $f(x) = x^7$.

(a) Fig. 14-1; (b) Fig. 14-2; (c) Fig. 14-3.

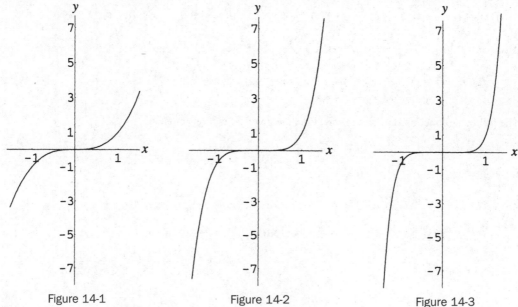

Figure 14-1          Figure 14-2          Figure 14-3

**EXAMPLE 14.2** Draw graphs of (a) $f(x) = x^4$; (b) $f(x) = x^6$; (c) $f(x) = x^8$.

(a) Fig. 14-4; (b) Fig. 14-5; (c) Fig. 14-6.

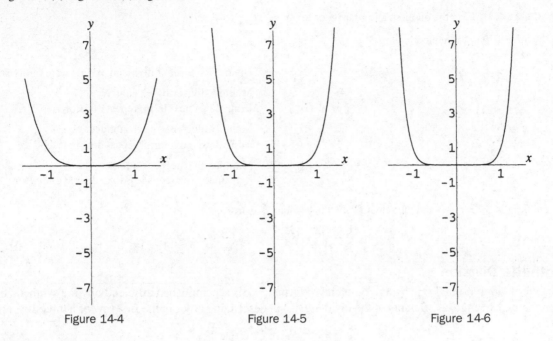

| Figure 14-4 | Figure 14-5 | Figure 14-6 |

## Zeros of Polynomials

If $f(c) = 0$, $c$ is called a *zero* of the polynomial $f(x)$.

## Division of Polynomials

If a polynomial $g(x)$ is a factor of another polynomial $f(x)$, then $f(x)$ is said to be *divisible* by $g(x)$. Thus $x^3 - 1$ is divisible both by $x - 1$ and by $x^2 + x + 1$. If a polynomial is not divisible by another, it is possible to apply the technique of long division to find a quotient and remainder, as in the following examples:

**EXAMPLE 14.3** Find the quotient and remainder for $(2x^4 - x^2 - 2)/(x^2 + 2x - 1)$.

Arrange the dividend and divisor in descending powers of the variable. Insert terms with zero coefficients and use the long division scheme.

$$
\begin{array}{r}
2x^2 - 4x + 9 \\
x^2 + 2x - 1 \overline{\smash{\big)}\ 2x^4 + 0x^3 - x^2 + 0x - 2} \\
\underline{2x^4 + 4x^3 - 2x^2} \\
-4x^3 + x^2 + 0x \\
\underline{-4x^3 - 8x^2 + 4x} \\
9x^2 - 4x - 2 \\
\underline{9x^2 + 18x - 9} \\
-22x + 7
\end{array}
$$

Divide first term of dividend by first term of divisor

Multiply divisor by $2x^2$; subtract
Bring down next term; repeat division step

Multiply divisor by $-4x$; subtract
Bring down next term; repeat division step

Multiply divisor by 9; subtract
Remainder; degree is less than degree of divisor

The quotient is $2x^2 - 4x + 9$ and the remainder is $-22x + 7$. Thus:

$$\frac{2x^4 - x^2 - 2}{x^2 + 2x - 1} = 2x^2 - 4x + 9 + \frac{-22x + 7}{x^2 + 2x - 1}$$

## Division Algorithm for Polynomials

If $f(x)$ and $g(x)$ are polynomials, with $g(x) \neq 0$, then there exist unique polynomials $q(x)$ and $r(x)$ such that

$$f(x) = g(x)q(x) + r(x) \quad \text{and} \quad \frac{f(x)}{g(x)} = q(x) + \frac{r(x)}{g(x)}$$

Either $r(x) = 0$ ($f(x)$ is divisible by $g(x)$) or the degree of $r(x)$ is less than the degree of $g(x)$. Therefore, if the degree of $g(x)$ is 1, the degree of $r(x)$ is 0, and the remainder is a constant polynomial $r$.

**EXAMPLE 14.4** Find the quotient and remainder for $(x^3 - 5x^2 + 7x - 9)/(x - 4)$.

Use the long division scheme:

$$
\begin{array}{r}
x^2 - x + \phantom{0}3 \\
x - 4 \,\overline{)\, x^3 - 5x^2 + 7x - \phantom{0}9} \\
\underline{x^3 - 4x^2} \phantom{xxxxxxxx} \\
-x^2 + 7x \phantom{xxx} \\
\underline{-x^2 + 4x} \phantom{xx} \\
3x - \phantom{0}9 \\
\underline{3x - 12} \\
3
\end{array}
$$

Divide first term of dividend by first term of divisor

Multiply divisor by $x^2$; subtract
Bring down next term; repeat division step

Multiply divisor by $-x$; subtract
Bring down next term; repeat division step

Multiply divisor by 3; subtract
Remainder; degree is less than degree of divisor

The quotient is $x^2 - x + 3$ and the remainder is the constant 3. Thus

$$\frac{x^3 - 5x^2 + 7x - 9}{x - 4} = x^2 - x + 3 + \frac{3}{x - 4}$$

## Synthetic Division

Division of a polynomial $f(x)$ by a polynomial of form $x - c$ is accomplished efficiently by the synthetic division scheme. Arrange coefficients of the dividend $f(x)$ in descending order in the first row of a three-row array.

$$c \mid a_n\, a_{n-1} \cdots a_1\, a_0$$

The third row is formed by bringing down the first coefficient of $f(x)$, then successively multiplying each coefficient in the third row by $c$, placing the result in the second row, adding this to the corresponding coefficient in the first row, and placing the result in the next position in the third row.

$$
\begin{array}{c|ccccc}
c & a_n & a_{n-1} & \cdots & a_1 & a_0 \\
 & & ca_n & cb_1 & \cdots cb_{n-2} & cb_{n-1} \\
\hline
 & a_n & b_1 & \cdots & b_{n-1} & r
\end{array}
$$

The last coefficient in the third row is the constant remainder; the other coefficients are the coefficients of the quotient, in descending order.

**EXAMPLE 14.5** Use synthetic division to find the quotient and remainder in the previous example.

In this case, $c = 4$. Arrange the coefficients of $x^3 - 5x^2 + 7x - 9$ in the first row of a three-row array; proceed to bring down the first coefficient, 1, then multiply by 4, place the result in the second row, add to $-5$, place the result in the third row. Continue to the last coefficient of the array.

$$
\begin{array}{c|rrrr}
4 & 1 & -5 & 7 & -9 \\
 & & 4 & -4 & 12 \\
\hline
 & 1 & -1 & 3 & 3
\end{array}
$$

As before, the quotient is $x^2 - x + 3$ and the remainder is 3.

## Remainder Theorem

When the polynomial $f(x)$ is divided by $x - c$, the remainder is $f(c)$.

**EXAMPLE 14.6** Verify the remainder theorem for the polynomial $f(x) = x^3 - 5x^2 + 7x - 9$ divided by $x - 4$.

Calculate $f(4) = 4^3 - 5 \cdot 4^2 + 7 \cdot 4 - 9 = 3$. The remainder in division has already been shown to be 3; thus, the conclusion of the theorem holds.

## Factor Theorem

A polynomial $f(x)$ has a factor of $x - c$ if and only if $f(c) = 0$. Thus, $x - c$ is a factor of a polynomial if and only if $c$ is a zero of the polynomial.

**EXAMPLE 14.7** Use the factor theorem to verify that $x + 2$ is a factor of $x^5 + 32$.

Let $f(x) = x^5 + 32$; then $f(-2) = (-2)^5 + 32 = 0$; hence $x - (-2) = x + 2$ is a factor of $f(x)$.

## Fundamental Theorem of Algebra

Every polynomial of positive degree with complex coefficients has at least one complex zero.

## Corollaries of the Fundamental Theorem

1. Every polynomial of positive degree n has a factorization of the form

$$P(x) = a_n (x - r_1)(x - r_2) \cdots (x - r_n)$$

   where the $r_i$ are not necessarily distinct. If in the factorization $x - r_i$ occurs $m$ times, $r_i$ is called a zero of multiplicity $m$. However, it is not necessarily possible to find the factorization using exact algebraic methods.
2. A polynomial of degree $n$ has at most $n$ complex zeros. If a zero of multiplicity $m$ is counted as $m$ zeros, then a polynomial of degree $n$ has exactly $n$ zeros.

## Further Theorems about Zeros

Further theorems about zeros of polynomials:

1. If $P(x)$ is a polynomial with *real* coefficients, and if $z$ is a complex zero of $P(x)$, then the complex conjugate $\bar{z}$ is also a zero of $P(x)$. That is, complex zeros of polynomials with real coefficients occur in complex conjugate pairs.
2. Any polynomial of degree $n > 0$ with real coefficients has a complete factorization using linear and quadratic factors, multiplied by the leading coefficient of the polynomial. However, it is not necessarily possible to find the factorization using exact algebraic methods.
3. If $P(x) = a_n x^n + a_{n-1} x^{n-1} + \cdots + a_1 x + a_0$ is a polynomial with *integral* coefficients and $r = p/q$ is a *rational* zero of $P(x)$ in lowest terms, then $p$ must be a factor of the constant term $a_0$ and $q$ must be a factor of the leading coefficient $a_n$.

**EXAMPLE 14.8** Find a polynomial of least degree with real coefficients and zeros 2 and $1 - 3i$.

By the factor theorem, $c$ is a zero of a polynomial only if $x - c$ is a factor. By the theorem on zeros of polynomials with real coefficients, if $1 - 3i$ is a zero of this polynomial, then so is $1 + 3i$. Hence the polynomial can be written as

$$P(x) = a(x - 2)[x - (1 - 3i)][x - (1 + 3i)]$$

Simplifying yields:

$$P(x) = a(x - 2)[(x - 1) + 3i][(x - 1) - 3i]$$
$$= a(x - 2)[(x - 1)^2 - (3i)^2]$$
$$= a(x - 2)(x^2 - 2x + 10)$$
$$= a(x^3 - 4x^2 + 14x - 20)$$

**EXAMPLE 14.9** List the possible rational zeros of $3x^2 + 5x - 8$.

From the theorem on rational zeros of polynomials with integer coefficients, the possible rational zeros are:

$$\frac{\text{Factors of } -8}{\text{Factors of } 3} = \frac{\pm 1, \pm 2, \pm 4, \pm 8}{\pm 1, \pm 3} = \pm 1, \pm 2, \pm 4, \pm 8, \pm \frac{1}{3}, \pm \frac{2}{3}, \pm \frac{4}{3}, \pm \frac{8}{3}$$

Note that the actual zeros are 1 and $-\frac{8}{3}$.

## Theorems Used In Locating Zeros

Theorems used in locating zeros of polynomials:

1. **INTERMEDIATE VALUE THEOREM:** Given a polynomial $f(x)$ with $a < b$, if $f(a) \neq f(b)$, then $f(x)$ takes on every value $c$ between $a$ and $b$ in the interval $(a,b)$.
2. **COROLLARY:** For a polynomial $f(x)$, if $f(a)$ and $f(b)$ have opposite signs, then $f(x)$ has at least one zero between $a$ and $b$.
3. **DESCARTES' RULE OF SIGNS:** If $f(x)$ is a polynomial with terms arranged in descending order, then the number of positive real zeros of $f(x)$ is either equal to the number of sign changes between successive terms of $f(x)$ or is less than this number by an even number. The number of negative real zeros of $f(x)$ is found by applying this rule to $f(-x)$.
4. If the third line of a synthetic division of $f(x)$ by $x - r$ is all positive for some $r > 0$, then $r$ is an upper bound for the zeros of $f(x)$; that is, there are no zeros greater than $r$. If the terms in the third line of a synthetic division of $f(x)$ by $x - r$ alternate in sign for some $r < 0$, then $r$ is a lower bound for the zeros of $f(x)$; that is, there are no zeros less than $r$. (0 may be regarded as positive or negative for the purpose of this theorem.)

## Solving Polynomial Equations

Solving polynomial equations and graphing polynomials:
   The following statements are equivalent:

1. $c$ is a zero of $P(x)$.
2. $c$ is a solution of the equation $P(x) = 0$.
3. $x - c$ is a factor of $P(x)$.
4. For real $c$, the graph of $y = P(x)$ has an $x$-intercept at $c$.

## Graphing a Polynomial

To graph a polynomial function for which all factors can be found:

1. Write the polynomial in factored form.
2. Determine the sign behavior of the polynomial from the signs of the factors.
3. Enter the $x$-intercepts of the polynomial on the $x$-axis.
4. If desired, form a table of values.
5. Sketch the graph of the polynomial as a smooth curve.

**EXAMPLE 14.10**   Sketch a graph of $y = 2x(x - 3)(x + 2)$.

The polynomial is already in factored form. Use the methods of Chapter 6 to obtain the sign chart shown in Fig. 14-7.

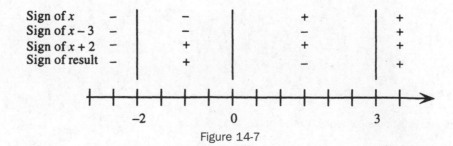

Figure 14-7

The graph has $x$-intercepts $-2, 0, 3$ and is below the $x$-axis on the intervals $(-\infty, -2)$ and $(0,3)$ and above the $x$-axis on the intervals $(-2,0)$ and $(3,\infty)$. Form a table of values as shown and sketch the graph as a smooth curve (Fig. 14-8).

| $x$ | –3 | –2 | –1 | 0 | 1 | 2 | 3 | 4 |
|---|---|---|---|---|---|---|---|---|
| $y$ | –36 | 0 | 8 | 0 | –12 | –16 | 0 | 48 |

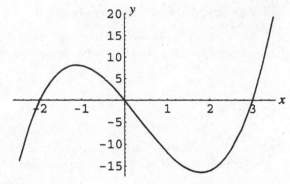

Figure 14-8

## SOLVED PROBLEMS

**14.1.** Prove the remainder theorem.

By the division algorithm, there exist polynomials $q(x)$ and $r(x)$ such that $f(x) = q(x)(x - c) + r(x)$. Since the degree of $r(x)$ is less than the degree of $x - c$, that is, less than 1, the degree of $r(x)$ must be zero. Thus $r(x)$ is a constant; call it $r$. Thus, for all $x$,

$$f(x) = q(x)(x - c) + r$$

In particular, let $x = c$. Then $f(c) = q(c)(c - c) + r$, that is, $f(c) = r$. Thus,

$$f(x) = q(x)(x - c) + f(c)$$

In other words, the remainder when $f(x)$ is divided by $c$ is $f(c)$.

**14.2.** Find the quotient and remainder when $2x^3 + 3x^2 - 13x + 5$ is divided by $2x - 3$. Use the long division scheme:

$$
\begin{array}{r}
x^2 + 3x - 2 \\
2x - 3 \overline{\smash{\big)}\, 2x^3 + 3x^2 - 13x + 5} \\
\end{array}
$$

| | |
|---|---|
| $\underline{2x^3 - 3x^2}$ | Multiply divisor by $x^2$; subtract |
| $6x^2 - 13x$ | Bring down next term; repeat division step |
| $\underline{6x^2 - \phantom{0}9x}$ | Multiply divisor by $3x$; subtract |
| $-4x + 5$ | Bring down next term; repeat division step |
| $\underline{-4x + 6}$ | Multiply divisor by $-2$; subtract |
| $-1$ | Remainder; degree is less than degree of divisor |

Divide first term of dividend by first term of divisor

The quotient is $x^2 + 3x - 2$ and the remainder is $-1$.

**14.3.** Find the quotient and remainder when $3x^5 - 7x^3 + 5x^2 + 6x - 6$ is divided by $x^3 - x + 2$.

Use the long division scheme:

$$
\begin{array}{r}
3x^2 \phantom{xxx} - 4 \\
x^3 - x + 2 \overline{\smash{\big)}\, 3x^5 \phantom{xx} - 7x^3 + 5x^2 + 6x - 6} \\
\end{array}
$$

| | |
|---|---|
| | Divide first term of dividend by first term of divisor |
| $\underline{3x^5 \phantom{xx} -3x^3 + 6x^2}$ | Multiply divisor by $3x^2$; subtract |
| $-4x^3 - \phantom{0}x^2 + 6x - 6$ | Bring down next term; repeat division step |
| $\underline{-4x^3 \phantom{xxxxx} + 4x - 8}$ | Multiply divisor by $-4$; subtract |
| $-x^2 + 2x + 2$ | Remainder; degree is less than degree of divisor |

The quotient is $3x^2 - 4$ and the remainder is $-x^2 + 2x + 2$.

**14.4.** Find the quotient and remainder when $2x^3 + 5x^2 - 10x + 9$ is divided by $x + 2$.

Use the synthetic division scheme. Note that in dividing by $x - c$, the coefficient $c$ is placed in the upper-left corner and used to multiply the numbers generated in the third line. In division by $x + 2$, that is, $x - (-2)$, use $c = -2$.

$$
\begin{array}{r|rrrr}
-2 & 2 & 5 & -10 & 9 \\
& & -4 & -2 & 24 \\
\hline
& 2 & 1 & -12 & 33
\end{array}
$$

The quotient is $2x^2 + x - 12$ and the remainder is 33.

**14.5.** Find the quotient and remainder when $-3t^5 + 10t^4 + 15t^2 + 18t - 6$ is divided by $t - 4$.

Use the scheme for synthetic division by $t - c$, with $c = 4$. Enter a zero for the missing coefficient of $t^3$.

$$
\begin{array}{r|rrrrrr}
4 & -3 & 10 & 0 & 15 & 18 & -6 \\
& & -12 & -8 & -32 & -68 & -200 \\
\hline
& -3 & -2 & -8 & -17 & -50 & -206
\end{array}
$$

The quotient is $-3t^4 - 2t^3 - 8t^2 - 17t - 50$ and the remainder is $-206$.

**14.6.** Find the quotient and remainder when $2x^3 - 5x^2 + 6x - 3$ is divided by $x - \frac{1}{2}$.

Use the scheme for synthetic division by $x - c$, with $c = \frac{1}{2}$.

$$
\begin{array}{r|rrrr}
\frac{1}{2} & 2 & -5 & 6 & -3 \\
& & 1 & -2 & 2 \\
\hline
& 2 & -4 & 4 & -1
\end{array}
$$

The quotient is $2x^2 - 4x + 4$ and the remainder is $-1$.

**14.7.** Find the quotient and remainder when $3x^4 + 8x^3 - x^2 + 7x + 2$ is divided by $x + \frac{2}{3}$.

Use the scheme for synthetic division by $x - c$, with $c = -\dfrac{2}{3}$.

$$
\begin{array}{r|rrrrr}
-\frac{2}{3} & 3 & 8 & -1 & 7 & 2 \\
& & -2 & -4 & \frac{10}{3} & -\frac{62}{9} \\
\hline
& 3 & 6 & -5 & \frac{31}{3} & -\frac{44}{9}
\end{array}
$$

The quotient is $3x^3 + 6x^2 - 5x + \frac{31}{3}$ and the remainder is $-\frac{44}{9}$.

**14.8.** Prove the factor theorem.

By the remainder theorem, when $f(x)$ is divided by $x - c$, the remainder is $f(c)$.
Assume $c$ is a zero of $f(x)$; then $f(c) = 0$. Therefore, $f(x) = q(x)(x - c) + f(c) = q(x)(x - c)$, that is, $x - c$ is a factor of $f(x)$.
Conversely, assume $x - c$ is a factor of $f(x)$; then the remainder when $f(x)$ is divided by $x - c$ must be zero. By the remainder theorem, this remainder is $f(c)$; hence, $f(c) = 0$.

**14.9.** Show that $x - a$ is a factor of $x^n - a^n$ for all integers $n$.

Let $f(x) = x^n - a^n$; then $f(a) = a^n - a^n = 0$. By the factor theorem, since $a$ is a zero of $f(x)$, $x - a$ is a factor.

**14.10.** Use the quadratic formula and the factor theorem to factor (a) $x^2 - 12x + 3$; (b) $x^2 - 4x + 13$.

(a) The zeros of $x^2 - 12x + 3$, that is, the solutions of $x^2 - 12x + 3 = 0$, are found from the quadratic formula. Using $a = 1, b = -12, c = 3$ yields

$$x = \frac{-(-12) \pm \sqrt{(-12)^2 - 4(1)(3)}}{2(1)} = \frac{12 \pm \sqrt{132}}{2} = 6 \pm \sqrt{33}$$

Since the zeros are $6 \pm \sqrt{33}$, the factors are $x - (6 + \sqrt{33})$ and $x - (6 - \sqrt{33})$. Thus

$$x^2 - 12x + 3 = [x - (6 + \sqrt{33})][x - (6 - \sqrt{33})] \text{ or } [(x - 6) - \sqrt{33}][(x - 6) + \sqrt{33}]$$

(b) Proceeding as in (a), use the quadratic formula with $a = 1, b = -4, c = 13$ to obtain

$$x = \frac{-(-4) \pm \sqrt{(-4)^2 - 4(1)(13)}}{2(1)} = \frac{4 \pm \sqrt{-36}}{2} = 2 \pm 3i$$

Since the zeros are $2 \pm 3i$, the factors are $x - (2 + 3i)$ and $x - (2 - 3i)$. Thus

$$x^2 - 4x + 13 = [x - (2 + 3i)][x - (2 - 3i)] \text{ or } [(x - 2) - 3i][(x - 2) + 3i]$$

**14.11.** Write the polynomial $P(x) = x^4 - 7x^3 + 13x^2 + 3x - 18$ as a product of first-degree factors, given that 3 is a zero of multiplicity 2.

Since 3 is a zero of multiplicity 2, there exists a polynomial $g(x)$ with $P(x) = (x - 3)^2 g(x)$. To find $g(x)$, use the scheme for synthetic division by $x - c$, with $c = 3$, twice:

$$
\begin{array}{r|rrrrr}
3 & 1 & -7 & 13 & 3 & -18 \\
  &   & 3 & -12 & 3 & 18 \\
\hline
3 & 1 & -4 & 1 & 6 & 0 \\
  &   & 3 & -3 & -6 & \\
\hline
  & 1 & -1 & -2 & 0 & \\
\end{array}
$$

Thus

$$P(x) = (x - 3)(x - 3)(x^2 - x - 2)$$
$$= (x - 3)(x - 3)(x - 2)(x + 1)$$

**14.12.** Write the polynomial $P(x) = 2x^3 + 2x^2 - 40x - 100$ as a product of first-degree factors, given that $-3 - i$ is a zero. Find all zeros of $P(x)$.

Since $P(x)$ has real coefficients and $-3 - i$ is a zero, $-3 + i$ is also a zero. Therefore, there exists a polynomial $g(x)$ with $P(x) = [x - (-3 - i)][x - (-3 + i)]g(x)$. To find $g(x)$, use the scheme for synthetic division by $x - c$ with, in turn, $c = -3 - i$ and $c = -3 + i$.

$$
\begin{array}{r|rrrr}
-3 - i & 2 & 2 & -40 & -100 \\
       &   & -6 - 2i & 10 + 10i & 100 \\
\hline
-3 + i & 2 & -4 - 2i & -30 + 10i & 0 \\
       &   & -6 + 2i & 30 - 10i & \\
\hline
       & 2 & -10 & 0 & \\
\end{array}
$$

Thus

$$P(x) = [x - (-3 - i)][x - (-3 + i)](2x - 10)$$

and the zeros of $P(x)$ are $-3 \pm i$ and 5.

**14.13.** Find a polynomial $P(x)$ of lowest degree, with real coefficients, such that 4 is a zero of multiplicity 3, $-2$ is a zero of multiplicity 2, 0 is a zero, and $5 + 2i$ is a zero.

Since $P(x)$ has real coefficients and $5 + 2i$ is a zero, $5 - 2i$ is also a zero. Thus, write

$$P(x) = a(x - 4)^3[x - (-2)]^2(x - 0)[x - (5 + 2i)][x - (5 - 2i)]$$
$$= a(x - 4)^3(x + 2)^2x[(x - 5) - 2i][(x - 5) + 2i]$$
$$= a(x - 4)^3(x + 2)^2x(x^2 - 10x + 29)$$

Here $a$ can be any real number.

**14.14.** Find a polynomial $P(x)$ of lowest degree, with integer coefficients, such that $\frac{2}{3}, \frac{3}{4}$, and $-\frac{1}{2}$ are zeros.

Write

$$P(x) = a\left(x - \frac{2}{3}\right)\left(x - \frac{3}{4}\right)\left[x - \left(-\frac{1}{2}\right)\right]$$
$$= a\left(\frac{3x - 2}{3}\right)\left(\frac{4x - 3}{4}\right)\left(\frac{2x + 1}{2}\right)$$
$$= 24b\left(\frac{3x - 2}{3}\right)\left(\frac{4x - 3}{4}\right)\left(\frac{2x + 1}{2}\right)$$
$$= b(3x - 2)(4x - 3)(2x + 1)$$

Here $b$ can be any integer.

**14.15.** Show that $f(x) = x^3 - 5$ has a zero between 1 and 2.

Since $f(1) = 1^3 - 5 = -4$ and $f(2) = 2^3 - 5 = 3$, $f(1)$ and $f(2)$ have opposite signs. Hence the polynomial has at least one zero between 1 and 2.

**14.16.** Show that $f(x) = 2x^4 + 3x^3 + x^2 - 2x - 8$ has a zero between $-2$ and $-1$.

Use the scheme for synthetic division with $c = -2$ and $c = -1$.

$$
\begin{array}{r|rrrrr}
-2 & 2 & 3 & 1 & -2 & -8 \\
   &   & -4 & 2 & -6 & 16 \\
\hline
   & 2 & -1 & 3 & -8 & 8
\end{array}
\qquad
\begin{array}{r|rrrrr}
-1 & 2 & 3 & 1 & -2 & -8 \\
   &   & -2 & -1 & 0 & 2 \\
\hline
   & 2 & 1 & 0 & -2 & -6
\end{array}
$$

Since $f(-2) = 8$ and $f(-1) = -6$, $f(-2)$ and $f(-1)$ have opposite signs. Hence the polynomial has at least one zero between $-2$ and $-1$.

**14.17.** Use Descartes' rule of signs to analyze the possible combinations of positive, negative, and imaginary zeros for $f(x) = x^3 - 3x^2 + 2x + 8$.

The coefficients of $f(x)$ exhibit two changes of sign. Thus there could be two or zero positive real zeros for $f$. To find the possible number of negative zeros, consider $f(-x)$.

$$f(-x) = (-x)^3 - 3(-x)^2 + 2(-x) + 8 = -x^3 - 3x^2 - 2x + 8$$

The coefficients of $f(-x)$ exhibit one change of sign. Thus there must be one negative real zero for $f$. Since there are either three or one real zeros, there can be either no or two imaginary zeros. The table indicates the possible combinations of zeros:

| POSITIVE | NEGATIVE | IMAGINARY |
|---|---|---|
| 2 | 1 | 0 |
| 0 | 1 | 2 |

**14.18.** Use Descartes' rule to signs of analyze the possible combinations of positive, negative, and imaginary zeros for $f(x) = -2x^6 + 3x^5 - 3x^3 + 5x^2 - 6x + 9$.

The coefficients of $f(x)$ exhibit five changes of sign. Thus there could be five or three positive real zeros for $f$ or one positive real zero for $f$.
To find the possible number of negative zeros, consider $f(-x)$.

$$f(-x) = -2(-x)^6 + 3(-x)^5 - 3(-x)^3 + 5(-x)^2 - 6(-x) + 9$$
$$= -2x^6 - 3x^5 + 3x^3 + 5x^2 + 6x + 9$$

The coefficients of $f(-x)$ exhibit one change of sign. Thus there must be one negative real zero for $f$. Since there are either six, four, or two real zeros, there can be either no, two, or four imaginary zeros. The table indicates the possible combinations of zeros:

| POSITIVE | NEGATIVE | IMAGINARY |
|---|---|---|
| 5 | 1 | 0 |
| 3 | 1 | 2 |
| 1 | 1 | 4 |

**14.19.** Use Descartes' rule of signs to show that $f(x) = x^3 + 7$ has no positive real zeros and must have a real negative zero.

Since $f(x)$ exhibits no changes of sign, there can be no positive real zeros. To find the possible number of negative zeros, consider $f(-x)$.

$$f(-x) = (-x)^3 + 7 = -x^3 + 7$$

The coefficients of $f(-x)$ exhibit one change of sign. Thus there must be one negative real zero for $f$.

**14.20.** Use Descartes' rule of signs to show that $f(x) = x^4 + 2x^2 + 1$ has no real zeros.

Since $f(x)$ exhibits no changes of sign, there can be no positive real zeros. To find the possible number of negative zeros, consider $f(-x)$.

$$f(-x) = (-x)^4 + 2(-x)^2 + 1 = x^4 + 2x^2 + 1$$

Since $f(-x)$ exhibits no changes of sign, there can be no negative real zeros. Since 0 is not a zero, there can be no real zeros of $f(x)$.

**14.21.** Find the smallest positive integer and the largest negative integer that are, respectively, upper and lower bounds for the zeros of $f(x) = x^3 + 2x^2 - 3x - 5$.

Use the scheme for synthetic division by $x - c$, with $c$ = successive positive integers (only the last line in the synthetic division is shown).

| | 1 | 2 | −3 | −5 |
|---|---|---|---|---|
| 1 | 1 | 3 | 0 | −5 |
| 2 | 1 | 4 | 5 | 5 |

Since the last line in synthetic division by $x - 2$ is all positive and the last line in synthetic division by $x - 1$ is not, 2 is the smallest positive integer that is an upper bound for the zeros of $f$.

Use the scheme for synthetic division by $x - c$, with $c$ = successive negative integers (only the last line in the synthetic division is shown).

| | 1 | 2 | -3 | -5 |
|---|---|---|---|---|
| -1 | 1 | 1 | -4 | -1 |
| -2 | 1 | 0 | -3 | 1 |
| -3 | 1 | -1 | 0 | -5 |

Since the last line in synthetic division by $x + 3$ alternates in sign (recall that 0 can be regarded as positive or negative in this context) and the last line in synthetic division by $x + 2$ does not, $-3$ is the largest negative integer that is a lower bound for the zeros of $f$.

**14.22.** Use the corollary of the intermediate value theorem to locate, between successive integers, the zeros of $f$ in the previous problem.

From the synthetic divisions carried out in the previous problem, since $f(1)$ and $f(2)$ have opposite signs, there is a zero of $f$ between 1 and 2. Similarly, since $f(-1)$ and $f(-2)$ have opposite signs, there is a zero of $f$ between $-1$ and $-2$. Finally, since $f(-2)$ and $f(-3)$ have opposite signs, there is a zero of $f$ between $-2$ and $-3$.

**14.23.** Use the corollary of the intermediate value theorem to locate, between successive integers, the zeros of $f(x) = x^4 - 3x^3 - 6x^2 + 33x - 35$.

Use the scheme for synthetic division by $x - c$, with $c$ = successive positive integers, then 0, then successive negative integers (only the last line in the synthetic divisions is shown).

| | 1 | -3 | -6 | 33 | -35 |
|---|---|---|---|---|---|
| 1 | 1 | -2 | -8 | 25 | -10 |
| 2 | 1 | -1 | -8 | 17 | -1 |
| 3 | 1 | 0 | -6 | 15 | 10 |
| 4 | 1 | 1 | -2 | 25 | 65 |
| 5 | 1 | 2 | 4 | 53 | 230 |
| 0 | 1 | -3 | -6 | 33 | -35 |
| -1 | 1 | -4 | -2 | 35 | -70 |
| -2 | 1 | -5 | 4 | 25 | -85 |
| -3 | 1 | -6 | 12 | -3 | -26 |
| -4 | 1 | -7 | 22 | -55 | 185 |

Since $f(2)$ and $f(3)$ have opposite signs, there is a zero of $f$ between 2 and 3. No other positive real zeros can be isolated from the data in the table (5 is an upper bound for the positive real zeros).

Since $f(-3)$ and $f(-4)$ have opposite signs, there is a zero of $f$ between $-3$ and $-4$. No other negative real zeros can be isolated from the data in the table ($-4$ is a lower bound for the negative real zeros).

**14.24.** List the possible rational zeros of $x^3 - 5x^2 + 7x - 12$.

From the theorem on rational zeros of polynomials with integer coefficients, the possible rational zeros are:

$$\frac{\text{Factors of } -12}{\text{Factors of } 1} = \frac{\pm 1, \pm 2, \pm 3, \pm 4, \pm 6, \pm 12}{\pm 1} = \pm 1, \pm 2, \pm 3, \pm 4, \pm 6, \pm 12$$

**14.25.** List the possible rational zeros of $4x^3 + 5x^2 + 7x - 18$.

From the theorem on rational zeros of polynomials with integer coefficients, the possible rational zeros are:

$$\frac{\text{Factors of } -18}{\text{Factors of } 4} = \frac{\pm 1, \pm 2, \pm 3, \pm 6, \pm 9, \pm 18}{\pm 1, \pm 2, \pm 4} = \pm 1, \pm 2, \pm 3, \pm 6 \ \pm 9, \pm 18, \pm \frac{1}{2}, \pm \frac{3}{2}, \pm \frac{9}{2}, \pm \frac{1}{4}, \pm \frac{3}{4}, \pm \frac{9}{4}$$

**14.26.** Find all zeros of $f(x) = x^3 + 3x^2 - 10x - 24$.

From Descartes' rule of signs, the following combinations of positive, negative, and imaginary zeros are possible.

| POSITIVE | NEGATIVE | IMAGINARY |
|---|---|---|
| 1 | 2 | 0 |
| 1 | 0 | 2 |

From the theorem on rational zeros of polynomials with integer coefficients, the possible rational zeros are $\pm 1, \pm 2, \pm 3, \pm 4 \ \pm 6, \pm 8, \pm 12, \pm 24$.

Use the scheme for synthetic division by $x - c$, with $c =$ successive positive integers from this list (only the last line in the synthetic division is shown).

| | 1 | 3 | −10 | −24 |
|---|---|---|---|---|
| 1 | 1 | 4 | −6 | −30 |
| 2 | 1 | 5 | 0 | −24 |
| 3 | 1 | 6 | 8 | 0 |

Thus, 3 is a zero and the polynomial can be factored as follows:

$$f(x) = (x - 3)(x^2 + 6x + 8)$$
$$= (x - 3)(x + 2)(x + 4)$$

Hence the zeros are 3, −2, and −4.

**14.27.** Find all zeros of $f(x) = 3x^4 + 16x^3 + 20x^2 - 9x - 18$.

From Descartes' rule of signs, the following combinations of positive, negative, and imaginary zeros are possible.

| POSITIVE | NEGATIVE | IMAGINARY |
|---|---|---|
| 1 | 3 | 0 |
| 1 | 1 | 2 |

From the theorem on rational zeros of polynomials with integer coefficients, the possible rational zeros are

$$\frac{\text{Factors of } -18}{\text{Factor of } 3} = \frac{\pm 1, \pm 2, \pm 3, \pm 6, \pm 9, \pm 18}{\pm 1, \pm 3} = \pm 1, \pm 2, \pm 3, \pm 6, \pm 9, \pm 18, \pm \frac{1}{3}, \pm \frac{2}{3}$$

Use the scheme for synthetic division by $x - c$, with $c =$ successive positive integers from this list (only the last line in the synthetic division is shown).

| | 3 | 16 | 20 | −9 | −18 |
|---|---|---|---|---|---|
| 1 | 3 | 19 | 39 | 30 | 12 |

Since $f(0)$ and $f(1)$ have opposite signs, the positive zero is between 0 and 1.

Now use the scheme for synthetic division by $x - c$, with $c =$ successive negative integers from the list (only the last line in the synthetic division is shown).

| | 3 | 16 | 20 | −9 | −18 |
|---|---|---|---|---|---|
| −1 | 3 | 13 | 7 | −16 | −2 |
| −2 | 3 | 10 | 0 | −9 | 0 |

Thus, −2 is a zero and the polynomial can be factored as follows:

$$f(x) = (x + 2)(3x^3 + 10x^2 - 9)$$

The possible rational zeros of the depressed polynomial $3x^3 + 10x^2 - 9$ that have not been eliminated are $-3$, $-9$, and $\pm\frac{1}{3}$. Synthetic division by $x - c$, with $c = -3$, yields (only the last line in the synthetic division is shown):

| | 3 | 10 | 0 | −9 |
|---|---|---|---|---|
| −3 | 3 | 1 | −3 | 0 |

Thus, −3 is a zero and the polynomial can be factored as follows:

$$f(x) = (x + 2)(x + 3)(3x^2 + x - 3)$$

The remaining zeros can be found by solving $3x^2 + x - 3 = 0$ by the quadratic formula to obtain $\dfrac{-1 \pm \sqrt{37}}{6}$ in addition to −2 and −3.

**14.28.** Find all zeros of $f(x) = 4x^4 - 4x^3 - 7x^2 - 6x + 18$.

From Descartes' rule of signs, the following combinations of positive, negative, and imaginary zeros are possible.

| POSITIVE | NEGATIVE | IMAGINARY |
|---|---|---|
| 2 | 2 | 0 |
| 2 | 0 | 2 |
| 0 | 2 | 2 |
| 0 | 0 | 4 |

From the theorem on rational zeros of polynomials with integer coefficients, the possible rational zeros are $\pm 1$, $\pm 2$, $\pm 3$, $\pm 6$, $\pm 9$, $\pm 18$, $\pm\frac{1}{2}$, $\pm\frac{3}{2}$, $\pm\frac{9}{2}$, $\pm\frac{1}{4}$, $\pm\frac{3}{4}$, $\pm\frac{9}{4}$.

Use the scheme for synthetic division by $x - c$, with $c =$ successive positive integers from this list (only the last line in the synthetic division is shown).

| | 4 | −4 | −7 | −6 | 18 |
|---|---|---|---|---|---|
| 1 | 4 | 0 | −7 | −13 | 5 |
| 2 | 4 | 4 | 1 | −4 | 10 |
| 3 | 4 | 8 | 17 | 45 | 153 |

3 is an upper bound for the positive zeros of $f$. Now use the scheme for synthetic division by $x - c$, with $c =$ successive positive rational numbers from this list (only the last line in the synthetic division is shown).

$$
\begin{array}{c|ccccc}
 & 4 & -4 & -7 & -6 & 18 \\
\hline
\frac{1}{2} & 4 & -2 & -8 & -10 & 13 \\
\\
\frac{3}{2} & 4 & 2 & -4 & -12 & 0 \\
\end{array}
$$

Thus, $\frac{3}{2}$ is a zero and the polynomial can be factored as follows:

$$
f(x) = \left(x - \frac{3}{2}\right)(4x^3 + 2x^2 - 4x - 12) = (2x - 3)(2x^3 + x^2 - 2x - 6)
$$

The only possible rational zero of the depressed polynomial $2x^3 + x^2 - 2x - 6$ that has not been eliminated from consideration is $\frac{3}{2}$. Synthetic division by $x - c$, with $c = \frac{3}{2}$, yields (only the last line in the synthetic division is shown):

$$
\begin{array}{c|cccc}
 & 2 & 1 & -2 & -6 \\
\hline
\frac{3}{2} & 2 & 4 & 4 & 0 \\
\end{array}
$$

Thus, $\frac{3}{2}$ is a double zero of the original polynomial, which can be factored as follows:

$$
f(x) = (2x - 3)\left(x - \frac{3}{2}\right)(2x^2 + 4x + 4) = (2x - 3)^2(x^2 + 2x + 2)
$$

The remaining zeros can be found by solving $x^2 + 2x + 2 = 0$ by the quadratic formula to obtain $-1 \pm i$ in addition to the double zero $\frac{3}{2}$.

**14.29.** Sketch the graphs of the following polynomial functions:

(a) $f(x) = 2x^3 - 9$  (b) $f(x) = \frac{1}{3}(x + 1)^4$

(c) $f(x) = -\frac{1}{2}(x + 3)^3 + 4$  (d) $f(x) = x^3 + 3x^2 - 10x - 24$

(e) $f(x) = 3x^4 + 16x^3 + 20x^2 - 9x - 18$  (f) $f(x) = 4x^4 - 4x^3 - 7x^2 - 6x + 18$

(a) The graph of $f(x) = 2x^3 - 9$ is the same as the graph of $f(x) = x^3$ stretched by a factor of 2 with respect to the $y$-axis and shifted down 9 units (see Fig. 14-9).

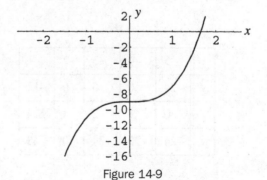

Figure 14-9

(b) The graph of $f(x) = \frac{1}{3}(x + 1)^4$ is the same as the graph of $f(x) = x^4$ shifted 1 unit to the left and compressed by a factor of 3 (see Fig. 14-10).

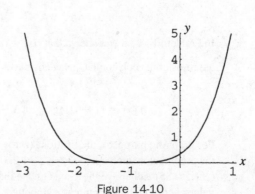

Figure 14-10

(c) The graph of $f(x) = -\frac{1}{2}(x + 3)^3 + 4$ is the
same as the graph of $f(x) = x^3$ shifted
3 units to the left, compressed by a factor
of 2, reflected with respect to the $y$-axis,
and shifted up 4 units (see Fig. 14-11).

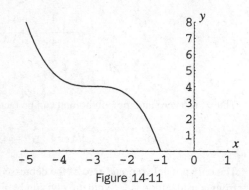

Figure 14-11

(d) In Problem 14-26 it was shown that $f(x) = x^3 + 3x^2 - 10x - 24 = (x - 3)(x + 2)(x + 4)$. Use the methods of Chapter 6 to obtain the sign chart shown in Fig. 14-12.

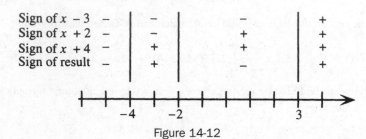

Figure 14-12

The graph has $x$-intercepts $-4$, $-2$, 3 and is below the $x$-axis on the intervals $(-\infty, -4)$ and $(-2, 3)$ and above the $x$-axis on the intervals $(-4, -2)$ and $(3, \infty)$. Form a table of values and sketch the graph as a smooth curve.
See Fig. 14-13 and the accompanying table.

| $x$ | −5 | −4 | −3 | −2 | −1 |
|---|---|---|---|---|---|
| $y$ | −24 | 0 | 6 | 0 | −12 |
| $x$ | 0 | 1 | 2 | 3 | 4 |
| $y$ | −24 | −30 | −24 | 0 | 48 |

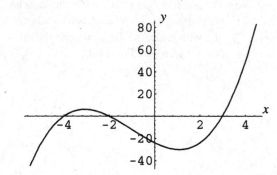

Figure 14-13

(e) In Problem 14-27 it was shown that $f(x) = (x + 2)(x + 3)(3x^2 + x - 3)$. It was further shown that $\dfrac{-1 \pm \sqrt{37}}{6}$ are zeros of the polynomial, hence; by the factor theorem, $f(x)$ can be completely factored as

$$f(x) = (x + 2)(x + 3)\left(x - \frac{-1 + \sqrt{37}}{6}\right)\left(x - \frac{-1 - \sqrt{37}}{6}\right)3$$

For graphing purposes, the irrational zeros may be approximated as 0.85 and $-1.2$. A sign chart shows that the graph has $x$-intercepts $-3$, $-2$, $-1.2$, 0.85 and is below the $x$-axis on the intervals $(-3,-2)$ and $(-1.2, 0.85)$ and above the $x$-axis on the intervals $(-\infty, -3)$, $(-2, -1.2)$, and $(0.85, \infty)$. Form a table of values and sketch the graph as a smooth curve.

See Fig. 14-14 and the accompanying table.

| x | −4 | −3 | −2 | −1 | 0 | 1 |
|---|---|---|---|---|---|---|
| y | 82 | 0 | 0 | −2 | −18 | 12 |

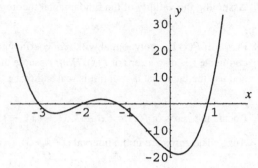

Figure 14-14

(f) In Problem 14-28 it was shown that $f(x) = (2x − 3)^2 (x^2 + 2x + 2)$. Thus the graph of the polynomial has an $x$-intercept at $x = \frac{3}{2}$ and is above the $x$-axis for all other values of $x$. Form a table of values and sketch the graph as a smooth curve.

See Fig. 14-15 and the accompanying table.

| x | −1.5 | −1 | −0.5 | 0 |
|---|---|---|---|---|
| y | 45 | 25 | 20 | 18 |
| x | 0.5 | 1 | 1.5 | 2 |
| y | 13 | 5 | 0 | 10 |

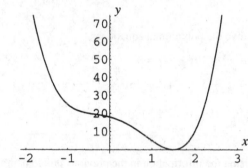

Figure 14-15

## SUPPLEMENTARY PROBLEMS

**14.30.** Find the quotient and remainder for the following:

(a) $(5x^4 + x^2 − 8x + 2)/(x^2 − 3x + 1)$   (b) $(x^5 + x^4 + 3x^3 − x^2 − x − 3)/(x^2 + x + 1)$

(c) $(x^3 − 3x^2 + 8x − 7)/(2x − 5)$   (d) $(x^6 − x^4 − 8x^3 + x + 2)/(x + 3)$

*Ans.*   (a) Quotient: $5x^2 + 15x + 41$, remainder: $100x − 39$; (b) quotient: $x^3 + 2x − 3$, remainder: 0;

(c) quotient: $\frac{1}{2}x^2 − \frac{1}{4}x + \frac{27}{8}$, remainder: $\frac{79}{8}$; (d) quotient: $x^5 − 3x^4 + 8x^3 − 32x^2 + 96x − 287$, remainder: 863

**14.31.** Given $f(x) = x^4 + 2x^3 + 6x^2 + 8x + 8$, find (a) $f(−3)$; (b) $f(2i)$; (c) $f(3 − i)$; (d) $f(−1 + i)$.

*Ans.*   (a) 65; (b) 0; (c) $144 − 192i$; (d) 0.

**14.32.** Find a polynomial $P(x)$ of lowest degree, with integer coefficients, such that $\frac{3}{5}$ and $−3 − 2i$ are zeros.

*Ans.*   $P(x) = a(5x^3 + 27x^2 + 47x − 39)$, where $a$ is any integer.

**14.33.** Show that $x + a$ is a factor of $x^n + a^n$ for all odd $n$.

**14.34.** Show that $x + a$ is a factor of $x^n - a^n$ for all even $n$.

**14.35.** Assuming the validity of the fundamental theorem of algebra, prove the first corollary stated above.

**14.36.** Prove: If $P(x)$ is a polynomial with *real* coefficients, then if $z$ is a complex zero of $P(x)$, then the complex conjugate $\bar{z}$ is also a zero of $P(x)$. *Hint:* Assume that $z$ is a zero of $P(x) = a_n x^n + a_{n-1} x^{n-1} + \cdots + a_1 x + a_0$, and use the facts that $\bar{a} = a$ if $a$ is real and that $\overline{z + w} = \bar{z} + \bar{w}$ and $\overline{zw} = \bar{z}\bar{w}$ for all complex numbers.

**14.37.** Locate the zeros of $f(x) = 6x^3 + 32x^2 + 41x + 12$ between successive integers.

*Ans.* The zeros are in the intervals $(-4,-3)$, $(-2,-1)$, and $(-1,0)$.

**14.38.** Find all zeros exactly for the following polynomials:

(a) $2x^3 - 5x^2 - 2x + 2$; (b) $x^4 + 2x^3 - 2x^2 - 6x - 3$; (c) $x^4 - x^3 - 3x^2 + 17x - 30$;

(d) $x^5 + 5x^3 + 6x$; (e) $3x^5 - 2x^4 - 9x^3 + 6x^2 - 12x + 8$

*Ans.* (a) $\{\frac{1}{2}, 1 \pm \sqrt{3}\}$; (b) $\{-1(\text{double}), \pm\sqrt{3}\}$; (c) $\{2, -3, 1\pm 2i\}$;

(d) $\{0, \pm i\sqrt{2}, \pm i\sqrt{3}\}$; (e) $\{\pm 2, \frac{2}{3}, \pm i\}$

**14.39.** Solve the polynomial equations:

(a) $x^3 - 19x - 30 = 0$ 　　　　　　　(b) $4x^3 + 40x = 22x^2 + 25$

(c) $x^5 - 5x^4 - 4x^3 + 36x^2 + 27x - 135 = 0$ 　　(d) $-12x^4 - 8x^3 + 49x^2 + 39x - 18 = 0$

*Ans.* (a) $\{-3, -2, 5\}$; (b) $\left\{\dfrac{5}{2}, \dfrac{3 \pm i}{2}\right\}$; (c) $\{3, -2 \pm i\}$; (d) $\{2, \frac{1}{3}, -\frac{3}{2}\}$

**14.40.** Using the information in the previous problem, draw graphs of

(a) $f(x) = x^3 - 19x - 30$ 　　　　　　(b) $f(x) = 4x^3 - 22x^2 + 40x - 25$

(c) $f(x) = x^5 - 5x^4 - 4x^3 + 36x^2 + 27x - 135$ 　(d) $f(x) = -12x^4 - 8x^3 + 49x^2 + 39x - 18$

*Ans.* (a) Fig. 14-16; (b) Fig. 14-17; (c) Fig. 14-18; (d) Fig. 14-19.

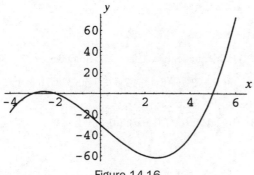

Figure 14-16

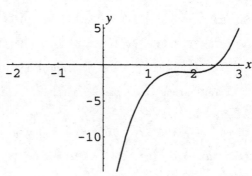

Figure 14-17

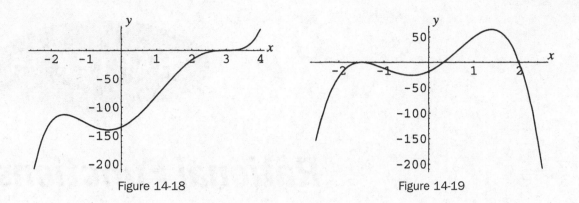

Figure 14-18          Figure 14-19

**14.41.** From a square piece of cardboard 20 inches on a side, an open box is to be made by removing squares of side $x$ and turning up the sides. Find the possible values of $x$ if the box is to have volume 576 cubic inches. (See Fig. 14-20.)

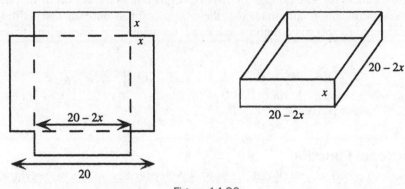

Figure 14-20

*Ans.*   4 inches, or $8 - \sqrt{28} \approx 2.7$ inches.

**14.42.** A silo is to be built in the shape of a right circular cylinder with a hemispherical top (see Fig. 14-21). If the total height of the silo is 30 feet and the total volume is $1008\pi$ cubic feet, find the radius of the cylinder.

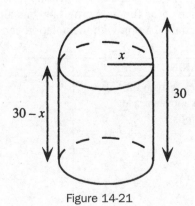

Figure 14-21

*Ans.*   6 feet.

CHAPTER 15

# Rational Functions

## Definition of Rational Function

A rational function is any function which can be specified by a rule written as $f(x) = \dfrac{P(x)}{Q(x)}$, where $P(x)$ and $Q(x)$ are polynomial functions. The domain of a rational function is the set of all real numbers for which $Q(x) \neq 0$. The assumption is normally made that the rational expression $P(x)/Q(x)$ is in lowest terms, that is, $P(x)$ and $Q(x)$ have no factors in common. (See below for analysis of cases where this assumption is not made.)

**EXAMPLE 15.1** $f(x) = \dfrac{12}{x}$, $g(x) = \dfrac{x^2}{x^2 - 9}$, $h(x) = \dfrac{(x + 1)(x - 4)}{x(x - 2)(x + 3)}$, and $k(x) = \dfrac{3x}{x^2 + 4}$, are examples of rational functions. The domains are, respectively, for $f, \{x \in \mathbf{R} | x \neq 0\}$, for $g$, $\{x \in \mathbf{R} | x \neq \pm3\}$, for $h$, $\{x \in \mathbf{R} | x \neq 0,2,-3\}$, and for $k$, $\mathbf{R}$ (since the denominator polynomial is never 0).

## Graph of a Rational Function

The graph of a rational function is analyzed in terms of the symmetry, intercepts, asymptotes, and sign behavior of the function.

1. If $Q(x)$ has no real zeros, the graph of $P(x)/Q(x)$ is a smooth curve for all real $x$.
2. If $Q(x)$ has real zeros, the graph of $P(x)/Q(x)$ consists of smooth curves on each open interval that does not include a zero. The graph has *vertical asymptotes* at each zero of $Q(x)$.

## Vertical Asymptotes

The line $x = a$ is a vertical asymptote for the graph of a function $f$ if, as $x$ approaches $a$ through values that are greater than or less than $a$, the value of the function grows beyond all bounds, either positive or negative. The cases are shown in the following table, along with the notation generally used:

| NOTATION | MEANING | GRAPH |
|---|---|---|
| $\displaystyle\lim_{x \to a^-} f(x) = \infty$ | As $x$ approaches $a$ from the left, $f(x)$ is positive and increases beyond all bounds. | *(graph)* Figure 15-1 |

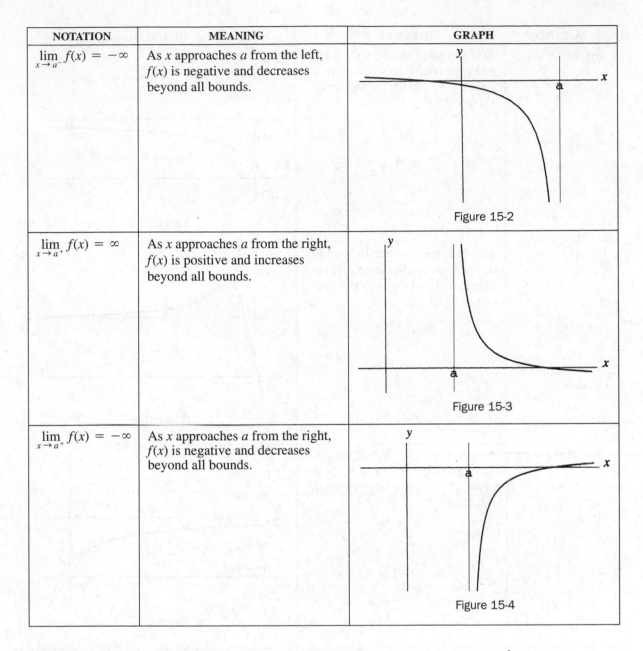

| NOTATION | MEANING | GRAPH |
|---|---|---|
| $\lim\limits_{x \to a^-} f(x) = -\infty$ | As $x$ approaches $a$ from the left, $f(x)$ is negative and decreases beyond all bounds. | Figure 15-2 |
| $\lim\limits_{x \to a^+} f(x) = \infty$ | As $x$ approaches $a$ from the right, $f(x)$ is positive and increases beyond all bounds. | Figure 15-3 |
| $\lim\limits_{x \to a^+} f(x) = -\infty$ | As $x$ approaches $a$ from the right, $f(x)$ is negative and decreases beyond all bounds. | Figure 15-4 |

**EXAMPLE 15.2**   Explain why the line $x = 2$ is a vertical asymptote for the graph of $f(x) = \dfrac{3}{x - 2}$.

Consider the values of $y = f(x)$ near $x = 2$, as shown in the table:

| $x$ | 1 | 1.9 | 1.99 | 1.999 | 3 | 2.1 | 2.01 | 2.001 |
|---|---|---|---|---|---|---|---|---|
| $y$ | $-3$ | $-30$ | $-300$ | $-3000$ | 3 | 30 | 300 | 3000 |

Clearly, as $x$ approaches 2 from the left, $f(x)$ is negative and decreases beyond all bounds, and, as $x$ approaches 2 from the right, $f(x)$ is positive and increases beyond all bounds, that is, $\lim\limits_{x \to 2^-} f(x) = -\infty$ and $\lim\limits_{x \to 2^+} f(x) = \infty$. Thus $x = 2$ is a vertical asymptote for the graph.

## Horizontal Asymptotes

The line $y = a$ is a horizontal asymptote for the graph of a function $f$ if, as $x$ grows beyond all bounds, either positive or negative, $f(x)$ approaches the value $a$. The cases are shown in the following table, along with the notation generally used:

| NOTATION | MEANING | GRAPH |
|---|---|---|
| $\lim_{x \to \infty} f(x) = a$ | As $x$ increases beyond all bounds, $f(x)$ approaches the value $a$. [In the figure, $f(x) < a$ for large positive values of $x$.] | 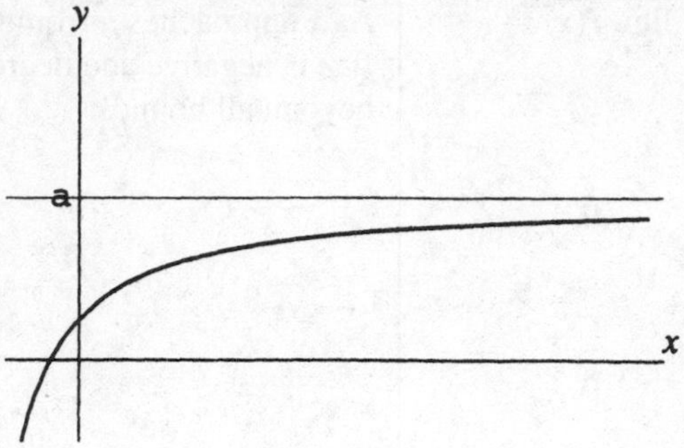 Figure 15-5 |
| $\lim_{x \to \infty} f(x) = a$ | As $x$ increases beyond all bounds, $f(x)$ approaches the value $a$. [In the figure, $f(x) > a$ for large positive values of $x$.] | 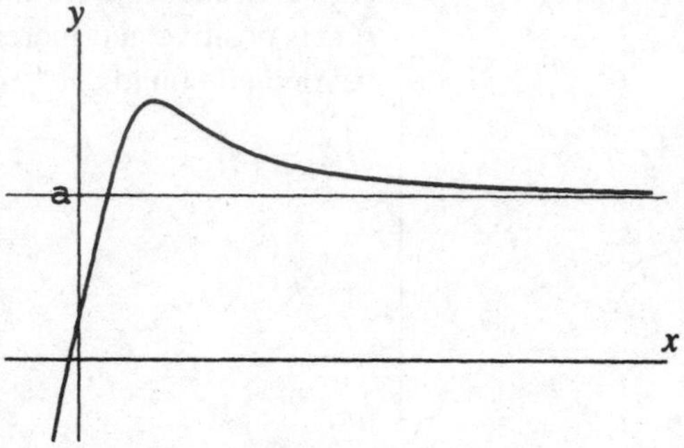 Figure 15-6 |
| $\lim_{x \to -\infty} f(x) = a$ | As $x$ decreases beyond all bounds, $f(x)$ approaches the value $a$. [In the figure, $f(x) < a$ for large negative values of $x$.] | 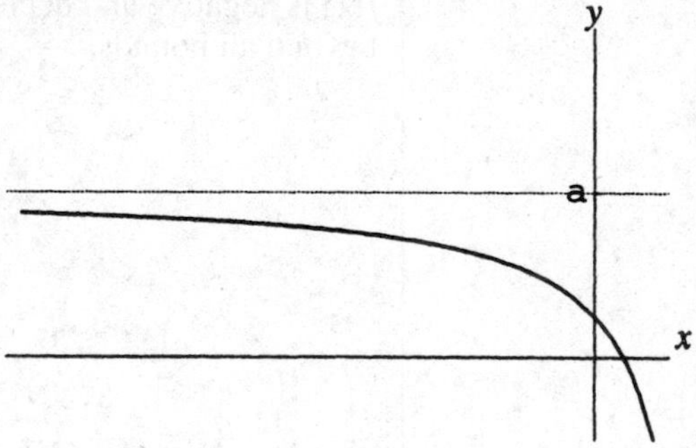 Figure 15-7 |
| $\lim_{x \to -\infty} f(x) = a$ | As $x$ decreases beyond all bounds, $f(x)$ approaches the value $a$. [In the figure, $f(x) > a$ for large negative values of $x$.] | 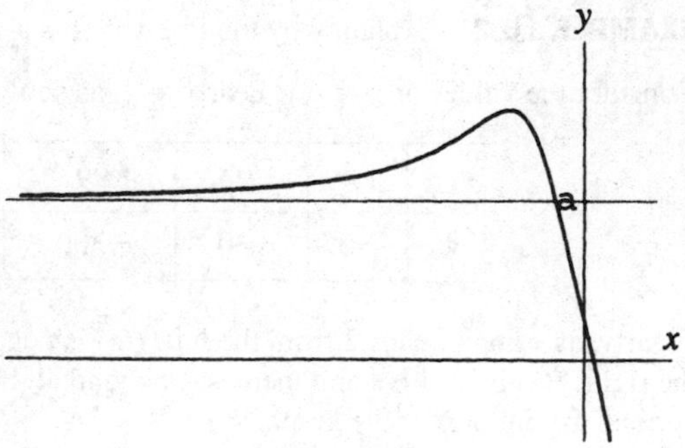 Figure 15-8 |

## Finding Horizontal Asymptotes

Let

$$f(x) = \frac{P(x)}{Q(x)} = \frac{a_n x^n + \cdots + a_1 x + a_0}{b_m x^m + \cdots + b_1 x + b_0}$$

with $a_n \neq 0$ and $b_m \neq 0$. Then

1. If $n < m$, the $x$ axis is a horizontal asymptote for the graph of $f$.
2. If $n = m$, the line $y = a_n/b_m$ is a horizontal asymptote for the graph of $f$.
3. If $n > m$, there is no horizontal asymptote for the graph of $f$. Instead, as $x \to \infty$ and as $x \to -\infty$, either $f(x) \to \infty$ or $f(x) \to -\infty$.

**EXAMPLE 15.3** Find the horizontal asymptotes, if any, for $f(x) = \dfrac{2x + 1}{x - 5}$.

Since the numerator and denominator both have degree 1, the quotient can be written as

$$f(x) = \frac{2x + 1}{x} \div \frac{x - 5}{x} = \frac{2 + \dfrac{1}{x}}{1 - \dfrac{5}{x}}$$

For large positive or negative values of $x$, this is very close to $\frac{2}{1}$, the ratio of the leading coefficients, thus $f(x) \to 2$. The line $y = 2$ is a horizontal asymptote.

## Oblique Asymptotes

Let

$$f(x) = \frac{P(x)}{Q(x)} = \frac{a_n x^n + \cdots + a_1 x + a_0}{b_m x^m + \cdots + b_1 x + b_0}$$

with $a_n \neq 0$ and $b_m \neq 0$. Then, if $n = m + 1$, $f(x)$ can be expressed using long division (see Chapter 14) in the form:

$$f(x) = ax + b + \frac{R(x)}{Q(x)}$$

where the degree of $R(x)$ is less than the degree of $Q(x)$. Then, as $x \to \infty$ or $x \to -\infty$, $f(x) \to ax + b$ and the line $y = ax + b$ is an oblique asymptote for the graph of the function.

**EXAMPLE 15.4** Find the oblique asymptote for the graph of the function $f(x) = \dfrac{x^3 + 1}{x^2 + x - 2}$.

Use the long division scheme to write $f(x) = x - 1 + \dfrac{3x - 1}{x^2 + x - 2}$. Hence, as $x \to \infty$ or $x \to -\infty$, $f(x) \to x - 1$, and the line $y = x - 1$ is an oblique asymptote for the graph of the function.

## Graphing a Rational Function

To sketch the graph of a rational function $y = f(x) = \dfrac{P(x)}{Q(x)}$:

1. Find any $x$-intercepts for the graph [the real zeros of $P(x)$] and plot the corresponding points. Find the $y$-intercept [$f(0)$, assuming 0 is in the domain of $f$] and plot the point $(0, f(0))$. Analyze the function for any symmetry with respect to the axes or the origin.
2. Find any real zeros of $Q(x)$ and enter any vertical asymptotes for the graph on the sketch.
3. Find any horizontal or oblique asymptote for the graph and enter this on the sketch.

4. Determine whether the graph intersects the horizontal or oblique asymptote. The graphs of $y = f(x)$ and $y = ax + b$ will intersect at real solutions of $f(x) = ax + b$.
5. Determine, from a sign chart if necessary, the intervals in which the function is positive and negative. Determine the behavior of the function near the asymptotes.
6. Sketch the graph of $f$ in each of the regions found in step 5.

**EXAMPLE 15.5**    Sketch the graph of the function $f(x) = -12/x$.

1. The graph has no $x$-intercepts or $y$-intercepts. Since $f(-x) = -f(x)$, the function is odd and the graph has origin symmetry.
2. Since 0 is the only zero of the denominator, the $y$-axis, $x = 0$, is the only vertical asymptote.
3. Since the degree of the denominator is greater than the degree of the numerator, the $x$-axis, $y = 0$, is the horizontal asymptote.
4. Since there is no solution to the equation $-12/x = 0$, the graph does not intersect the horizontal asymptote.
5. If $x$ is negative, $f(x)$ is positive. If $x$ is positive, $f(x)$ is negative. Hence, $\lim_{x \to 0^-} f(x) = \infty$ and $\lim_{x \to 0^+} f(x) = -\infty$.
6. Sketch the graph (Fig. 15-9).

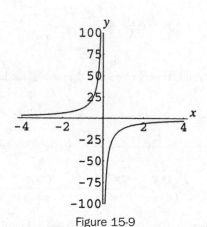

Figure 15-9

## SOLVED PROBLEMS

**15.1.** Find any vertical asymptotes for the graph of

   (a) $f(x) = \dfrac{x}{x^2 - 4}$      (b) $f(x) = \dfrac{2x}{x^2 + 4}$      (c) $f(x) = \dfrac{2x - 1}{x^2 - x - 2}$      (d) $f(x) = \dfrac{3}{x^3 + 8}$

   (a) Since the real zeros of $x^2 - 4$ are $\pm 2$, the vertical asymptotes are $x = \pm 2$.

   (b) Since $x^2 + 4$ has no real zeros, there are no vertical asymptotes.

   (c) Since the real zeros of $x^2 - x - 2$ are 2 and $-1$, the vertical asymptotes are $x = 2$ and $x = -1$.

   (d) Since the only real zero of $x^3 + 8$ is $-2$, the only vertical asymptote is $x = -2$.

**15.2.** Find any vertical asymptotes for the graph of $f(x) = \dfrac{x^2 - x}{x^2 - 1}$.

It would seem as if the graph has vertical asymptotes $x = \pm 1$, since these are the real zeros of the denominator polynomial. However, the expression for the function is not in lowest terms, in fact,

$$f(x) = \frac{x(x - 1)}{(x + 1)(x - 1)} = \frac{x}{x + 1} \text{ if } x \neq 1$$

Since, as $x \to 1^+$ or $x \to 1^-$, the function value does not increase or decrease beyond all bounds, the line $x = 1$ is not a vertical asymptote, and the only vertical asymptote is $x = -1$.

**15.3.** Find any horizontal asymptotes for the graph of

(a) $f(x) = \dfrac{4x^2}{x^2 + 4}$     (b) $f(x) = \dfrac{x^2}{x + 4}$     (c) $f(x) = \dfrac{2x}{x^2 - 4}$     (d) $f(x) = \dfrac{3x^2 + 5x + 2}{4x^2 + 1}$

(a) Since numerator and denominator both have degree 2, the quotient can be written as

$$f(x) = \frac{4x^2}{x^2} \div \frac{x^2 + 4}{x^2} = \frac{4}{1 + \dfrac{4}{x^2}}$$

For large positive or negative values of $x$, this is very close to $\frac{4}{1}$, the ratio of the leading coefficients, thus $f(x) \to x - 4$. The line $y = 4$ is a horizontal asymptote.

(b) Since the degree of the numerator is greater than the degree of the denominator, the graph has no horizontal asymptote.

(c) Since the degree of the numerator is less than the degree of the denominator, the $x$-axis, $y = 0$, is the horizontal asymptote.

(d) Since numerator and denominator both have degree 2, the quotient can be written as

$$f(x) = \frac{3x^2 + 5x + 2}{x^2} \div \frac{4x^2 + 1}{x^2} = \frac{3 + \dfrac{5}{x} + \dfrac{2}{x^2}}{4 + \dfrac{1}{x^2}}$$

For large positive or negative values of $x$, this is very close to $\frac{3}{4}$, the ratio of the leading coefficients; thus $f(x) \to \frac{3}{4}$. The line $y = \frac{3}{4}$ is a horizontal asymptote.

**15.4.** Find any oblique asymptotes for the graph of

(a) $f(x) = \dfrac{x^2}{x + 4}$     (b) $f(x) = \dfrac{x^3}{x + 4}$     (c) $f(x) = \dfrac{x^2 - 5x + 3}{2x - 5}$     (d) $f(x) = \dfrac{2x^3 - x}{x^2 + 2x + 1}$

(a) Use the synthetic division scheme to write $f(x) = x - 4 + \dfrac{16}{x + 4}$. Hence, as $x \to \infty$ or $x \to -\infty$, $f(x) \to x - 4$, and the line $y = x - 4$ is an oblique asymptote for the graph of the function.

(b) Since the degree of the numerator is not equal to 1 more than the degree of the denominator, the graph does not have an oblique asymptote. However, if the synthetic division scheme is used to write

$$f(x) = x^2 - 4x + 16 + \frac{-64}{x + 4}$$

then, as $x \to \infty$ or $x \to -\infty$, $f(x) \to x^2 - 4x + 16$. The graph of $f$ then is said to *approach asymptotically* the curve $y = x^2 - 4x + 16$.

(c) Use the long division scheme to write $f(x) = \dfrac{1}{2}x - \dfrac{5}{4} + \dfrac{-\frac{13}{4}}{2x - 5}$. Hence, as $x \to \infty$ or $x \to -\infty$, $f(x) \to \frac{1}{2}x - \frac{5}{4}$, and the line $y = \frac{1}{2}x - \frac{5}{4}$ is an oblique asymptote for the graph of the function.

(d) Use the long division scheme to write $f(x) = 2x - 4 + \dfrac{5x + 4}{x^2 + 2x + 1}$. Hence, as $x \to \infty$ or $x \to -\infty$, $f(x) \to 2x - 4$, and the line $y = 2x - 4$ is an oblique asymptote for the graph of the function.

**15.5.** Sketch a graph of $f(x) = \dfrac{4}{x + 2}$.

Apply the steps listed above for sketching the graph of a rational function.
Since $f(0) = 2$, the $y$-intercept is 2. Since $f(x)$ is never 0, there is no $x$-intercept. The graph has no symmetry with respect to axes or origin.
Since $x + 2 = 0$ when $x = -2$, this line is the only vertical asymptote.

Since the degree of the denominator is greater than the degree of the numerator, the $x$-axis is the horizontal asymptote.

Since $f(x) = 0$ has no solutions, the graph does not cross its horizontal asymptote.

A sign chart shows that the values of the function are negative on $(-\infty, -2)$ and positive on $(-2, \infty)$. Thus, $\lim\limits_{x \to -2^-} f(x) = -\infty$ and $\lim\limits_{x \to -2^+} f(x) = \infty$.

The graph is shown in Fig. 15-10.

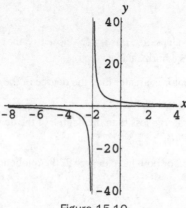

Figure 15-10

**15.6.** Sketch a graph of $f(x) = -\dfrac{3}{x^2}$.

The graph has no $x$- intercepts or $y$-intercepts. Since $f(-x) = f(x)$, the function is even and the graph has $y$-axis symmetry.

Since $x^2 = 0$ when $x = 0$, the $y$-axis is the only vertical asymptote.

Since the degree of the denominator is greater than the degree of the numerator, the $x$-axis is the horizontal asymptote.

Since $f(x) = 0$ has no solutions, the graph does not cross its horizontal asymptote.

Since $x^2$ is never negative, the function values are negative throughout the domain. Thus, $\lim\limits_{x \to 0^-} f(x) = -\infty$ and $\lim\limits_{x \to 0^+} f(x) = -\infty$.

The graph is shown in Fig. 15-11.

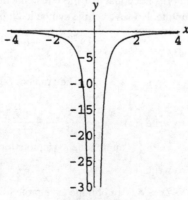

Figure 15-11

**15.7.** Sketch the graph of $f(x) = \dfrac{x + 3}{x - 2}$.

Since $f(0) = -\dfrac{3}{2}$, the $y$-intercept is $-\dfrac{3}{2}$. Since $f(x) = 0$ if $x = -3$, the $x$-intercept is $-3$. The graph has no symmetry with respect to axes or origin.

Since $x - 2 = 0$ when $x = 2$, this line is the only vertical asymptote.

Since the numerator and denominator both have degree 1, and the ratio of leading coefficients is $\frac{1}{1}$, or 1, the line $y = 1$ is the horizontal asymptote.

Since $f(x) = 1$ has no solutions, the graph does not cross its horizontal asymptote.

A sign chart shows that the values of the function are positive on $(-\infty, -3)$ and $(2, \infty)$ and negative on $(-3, 2)$. Thus, $\lim\limits_{x \to 2^-} f(x) = -\infty$ and $\lim\limits_{x \to 2^+} f(x) = \infty$.

The graph is shown in Fig. 15-12.

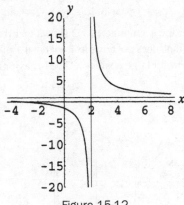

Figure 15-12

**15.8.** Sketch the graph of $f(x) = \dfrac{2x}{x^2 - 4}$.

Since $f(0) = 0$, and this is the only zero of the function, the $x$-intercept and the $y$-intercept are both 0, that is, the graph passes through the origin. Since $f(-x) = -f(x)$, the function is odd and the graph has origin symmetry.

Since $x^2 - 4 = 0$ when $x = \pm 2$, these lines are vertical asymptotes for the graph.

Since the degree of the denominator is greater than the degree of the numerator, the $x$-axis is the horizontal asymptote.

Since $f(x) = 0$ has the solution 0, the graph crosses its horizontal asymptote at the origin.

A sign chart shows that the values of the function are positive on both $(-2, 0)$ and $(2, \infty)$ and negative on both $(-\infty, -2)$ and $(0, 2)$. Thus, $\lim\limits_{x \to -2^-} f(x) = -\infty$ and $\lim\limits_{x \to -2^+} f(x) = \infty$; also, $\lim\limits_{x \to 2^-} f(x) = -\infty$ and $\lim\limits_{x \to 2^+} f(x) = \infty$.

The graph is shown in Fig. 15-13.

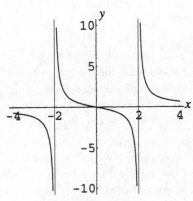

Figure 15-13

**15.9.** Sketch the graph of $f(x) = \dfrac{-2x^2}{x^2 - 4}$.

Since $f(0) = 0$, the $x$-intercept and the $y$-intercept are both 0, that is, the graph passes through the origin. Since $f(-x) = f(x)$, the function is even and the graph has $y$-axis symmetry.

Since $x^2 - 4 = 0$ when $x = \pm 2$, these lines are vertical asymptotes for the graph.

Since the numerator and denominator both have degree 2, and the ratio of leading coefficients is $-\frac{2}{1}$, or $-2$, the line $y = -2$ is the horizontal asymptote.

Since $f(x) = -2$ has no solutions, the graph does not cross its horizontal asymptote.

A sign chart shows that the values of the function are positive on $(-2, 2)$ and negative on $(-\infty, -2)$ and $(2, \infty)$. Thus, $\lim\limits_{x \to -2^-} f(x) = -\infty$ and $\lim\limits_{x \to -2^+} f(x) = \infty$, also $\lim\limits_{x \to 2^-} f(x) = \infty$ and $\lim\limits_{x \to 2^+} f(x) = -\infty$.

Moreover, since the behavior near the asymptote $x = 2$ shows that the function values are large and negative for $x$ greater than 2, and since the graph does not cross its horizontal asymptote, the graph must therefore approach the horizontal asymptote from *below* for large positive $x$. The behavior for large negative $x$ is the same, since the function is even.

The graph is shown in Fig. 15-14.

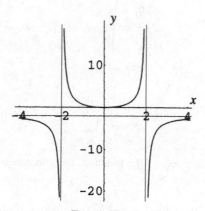

Figure 15-14

**15.10.** Sketch the graph of $f(x) = \dfrac{x^3}{x^2 - 4}$.

Since $f(0) = 0$, and this is the only zero of the function, the $x$-intercept and the $y$-intercept are both 0; that is, the graph passes through the origin. Since $f(-x) = -f(x)$, the function is odd and the graph has origin symmetry.

Since $x^2 - 4 = 0$ when $x = \pm 2$, these lines are vertical asymptotes for the graph.

Since the degree of the numerator is 1 more than the degree of the denominator, the graph has an oblique asymptote. Long division shows that

$$f(x) = \frac{x^3}{x^2 - 4} = x + \frac{4x}{x^2 - 4}$$

Thus, as $x \to \infty$, $f(x) \to x$, and the line $y = x$ is the oblique asymptote.

Since $f(x) = x$ has the solution 0, the graph crosses the oblique asymptote at the origin.

A sign chart shows that the values of the function are positive on $(-2, 0)$ and $(2, \infty)$ and negative on $(-\infty, -2)$ and $(0, 2)$: $\lim\limits_{x \to -2^-} f(x) = -\infty$ and $\lim\limits_{x \to -2^+} f(x) = \infty$; also, $\lim\limits_{x \to 2^-} f(x) = -\infty$ and $\lim\limits_{x \to 2^+} f(x) = \infty$.

Moreover, since the behavior near the asymptote $x = 2$ shows that the function values are large and positive for $x$ greater than 2, and since the graph does not cross its oblique asymptote here, the graph must therefore approach the oblique asymptote from *above* for large positive $x$. Since the function is odd, the graph must therefore approach the oblique asymptote from *below* for large negative $x$.

The graph is shown in Fig. 15-15.

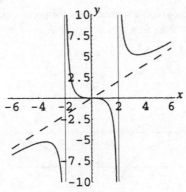

Figure 15-15

**15.11.** Sketch a graph of $f(x) = \dfrac{x^2 + x}{x^2 - 3x + 2}$.

Since $f(0) = 0$, the y-intercept is 0. Since $x^2 + x = 0$ when $x = 0$ and $-1$, these are both x-intercepts. The graph passes through the origin. There is no obvious symmetry.

Since $x^2 - 3x + 2 = 0$ when $x = 1$ and 2, these lines are vertical asymptotes.

Since the numerator and denominator both have degree 2, and the ratio of leading coefficients is $\frac{1}{1}$, or 1, the line $y = 1$ is the horizontal asymptote.

Since $f(x) = 1$ has the solution $\frac{1}{2}$, the graph crosses the horizontal asymptote at $\left(\frac{1}{2}, 1\right)$.

A sign chart shows that the values of the function are negative on $(-1, 0)$ and $(1, 2)$, and positive on $(-\infty, -1)$, $(0, 1)$, and $(2, \infty)$. Thus, $\lim\limits_{x \to 1^-} f(x) = \infty$ and $\lim\limits_{x \to 1^+} f(x) = -\infty$; also, $\lim\limits_{x \to 2^-} f(x) = -\infty$ and $\lim\limits_{x \to 2^+} f(x) = \infty$.

Moreover, since the behavior near the asymptote $x = 2$ shows that the function values are large and positive for $x$ greater than 2, and since the graph does not cross its horizontal asymptote here, the graph must therefore approach the horizontal asymptote from *above* for large positive $x$. Similarly, the graph must approach the horizontal asymptote from *below* for large negative $x$.

The graph is shown in Fig. 15-16.

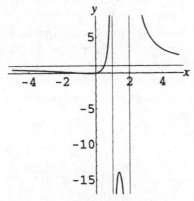

Figure 15-16

**15.12.** Sketch a graph of $f(x) = \dfrac{x^2 - 9}{x^2 + 4}$.

Since $f(0) = -\frac{9}{4}$, the y-intercept is $-\frac{9}{4}$. Since $x^2 - 9 = 0$ when $x = \pm 3$, these are both x-intercepts. Since $f(-x) = f(x)$, the function is even and the graph has y-axis symmetry.

Since $x^2 + 4$ has no real zeros, the graph has no vertical asymptotes.

Since the numerator and denominator both have degree 2, and the ratio of leading coefficients is $\frac{1}{1}$, or 1, the line $y = 1$ is the horizontal asymptote.

Since $f(x) = 1$ has no solutions, the graph does not cross the horizontal asymptote.
A sign chart shows that the values of the function are positive on $(-\infty, -3)$ and $(3, \infty)$ and negative on $(-3, 3)$.
The graph is shown in Fig. 15-17.

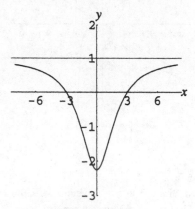

Figure 15-17

**15.13.** To understand an example of the special case when the numerator and denominator of a rational expression have factors in common, analyze and sketch a graph of $f(x) = \dfrac{x^2 - 4}{x^2 - 3x + 2}$.

Factoring numerator and denominator yields

$$f(x) = \frac{(x - 2)(x + 2)}{(x - 2)(x - 1)} = \frac{x + 2}{x - 1} \quad \text{for } x \neq 2$$

Thus, the graph of the function is identical with the graph of $g(x) = (x + 2)/(x - 1)$, except that 2 is not in the domain of $f$. Draw a graph of $y = g(x)$. The graph of $y = f(x)$ is conventionally shown as the graph of $g$ with a small circle centered at $(2,4)$ to indicate that this point is not on the graph.
Since $g(0) = -2$, the $y$-intercept is $-2$. Since $g(x) = 0$ if $x = -2$, the $x$-intercept is $-2$. The graph has no symmetry with respect to axes or origin.
Since $x - 1 = 0$ when $x = 1$, this line is the only vertical asymptote.
Since the numerator and denominator both have degree 1, and the ratio of leading coefficients is $\frac{1}{1}$, or 1, the line $y = 1$ is the horizontal asymptote.
Since $g(x) = 1$ has no solutions, the graph does not cross its horizontal asymptote.
A sign chart shows that the values of the function are positive on $(-\infty, -2)$ and $(1, \infty)$ and negative on $(-2, 1)$.
Thus, $\lim\limits_{x \to 1^-} g(x) = -\infty$ and $\lim\limits_{x \to 1^+} g(x) = \infty$.
The graph is shown in Fig. 15-18.

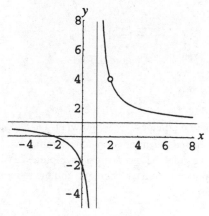

Figure 15-18

**15.14.** Find all intercepts for the graphs of the following rational functions:

(a) $f(x) = \dfrac{2x}{x + 4}$; (b) $f(x) = \dfrac{4x^2 - 1}{x^2 + 4}$; (c) $f(x) = \dfrac{x - 1}{x^2 - 4x}$; (d) $f(x) = \dfrac{x^3 + 27}{x^4 - 5x^2 + 4}$

*Ans.*   (a)  $x$-intercept: 0, $y$-intercept: 0; (b)  $x$-intercept: $\pm\frac{1}{2}$, $y$-intercept: $-\frac{1}{4}$;

(c)  $x$-intercept: 1, $y$-intercept: none; (d)  $x$-intercept: $-3$, $y$-intercept: $\frac{27}{4}$

**15.15.** Find all horizontal and vertical asymptotes for the graphs in the previous problem.

(a) horizontal: $y = 2$, vertical: $x = -4$; (b) horizontal: $y = 4$, vertical: none;

(c) horizontal: $y = 0$, vertical: $x = 0$, $x = 4$; (d) horizontal: $y = 0$, vertical: $x = \pm 1$, $x = \pm 2$

**15.16.** (a)  State intercepts and asymptotes and sketch the graph of $f(x) = \dfrac{2x}{x - 2}$.

(b)  Show that $f$ is one-to-one on its domain and that $f(x) = f^{-1}(x)$.

*Ans.*   (a)  Intercepts: the origin.
Asymptotes: $x = 2$, $y = 2$.
The graph is shown in Fig. 15-19.

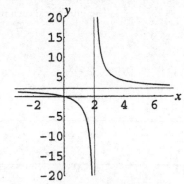

Figure 15-19

**15.17.** State intercepts and asymptotes and sketch the graph of $f(x) = \dfrac{2x^2}{x - 2}$.

*Ans.*   Intercepts: the origin.
Asymptotes: $x = 2$, $y = 2x + 4$.
The graph is shown in Fig. 15-20.

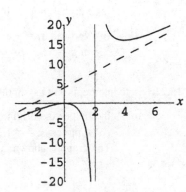

Figure 15-20

**15.18.** State intercepts and asymptotes and sketch the graph of $f(x) = \dfrac{2}{(x-2)^2}$.

Ans. Intercepts: $(0, \frac{1}{2})$.
Asymptotes: $x = 2$, $y = 0$.
The graph is shown in Fig. 15-21.

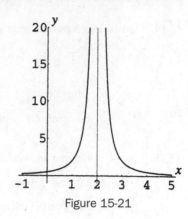

Figure 15-21

**15.19.** State intercepts and asymptotes and sketch the graph of $f(x) = \dfrac{2}{x^2 - 1}$.

Ans. Intercepts: $(0, -2)$.
Asymptotes: $x = \pm 1$, $y = 0$.
The graph is shown in Fig. 15-22.

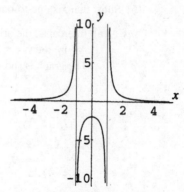

Figure 15-22

**15.20.** Find all vertical and oblique asymptotes for the graphs of the following rational functions:

(a) $f(x) = \dfrac{x^2}{x+2}$; (b) $f(x) = \dfrac{x^2 - 4x}{x - 1}$; (c) $f(x) = \dfrac{8x^3 - 1}{x^2 + 4}$; (d) $f(x) = \dfrac{x^4 - 5x^2 + 6}{x^3 + x^2}$; (e) $f(x) = \dfrac{x^3 - 2x}{x + 6}$

Ans. (a) vertical: $x = -2$, oblique: $y = x - 2$; (b) vertical: $x = 1$, oblique: $y = x - 3$;

(c) vertical: none, oblique: $y = 8x$; (d) vertical: $x = 0$, $x = -1$, oblique: $y = x - 1$;

(e) vertical: $x = -6$, oblique: none, however, the graph approaches asymptotically the graph
of $y = x^2 - 6x + 34$

**15.21.** State intercepts and asymptotes and sketch the graph of $f(x) = \dfrac{x^3}{x^2 - 1}$.

Ans. Intercepts: the origin.
Asymptotes: $x = \pm 1$, $y = x$.
The graph is shown in Fig. 15-23.

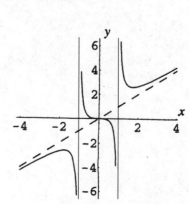

Figure 15-23

**15.22.** State intercepts and asymptotes and sketch the graph of $f(x) = \dfrac{3x}{x^2 + 1}$.

*Ans.* Intercepts: the origin.
Asymptotes: $y = 0$.
The graph is shown in Fig. 15-24.

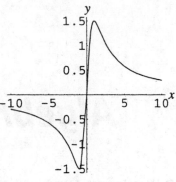

Figure 15-24

**15.23.** State intercepts and asymptotes and sketch the graph of $f(x) = \dfrac{x^3 - x^2 - x + 1}{x^2 + 1}$.

*Ans.* Intercepts: $(0,1)$, $(1,0)$, $(-1,0)$.
Asymptotes: $y = x - 1$.
The graph is shown in Fig. 15-25.

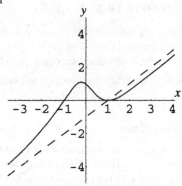

Figure 15-25

**15.24.** A field is to be marked off in the shape of a rectangle of area 144 square feet.

(a) Write an expression for the perimeter $P$ as a function of the length $x$.

(b) Sketch a graph of the perimeter function and determine approximately from the graph the dimensions for which the perimeter is a minimum.

*Ans.* (a) $P(x) = 2x + \dfrac{288}{x}$

(b) See Fig. 15-26. Dimensions: 12 feet by 12 feet.

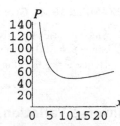

Figure 15-26

# CHAPTER 16

# Algebraic Functions; Variation

## Definition of Algebraic Function

An algebraic function is any function for which the rule is a polynomial or can be derived from polynomials by addition, subtraction, multiplication, division, or raising to an integer or rational exponent.

**EXAMPLE 16.1**   Examples of algebraic functions include:

(a) Polynomial functions, such as $f(x) = 5x^2 - 3x$
(b) Rational functions, such as $f(x) = 12/x^2$
(c) Absolute value functions, such as $f(x) = |x - 3|$, since $|x - 3| = \sqrt{(x - 3)^2}$
(d) Other functions involving rational powers, such as $f(x) = \sqrt{x}$, $f(x) = \sqrt[3]{x}$, $f(x) = 1/\sqrt{x}$, $f(x) = \sqrt{1 - x^2}$, and so on

## Variation

The term *variation* is used to describe many forms of simple functional dependence. The general pattern is that one variable, called the *dependent* variable, is said to vary as a result of changes in one or more other variables, called the *independent* variables. Variation statements always include a nonzero constant multiple, referred to as the *constant of variation,* or *constant of proportionality,* and often denoted $k$.

## Direct Variation

To describe a relation of the form $y = kx$, the following language is used:

1. $y$ varies directly as $x$ (occasionally, $y$ varies as $x$).
2. $y$ is directly proportional to $x$.

**EXAMPLE 16.2**   Given that $p$ varies directly as $q$, find an expression for $p$ in terms of $q$ if $p = 300$ when $q = 12$.

1. Since $p$ varies directly as $q$, write $p = kq$.
2. Since $p = 300$ when $q = 12$, substitute these values to obtain $300 = k(12)$, or $k = 25$.
3. Hence $p = 25q$ is the required expression.

## Inverse Variation

To describe a relation of the form $xy = k$, or $y = k/x$, the following language is used:

1. $y$ varies inversely as $x$.
2. $y$ is inversely proportional to $x$.

**EXAMPLE 16.3**   Given that $s$ varies inversely as $t$, find an expression for $s$ in terms of $t$ if $s = 5$ when $t = 8$.

1. Since $s$ varies inversely as $t$, write $s = k/t$.
2. Since $s = 5$ when $t = 8$, substitute these values to obtain $5 = k/8$, or $k = 40$.
3. Hence $s = 40/t$ is the required expression.

## Joint Variation

To describe a relation of the form $z = kxy$, the following language is used:

1. $z$ varies jointly as $x$ and $y$.
2. $z$ varies directly as the product of $x$ and $y$.

**EXAMPLE 16.4**    Given that $z$ varies jointly as $x$ and $y$ and $z = 3$ when $x = 4$ and $y = 5$, find an expression for $z$ in terms of $x$ and $y$.

1. Since $z$ varies jointly as $x$ and $y$, write $z = kxy$.
2. Since $z = 3$ when $x = 4$ and $y = 5$, substitute these values to obtain $3 = k \cdot 4 \cdot 5$, or $k = \frac{3}{20}$.
3. Hence $z = \frac{3}{20} xy$.

## Combined Variation

These types of variation can also be combined.

**EXAMPLE 16.5**    Given that $z$ varies directly as the square of $x$ and inversely as $y$ and $z = 5$ when $x = 3$ and $y = 12$, find an expression for $z$ in terms of $x$ and $y$.

1. Write $z = \frac{kx^2}{y}$.
2. Since $z = 5$ when $x = 3$ and $y = 12$, substitute these values to obtain $5 = k \cdot \frac{3^2}{12}$ or $k = 20/3$.
3. Hence $z = \frac{20x^2}{3y}$.

## SOLVED PROBLEMS

**16.1.** State the domain and range, and sketch a graph for:

     (a) $f(x) = \sqrt{x}$; (b) $f(x) = \sqrt[3]{x}$; (c) $f(x) = \sqrt[4]{x}$; (d) $f(x) = \sqrt[5]{x}$

     (a) Domain $[0, \infty)$
         Range: $[0, \infty)$
         The graph is shown in Fig. 16-1.

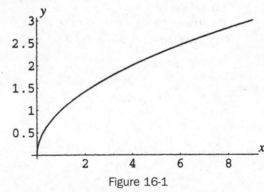

Figure 16-1

     (b) Domain: $R$
         Range: $R$
         The graph is shown in Fig. 16-2.

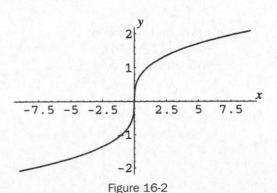

Figure 16-2

(c) Domain: $[0, \infty)$
    Range: $[0, \infty)$
    The graph is shown in Fig. 16-3.

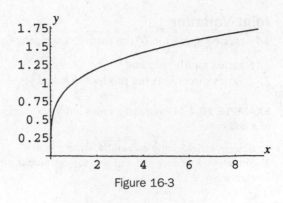

Figure 16-3

(d) Domain: **R**
    Range: **R**
    The graph is shown in Fig. 16-4.

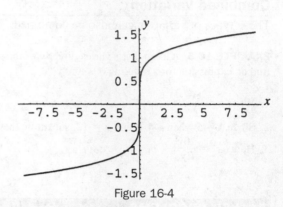

Figure 16-4

**16.2.** State the domain and range, and sketch a graph for:

(a) $f(x) = 1/\sqrt{x}$     (b) $f(x) = 1/\sqrt[3]{x}$

(a) Domain: $(0, \infty)$
    Range: $(0, \infty)$
    The graph is shown in Fig. 16-5.

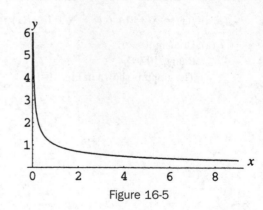

Figure 16-5

(b) Domain: $\{x \in \mathbf{R} | x \neq 0\}$
    Range: $\{y \in \mathbf{R} | y \neq 0\}$
    The graph is shown in Fig. 16-6.

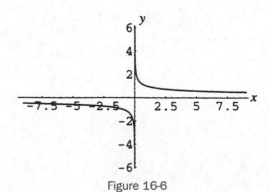

Figure 16-6

**16.3.** Analyze and sketch a graph for (a) $f(x) = \sqrt{9 - x^2}$; (b) $f(x) = -\sqrt{9 - x^2}$.

(a) If $y = \sqrt{9 - x^2}$, then $x^2 + y^2 = 9$, $y \geq 0$.
Thus, the graph of the function is the
upper half (semicircle) of the graph of
$x^2 + y^2 = 9$.
The domain is $\{x \in R| -3 \leq x \leq 3\}$ and
the range is $\{y \in R| 0 \leq y \leq 3\}$.
The graph is shown in Fig. 16-7.

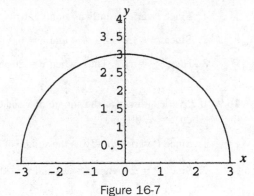

Figure 16-7

(b) If $y = -\sqrt{9 - x^2}$, then $x^2 + y^2 = 9$,
$y \leq 0$. Thus, the graph of the function
is the lower half (semicircle) of the
graph of $x^2 + y^2 = 9$.
The domain is $\{x \in R| -3 \leq x \leq 3\}$
and the range is $\{y \in R| -3 \leq y \leq 0\}$.
The graph is shown in Fig. 16-8.

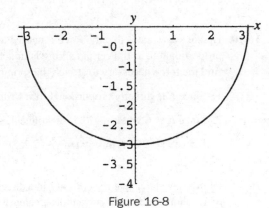

Figure 16-8

**16.4.** If $s$ varies directly as the square of $x$ and $s = 5$ when $x = 4$, find $s$ when $x = 20$.

1. Since $s$ varies directly as the square of $x$, write $s = kx^2$.

2. Since $s = 5$ when $x = 4$, substitute these values to obtain $5 = k \cdot 4^2$, or $k = \frac{5}{16}$.

3. Hence $s = 5x^2/16$. Thus, when $x = 20$, $s = 5(20)^2/16 = 125$.

**16.5.** If $y$ is directly proportional to the cube root of $x$ and $y = 12$ when $x = 64$, find $y$ when $x = \frac{1}{8}$.

1. Since $y$ is directly proportional to the cube root of $x$, write $y = k\sqrt[3]{x}$.

2. Since $y = 12$ when $x = 64$, substitute these values to obtain $12 = k\sqrt[3]{64} = 4k$, or $k = 3$.

3. Hence $y = 3\sqrt[3]{x}$. Thus, when $x = \frac{1}{8}$, $y = 3\sqrt[3]{1/8} = \frac{3}{2}$.

**16.6.** If $I$ is inversely proportional to the square of $t$, and $I = 100$ when $t = 15$, find $I$ when $t = 12$.

1. Since $I$ is inversely proportional to the square of $t$, write $I = \dfrac{k}{t^2}$.

2. Since $I = 100$ when $t = 15$, substitute these values to obtain $100 = \dfrac{k}{15^2}$, or $k = 22{,}500$.

3. Hence $I = \dfrac{22{,}500}{t^2}$. Thus, when $t = 12$, $I = \dfrac{22{,}500}{12^2} = 156.25$.

**16.7.** If $u$ varies inversely as the cube root of $x$, and $u = 56$ when $x = -8$, find $u$ when $x = 1000$.

1. Since $u$ varies inversely as the cube root of $x$, write $u = \dfrac{k}{\sqrt[3]{x}}$.

2. Since $u = 56$ when $x = -8$, substitute these values to obtain $56 = \dfrac{k}{\sqrt[3]{-8}}$, or $k = -112$.

3. Hence $u = \dfrac{-112}{\sqrt[3]{x}}$. Thus, when $x = 1000$, $u = \dfrac{-112}{\sqrt[3]{1000}} = -11.2$.

**16.8.** If $z$ varies jointly as $x$ and $y$, and $z = 3$ when $x = 4$ and $y = 6$, find $z$ when $x = 20$ and $y = 9$.

   1. Since $z$ varies jointly as $x$ and $y$, write $z = kxy$.

   2. Since $z = 3$ when $x = 4$ and $y = 6$, substitute these values to obtain $3 = k \cdot 4 \cdot 6$ or $k = \frac{1}{8}$.

   3. Hence $z = xy/8$. Thus, when $x = 20$ and $y = 9$, $z = (20 \cdot 9)/8 = 22.5$.

**16.9.** If $P$ varies jointly as the square of $x$ and the fourth root of $y$, and $P = 24$ when $x = 12$ and $y = 81$, find $P$ when $x = 1200$ and $y = \frac{1}{16}$.

   1. Since $P$ varies jointly as the square of $x$ and the fourth root of $y$, write $P = kx^2 \sqrt[4]{y}$.

   2. Since $P = 24$ when $x = 12$ and $y = 81$, substitute these values to obtain $24 = k \cdot 12^2 \sqrt[4]{81}$ or $k = \frac{1}{18}$.

   3. Hence $P = \dfrac{x^2 \sqrt[4]{y}}{18}$. Thus, when $x = 1200$ and $y = \frac{1}{16}$, $P = \dfrac{(1200)^2 \sqrt[4]{1/16}}{18} = 40{,}000$.

**16.10.** Hooke's law states that the force $F$ needed to stretch a spring $x$ units beyond its natural length is directly proportional to $x$. If a certain spring is stretched 0.5 inches from its natural length by a force of 6 pounds, find the force necessary to stretch the spring 2.25 inches.

   1. Since $F$ is directly proportional to $x$, write $F = kx$.

   2. Since $F = 6$ when $x = 0.5$, substitute these values to obtain $6 = k(0.5)$ or $k = 12$.

   3. Hence $F = 12x$. Thus, when $x = 2.25$, $F = 12(2.25) = 27$ pounds.

**16.11.** Ohm's law states that the current $I$ in a direct-current circuit varies inversely as the resistance $R$. If a resistance of 12 ohms produces a current of 3.5 amperes, find the current when the resistance is 2.4 ohms.

   1. Since $I$ varies inversely as $R$, write $I = k/R$.

   2. Since $I = 3.5$ when $R = 12$, substitute these values to obtain $3.5 = k/12$ or $k = 42$.

   3. Hence $I = 42/R$. Thus, when $R = 2.4$, $I = 42/2.4 = 17.5$ amperes.

**16.12.** The pressure $P$ of wind on a wall varies jointly as the area $A$ of the wall and the square of the velocity $v$ of the wind. If $P = 100$ pounds when $A = 80$ square feet and $v = 40$ miles per hour, find $P$ if $A = 120$ square feet and $v = 50$ miles per hour.

   1. Since $P$ varies jointly as $A$ and $v$, write $P = kAv^2$.

   2. Since $P = 100$ when $A = 80$ and $v = 40$, substitute these values to obtain $100 = k \cdot 80 \cdot 40^2$ or $k = 1/1280$.

   3. Hence $P = Av^2/1280$. Thus, when $A = 120$ and $v = 50$, $P = 120 \cdot 50^2/1280 = 234.375$ pounds.

**16.13.** The weight $w$ of an object on or above the surface of the earth varies inversely as the square of the distance $d$ of the object from the center of the earth. If an astronaut weighs 120 pounds at the surface of the earth, how much (to the nearest pound) would she weigh in a satellite 400 miles above the surface? (Use 4000 miles as the radius of the earth.)

   1. Since $w$ varies inversely as the square of $d$, write $w = k/d^2$.

   2. Since $w = 120$ at the surface of the earth, when $d = 4000$, substitute these values to obtain $120 = k/4000^2$, or $k = 1.92 \times 10^9$.

   3. Hence $w = 1.92 \times 10^9/d^2$. Thus, when $d = 4000 + 400 = 4400$, $w = 1.92 \times 10^9/4400^2$, or approximately 99 pounds.

**16.14.** The volume $V$ of a given mass of gas varies directly as the temperature $T$ and inversely as the pressure $P$. If a gas has volume 16 cubic inches when the temperature is 320°K and the pressure is 300 pounds per square inch, find the volume when the temperature is 350°K and the pressure is 280 pounds per square inch.

    1. Since $V$ varies directly as $T$ and inversely as $P$, write $V = kT/P$.

    2. Since $V = 16$ when $T = 320$ and $P = 300$, substitute these values to obtain $16 = k \cdot 320/300$ or $k = 15$.

    3. Hence $V = 15T/P$. Thus, when $T = 350$ and $P = 280$, $V = 15 \cdot 350/280 = 18.75$ cubic inches.

**16.15.** If $y$ varies directly as the square of $x$, what is the effect on $y$ of doubling $x$?

    1. Since $y = kx^2$, write $k = y/x^2$.

    2. While $x$ and $y$ vary, $k$ remains constant; hence for different $x$ and $y$ values, $y_1/x_1^2 = y_2/x_2^2$, or $y_2 = y_1 x_2^2/x_1^2$.

    3. Hence, if $x_2 = 2x_1$, $y_2 = y_1(2x_1)^2/x_1^2 = 4y_1$. Thus, if $x$ is doubled, $y$ is multiplied by 4.

**16.16.** If $y$ varies inversely as the cube of $x$, what is the effect on $y$ of doubling $x$?

    1. Since $y = k/x^3$, write $k = x^3 y$.

    2. While $x$ and $y$ vary, $k$ remains constant; hence for different $x$ and $y$ values, $x_1^3 y_1 = x_2^3 y_2$, or $y_2 = y_1 x_1^3/x_2^3$.

    3. Hence, if $x_2 = 2x_1$, $y_2 = y_1 x_1^3/(2x_1)^3 = y_1/8$. Thus, if $x$ is doubled, $y$ is divided by 8.

**16.17.** The strength $W$ of a rectangular beam of wood varies jointly as the width $w$ and the square of the depth $d$, and inversely as the length $L$ of the beam. What would be the effect on $W$ of doubling $w$ and $d$ while decreasing $L$ by a factor of 20%?

    1. Since $W = kwd^2/L$, write $k = WL/(wd^2)$.

    2. For different values of the variables, $k$ remains constant, hence $W_1 L_1/(w_1 d_1^2) = W_2 L_2/(w_2 d_2^2)$.

    3. Hence, if $w_2 = 2w_1$, $d_2 = 2d_1$, and $L_2 = L_1 - 0.2L_1 = 0.8L_1$, write:

$$\frac{W_1 L_1}{w_1 d_1^2} = \frac{W_2(0.8L_1)}{(2w_1)(2d_1)^2} \text{ and solve for } W_2 \text{ to obtain } W_2 = \frac{W_1 L_1}{w_1 d_1^2} \cdot \frac{(2w_1)(2d_1)^2}{0.8L_1} = 10W_1. \text{ Thus, } W \text{ would be}$$

multiplied by 10.

## SUPPLEMENTARY PROBLEMS

**16.18.** State the domain and range and sketch a graph of the following functions:

    (a) $f(x) = \sqrt[3]{x - 2}$       (b) $f(x) = -1/\sqrt{x + 3}$       (c) $f(x) = \sqrt{4 - (x + 2)^2}$

*Ans.*   (a) Domain: $R$, Range $R$

       The graph is shown in Fig. 16-9.

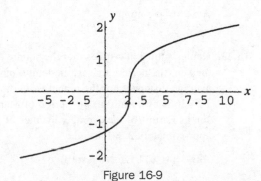

Figure 16-9

(b) Domain: $\{x \in R | x > -3\}$
Range: $\{y \in R | y < 0\}$
The graph is shown in Fig. 16-10.

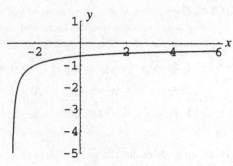

Figure 16-10

(c) Domain: $\{x \in R | -4 \leq x \leq 0\}$
Range: $\{y \in R | 0 \leq y \leq 2\}$
The graph is shown in Fig. 16-11.

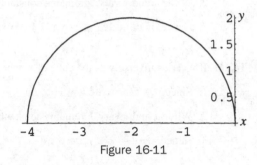

Figure 16-11

**16.19.** If $y$ varies directly as the fourth power of $x$, and $y = 2$ when $x = \frac{1}{2}$, find $y$ when $x = 2$.

*Ans.* 512

**16.20.** If $y$ varies inversely as the square root of $x$, and $y = 2$ when $x = \frac{1}{2}$, find $y$ when $x = 2$.

*Ans.* 1

**16.21.** If a spring of natural length 5 centimeters is displaced 0.3 centimeter from its natural length by a weight of 6 pounds, use Hooke's law (Problem 16-10) to determine the weight necessary to displace the spring 1 centimeter.

*Ans.* 20 pounds

**16.22.** Newton's law of cooling states that the rate $r$ at which a body cools is directly proportional to the difference between the temperature $T$ of the body and the temperature $T_0$ of its surroundings. If a cup of hot coffee at temperature $140°$ is in a room at temperature $68°$ and is cooling at the rate of $9°$ per minute, find the rate at which it will be cooling when its temperature has dropped to $116°$.

*Ans.* $6°$ per minute.

**16.23.** Kepler's third law states that the square of the time $T$ required for a planet to complete one orbit around the sun (the period, that is, the length of one planetary year) is directly proportional to the cube of the average distance $d$ of the planet from the sun. For the planet Earth, assume $d = 93 \times 10^6$ miles and $T = 365$ days. Find (a) the period of Mars, given that Mars is approximately 1.5 times as distant from the sun as Earth; (b) the average distance of Venus from the sun, given that the period of Venus is approximately 223 Earth days.

*Ans.* (a) 671 Earth days; (b) $67 \times 10^6$ miles

**16.24.** The resistance $R$ of a wire varies directly as the length $L$ and inversely as the square of the diameter $d$. A 4-meter-long piece of wire with a diameter of 6 millimeters has a resistance of 600 ohms. What diameter should be used if a 5-meter piece of this wire is to have a resistance of 1000 ohms?

*Ans.*   $\sqrt{27} \approx 5.2$ millimeters

**16.25.** Coulomb's law states that the force $F$ of attraction between two oppositely charged particles varies jointly as the magnitudes $q_1$ and $q_2$ of their electrical charges and inversely as the square of the distance $d$ between the particles. What is the effect on $F$ of doubling the magnitude of the charges and halving the distance between them?

*Ans.*   The force is multiplied by a factor of 16.

# CHAPTER 17

# Exponential Functions

## Definition of Exponential Function

An exponential function is any function for which the rule specifies the independent variable in an exponent. A *basic* exponential function has the form $F(x) = a^x$, $a > 0$, $a \neq 1$. The domain of a basic exponential function is considered to be the set of all real numbers, unless otherwise specified.

**EXAMPLE 17.1**   The following are examples of exponential functions:

(a) $f(x) = 2^x$; (b) $f(x) = \left(\frac{1}{2}\right)^x$; (c) $f(x) = 4^{-x}$; (d) $f(x) = 2^{-x^2}$

## Properties of Exponents

Properties of exponents can be restated for convenience in terms of variable exponents. Assuming $a, b > 0$, then for all real $x$ and $y$:

$$a^x a^y = a^{x+y} \qquad (ab)^x = a^x b^x$$

$$\frac{a^x}{a^y} = a^{x-y} \qquad \left(\frac{a}{b}\right)^x = \frac{a^x}{b^x}$$

$$(a^p)^x = a^{px}$$

## The Number $e$

The number $e$ is called the natural exponential base. It is defined as $\lim_{n \to \infty} \left(1 + \frac{1}{n}\right)^n$. $e$ is an irrational number with a value approximately $2.718\ 281\ 828\ 459\ 045.\ldots$

## Exponential Growth and Decay

Applications generally distinguish between exponential *growth* and *decay*. A basic exponential growth function is an increasing exponential function; an exponential decay function is a decreasing exponential function.

## Compound Interest

If a principal of $P$ dollars is invested at an annual rate of interest $r$, and the interest is compounded $n$ times per year, then the amount of money $A(t)$ generated at time $t$ is given by the formula:

$$A(t) = P\left(1 + \frac{r}{n}\right)^{nt}$$

## Continuous Compound Interest

If a principal of $P$ dollars is invested at an annual rate of interest $r$, and the interest is compounded *continuously*, then the amount of money $A(t)$ available at any later time $t$ is given by the formula:

$$A(t) = Pe^{rt}$$

## Unlimited Population Growth

If a population consisting initially of $N_0$ individuals also is modeled as growing without limit, the population $N(t)$ at any later time $t$ is given by the formula ($k$ is a constant to be determined):

$$N(t) = N_0 e^{kt}$$

Alternatively, a different base can be used.

## Logistic Population Growth

If a population consisting initially of $N_0$ individuals is modeled as growing with a limiting population (due to limited resources) of $P$ individuals, the population $N(t)$ at any later time $t$ is given by the formula ($k$ is a constant to be determined):

$$N(t) = \frac{N_0 P}{N_0 + (P - N_0)e^{-kt}}$$

## Radioactive Decay

If an amount $Q_0$ of a radioactive substance is present at time $t = 0$, then the amount $Q(t)$ of the substance present at any later time $t$ is given by the formula ($k$ is a constant to be determined):

$$Q(t) = Q_0 e^{-kt}$$

Alternatively, a different base can be used.

### SOLVED PROBLEMS

**17.1** Explain why the domain of a basic exponential function is considered to be $R$. What is the range of the function?

Consider, for example, the function $f(x) = 2^x$. The quantity $2^x$ is defined for all integer $x$; for example, $2^3 = 8$, $2^{-3} = \frac{1}{8}$, $2^0 = 1$, and so on. Moreover, the quantity $2^x$ is defined for all noninteger rational $x$, for example, $2^{1/2} = \sqrt{2}$, $2^{5/3} = \sqrt[3]{2^5}$, $2^{-3/4} = 1/\sqrt[4]{2^3}$, and so on.

To define the quantity $2^x$ for $x$ an irrational number, for example, $2^{\sqrt{2}}$, use the nonterminating decimal representing $\sqrt{2}$, that is, $1.4142 \ldots$, and consider the rational powers $2^1$, $2^{1.4}$, $2^{1.41}$, $2^{1.414}$, $2^{1.4142}$, and so on. It can be shown in calculus that each successive power gets closer to a real number, which is defined as $2^{\sqrt{2}}$. This process can be applied to define the quantity $2^x$ for $x$ any irrational number, hence, $2^x$ is defined for all real numbers $x$. The domain of $f(x) = 2^x$ is considered to be $R$, and similarly for any exponential function $f(x) = a^x$, $a > 0$, $a \neq 1$.

Since $2^x$ is positive for all real $x$, the range of the function is the positive numbers, $(0, \infty)$.

**17.2.** Analyze and sketch the graph of a basic exponential function of form $f(x) = a^x$, $a > 1$.

The graph has no obvious symmetry. Since $a^0 = 1$, the graph passes through the point $(0, 1)$. Since $a^1 = a$, the graph passes through the point $(1, a)$.

It can be shown that, if $x_1 < x_2$, then $a^{x_1} < a^{x_2}$, that is, the function is increasing on $R$; hence the term exponential growth function. Therefore, the basic exponential function is a one-to-one function.

It can further be shown that as $x \to \infty$, $a^x \to \infty$, and as $x \to -\infty$, $a^x \to 0$. Thus, the negative $x$-axis is a horizontal asymptote for the graph.

Since $a^x$ is positive for all real $x$ (see the previous problem), the range of the function is $(0,\infty)$.

The graph is shown in Fig. 17-1.

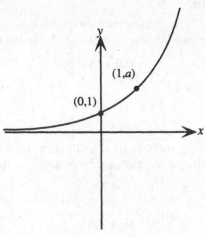

Figure 17-1

**17.3.** Analyze and sketch the graph of a basic exponential function of form $f(x) = a^x$, $a < 1$.

The graph has no obvious symmetry. Since $a^0 = 1$, the graph passes through the point $(0,1)$. Since $a^1 = a$, the graph passes through the point $(1,a)$.

It can be shown that, if $x_1 < x_2$, then $a^{x_1} > a^{x_2}$, that is, the function is decreasing on $R$; hence the term exponential decay function.

It can further be shown, that as $x \to \infty$, $a^x \to 0$, and as $x \to -\infty$, $a^x \to \infty$. Thus, the positive $x$-axis is a horizontal asymptote for the graph.

Since $a^x$ is positive for all real $x$, the range of the function is $(0,\infty)$.

The graph is shown in Fig. 17-2.

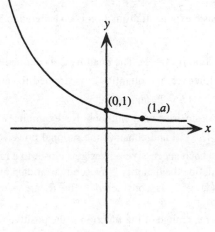

Figure 17-2

**17.4.** Show that the graph of $f(x) = a^{-x}$, $a > 1$, is an exponential decay curve.

Let $b = 1/a$. Then, since $a > 1$, it follows that $b < 1$. Moreover, $a^{-x} = (1/b)^{-x} = b^x$. Since the graph of $f(x) = b^x$, $b < 1$, is an exponential decay curve, so is the graph of $f(x) = a^{-x}$, $a > 1$.

**17.5.** Sketch a graph of (a) $f(x) = 2^x$ (b) $f(x) = 2^{-x}$.

(a) Form a table of values.

| $x$ | $y$ |
|---|---|
| $-2$ | $\frac{1}{4}$ |
| $-1$ | $\frac{1}{2}$ |
| $0$ | $1$ |
| $1$ | $2$ |
| $2$ | $4$ |
| $3$ | $8$ |

Domain: $R$, Range: $(0,\infty)$
Asymptote: negative $x$ axis.
The graph is shown in Fig. 17-3.

(b) Form a table of values.

| $x$ | $y$ |
|---|---|
| $-3$ | $8$ |
| $-2$ | $4$ |
| $-1$ | $2$ |
| $0$ | $1$ |
| $1$ | $\frac{1}{2}$ |
| $2$ | $\frac{1}{4}$ |

Domain: $R$, Range: $(0,\infty)$
Asymptote: positive $x$ axis.
The graph is shown in Fig. 17-4.

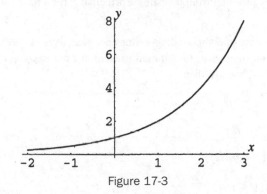

Figure 17-3

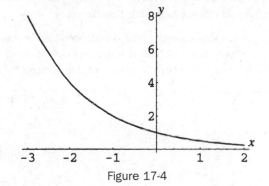

Figure 17-4

**17.6.** Explain the definition of the natural exponential base $e$.

Consider the following table of values for the quantity $\left(1 + \frac{1}{n}\right)^n$.

| $n$ | 1 | 10 | 100 | 1000 | 10,000 | 100,000 | 1,000,000 |
|---|---|---|---|---|---|---|---|
| $\left(1 + \frac{1}{n}\right)^n$ | 2 | 2.59374246 | 2.70481383 | 2.71692393 | 2.71814593 | 2.71826824 | 2.71828047 |

As $n \to \infty$, the quantity $\left(1 + \frac{1}{n}\right)^n$ does not increase beyond all bounds, but seems to approach a value.

In calculus it is shown that this value is an irrational number, called $e$, with a decimal approximation of 2.718 281 828 459 045. . . In calculus this number and the exponential functions $f(x) = e^x$, $f(x) = e^{-x}$, and so on, are shown to have special properties.

**17.7.** Derive the formula $A(t) = P(1 + r/n)^{nt}$ for the amount of money resulting from investing a principal $P$ for a time $t$ at an annual rate $r$, compounded $n$ times per year.

First assume that the amount $P$ is invested for one year at the simple interest rate of $r$. Then the interest after one year is $I = Prt = Pr(1) = Pr$. The amount of money present after one year is then

$$A = P + I = P + Pr = P(1 + r).$$

If this amount is then invested for a second year at the simple interest rate of $r$, then the interest after the second year is $P(1 + r)r(1) = P(1 + r)r$. The amount of money present after two years is then

$$A = P(1 + r) + P(1 + r)r = P(1 + r)(1 + r) = P(1 + r)^2$$

Thus the amount present at the end of each year is multiplied by a factor of $1 + r$ during the next year. Generalizing, the amount present at time $t$, assuming compounding once per year, is

$$A(t) = P(1 + r)^t$$

Now assume that interest is compounded $n$ times per year. The interest after one compounding period is then $I = Pr/n$. The amount of money present after one compounding period is

$$A = P + I = P + Pr/n = P(1 + r/n)$$

Thus, the amount present at the end of each compounding period is multiplied by a factor of $1 + r/n$ during the next period. Hence the amount present after one year, $n$ compounding periods, is

$$A = P(1 + r/n)^n$$

and the amount present at time $t$ is given by

$$A(t) = P((1 + r/n)^n)^t = P(1 + r/n)^{nt}$$

**17.8.** Derive the formula $A(t) = Pe^{rt}$ for the amount of money resulting from investing a principal $P$ for a time $t$ at an annual rate $r$, compounded continuously.

Continuous compounding is understood as the limiting case of compounding $n$ times per year, as $n \to \infty$. From the previous problem, if interest is compounded $n$ times per year, the amount present at time $t$ is given by $A(t) = P(1 + r/n)^{nt}$. If $n$ is allowed to increase beyond all bounds, then

$$A(t) = \lim_{n \to \infty} P(1 + r/n)^{nt}$$
$$= \lim_{n \to \infty} P(1 + r/n)^{(n/r)rt}$$
$$= \lim_{n \to \infty} P[(1 + r/n)^{n/r}]^{rt}$$
$$= \lim_{n/r \to \infty} P[(1 + r/n)^{n/r}]^{rt}$$
$$= P[\lim_{n/r \to \infty} (1 + r/n)^{n/r}]^{rt}$$
$$= Pe^{rt}$$

**17.9.** Calculate the amount of money present if $1000 is invested at 5% interest for seven years, compounded

(a) yearly; (b) quarterly; (c) monthly; (d) daily; (e) continuously.

(a) Use $A(t) = P(1 + r/n)^{nt}$ with $P = 1000$, $r = 0.05$, $t = 7$, and $n = 1$.
$$A(7) = 1000(1 + 0.05/1)^{1 \cdot 7} = \$1407.10$$

(b) Use $A(t) = P(1 + r/n)^{nt}$ with $P = 1000$, $r = 0.05$, $t = 7$, and $n = 4$.
$$A(7) = 1000(1 + 0.05/4)^{4 \cdot 7} = \$1415.99$$

(c) Use $A(t) = P(1 + r/n)^{nt}$ with $P = 1000$, $r = 0.05$, $t = 7$, and $n = 12$.
$$A(7) = 1000(1 + 0.05/12)^{12 \cdot 7} = \$1418.04$$

(d) Use $A(t) = P(1 + r/n)^{nt}$ with $P = 1000$, $r = 0.05$, $t = 7$, and $n = 365$.
$$A(7) = 1000(1 + 0.05/365)^{365 \cdot 7} = \$1419.03$$

(e) Use $A(t) = Pe^{rt}$ with $P = 1000$, $r = 0.05$, and $t = 7$.
$$A(7) = 1000 \cdot e^{0.05 \cdot 7} = \$1419.07$$

Note that the difference in interest that results from increasing the frequency of compounding from daily to continuously is quite small.

**17.10.** Simplify the expressions:

(a) $\left(\dfrac{e^x + e^{-x}}{2}\right)^2 - \left(\dfrac{e^x - e^{-x}}{2}\right)^2$

(b) $\dfrac{(e^x + e^{-x})(e^x + e^{-x}) - (e^x - e^{-x})(e^x - e^{-x})}{(e^x + e^{-x})^2}$

(a) $\left(\dfrac{e^x + e^{-x}}{2}\right)^2 - \left(\dfrac{e^x - e^{-x}}{2}\right)^2 = \dfrac{(e^x)^2 + 2e^x e^{-x} + (e^{-x})^2}{4} - \dfrac{(e^x)^2 - 2e^x e^{-x} + (e^{-x})^2}{4}$

$$= \dfrac{e^{2x} + 2 + e^{-2x} - e^{2x} + 2 - e^{-2x}}{4}$$

$$= \dfrac{4}{4} = 1$$

(b) $\dfrac{(e^x + e^{-x})(e^x + e^{-x}) - (e^x - e^{-x})(e^x - e^{-x})}{(e^x + e^{-x})^2} = \dfrac{e^{2x} + 2e^x e^{-x} + e^{-2x} - e^{2x} + 2e^x e^{-x} - e^{-2x}}{(e^x + e^{-x})^2}$

$$= \dfrac{4}{(e^x + e^{-x})^2}$$

For an alternate form, regard the last expression as a complex fraction and multiply numerator and denominator by $e^{2x}$ to obtain

$$\dfrac{4}{(e^x + e^{-x})^2} = \dfrac{4e^{2x}}{e^{2x}(e^x + e^{-x})^2} = \dfrac{4e^{2x}}{(e^{2x} + 1)^2}$$

**17.11.** Find the zeros of the function $f(x) = xe^{-x} - e^{-x}$.

Solve $xe^{-x} - e^{-x} = 0$ by factoring to obtain

$$e^{-x}(x - 1) = 0$$
$$e^{-x} = 0 \text{ or } x - 1 = 0$$
$$x = 1$$

Since $e^{-x}$ is never 0, the only zero of the function is 1.

**17.12.** The number of bacteria in a culture is counted as 400 at the start of an experiment. If the number of bacteria doubles every 3 hours, the number of individuals can be represented by the formula $N(t) = 400(2)^{t/3}$. Find the number of bacteria present in the culture after 24 hours.

$$N(24) = 400(2)^{24/3} = 400 \cdot 2^8 = 102{,}400 \text{ individuals}$$

**17.13.** Human populations can be modeled over short periods by unlimited exponential growth functions. If a country has a population of 22 million in 2000 and maintains a population growth rate of 1% per year, then its population in millions at a later time, taking $t = 0$ in 2000, can be modeled as $N(t) = 22e^{0.01t}$. Estimate the population in the year 2010.

In the year 2010, $t = 10$. Hence $N(10) = 22e^{0.01(10)} = 24.3$. Hence the population is estimated to be 24.3 million.

**17.14.** A herd of deer is introduced onto an island. The initial population is 500 individuals, and it is estimated that the long-term sustainable population is 2000 individuals. If the size of the population is given by the logistic growth function

$$N(t) = \dfrac{2000}{1 + 3e^{-0.05t}}$$

estimate the number of deer present after (a) 1 year; (b) 20 years; (c) 50 years.
Use the given formula with the given values of $t$.

(a) $t = 1$: $N(1) = \dfrac{2000}{1 + 3e^{-0.05(1)}} \approx 520$ individuals

(b) $t = 20$: $N(20) = \dfrac{2000}{1 + 3e^{-0.05(20)}} \approx 950$ individuals

(c) $t = 50$: $N(50) = \dfrac{2000}{1 + 3e^{-0.05(50)}} \approx 1600$ individuals

**17.15.** Draw a graph of the function $N(t)$ from the previous problem.

Use the calculated values. Note also that $N(0)$ is given as 500, and that as $t \to \infty$, since $e^{-0.05t} \to 0$, the value of the function approaches 2000 asymptotically. The graph is shown in Fig. 17-5.

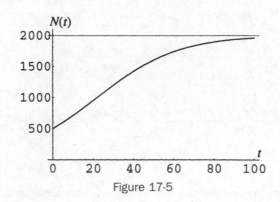

Figure 17-5

**17.16.** A certain radioactive isotope decays according to the formula $Q(t) = Q_0 e^{-0.034t}$, where $t$ is the time in years and $Q_0$ is the number of grams present initially. If 20 grams are present initially, approximate to the nearest tenth of a gram the amount present after 10 years.

Use the given formula with $Q_0 = 20$ and $t = 10$: $Q(10) = 20 \cdot e^{-0.034(10)} = 14.2$ grams.

**17.17.** If a radioactive isotope decays according to the formula $Q(t) = Q_0 \cdot 2^{-t/T}$, where $t$ is the time in years and $Q_0$ is the number of grams present initially, show that the amount present at time $t = T$ is $Q_0/2$. ($T$ is called the *half-life* of the isotope.)

Use the given formula with $t = T$. Then $Q(T) = Q_0 \cdot 2^{-T/T} = Q_0 \cdot 2^{-1} = Q_0/2$.

## SUPPLEMENTARY PROBLEMS

**17.18.** Sketch a graph of the functions (a) $f(x) = 1 - e^{-x}$ and (b) $f(x) = 2^{-x^2/2}$

*Ans.* (a) Fig. 17-6; (b) Fig. 17-7.

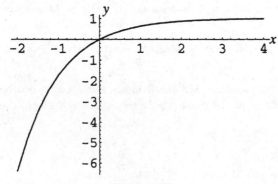

Figure 17-6

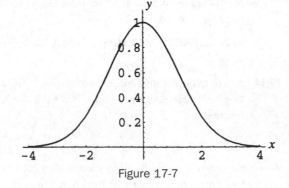

Figure 17-7

**17.19.** Simplify the expression $\left(\dfrac{e^x + e^{-x}}{2}\right)^2 + \left(\dfrac{e^x - e^{-x}}{2}\right)^2$.

*Ans.* $\dfrac{e^{2x} + e^{-2x}}{2}$ or $\dfrac{e^{4x} + 1}{2e^{2x}}$

**17.20.** Prove that the difference quotient (see Chapter 9) for $f(x) = e^x$ can be written as

$$\frac{e^h - 1}{h} e^x$$

**17.21.** Find the zeros of the function $f(x) = -x^2 e^{-x} + 2xe^{-x}$.

*Ans.*   $0, 2$

**17.22.** $8000 is invested in an account yielding 5.5% interest. Find the amount of money in the account after one year if interest is compounded (a) quarterly; (b) daily; (c) continuously.

*Ans.*   (a) $8449.16; (b) $8452.29; (c) $8452.32

**17.23.** In the previous problem, find the annualized percentage rate for the account (this is the equivalent rate without compounding that would yield the same amount of interest).

*Ans.*   (a) 5.61%; (b) 5.65%; (c) 5.65%

**17.24.** How much would have to be invested at 5.5% compounded continuously to obtain $5000 after 10 years?

*Ans.*   $2884.75

**17.25.** A family has just had a new child. How much would have to be invested at 6%, compounded daily, in order to have $60,000 for her college education in 17 years?

*Ans.*   $21,637.50

**17.26.** If the number of bacteria in a culture is given by the formula $Q(t) = 250 \cdot 3^{t/4}$, where $t$ is measured in days, estimate (a) the initial population; (b) the population after 4 days; (c) the population after 14 days.

*Ans.*   (a) 250; (b) 750; (c) 11,700

**17.27.** If the population of trout in a lake is given by the formula $N(t) = \dfrac{8000}{2 + 3e^{-0.037t}}$, where $t$ is measured in years, estimate (a) the initial population; (b) the population after 10 years; (c) the long-term limiting value of the population.

*Ans.*   (a) 1600; (b) 1960; (c) 4000

**17.28.** If a radioactive isotope decays according to the formula $Q(t) = Q_0 \cdot 2^{-t/12}$, where $t$ is measured in years, find the portion of an initial amount remaining after (a) 1 year; (b) 12 years; (c) 100 years.

*Ans.*   (a) $0.94Q_0$; (b) $0.5Q_0$; (c) $0.003Q_0$

**17.29.** The half-life (see Problem 17.17) of Carbon-14 is 5730 years.

(a)  If 100 grams of Carbon-14 were present initially, how much would remain after 3000 years?

(b)  If a sample contains 38 grams of Carbon-14, how much was present 4500 years ago?

*Ans.*   (a) 69.6 grams; (b) 65.5 grams

# CHAPTER 18

# Logarithmic Functions

## Definition of Logarithmic Function

A logarithmic function, $f(x) = \log_a x$, $a > 0$, $a \neq 1$, is the inverse function to an exponential function $F(x) = a^x$. Thus, if $y = \log_a x$, then $x = a^y$. That is, the logarithm of $x$ to the base $a$ is the exponent to which $a$ must be raised to obtain $x$. Conversely, if $x = a^y$, then $y = \log_a x$.

**EXAMPLE 18.1**   The function $f(x) = \log_2 x$ is defined as $f: y = \log_2 x$ if $2^y = x$. Since $2^4 = 16$, 4 is the exponent to which 2 must be raised to obtain 16, and $\log_2 16 = 4$.

**EXAMPLE 18.2**   The statement $10^3 = 1000$ can be rewritten in terms of the logarithm to the base 10. Since 3 is the exponent to which 10 must be raised to obtain 1000, $\log_{10} 1000 = 3$.

## Relation between Logarithmic and Exponential Functions

$$\log_a a^x = x \qquad a^{\log_a x} = x$$

**EXAMPLE 18.3**   $\log_5 5^3 = 3$; $5^{\log_5 25} = 25$

## Properties of Logarithms

($M, N$ positive real numbers)

$$\log_a 1 = 0 \qquad\qquad \log_a a = 1$$
$$\log_a (MN) = \log_a M + \log_a N \qquad \log_a (M^p) = p\log_a M$$
$$\log_a \left(\frac{M}{N}\right) = \log_a M - \log_a N$$

**EXAMPLE 18.4**   (a) $\log_5 1 = 0$ (since $5^0 = 1$)   (b) $\log_4 4 = 1$ (since $4^1 = 4$)

(c) $\log_6 6x = \log_6 6 + \log_6 x = 1 + \log_6 x$   (d) $\log_6 x^6 = 6 \log_6 x$

(e) $\log_{1/2}(2x) = \log_{1/2}\frac{x}{1/2} = \log_{1/2} x - \log_{1/2}\left(\frac{1}{2}\right) = \log_{1/2} x - 1$

## Special Logarithmic Functions

$\log_{10} x$ is abbreviated as $\log x$ (common logarithm).
$\log_e x$ is abbreviated as $\ln x$ (natural logarithm).

## SOLVED PROBLEMS

**18.1.**   Write the following in exponential form:

(a) $\log_2 8 = 3$; (b) $\log_{25} 5 = \frac{1}{2}$; (c) $\log_{10} \frac{1}{100} = -2$;

(d) $\log_8 \frac{1}{4} = -\frac{2}{3}$; (e) $\log_b c = d$; (f) $\log_e (x^2 + 5x - 6) = y - C$

(a) If $y = \log_a x$, then $x = a^y$. Hence, if $3 = \log_2 8$, then $8 = 2^3$.

(b) If $y = \log_a x$, then $x = a^y$. Hence, if $\frac{1}{2} = \log_{25} 5$, then $5 = 25^{1/2}$.

(c) If $-2 = \log_{10} \frac{1}{100}$, then $\frac{1}{100} = 10^{-2}$.

(d) If $-\frac{2}{3} = \log_8 \frac{1}{4}$, then $8^{-2/3} = \frac{1}{4}$.

(e) If $d = \log_b c$, then $b^d = c$.

(f) If $y - C = \log_e(x^2 + 5x - 6)$, then $e^{y-C} = x^2 + 5x - 6$.

**18.2.** Write the following in logarithmic form:

(a) $3^5 = 243$; (b) $6^{-3} = \frac{1}{216}$; (c) $256^{3/4} = 64$;

(d) $\left(\frac{1}{2}\right)^{-5} = 32$; (e) $u^m = p$; (f) $e^{at+b} = y - C$

(a) If $x = a^y$, then $y = \log_a x$. Hence if $243 = 3^5$, then $5 = \log_3 243$.

(b) If $x = a^y$, then $y = \log_a x$. Hence if $\frac{1}{216} = 6^{-3}$, then $-3 = \log_6 \frac{1}{216}$.

(c) If $64 = 256^{3/4}$, then $\log_{256} 64 = \frac{3}{4}$.

(d) If $32 = \left(\frac{1}{2}\right)^{-5}$, then $\log_{1/2} 32 = -5$.

(e) If $p = u^m$, then $\log_u p = m$.

(f) If $y - C = e^{at+b}$, then $\log_e(y - C) = at + b$.

**18.3.** Evaluate the following logarithms:

(a) $\log_7 49$; (b) $\log_4 256$; (c) $\log_{10} 0.000001$; (d) $\log_{27} \frac{1}{9}$; (e) $\log_{1/5} 125$

(a) The logarithm to base 7 of 49 is the exponent to which 7 must be raised to obtain 49. This exponent is 2; hence $\log_7 49 = 2$.

(b) The logarithm to base 4 of 256 is the exponent to which 4 must be raised to obtain 256. This exponent is 4; hence $\log_4 256 = 4$.

(c) Set $\log_{10} 0.000001 = x$. Then $\log_{10} 10^{-6} = x$. Rewritten in exponential form, $10^x = 10^{-6}$. Since the exponential function is a one-to-one function, $x = -6$; hence $\log_{10} 0.000001 = -6$.

(d) Set $\log_{27} \frac{1}{9} = x$. Rewritten in exponential form, $27^x = \frac{1}{9}$, or $(3^3)^x = 3^{3x} = 3^{-2}$. Since the exponential function is a one-to-one function, $3x = -2$, $x = -\frac{2}{3}$, hence $\log_{27} \frac{1}{9} = -\frac{2}{3}$.

(e) Set $\log_{1/5} 125 = x$. Rewritten in exponential form, $\left(\frac{1}{5}\right)^x = 125$, or $(5^{-1})^x = 5^{-x} = 5^3$. Since the exponential function is a one-to-one function, $-x = 3$, $x = -3$; hence $\log_{1/5} 125 = -3$.

**18.4.** (a) Determine the domain and range of the logarithm function to base $a$.

(b) Evaluate $\log_5(-25)$.

(a) Since the logarithm function is the inverse function to the exponential function with base $a$, and since the exponential function has domain $R$ and range $(0,\infty)$, the logarithm function must have domain $(0,\infty)$ and range $R$.

(b) Since $-25$ is not in the domain of the logarithm function, $\log_5(-25)$ is undefined.

**18.5.** Sketch a graph of $f(x) = a^x$, $a > 1$, $f^{-1}(x) = \log_a x$, and the line $y = x$ on the same Cartesian coordinate system.

*Note*: The graph is shown in Fig. 18-1.

The domain of $f$ is $R$ and the range of $f$ is $(0,\infty)$.
The points $(0,1)$ and $(1,a)$ are on the graph of $f$.
The negative $x$ axis is an asymptote.
The domain of $f^{-1}$ is $(0,\infty)$ and the range of $f^{-1}$ is $R$.
The points $(1,0)$ and $(a,1)$ are on the graph of $f^{-1}$.
The negative $y$ axis is an asymptote.

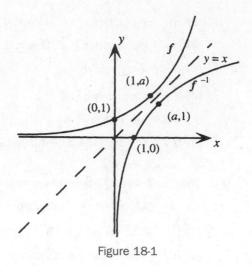

Figure 18-1

**18.6.** Sketch a graph of

(a) $f(x) = \log_5 x$

(b) $g(x) = \log_{1/4} x$

(a) Form a table of values.

(b) Form a table of values.

| $x$ | $y$ |
|---|---|
| $\frac{1}{5}$ | $-1$ |
| $1$ | $0$ |
| $5$ | $1$ |
| $25$ | $2$ |

| $x$ | $y$ |
|---|---|
| $\frac{1}{4}$ | $1$ |
| $1$ | $0$ |
| $4$ | $-1$ |
| $16$ | $-2$ |

Domain: $(0,\infty)$, range: $R$
Asymptote: negative $y$-axis.
The graph is shown in Fig. 18-2.

Domain: $(0,\infty)$, range: $R$
Asymptote: positive $y$-axis.
The graph is shown in Fig. 18-3.

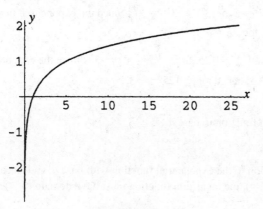

Figure 18-2

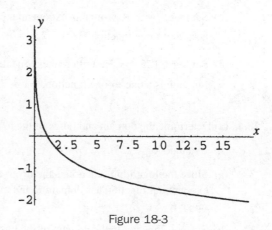

Figure 18-3

**18.7.** Prove the logarithmic–exponential function relations.

(a) If $y = \log_a x$, then $x = a^y$. Hence $x = a^y = a^{\log_a x}$.

(b) Similarly, reversing the letters, if $x = \log_a y$, then $y = a^x$. Hence $x = \log_a y = \log_a a^x$.

**18.8.** Show that if $\log_a u = \log_a v$, then $u = v$.

Since the exponential function, $f(x) = a^x$, is a one-to-one function, its inverse function, $f^{-1}(x) = \log_a x$, is also a one-to-one function and $(f \circ f^{-1})(x) = f(f^{-1}(x)) = f(\log_a x) = a^{\log_a x} = x$.

Hence, if $\log_a u = \log_a v$, then $a^{\log_a u} = a^{\log_a v}$ and $u = v$.

**18.9.** Evaluate, using the logarithmic–exponential function relations:

(a) $\log_3 3^5$; (b) $\log_2 256$; (c) $\log_a \sqrt[3]{a^2}$; (d) $\log 0.00001$;

(e) $5^{\log_5 3}$; (f) $e^{\ln \pi}$; (g) $a^{\log_a (x^2 - 5x + 6)}$; (h) $36^{\log_6 7}$

*Ans.* (a) $\log_3 3^5 = 5$; (b) $\log_2 256 = \log_2 2^8 = 8$;

(c) $\log_a \sqrt[3]{a^2} = \log_a a^{2/3} = \dfrac{2}{3}$; (d) $\log 0.00001 = \log_{10} 10^{-5} = -5$;

(e) $5^{\log_5 3} = 3$; (f) $e^{\ln \pi} = e^{\log_e \pi} = \pi$;

(g) $a^{\log_a (x^2 - 5x + 6)} = x^2 - 5x + 6$; (h) $36^{\log_6 7} = (6^2)^{\log_6 7} = 6^{2\log_6 7} = (6^{\log_6 7})^2 = 7^2 = 49$

**18.10.** Prove the properties of logarithms.

The properties $\log_a 1 = 0$ and $\log_a a = 1$ follow directly from the logarithmic–exponential relations, since $\log_a 1 = \log_a a^0 = 0$ and $\log_a a = \log_a a^1 = 1$.

To prove the other properties, let $u = \log_a M$ and $v = \log_a N$. Then $M = a^u$ and $N = a^v$.
Therefore $MN = a^u a^v$; thus $MN = a^{u+v}$. Rewriting in logarithmic form, $\log_a MN = u + v$.
Hence $\log_a MN = \log_a M + \log_a N$.

Similarly, $\dfrac{M}{N} = \dfrac{a^u}{a^v}$, thus $\dfrac{M}{N} = a^{u-v}$. Rewriting in logarithmic form, $\log_a \dfrac{M}{N} = u - v$.

Hence $\log_a\left(\dfrac{M}{N}\right) = \log_a M - \log_a N$.

Finally $M^p = (a^u)^p = a^{up}$; thus $M^p = a^{pu}$. Rewriting in logarithmic form, $\log_a M^p = pu$.
Hence $\log_a M^p = p \log_a M$.

**18.11.** Use the properties of logarithms to rewrite in terms of logarithms of simpler expressions:

(a) $\log_a \dfrac{xy}{z}$; (b) $\log_a(x^2 - 1)$; (c) $\log_a \dfrac{x^3(x + 5)}{(x - 4)^2}$; (d) $\log_a \sqrt{\dfrac{x^2 + y^2}{xy}}$; (e) $\ln(Ce^{5x + 1})$

*Ans.* (a) $\log_a \dfrac{xy}{z} = \log_a xy - \log_a z = \log_a x + \log_a y - \log_a z$

(b) $\log_a (x^2 - 1) = \log_a[(x - 1)(x + 1)] = \log_a(x - 1) + \log_a (x + 1)$ *Note:* The properties of logarithms can be used to transform expressions involving logarithms of products, quotients, and powers. They do not allow simplification of logarithms of sums or differences.

(c) $\log_a \dfrac{x^3(x + 5)}{(x - 4)^2} = \log_a x^3 + \log_a (x + 5) - \log_a (x - 4)^2 = 3 \log_a x + \log_a(x + 5) - 2 \log_a(x - 4)$

(d) $\log_a \sqrt{\dfrac{x^2 + y^2}{xy}} = \dfrac{1}{2}\left[\log_a \dfrac{x^2 + y^2}{xy}\right] = \dfrac{1}{2}\left[\log_a(x^2 + y^2) - \log_a(xy)\right] = \dfrac{1}{2}\left[\log_a(x^2 + y^2) - \log_a x - \log_a y\right]$

(e) $\ln(Ce^{5x + 1}) = \ln C + \ln e^{5x + 1} = \ln C + 5x + 1$

**18.12.** Write as one logarithm:

(a) $3 \log_a u - \log_a v$; (b) $\dfrac{1}{3} \log_a 5 - 3 \log_a x - 4 \log_a y$; (c) $\dfrac{1}{3} \log_a (x - 3) + 3 \log_a x + 2 \log_a (1 + x)$;

(d) $\dfrac{1}{2}[\log_a x + 3 \log_a y - 5 \log_a (z - 2)]$; (e) $\dfrac{1}{2}\ln(x + 1) - \dfrac{1}{2}\ln (x - 1) + \ln C$

*Ans.* (a) $3 \log_a u - \log_a v = \log_a u^3 - \log_a v = \log_a \dfrac{u^3}{v}$

(b) $\dfrac{1}{3} \log_a 5 - 3 \log_a x - 4 \log_a y = \log_a \sqrt[3]{5} - \log_a (x^3 y^4) = \log_a \dfrac{\sqrt[3]{5}}{x^3 y^4}$

(c) $\frac{1}{3}\log_a (x - 3) + 3 \log_a x + 2 \log_a (1 + x) = \log_a(x - 3)^{1/3} + \log_a x^3 + \log_a(1 + x)^2$

$$= \log_a \sqrt[3]{x - 3} + \log_a x^3(1 + x)^2$$

$$= \log_a[x^3(1 + x)^2 \sqrt[3]{x - 3}]$$

(d) $\frac{1}{2}[\log_a x + 3 \log_a y - 5 \log_a(z - 2)] = \frac{1}{2}[\log_a xy^3 - \log_a(z - 2)^5]$

$$= \frac{1}{2}\left[\log_a \frac{xy^3}{(z - 2)^5}\right] = \log_a \sqrt{\frac{xy^3}{(z - 2)^5}}$$

(e) $\frac{1}{2}\ln(x + 1) - \frac{1}{2}\ln(x - 1) + \ln C = \frac{1}{2}\ln\left(\frac{x - 1}{x + 1}\right) + \ln C$

$$= \ln\sqrt{\frac{x - 1}{x + 1}} + \ln C = \ln C\sqrt{\frac{x - 1}{x + 1}}$$

## SUPPLEMENTARY PROBLEMS

**18.13.** Write in exponential form:

(a) $\log_{1000} 10 = \frac{1}{3}$; (b) $\log_7 \frac{1}{49} = -2$; (c) $\log_u \frac{1}{\sqrt{u}} = -\frac{1}{2}$

*Ans.* (a) $1000^{1/3} = 10$; (b) $7^{-2} = \frac{1}{49}$; (c) $u^{-1/2} = \frac{1}{\sqrt{u}}$

**18.14.** Write in logarithmic form:

(a) $\left(\frac{1}{4}\right)^{-3} = 64$; (b) $e^{2/5} = \sqrt[5]{e^2}$; (c) $m^{-p} = T$

*Ans.* (a) $\log_{1/4} 64 = -3$; (b) $\ln \sqrt[5]{e^2} = \frac{2}{5}$; (c) $\log_m T = -p$

**18.15.** Evaluate:

(a) $\ln Ce^{-at}$; (b) $\log_4 \frac{1}{8}$; (c) $\log_{10}(-100)$; (d) $\log_{1/256} \frac{1}{2}$

*Ans.* (a) $\ln C - at$; (b) $-\frac{3}{2}$; (c) undefined; (d) $\frac{1}{8}$

**18.16.** (a) Evaluate $\log_3 81$; (b) Evaluate $\log_3 \frac{1}{81}$; (c) Show that $\log_a \frac{1}{N} = -\log_a N$

*Ans.* (a) 4; (b) −4

**18.17.** (a) Evaluate $\log_5 125$; (b) Evaluate $\log_{125} 5$; (c) Show that $\log_a b = \frac{1}{\log_b a}$

*Ans.* (a) 3; (b) $\frac{1}{3}$

**18.18.** Write in terms of logarithms of simpler expressions:

(a) $\log_a a(x - r)(x - s)$; (b) $\log_a \frac{a^2}{x^3 y^4}$; (c) $\ln \frac{a + \sqrt{a^2 - x^2}}{a - \sqrt{a^2 - x^2}}$; (d) $\ln \frac{e^x - e^{-x}}{2}$

*Ans.* (a) $1 + \log_a(x - r) + \log_a(x - s)$; (b) $2 - 3 \log_a x - 4 \log_a y$

(c) $\ln(a + \sqrt{a^2 - x^2}) - \ln(a - \sqrt{a^2 - x^2})$, or (after rationalizing the denominator)

$2 \ln(a + \sqrt{a^2 - x^2}) - 2 \ln x$;

(d) $\ln(e^x - e^{-x}) - \ln 2$

**18.19.** Evaluate (a) $10^{(1/2)\log 3}$; (b) $5^{3\log_5 7}$; (c) $2^{-3 \log_2 5}$

*Ans.* (a) $\sqrt{3}$; (b) 343; (c) $\frac{1}{125}$

**18.20.** Write as one logarithm:

(a) $2 \ln x - 8 \ln y + 4 \ln z$; (b) $\log(1 - x) + \log(x - 3)$;

(c) $\dfrac{\ln(x + h) - \ln x}{h}$; (d) $x \ln x - (x - 1)\ln(x - 1)$;

(e) $\log_c \dfrac{-b + \sqrt{b^2 - 4ac}}{2a} + \log_c \dfrac{-b - \sqrt{b^2 - 4ac}}{2a}$

*Ans.*  (a) $\ln \dfrac{x^2 z^4}{y^8}$; (b) undefined (there is no value of $x$ for which both logarithms are defined);

(c) $\ln \left(\dfrac{x + h}{x}\right)^{1/h}$; (d) $\ln \dfrac{x^x}{(x - 1)^{x - 1}}$; (e) $1 - \log_c a$

**18.21.** Given $\log_a 2 = 0.69$, $\log_a 3 = 1.10$, and $\log_a 5 = 1.61$, use the properties of logarithms to evaluate:

(a) $\log_a 30$; (b) $\log_a \dfrac{6}{5}$; (c) $\log_a \dfrac{1}{\sqrt{15}}$; (d) $\log_a \left(-\dfrac{5}{6}\right)$

*Ans.*  (a) 3.40; (b) 0.18; (c) $-1.36$; (d) undefined

**18.22.** Sketch graphs of:

(a) $f(x) = \log_3(x + 2)$; (b) $F(x) = 3 - \log_2 x$; (c) $g(x) = \ln |x|$; (d) $G(x) = -\ln(-x)$;

*Ans.*  (a) Fig. 18-4; (b) Fig 18-5; (c) Fig. 18-6; (d) Fig. 18-7

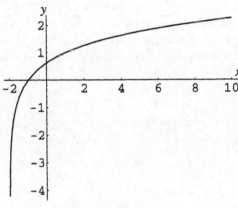

Figure 18-4

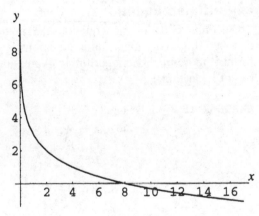

Figure 18-5

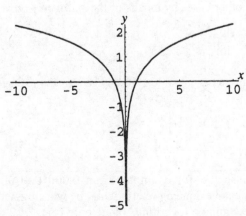

Figure 18-6

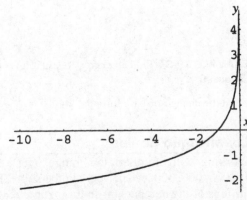

Figure 18-7

# CHAPTER 19

# Exponential and Logarithmic Equations

## Exponential Equations

Exponential equations are equations that involve a variable in an exponent. The crucial step in solving exponential equations is generally to take the logarithm of both sides to an appropriate base, commonly base 10 or base $e$.

**EXAMPLE 19.1**  Solve $e^x = 2$.

$$e^x = 2 \qquad \text{Take logarithms of both sides}$$
$$\ln(e^x) = \ln(2) \qquad \text{Apply the function-inverse function relation}$$
$$x = \ln 2$$

## Logarithmic Equations

Logarithmic equations are equations that involve the logarithm of a variable or variable expression. The crucial step in solving logarithmic equations is generally to rewrite the logarithmic statement in exponential form. If more than one logarithmic expression is present, these can be combined into one by using properties of logarithms.

**EXAMPLE 19.2**  Solve $\log_2 (x - 3) = 4$

$$\log_2 (x - 3) = 4 \qquad \text{Rewrite in exponential form}$$
$$2^4 = x - 3 \qquad \text{Isolate the variable}$$
$$x = 2^4 + 3$$
$$x = 19$$

## Change-of-Base Formula

Logarithmic expressions can be rewritten in terms of other bases by means of the *change-of-base* formula:

$$\log_a x = \frac{\log_b x}{\log_b a}$$

**EXAMPLE 19.3**  Find an expression, in terms of logarithms to base $e$, for $\log_5 10$, and give an approximate value for the quantity.

From the change-of-base formula, $\log_5 10 = \dfrac{\ln 10}{\ln 5} \approx 1.43$.

## Logarithmic Scales

Working with numbers that range over very wide scales, for example, from $0.000\ 000\ 000\ 001$ to $10,000,000,000$, can be very cumbersome. The work can be done more efficiently by working with the logarithms of the numbers (as in this example, where the common logarithms range only from $-12$ to $+10$).

## Examples of Logarithmic Scales

1. **SOUND INTENSITY:** The decibel scale for measuring sound intensity is defined as follows:

$$D = 10 \log \frac{I}{I_0}$$

where $D$ is the decibel level of the sound, $I$ is the intensity of the sound (measured in watts per square meter), and $I_0$ is the intensity of the smallest audible sound.

2. **EARTHQUAKE INTENSITY:** There is more than one logarithmic scale, called a Richter scale, used to measure the destructive power of an earthquake. A commonly used Richter scale is defined as follows:

$$R = \frac{2}{3} \log \frac{E}{E_0}$$

where $R$ is called the (Richter) magnitude of the earthquake, $E$ is the energy released by the earthquake (measured in joules), and $E_0$ is the energy released by a very small reference earthquake.

## SOLVED PROBLEMS

**19.1.** Prove the change-of-base formula.

Let $y = \log_a x$. Then, rewritten in exponential form, $x = a^y$. Taking logarithms of both sides to the base $b$ yields:

$$\log_b x = \log_b a^y$$
$$= y \log_b a \quad \text{by the properties of logarithms}$$

Hence,

$$y = \frac{\log_b x}{\log_b a}, \text{ that is, } \log_a x = \frac{\log_b x}{\log_b a}$$

**19.2.** Solve $2^x = 6$.

Take logarithms of both sides to base $e$ (base 10 could equally well be used, but base $e$ is standard in most calculus situations).

$$\ln 2^x = \ln 6$$
$$x \ln 2 = \ln 6$$
$$x = \frac{\ln 6}{\ln 2} \quad \text{Exact answer}$$
$$x \approx 2.58 \quad \text{Approximate answer}$$

Alternatively, take logarithms of both sides to base 2 and apply the change of base formula:

$$\log_2 2^x = \log_2 6$$
$$x = \log_2 6$$
$$x = \frac{\ln 6}{\ln 2} \quad \text{by the-of-base formula}$$

**19.3.** Solve $2^{3x-4} = 15$.

Proceed as in the previous problem.

$$\ln 2^{3x-4} = \ln 15$$
$$(3x - 4)\ln 2 = \ln 15$$
$$3x \ln 2 - 4 \ln 2 = \ln 15$$
$$3x \ln 2 = \ln 15 + 4 \ln 2$$
$$x = \frac{\ln 15 + 4 \ln 2}{3 \ln 2} \quad \text{Exact answer}$$
$$x = 2.64 \quad \text{Approximate answer}$$

**19.4.**   Solve $5^{4-x} = 7^{3x+1}$.

Proceed as in the previous problem.

$$\ln 5^{4-x} = \ln 7^{3x+1}$$
$$(4-x)\ln 5 = (3x+1)\ln 7$$
$$4\ln 5 - x\ln 5 = 3x\ln 7 + \ln 7$$
$$4\ln 5 - \ln 7 = x\ln 5 + 3x\ln 7$$
$$x = \frac{4\ln 5 - \ln 7}{\ln 5 + 3\ln 7} \qquad \text{Exact answer}$$
$$x \approx 0.60 \qquad\qquad \text{Approximate answer}$$

**19.5.**   Solve $2^x - 2^{-x} = 1$.

Before taking logarithms of both sides it is crucial to isolate the exponential form:

$$2^x - \frac{1}{2^x} = 1 \quad \text{Multiply both sides by } 2^x$$

$$2^x \cdot 2^x - 2^x \cdot \frac{1}{2^x} = 2^x$$

$$(2^x)^2 - 1 = 2^x$$
$$(2^x)^2 - 2^x - 1 = 0$$

This equation is in quadratic form. Introduce the substitution $u = 2^x$. Then $u^2 = (2^x)^2$ and the equation becomes:

$$u^2 - u - 1 = 0$$

Now apply the quadratic formula with $a = 1$, $b = -1$, $c = -1$.

$$u = \frac{-(-1) \pm \sqrt{(-1)^2 - 4(1)(-1)}}{2(1)}$$

$$= \frac{1 \pm \sqrt{5}}{2}$$

Now undo the substitution $2^x = u$ and take logarithms of both sides.

$$2^x = \frac{1 \pm \sqrt{5}}{2}$$

$$x \ln 2 = \ln \frac{1 \pm \sqrt{5}}{2}$$

$$x = \ln\left(\frac{1 + \sqrt{5}}{2}\right)\Big/ \ln 2 \qquad \text{or} \qquad \ln\left(\frac{1 - \sqrt{5}}{2}\right)\Big/ \ln 2$$

Note that since $\dfrac{1 - \sqrt{5}}{2}$ is negative, it is not in the domain of the logarithm function. Hence the only solution is $x = \ln\left(\dfrac{1 + \sqrt{5}}{2}\right)\Big/ \ln 2$ or, approximately, 0.69.

**19.6.**   Solve $\dfrac{e^x - e^{-x}}{e^x + e^{-x}} = y$ for $x$ in terms of $y$.

First note that the left side is a complex fraction (since $e^{-x} = 1/e^x$) and write it as a simple fraction.

$$\frac{e^x - 1/e^x}{e^x + 1/e^x} = y$$

$$\frac{e^x(e^x - 1/e^x)}{e^x(e^x + 1/e^x)} = y$$

$$\frac{(e^x)^2 - 1}{(e^x)^2 + 1} = y$$

$$\frac{e^{2x} - 1}{e^{2x} + 1} = y$$

Now, isolate the exponential form $e^{2x}$.

$$e^{2x} - 1 = y(e^{2x} + 1)$$
$$e^{2x} - 1 = e^{2x}y + y$$
$$e^{2x} - e^{2x}y = 1 + y$$
$$e^{2x}(1 - y) = 1 + y$$
$$e^{2x} = \frac{1 + y}{1 - y}$$

Taking logarithms of both sides yields:

$$\ln e^{2x} = \ln\frac{1 + y}{1 - y}$$

$$2x = \ln\frac{1 + y}{1 - y}$$

$$x = \frac{1}{2}\ln\frac{1 + y}{1 - y}$$

This is valid as long as the expression $\dfrac{1 + y}{1 - y}$ is positive, that is, for $-1 < y < 1$.

**19.7.** Solve $\log_2 (3x - 4) = 5$.

Rewrite the logarithm statement in exponential form, then isolate the variable.

$$2^5 = 3x - 4$$
$$32 = 3x - 4$$
$$x = 12$$

**19.8.** Solve $\log x + \log (x + 3) = 1$.

Use the properties of logarithms to combine the logarithmic expressions into one expression, then rewrite the logarithm statement in exponential form.

$$\log [x (x + 3)] = 1$$
$$10^1 = x (x + 3)$$
$$x^2 + 3x = 10$$

This quadratic equation is solved by factoring:

$$(x + 5)(x - 2) = 0$$
$$x = -5 \text{ or } x = 2$$

Since $-5$ is not in the domain of the logarithm function, the only solution is 2.

**19.9.** Solve for $y$ in terms of $x$ and $C$: $\ln (y + 2) = x + \ln C$.

Use the properties of logarithms to combine the logarithmic expressions into one expression, then rewrite the logarithm statement in exponential form.

$$\ln (y + 2) - \ln C = x$$
$$\ln\left(\frac{y + 2}{C}\right) = x$$
$$\frac{y + 2}{C} = e^x$$
$$y = Ce^x - 2$$

**19.10.** A certain amount of money $P$ is invested at an annual rate of interest of 4.5%. How many years (to the nearest tenth of a year) would it take for the amount of money to double, assuming interest is compounded quarterly?

Use the formula $A(t) = P\left(1 + \frac{r}{n}\right)^{nt}$ from Chapter 17, with $n = 4$ and $r = 0.045$, to find $t$ when $A(t) = 2P$.

$$2P = P\left(1 + \frac{0.045}{4}\right)^{4t}$$
$$2 = \left(1 + \frac{0.045}{4}\right)^{4t}$$

To isolate $t$, take logarithms of both sides to base $e$.

$$\ln 2 = 4t \ln\left(1 + \frac{0.045}{4}\right)$$
$$t = \frac{\ln 2}{4 \ln\left(1 + \frac{0.045}{4}\right)}$$
$$t \approx 15.5 \text{ years}$$

**19.11.** In the previous problem, how many years (to the nearest tenth of a year) would it take for the amount of money to double, assuming interest is compounded continuously?

Use the formula $A(t) = Pe^{rt}$ from Chapter 17, with $r = 0.045$, to find $t$ when $A(t) = 2P$.

$$2P = Pe^{0.045t}$$
$$2 = e^{0.045t}$$

To isolate $t$, take logarithms of both sides to base $e$.

$$\ln 2 = 0.045t$$
$$t = \frac{\ln 2}{0.045}$$
$$t \approx 15.4 \text{ years}$$

**19.12.** A radioactive isotope has a half-life of 35.2 years. How many years (to the nearest tenth of a year) would it take before an initial quantity of 1 gram decays to 0.01 gram?

Use the formula $Q(t) = Q_0 e^{-kt}$ from Chapter 17.
First, determine $k$ by using $t = 35.2$, $Q_0 = 1$, and $Q(35.2) = Q_0/2 = 1/2$.

$$1/2 = 1e^{-k(35.2)}$$

To isolate $k$, take logarithms of both sides to base $e$.

$$\ln(1/2) = -k(35.2)$$
$$k = -\frac{\ln(1/2)}{35.2}$$
$$= \frac{\ln 2}{35.2}$$

Thus, for this isotope, the quantity remaining after $t$ years is given by:

$$Q(t) = Q_0 e^{\frac{-t \ln 2}{35.2}}$$

To find the time required for the initial quantity to decay to 0.01 gram, use this formula with $Q(t) = 0.01$, $Q_0 = 1$ and solve for $t$.

$$0.01 = 1 e^{\frac{-t \ln 2}{35.2}}$$

To isolate $t$, take logarithms of both sides to base $e$.

$$\ln 0.01 = \frac{-t \ln 2}{35.2}$$

$$t = \frac{-35.2 \ln 0.01}{\ln 2}$$

$$t \approx 233.9 \text{ years}$$

**19.13.** (a) Calculate the decibel level of the smallest audible sound, $I_0 = 10^{-12}$ watts per square meter.

(b) Calculate the decibel level of a rock concert at an intensity of $10^{-1}$ watts per square meter.

(c) Calculate the intensity of a sound with decibel level 85.

Use the formula $D = 10 \log \dfrac{I}{I_0}$.

(a) Set $I = I_0$. Then $D = 10 \log \dfrac{I_0}{I_0} = 10 \log 1 = 0$.

(b) Set $I = 10^{-1}$ and $I_0 = 10^{-12}$. Then $D = 10 \log \dfrac{10^{-1}}{10^{-12}} = 10 \log 10^{11} = 10 \cdot 11 = 110$ decibels.

(c) Set $D = 85$ and $I_0 = 10^{-12}$. Then $85 = 10 \log \dfrac{I}{10^{-12}}$. Solving for $I$ yields:

$$8.5 = \log \frac{I}{10^{-12}}$$

$$\frac{I}{10^{-12}} = 10^{8.5}$$

$$I = 10^{-12} \cdot 10^{8.5}$$

$$I = 10^{-3.5}$$

$$I \approx 3.2 \times 10^{-4} \text{ watts/square meter}$$

**19.14.** (a) Find the Richter scale magnitude of an earthquake that releases energy of $1000E_0$. (b) Find the energy released by an earthquake that measures 5.0 on the Richter scale, given that $E_0 = 10^{4.40}$ joules. (c) What is the ratio in energy released between an earthquake that measures 8.1 on the Richter scale and an aftershock measuring 5.4 on the scale?

Use the formula $R = \dfrac{2}{3} \log \dfrac{E}{E_0}$.

(a) Set $E = 1000E_0$. Then $R = \dfrac{2}{3} \log \dfrac{1000E_0}{E_0} = \dfrac{2}{3} \log 1000 = \dfrac{2}{3} \cdot 3 = 2$.

(b) Set $R = 5$. Then $5 = \dfrac{2}{3} \log \dfrac{E}{E_0}$. Solving for $E$ yields:

$$\frac{15}{2} = \log \frac{E}{E_0}$$

$$\frac{E}{E_0} = 10^{15/2}$$

$$E = E_0 \cdot 10^{7.5}$$

$$= 10^{4.40} \cdot 10^{7.5}$$

$$\approx 7.94 \times 10^{11} \text{ joules}$$

(c) First, solve the formula for $E$ in terms of $R$ and $R_0$.

$$\log \frac{E}{E_0} = \frac{3R}{2}$$

$$\frac{E}{E_0} = 10^{3R/2}$$

$$E = E_0 10^{3R/2}$$

Then set $R_1 = 8.1$ and $R_2 = 5.4$ and find the ratio of the corresponding energies $E_1$ and $E_2$.

$$E_1 = E_0 10^{3R_1/2} \qquad E_2 = E_0 10^{3R_2/2}$$
$$E_1 = E_0 10^{3(8.1)/2} \qquad E_2 = E_0 10^{3(5.4)/2}$$
$$E_1/E_2 = (E_0 10^{3(8.1)/2})/(E_0 10^{3(5.4)/2})$$
$$E_1/E_2 = 10^{12.15}/10^{8.1}$$
$$E_1/E_2 = 10^{4.05}/1$$
$$E_1/E_2 \approx 11{,}200/1$$

The energy released by the earthquake is more than 11,000 times the energy released by the aftershock.

## SUPPLEMENTARY PROBLEMS

**19.15.** Show that $a^b = e^{b \ln a}$.

**19.16.** Solve (a) $e^{5x-3} = 10$; (b) $5^{3+x} = 20^{x-3}$; (c) $4^{x^2-2x} = 12$.

*Ans.* (a) $x = \dfrac{3 + \ln 10}{5} \approx 1.06$; (b) $x = \dfrac{3 \ln 5 + 3 \ln 20}{\ln 20 - \ln 5} \approx 9.97$;

(c) $x = 1 \pm \sqrt{1 + \dfrac{\ln 12}{\ln 4}}$; $x \approx 2.67, -0.67$

**19.17.** Solve in terms of logarithms to base 10: (a) $2^x - 6(2^{-x}) = 6$; (b) $\dfrac{10^x - 10^{-x}}{10^x + 10^{-x}} = \dfrac{1}{2}$.

*Ans.* (a) $x = \log(3 + \sqrt{15})/\log 2$; (b) $x = (\log 3)/2$

**19.18.** Solve: (a) $\log_3 (x - 2) + \log_3 (x - 4) = 2$; (b) $2 \ln x - \ln (x + 1) = 3$

*Ans.* (a) $x = 3 + \sqrt{10} \approx 6.16$; (b) $x = \dfrac{e^3 + \sqrt{e^6 + 4e^3}}{2} \approx 21.04$

**19.19.** Solve for $t$, using natural logarithms: (a) $Q = Q_0 e^{kt}$; (b) $A = P\left(1 + \dfrac{r}{n}\right)^{nt}$.

*Ans.* (a) $t = \dfrac{1}{k} \ln \dfrac{Q}{Q_0}$; (b) $t = \dfrac{\ln (A/P)}{n \ln (1 + r/n)}$

**19.20.** Solve for $t$, using natural logarithms: (a) $I = \dfrac{V}{R}\left(1 - e^{-Rt/L}\right)$; (b) $N = \dfrac{N_0 P}{N_0 + (P - N_0)e^{-kt}}$

*Ans.* (a) $t = \dfrac{L}{R} \ln \dfrac{V}{V - RI}$; (b) $t = \dfrac{1}{k} \ln \dfrac{N(P - N_0)}{N_0(P - N)}$

**19.21.** Solve for $x$ in terms of $y$: (a) $\dfrac{e^x + e^{-x}}{2} = y$; (b) $\dfrac{e^x - e^{-x}}{2} = y$.

*Ans.* (a) $x = \ln(y \pm \sqrt{y^2 - 1})$; (b) $x = \ln(y + \sqrt{y^2 + 1})$

**19.22.** How many years would it take an investment to triple at 6% interest compounded quarterly?

*Ans.* 18.4 years

**19.23.** At what rate of interest would an investment double in eight years, compounded continuously?

*Ans.* 8.66%

**19.24.** If a sample of a radioactive isotope decays from 400 grams to 300 grams in 5.3 days, find the half-life of this isotope.

*Ans.* 12.8 days

**19.25.** If the intensity of one sound is 1000 times the intensity of another sound, what is the difference in the decibel level of the two sounds?

*Ans.* 30 decibels

**19.26.** Newton's law of cooling states that the temperature $T$ of a body, initially at temperature $T_0$, placed in a surrounding medium at a lower temperature $T_m$, is given by the formula $T = T_m + (T_0 - T_m)e^{-kt}$. If a cup of coffee, at temperature 160° at 7 a.m., is brought outside into air at 40°, and cools to 140° by 7:05 a.m., (a) find its temperature at 7:10 a.m. (b) At what time will the temperature have fallen to 100°?

*Ans.* (a) 123°; (b) 7:19 a.m.

# Trigonometric Functions

## Unit Circle

The unit circle is the circle $U$ with center $(0,0)$ and radius 1. The equation of the unit circle is $x^2 + y^2 = 1$. The circumference of the unit circle is $2\pi$.

**EXAMPLE 20.1**  Draw a unit circle and indicate its intercepts (see Fig. 20-1).

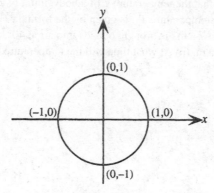

Figure 20-1

## Points on a Unit Circle

A unique point $P$ on a unit circle $U$ can be associated with any given real number $t$ in the following manner:

1. Associated with $t = 0$ is the point $(1,0)$.
2. Associated with any *positive* real number $t$ is the point $P(x,y)$ found by proceeding a distance $|t|$ in the *counterclockwise* direction from the point $(1,0)$ (see Fig. 20-2).
3. Associated with any *negative* real number $t$ is the point $P(x,y)$ found by proceeding a distance $|t|$ in the *clockwise* direction from the point $(1,0)$ (see Fig. 20-3).

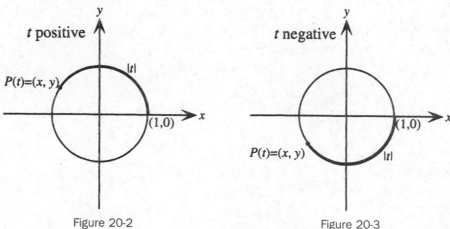

Figure 20-2                                      Figure 20-3

## Definition of the Trigonometric Functions

If $t$ is a real number and $P(x,y)$ is the point, referred to as $P(t)$, on the unit circle $U$ that corresponds to $P$, then the six *trigonometric functions* of $t$—sine, cosine, tangent, cosecant, secant, and cotangent, abbreviated sin, cos, tan, csc, sec, and cot, respectively—are defined as follows:

$$\sin t = y \qquad\qquad \csc t = \frac{1}{y} \text{ (if } y \neq 0)$$

$$\cos t = x \qquad\qquad \sec t = \frac{1}{x} \text{ (if } x \neq 0)$$

$$\tan t = \frac{y}{x} \text{ (if } x \neq 0) \qquad \cot t = \frac{x}{y} \text{ (if } y \neq 0)$$

**EXAMPLE 20.2**  If $t$ is a real number such that $P\left(\frac{3}{5}, -\frac{4}{5}\right)$ is the point on the unit circle that corresponds to $t$, find the six trigonometric functions of $t$.

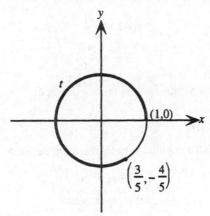

Figure 20-4

Since the $x$ coordinate of $P$ is $\frac{3}{5}$ and the $y$ coordinate of $P$ is $-\frac{4}{5}$, the six trigonometric functions of $t$ are as follows:

$$\sin t = y = -\frac{4}{5} \qquad \cos t = x = \frac{3}{5} \qquad \tan t = \frac{y}{x} = \frac{-4/5}{3/5} = -\frac{4}{3}$$

$$\csc t = \frac{1}{y} = \frac{1}{-4/5} = -\frac{5}{4} \qquad \sec t = \frac{1}{x} = \frac{1}{3/5} = \frac{5}{3} \qquad \cot t = \frac{x}{y} = \frac{3/5}{-4/5} = -\frac{3}{4}$$

## Symmetries of the Points on a Unit Circle

For any real number $t$, the following relations can be shown to hold:

1. $P(t + 2\pi) = P(t)$.
2. If $P(t) = (x,y)$, then $P(-t) = (x,-y)$.
3. If $P(t) = (x,y)$, then $P(t + \pi) = (-x,-y)$.

## Periodic Functions

A function $f$ is called *periodic* if there exists a real number $p$ such that $f(t + p) = f(t)$ for every real number $t$ in the domain of $f$. The smallest such real number is called *the period* of the function.

## Periodicity of the Trigonometric Functions

The trigonometric functions are all periodic. The following important relations can be shown to hold:

$$\sin(t + 2\pi) = \sin t \qquad \cos(t + 2\pi) = \cos t \qquad \tan(t + \pi) = \tan t$$
$$\csc(t + 2\pi) = \csc t \qquad \sec(t + 2\pi) = \sec t \qquad \cot(t + \pi) = \cot t$$

## Notation

Notation for exponents: The expressions for the squares of the trigonometric functions arise frequently. $(\sin t)^2$ is generally written $\sin^2 t$, $(\cos t)^2$ is generally written $\cos^2 t$, and so on. Similarly, $(\sin t)^3$ is generally written $\sin^3 t$, and so on.

## Identities

An identity is an equation that is true for all values of the variables it contains, as long as both sides are meaningful.

## Trigonometric Identities

1. **PYTHAGOREAN IDENTITIES.** For all $t$ for which both sides are defined:

$$\cos^2 t + \sin^2 t = 1 \qquad 1 + \tan^2 t = \sec^2 t \qquad \cot^2 t + 1 = \csc^2 t$$

$$\cos^2 t = 1 - \sin^2 t \qquad \tan^2 t = \sec^2 t - 1 \qquad \cot^2 t = \csc^2 t - 1$$

$$\sin^2 t = 1 - \cos^2 t \qquad 1 = \sec^2 t - \tan^2 t \qquad 1 = \csc^2 t - \cot^2 t$$

2. **RECIPROCAL IDENTITIES.** For all $t$ for which both sides are defined:

$$\sin t = \frac{1}{\csc t} \qquad\qquad \cos t = \frac{1}{\sec t} \qquad\qquad \tan t = \frac{1}{\cot t}$$

$$\csc t = \frac{1}{\sin t} \qquad\qquad \sec t = \frac{1}{\cos t} \qquad\qquad \cot t = \frac{1}{\tan t}$$

3. **QUOTIENT IDENTITIES.** For all $t$ for which both sides are defined:

$$\tan t = \frac{\sin t}{\cos t} \qquad\qquad\qquad \cot t = \frac{\cos t}{\sin t}$$

4. **IDENTITIES FOR NEGATIVES.** For all $t$ for which both sides are defined:

$$\sin(-t) = -\sin t \qquad\qquad \cos(-t) = \cos t \qquad\qquad \tan(-t) = -\tan t$$

$$\csc(-t) = -\csc t \qquad\qquad \sec(-t) = \sec t \qquad\qquad \cot(-t) = -\cot t$$

## SOLVED PROBLEMS

**20.1.** Find the domain and range of the sine and cosine functions.

For any real number $t$, a unique point $P(t) = (x,y)$ on the unit circle $x^2 + y^2 = 1$ is associated with $t$. Since $\sin t = y$ and $\cos t = x$ are defined for all $t$, the domain of the sine and cosine functions is $\mathbf{R}$. Since $y$ and $x$ are coordinates of points on the unit circle, $-1 \le y \le 1$ and $-1 \le x \le 1$, hence the range of the sine and cosine functions is given by $-1 \le \sin t \le 1$ and $-1 \le \cos t \le 1$, that is, $[-1,1]$.

**20.2.** For what values of $t$ is the $y$-coordinate of $P(t)$ equal to zero?

See Fig. 20-5.

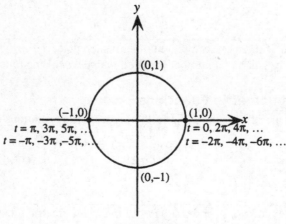

Figure 20-5

By definition, $P(0) = (1,0)$. Since the perimeter of the unit circle is $2\pi$, if $t$ is any positive or negative integer multiple of $2\pi$, then again $P(t) = (1,0)$.

Since $\pi$ is half the perimeter of the unit circle, $P(\pi)$ is halfway around the unit circle from $(1,0)$; that is, $P(\pi) = (-1,0)$. Furthermore, if $t$ is equal to $\pi$ plus any positive or negative integer multiple of $2\pi$, then again $P(t) = (-1,0)$.

Summarizing, the $y$-coordinate of $P(t)$ is equal to zero if $t$ is any integer multiple of $\pi$; thus, $n\pi$.

**20.3.** For what values of $t$ is the $x$-coordinate of $P(t)$ equal to zero?

See Fig. 20-6. Since the perimeter of the unit circle is $2\pi$, one-fourth of the perimeter is $\pi/2$. Thus $P(\pi/2)$ is one-fourth of the way around the unit circle from $(1,0)$; that is, $P(\pi/2) = (0,1)$. Also, if $t$ is equal to $\pi/2$ plus any positive or negative integer multiple of $2\pi$, then again $P(t) = (0,1)$.

Next, if $t = \pi + \pi/2$, or $3\pi/2$, then $t$ is three-fourths of the way around the unit circle from $(1,0)$; that is, $P(3\pi/2) = (0,-1)$. And if $t$ is equal to $3\pi/2$ plus any positive or negative integer multiple of $2\pi$, then again $P(t) = (0,-1)$.

Summarizing, the $x$-coordinate of $P(t)$ is equal to zero if $t$ is $\pi/2$ or $3\pi/2$ plus any integer multiple of $2\pi$; thus, $\pi/2 + 2\pi n$ or $3\pi/2 + 2\pi n$.

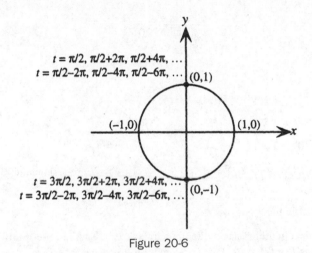

Figure 20-6

**20.4.** Find the domains of the tangent and secant functions.

For any real number $t$, a unique point $P(t) = (x, y)$ on the unit circle $x^2 + y^2 = 1$ is associated with $t$. Since $\tan t$ is defined as $y/x$ and $\sec t$ is defined as $1/x$, each function is defined for all values of $t$ except those for which $x = 0$. From Problem 20.3, these values are $\pi/2 + 2\pi n$ or $3\pi/2 + 2\pi n$, $n$ any integer. Thus, the domains of the tangent and secant functions are $\{t \in \mathbf{R} | t \neq \pi/2 + 2\pi n, 3\pi/2 + 2\pi n\}$, $n$ any integer.

**20.5.** Find the domains of the cotangent and cosecant functions.

For any real number $t$, a unique point $P(t) = (x, y)$ on the unit circle $x^2 + y^2 = 1$ is associated with $t$. Since $\cot t$ is defined as $x/y$ and $\csc t$ is defined as $1/y$, each function is defined for all values of $t$ except those for which $y = 0$. From Problem 20.2, these values are $n\pi$, for $n$ any integer. Thus, the domains of the cotangent and cosecant functions are $\{t \in \mathbf{R} | t \neq n\pi\}$, $n$ any integer.

**20.6** Find the ranges of the tangent, cotangent, secant, and cosecant functions.

For any real number $t$, a unique point $P(t) = (x, y)$ on the unit circle $x^2 + y^2 = 1$ is associated with $t$. Since $\tan t$ is defined as $y/x$ and $\cot t$ is defined as $x/y$, and, since, for various values of $t$, $x$ may be greater than $y$, less than $y$, or equal to $y$, $\tan t = y/x$ and $\cot t = x/y$ may assume any real value. Thus, the ranges of the tangent and cotangent functions are both $\mathbf{R}$.

Since $\sec t$ is defined as $1/x$ and $\csc t$ is defined as $1/y$, and for any point on the unit circle, $-1 \leq x \leq 1$ and $-1 \leq y \leq 1$, it follows that $|1/x| \geq 1$ and $|1/y| \geq 1$, that is, $|\sec t| \geq 1$ and $|\csc t| \geq 1$. Thus, the ranges of the secant and cosecant functions are both $(-\infty, -1] \cup [1, \infty)$.

**20.7.** Find the six trigonometric functions of 0.

See Fig. 20-7.

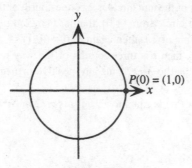

Figure 20-7

Since $P(0) = (1, 0) = (x, y)$, it follows that

$$\sin(0) = y = 0 \qquad\qquad \csc(0) = 1/y = 1/0 \text{ is undefined}$$

$$\cos(0) = x = 1 \qquad\qquad \sec(0) = 1/x = 1/1 = 1$$

$$\tan(0) = y/x = 0/1 = 0 \qquad\qquad \cot(0) = x/y = 1/0 \text{ is undefined}$$

**20.8.** Find the six trigonometric functions of $\pi/2$.

See Fig. 20-8. Since the circumference of the unit circle is $2\pi$, $P(\pi/2)$ is one-fourth of the way around the unit circle from $(1, 0)$. Thus $P(\pi/2) = (0, 1) = (x, y)$ and it follows that

$$\sin(\pi/2) = y = 1 \qquad\qquad \csc(\pi/2) = 1/y = 1/1 = 1$$

$$\cos(\pi/2) = x = 0 \qquad\qquad \sec(\pi/2) = 1/x = 1/0 \text{ is undefined}$$

$$\tan(\pi/2) = y/x = 1/0 \text{ is undefined} \qquad\qquad \cot(\pi/2) = x/y = 0/1 = 0$$

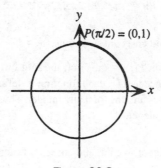

Figure 20-8

**20.9.** If $P(t)$ is in a quadrant, it is said that $t$ is in that quadrant. For $t$ in each of the four quadrants, derive the following table showing the signs of the six trigonometric functions of $t$.

|         | QUADRANT I | QUADRANT II | QUADRANT III | QUADRANT IV |
|---------|------------|-------------|--------------|-------------|
| $\sin t$ | + | + | − | − |
| $\cos t$ | + | − | − | + |
| $\tan t$ | + | − | + | − |
| $\csc t$ | + | + | − | − |
| $\sec t$ | + | − | − | + |
| $\cot t$ | + | − | + | − |

Since $\sin t = y$ and $\csc t = 1/y$, and $y$ is positive in quadrants I and II, and negative in quadrants III and IV, the signs of $\sin t$ and $\csc t$ are as shown.

Since $\cos t = x$ and $\sec t = 1/x$, and $x$ is positive in quadrants I and IV, and negative in quadrants II and III, the signs of $\cos t$ and $\sec t$ are as shown.

Since $\tan t = y/x$ and $\cot t = x/y$, and $x$ and $y$ have the same signs in quadrants I and III and opposite signs in quadrants II and IV, the signs of $\tan t$ and $\cot t$ are as shown.

**20.10.** Find the six trigonometric functions of $\pi/4$.

See Fig. 20-9. Since $\pi/4$ is one-half the way from 0 to $\pi/2$, the point $P(\pi/4) = (x, y)$ lies on the line $y = x$. Thus the coordinates $(x, y)$ satisfy both $x^2 + y^2 = 1$ and $y = x$. Substituting yields:

$$x^2 + x^2 = 1$$
$$2x^2 = 1$$
$$x^2 = 1/2$$
$$x = 1/\sqrt{2} \qquad \text{since } x \text{ is positive}$$

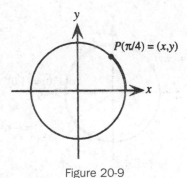

Figure 20-9

Hence $P(\pi/4) = (x, y) = \left(1/\sqrt{2}, 1/\sqrt{2}\right)$. Hence it follows that:

$$\sin\left(\frac{\pi}{4}\right) = y = \frac{1}{\sqrt{2}} \qquad\qquad \csc\left(\frac{\pi}{4}\right) = \frac{1}{y} = \frac{1}{1/\sqrt{2}} = \sqrt{2}$$

$$\cos\left(\frac{\pi}{4}\right) = x = \frac{1}{\sqrt{2}} \qquad\qquad \sec\left(\frac{\pi}{4}\right) = \frac{1}{x} = \frac{1}{1/\sqrt{2}} = \sqrt{2}$$

$$\tan\left(\frac{\pi}{4}\right) = \frac{y}{x} = \frac{1/\sqrt{2}}{1/\sqrt{2}} = 1 \qquad\qquad \cot\left(\frac{\pi}{4}\right) = \frac{x}{y} = \frac{1/\sqrt{2}}{1/\sqrt{2}} = 1$$

**20.11.** Prove the symmetry properties listed on page 177 for points on a unit circle.

    (a) For any real number $t$, $P(t + 2\pi) = P(t)$.

    (b) If $P(t) = (x,y)$, then $P(-t) = (x,-y)$.

    (c) If $P(t) = (x,y)$, then $P(t + \pi) = (-x,-y)$.

    (a) Let $P(t) = (x, y)$. Since the circumference of the unit circle is precisely $2\pi$, the point $P(t + 2\pi)$ is obtained by going exactly once around the unit circle from $P(t)$. Thus the coordinates of $P(t + 2\pi)$ are the same as those of $P(t)$.

    (b) See Fig. 20-10.

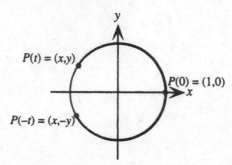

Figure 20-10

Let $P(t) = (x, y)$. Since $P(t)$ and $P(-t)$ are obtained by going the same distance around the unit circle from the same point, $P(0)$, the coordinates of the two points will be equal in absolute value. The $x$-coordinates of the two points will be the same; however, since the two points are reflections of each other with respect to the $x$-axis, the $y$-coordinates of the points will be opposite in sign. Hence the coordinates of $P(-t)$ are $(x,-y)$.

    (c) See Fig. 20-11.

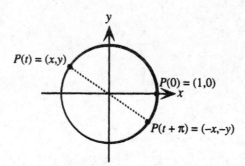

Figure 20-11

Let $P(t) = (x, y)$. Since $P(t + \pi)$ is obtained by going halfway around the unit circle from $P(t)$, the two points are at opposite ends of a diameter, hence they are reflections of each other with respect to the origin. Hence $P(t + \pi) = (-x, -y)$.

**20.12.** Find the six trigonometric functions of $5\pi/4$.

    Since $\dfrac{5\pi}{4} = \dfrac{\pi}{4} + \pi$, and $P\!\left(\dfrac{\pi}{4}\right) = \left(\dfrac{1}{\sqrt{2}}, \dfrac{1}{\sqrt{2}}\right)$, it follows that $P\!\left(\dfrac{5\pi}{4}\right) = \left(-\dfrac{1}{\sqrt{2}}, -\dfrac{1}{\sqrt{2}}\right)$. Hence the six trigonometric functions of $5\pi/4$ are:

$$\sin\!\left(\frac{5\pi}{4}\right) = y = -\frac{1}{\sqrt{2}} \qquad \csc\!\left(\frac{5\pi}{4}\right) = \frac{1}{y} = \frac{1}{-1/\sqrt{2}} = -\sqrt{2}$$

$$\cos\left(\frac{5\pi}{4}\right) = x = -\frac{1}{\sqrt{2}} \qquad \sec\left(\frac{5\pi}{4}\right) = \frac{1}{x} = \frac{1}{-1/\sqrt{2}} = -\sqrt{2}$$

$$\tan\left(\frac{5\pi}{4}\right) = \frac{y}{x} = \frac{-1/\sqrt{2}}{-1/\sqrt{2}} = 1 \qquad \cot\left(\frac{5\pi}{4}\right) = \frac{x}{y} = \frac{-1/\sqrt{2}}{-1/\sqrt{2}} = 1$$

**20.13.** Prove the periodicity properties for the sine, cosine, and tangent functions.

Let $P(t) = (x, y)$; then $P(t + 2\pi) = P(t) = (x, y)$. It follows immediately that $\sin(t + 2\pi) = y = \sin t$ and $\cos(t + 2\pi) = x = \cos t$.

Also, $P(t + \pi) = (-x, -y)$. Hence $\tan(t + \pi) = \frac{-y}{-x} = \frac{y}{x} = \tan t$.

**20.14.** Prove the reciprocal identities.

Let $P(t) = (x, y)$; then it follows that:

$$\csc t = \frac{1}{y} = \frac{1}{\sin t} \qquad \sec t = \frac{1}{x} = \frac{1}{\cos t} \qquad \cot t = \frac{x}{y} = 1 \div \frac{y}{x} = 1 \div (\tan t) = \frac{1}{\tan t}$$

Hence it follows by algebra that:

$$\sin t = \frac{1}{\csc t} \qquad \cos t = \frac{1}{\sec t} \qquad \tan t = \frac{1}{\cot t}$$

**20.15.** Prove the periodicity properties for the cosecant, secant, and cotangent functions.

Use the reciprocal identities and the periodicity properties for sine, cosine, and tangent.

$$\csc(t + 2\pi) = \frac{1}{\sin(t + 2\pi)} = \frac{1}{\sin t} = \csc t$$

$$\sec(t + 2\pi) = \frac{1}{\cos(t + 2\pi)} = \frac{1}{\cos t} = \sec t$$

$$\cot(t + \pi) = \frac{1}{\tan(t + \pi)} = \frac{1}{\tan t} = \cot t$$

**20.16.** Prove the quotient identities.

Let $P(t) = (x, y)$, then it follows that:

$$\tan t = \frac{y}{x} = \frac{\sin t}{\cos t} \qquad \text{and} \qquad \cot t = \frac{x}{y} = \frac{\cos t}{\sin t}$$

**20.17.** Find the six trigonometric functions of $5\pi/2$.

Since $\frac{5\pi}{2} = 2\pi + \frac{\pi}{2}$ and $P\left(\frac{\pi}{2}\right) = (0, 1)$, it follows that $P\left(\frac{5\pi}{2}\right) = (0, 1)$ and

$\sin(5\pi/2) = y = 1 \qquad\qquad \csc(5\pi/2) = 1/y = 1/1 = 1$

$\cos(5\pi/2) = x = 0 \qquad\qquad \sec(5\pi/2) = 1/x = 1/0$ is undefined

$\tan(5\pi/2) = y/x = 1/0$ is undefined $\qquad \cot(5\pi/2) = x/y = 0/1 = 0$

**20.18.** Prove the identities for negatives,

Let $P(t) = (x, y)$. Then $P(-t) = (x, -y)$ by the symmetry properties of points on a unit circle. It follows that:

$\sin(-t) = -y = -\sin t \qquad\qquad \csc(-t) = \dfrac{1}{\sin(-t)} = \dfrac{1}{-\sin t} = -\dfrac{1}{\sin t} = -\csc t$

$\cos(-t) = x = \cos t \qquad\qquad \sec(-t) = \dfrac{1}{\cos(-t)} = \dfrac{1}{\cos t} = \sec t$

$\tan(-t) = \dfrac{-y}{x} = -\dfrac{y}{x} = -\tan t \qquad\qquad \cot(-t) = \dfrac{x}{-y} = -\dfrac{x}{y} = -\cot t$

**20.19.** Find the six trigonometric functions of $-\pi/4$.

Use the identities for negatives and the results of Problem 20.10.

$$\sin\left(-\frac{\pi}{4}\right) = -\sin\left(\frac{\pi}{4}\right) = -\frac{1}{\sqrt{2}} \qquad \cos\left(-\frac{\pi}{4}\right) = \cos\left(\frac{\pi}{4}\right) = \frac{1}{\sqrt{2}} \qquad \tan\left(-\frac{\pi}{4}\right) = -\tan\left(\frac{\pi}{4}\right) = -1$$

$$\csc\left(-\frac{\pi}{4}\right) = -\csc\left(\frac{\pi}{4}\right) = -\sqrt{2} \qquad \sec\left(-\frac{\pi}{4}\right) = \sec\left(\frac{\pi}{4}\right) = \sqrt{2} \qquad \cot\left(-\frac{\pi}{4}\right) = -\cot\left(\frac{\pi}{4}\right) = -1$$

**20.20.** Prove the Pythagorean identity $\cos^2 t + \sin^2 t = 1$.

For any real number $t$, a unique point $P(t) = (x, y)$ on the unit circle $x^2 + y^2 = 1$ is associated with $t$. By definition, $\cos t = x$ and $\sin t = y$; hence for any $t$, $(\cos t)^2 + (\sin t)^2 = 1$, that is,

$$\cos^2 t + \sin^2 t = 1$$

**20.21.** Prove the Pythagorean identity $1 + \tan^2 t = \sec^2 t$.

Start with $\cos^2 t + \sin^2 t = 1$ and divide both sides by $\cos^2 t$. Then it follows that:

$$\frac{\cos^2 t}{\cos^2 t} + \frac{\sin^2 t}{\cos^2 t} = \frac{1}{\cos^2 t}$$

$$1 + \left(\frac{\sin t}{\cos t}\right)^2 = \left(\frac{1}{\cos t}\right)^2$$

$$1 + \tan^2 t = \sec^2 t$$

**20.22.** Given $\sin t = \frac{1}{2}$ and $t$ in quadrant II, find the other five trigonometric functions of $t$.

1. Cosine. From the Pythagorean identity, $\cos^2 t = 1 - \sin^2 t$. Since $t$ is specified in quadrant II, $\cos t$ must be negative (see Problem 20.9). Hence,

$$\cos t = -\sqrt{1 - \sin^2 t} = -\sqrt{1 - \left(\frac{1}{2}\right)^2} = -\sqrt{\frac{3}{4}} = -\frac{\sqrt{3}}{2}$$

2. Tangent. From the quotient identity,

$$\tan t = \frac{\sin t}{\cos t} = \frac{\frac{1}{2}}{-\sqrt{3}/2} = -\frac{1}{\sqrt{3}}$$

3. Cotangent. From the reciprocal identity,

$$\cot t = \frac{1}{\tan t} = \frac{1}{-1/\sqrt{3}} = -\sqrt{3}$$

4. Secant. From the reciprocal identity,

$$\sec t = \frac{1}{\cos t} = \frac{1}{-\sqrt{3}/2} = -\frac{2}{\sqrt{3}}$$

5. Cosecant. From the reciprocal identity,

$$\csc t = \frac{1}{\sin t} = \frac{1}{\frac{1}{2}} = 2$$

**20.23.** Given $\tan t = -2$ and $t$ in quadrant IV, find the other five trigonometric functions of $t$.

1. Secant. From the Pythagorean identity, $\sec^2 t = 1 + \tan^2 t$. Since $t$ is specified in quadrant IV, $\sec t$ must be positive (see Problem 20.9). Hence,

$$\sec t = \sqrt{1 + \tan^2 t} = \sqrt{1 + (-2)^2} = \sqrt{5}$$

2. Cosine. From the reciprocal identity,

$$\cos t = \frac{1}{\sec t} = \frac{1}{\sqrt{5}}$$

3. Sine. From the quotient identity, $\tan t = \frac{\sin t}{\cos t}$; hence,

$$\sin t = \tan t \cos t = (-2)\frac{1}{\sqrt{5}} = -\frac{2}{\sqrt{5}}$$

4. Cotangent. From the reciprocal identity,

$$\cot t = \frac{1}{\tan t} = \frac{1}{-2} = -\frac{1}{2}$$

5. Cosecant. From the reciprocal identity,

$$\csc t = \frac{1}{\sin t} = \frac{1}{-2/\sqrt{5}} = -\frac{\sqrt{5}}{2}$$

**20.24.** For an arbitrary value of $t$ express the other trigonometric functions in terms of $\sin t$.

1. Cosine. From the Pythagorean identity, $\cos^2 t = 1 - \sin^2 t$. Hence $\cos t = \pm\sqrt{1 - \sin^2 t}$.

2. Tangent. From the quotient identity, $\tan t = \frac{\sin t}{\cos t}$. Using the previous result, $\tan t = \pm\frac{\sin t}{\sqrt{1 - \sin^2 t}}$.

3. Cotangent. From the quotient identity, $\cot t = \frac{\cos t}{\sin t}$. Hence $\cot t = \pm\frac{\sqrt{1 - \sin^2 t}}{\sin t}$.

4. Secant. From the reciprocal identity, $\sec t = \frac{1}{\cos t}$. Hence $\sec t = \frac{1}{\pm\sqrt{1 - \sin^2 t}} = \pm\frac{1}{\sqrt{1 - \sin^2 t}}$.

5. Cosecant. From the reciprocal identity, $\csc t = \frac{1}{\sin t}$.

## SUPPLEMENTARY PROBLEMS

**20.25.** If $t$ is a point on the unit circle with coordinates $\left(-\frac{5}{13}, -\frac{12}{13}\right)$, find the six trigonometric functions of $t$.

*Ans.* $\sin t = -12/13$, $\cos t = -5/13$, $\tan t = 12/5$, $\cot t = 5/12$, $\sec t = -13/5$, $\csc t = -13/12$

**20.26.** If $t$ is a point on the unit circle with coordinates $\left(\frac{2}{\sqrt{5}}, -\frac{1}{\sqrt{5}}\right)$, find the six trigonometric functions of $t$.

*Ans.* $\sin t = -1/\sqrt{5}$, $\cos t = 2/\sqrt{5}$, $\tan t = -1/2$, $\cot t = -2$, $\sec t = \sqrt{5}/2$, $\csc t = -\sqrt{5}$

**20.27.** Find the six trigonometric functions of $\pi$.

*Ans.* $\sin \pi = 0$, $\cos \pi = -1$, $\tan \pi = 0$, $\cot \pi$ is undefined, $\sec \pi = -1$, $\csc \pi$ is undefined.

**20.28.** Find the six trigonometric functions of $-\pi/2$.

*Ans.* $\sin(-\pi/2) = -1$, $\cos(-\pi/2) = 0$, $\tan(-\pi/2)$ is undefined,

$\cot(-\pi/2) = 0$, $\sec(-\pi/2)$ is undefined, $\csc(-\pi/2) = -1$

**20.29.** Find the six trigonometric functions of $7\pi/4$.

*Ans.* $\sin(7\pi/4) = -1/\sqrt{2}$, $\cos(7\pi/4) = 1/\sqrt{2}$, $\tan(7\pi/4) = -1$,

$\cot(7\pi/4) = -1$, $\sec(7\pi/4) = \sqrt{2}$, $\csc(7\pi/4) = -\sqrt{2}$

**20.30.** Prove that for all $t$, $\sin(t + 2\pi n) = \sin t$ for any integer value of $n$.

**20.31.** Prove the Pythagorean identity $\cot^2 t + 1 = \csc^2 t$.

**20.32.** Given $\cos t = 2/5$ and $t$ in quadrant I, find the other five trigonometric functions of $t$.

    *Ans.*  $\sin t = \sqrt{21}/5$,  $\tan t = \sqrt{21}/2$,  $\cot t = 2/\sqrt{21}$,  $\sec t = 5/2$,  $\csc t = 5/\sqrt{21}$

**20.33.** Given $\tan t = -2/3$ and $t$ in quadrant IV, find the other five trigonometric functions of $t$.

    *Ans.*  $\sin t = -2/\sqrt{13}$,  $\cos t = 3/\sqrt{13}$,  $\cot t = -3/2$,  $\sec t = \sqrt{13}/3$,  $\csc t = -\sqrt{13}/2$

**20.34.** Given $\cot t = \sqrt{5}$ and $t$ in quadrant III, find the other five trigonometric functions of $t$.

    *Ans.*  $\sin t = -1/\sqrt{6}$,  $\cos t = -\sqrt{5}/\sqrt{6}$,  $\tan t = 1/\sqrt{5}$,  $\sec t = -\sqrt{6}/\sqrt{5}$,  $\csc t = -\sqrt{6}$

**20.35.** Given $\sec t = -\frac{13}{5}$ and $t$ in quadrant II, find the other five trigonometric functions of $t$.

    *Ans.*  $\sin t = \frac{12}{13}$,  $\cos t = -\frac{5}{13}$,  $\tan t = -\frac{12}{5}$,  $\cot t = -\frac{5}{12}$,  $\csc t = \frac{13}{12}$

**20.36.** Given $\sin t = a$ and $t$ in quadrant II, find the other five trigonometric functions of $t$.

    *Ans.*  $\cos t = -\sqrt{1 - a^2}$,  $\tan t = -\dfrac{a}{\sqrt{1 - a^2}}$,  $\cot t = \dfrac{-\sqrt{1 - a^2}}{a}$,  $\sec t = -\dfrac{1}{\sqrt{1 - a^2}}$,  $\csc t = \dfrac{1}{a}$

**20.37.** Given $\cos t = a$ and $t$ in quadrant IV, find the other five trigonometric functions of $t$.

    *Ans.*  $\sin t = -\sqrt{1 - a^2}$,  $\tan t = \dfrac{-\sqrt{1 - a^2}}{a}$,  $\cot t = -\dfrac{a}{\sqrt{1 - a^2}}$,  $\sec t = \dfrac{1}{a}$,  $\csc t = -\dfrac{1}{\sqrt{1 - a^2}}$

**20.38.** Given $\tan t = a$ and $t$ in quadrant II, find the other five trigonometric functions of $t$.

    *Ans.*  $\sin t = -\dfrac{a}{\sqrt{a^2 + 1}}$,  $\cos t = -\dfrac{1}{\sqrt{a^2 + 1}}$,  $\cot t = \dfrac{1}{a}$,  $\sec t = -\sqrt{a^2 + 1}$,  $\csc t = \dfrac{-\sqrt{a^2 + 1}}{a}$

**20.39.** For an arbitrary value of $t$, express the other trigonometric functions in terms of $\tan t$.

    *Ans.*  $\sin t = \pm\dfrac{\tan t}{\sqrt{1 + \tan^2 t}}$,  $\cos t = \pm\dfrac{1}{\sqrt{1 + \tan^2 t}}$,  $\cot t = \dfrac{1}{\tan t}$,

    $\sec t = \pm\sqrt{1 + \tan^2 t}$,  $\csc t = \pm\dfrac{\sqrt{1 + \tan^2 t}}{\tan t}$

**20.40.** For an arbitrary value of $t$, express the other trigonometric functions in terms of $\cos t$.

    *Ans.*  $\sin t = \pm\sqrt{1 - \cos^2 t}$,  $\tan t = \pm\dfrac{\sqrt{1 - \cos^2 t}}{\cos t}$,  $\cot t = \pm\dfrac{\cos t}{\sqrt{1 - \cos^2 t}}$,

    $\sec t = \dfrac{1}{\cos t}$,  $\csc t = \pm\dfrac{1}{\sqrt{1 - \cos^2 t}}$

**20.41.** Show that cosine and secant are even functions.

**20.42.** Show that sine, tangent, cotangent, and cosecant are odd functions.

# CHAPTER 21

# Graphs of Trigonometric Functions

## Graphs of Basic Sine and Cosine Functions

The domains of $f(t) = \sin t$ and $f(t) = \cos t$ are identical: all real numbers, $\boldsymbol{R}$. The ranges of these functions are also identical: the interval $[-1, 1]$. The graph of $u = \sin t$ is shown in Fig. 21-1.

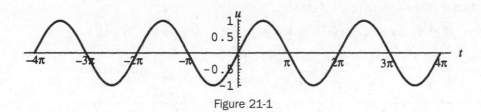

Figure 21-1

The graph of $u = \cos t$ is shown in Fig. 21-2.

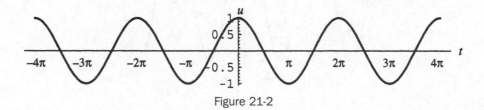

Figure 21-2

## Properties of the Basic Graphs

The function $f(t) = \sin t$ is periodic with period $2\pi$. Its graph repeats a *cycle*, regarded as the portion of the graph for $0 \le t \le 2\pi$. The graph is often referred to as the *basic sine curve*. The *amplitude* of the basic sine curve, defined as half the difference between the maximum and minimum values of the function, is 1. The function $f(t) = \cos t$ is also periodic with period $2\pi$. Its graph, called the *basic cosine curve*, also repeats a cycle, regarded as the portion of this graph for $0 \le t \le 2\pi$. The graph can also be thought of as a sine curve with amplitude 1, shifted left by an amount $\pi/2$.

## Graphs of Other Sine and Cosine Functions

The graphs of the following are variations of the basic sine and cosine curves.

1. **GRAPHS OF** $u = A \sin t$ **AND** $u = A \cos t$. The graph of $u = A \sin t$ for positive $A$ is a basic sine curve, but stretched by a factor of $A$, hence with amplitude $A$, referred to as a *standard* sine curve. The graph of

$u = A \sin t$ for negative $A$ is a standard sine curve with amplitude $|A|$, reflected with respect to the x-axis, called an *upside-down sine curve*. Similarly, the graph of $u = A \cos t$ for positive $A$ is a basic cosine curve with amplitude $|A|$, referred to as a *standard* cosine curve. The graph of $u = A \cos t$ for negative $A$ is a standard cosine curve with amplitude $|A|$, reflected with respect to the x-axis, called an *upside-down cosine curve*.

2. **GRAPHS OF** $u = \sin bt$ **AND** $u = \cos bt$ ($b$ positive). The graph of $u = \sin bt$ is a standard sine curve, compressed by a factor of $b$ with respect to the x-axis, hence with period $2\pi/b$. The graph of $u = \cos bt$ is a standard cosine curve with period $2\pi/b$.

3. **GRAPHS OF** $u = \sin(t - c)$ **AND** $u = \cos(t - c)$. The graph of $u = \sin(t - c)$ is a standard sine curve shifted to the right $|c|$ units if $c$ is positive, shifted to the left $|c|$ units if $c$ is negative. The graph of $u = \cos(t - c)$ is a standard cosine curve shifted to the right $|c|$ units if $c$ is positive, shifted to the left $|c|$ units if $c$ is negative. $c$ is referred to as the *phase shift*. (*Note*: The definition of phase shift is not universally agreed upon.)

4. **GRAPHS OF** $u = \sin t + d$ **AND** $u = \cos t + d$. The graph of $u = \sin t + d$ is a standard sine curve shifted up $|d|$ units if $d$ is positive, shifted down $|d|$ units if $d$ is negative. The graph of $u = \cos t + d$ is a standard cosine curve shifted up $|d|$ units if $d$ is positive, shifted down $|d|$ units if $d$ is negative.

5. **GRAPHS OF** $u = A \sin(bt - c) + d$ **AND** $u = A \cos(bt - c) + d$ display combinations of the above features. In general, assuming $A$, $b$, $c$, $d$ positive, the graphs are standard sine and cosine curves, respectively, with amplitude $A$, period $2\pi/b$, phase shift $c/b$, shifted up $d$ units.

**EXAMPLE 21.1**   Sketch a graph of $u = 3 \cos t$.

The graph (Fig. 21-3) is a standard cosine curve with amplitude 3 and period $2\pi$.

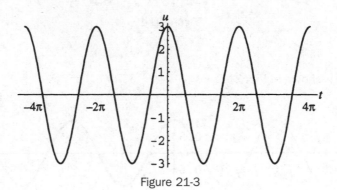

Figure 21-3

**EXAMPLE 21.2**   Sketch a graph of $u = -2 \sin 2t$.

The graph (Fig. 21-4) is an upside-down sine curve with amplitude $|-2| = 2$ and period $2\pi/2 = \pi$.

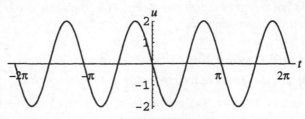

Figure 21-4

## Graphs of the Other Trigonometric Functions

1. **TANGENT.** The domain of the tangent function is $\{t \in R | t \neq \pi/2 + 2\pi n, 3\pi/2 + 2\pi n\}$ and the range is $R$. The graph is shown in Fig. 21-5.

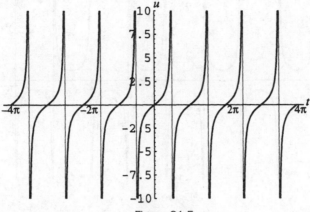

Figure 21-5

2. **SECANT.** The domain of the secant function is $\{t \in R | t \neq \pi/2 + 2\pi n, 3\pi/2 + 2\pi n\}$ and the range is $(-\infty, -1] \cup [1, \infty)$. The graph is shown in Fig. 21-6.

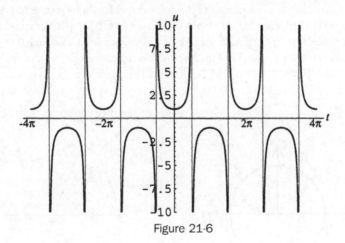

Figure 21-6

3. **COTANGENT.** The domain of the cotangent function is $\{t \in R | t \neq n\pi\}$ and the range is $R$. The graph is shown in Fig. 21-7.

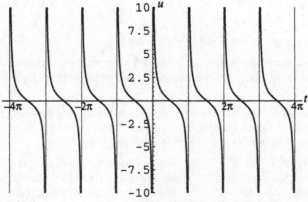

Figure 21-7

4. **COSECANT.** The domain of the cosecant function is $\{t \in R | t \neq n\pi\}$ and the range is $(-\infty, -1] \cup [1, \infty)$. The graph is shown in Fig. 21-8.

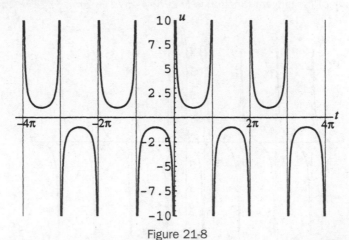

Figure 21-8

## SOLVED PROBLEMS

**21.1.** Explain the properties of the graph of the sine function.

Recall that $\sin t$ is defined as the $y$-coordinate of the point $P(t)$ obtained by proceeding a distance $|t|$ around the unit circle from the point $(1,0)$. (See Fig. 21-9.) As $t$ increases from 0 to $\pi/2$, the $y$-coordinate of $P(t)$ increases from 0 to 1; as $t$ increases from $\pi/2$ through $\pi$ to $3\pi/2$, $y$ decreases from 1 through 0 to $-1$; as $t$ increases from $3\pi/2$ to $2\pi$, $y$ increases from $-1$ to 0. (See Fig. 21-10.) This represents one cycle or period of the sine function; since the sine function is periodic with period $2\pi$, the cycle shown in Fig. 21-10 is repeated as $t$ increases from $2\pi$ to $4\pi$, $4\pi$ to $6\pi$, and so on. For negative $t$, the cycle is also repeated as $t$ increases from $-2\pi$ to 0, from $-4\pi$ to $-2\pi$, and so on.

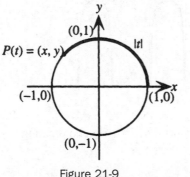

Figure 21-9

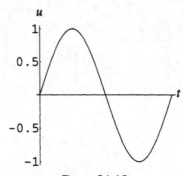

Figure 21-10

**21.2.** Explain how to sketch a graph of $u = A \sin(bt - c) + d$.

1. Determine amplitude and shape: Amplitude $= |A|$. If $A$ is positive, the curve is a standard sine curve; if $A$ is negative, the curve is an upside-down sine curve. The maximum height of the curve is $d + |A|$, the minimum is $d - |A|$.

2. Determine period and phase shift: Since $\sin T$ goes through one cycle in the interval $0 \leq T \leq 2\pi$, $\sin(bt - c)$ goes through one cycle in the interval $0 \leq bt - c \leq 2\pi$; that is, $c/b \leq t \leq (c + 2\pi)/b$. The graph is a standard (or upside-down) sine curve with period $2\pi/b$ and phase shift $c/b$.

3. Divide the interval from $c/b$ to $(c + 2\pi)/b$ into four equal subintervals and sketch one cycle of the curve. For positive $A$, the curve increases from a height of $d$ to its maximum height in the first subinterval, decreases to $d$ in the second and to its minimum height in the third, then increases to $d$ in the fourth. For negative $A$, the curve decreases from a height of $d$ to its minimum height in the first subinterval, increases to $d$ in the second and to its maximum height in the third, then decreases to $d$ in the fourth.

4. Show the behavior of the curve in further cycles as desired.

**21.3.** Explain the properties of the graph of the cosine function.

Recall that $\cos t$ is defined as the $x$-coordinate of the point $P(t)$ obtained by proceeding a distance $|t|$ around the unit circle from the point $(1,0)$. (See Fig. 21-11.) As $t$ increases from 0 through $\pi/2$ to $\pi$, the $x$-coordinate of $P(t)$ decreases from 1 through 0 to $-1$; as $t$ increases from $\pi$ through $3\pi/2$, to $2\pi$, $x$ increases from $-1$ through 0 to 1. (See Fig. 21-12). This represents one cycle or period of the cosine function; since the cosine function is periodic with period $2\pi$, the cycle shown in Fig. 21-12 is repeated as $t$ increases from $2\pi$ to $4\pi$, $4\pi$ to $6\pi$, and so on. For negative $t$, the cycle is also repeated as $t$ increases from $-2\pi$ to 0, from $-4\pi$ to $-2\pi$, and so on.

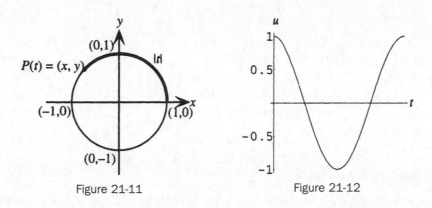

Figure 21-11                    Figure 21-12

**21.4.** Explain how to sketch a graph of $u = A\cos(bt - c) + d$.

1. Determine amplitude and shape: Amplitude $= |A|$. If $A$ is positive, the curve is a standard cosine curve; if $A$ is negative, the curve is an upside down cosine curve. The maximum height of the curve is $d + |A|$, the minimum is $d - |A|$.

2. Determine period and phase shift: Since $\cos T$ goes through one cycle in the interval $0 \le T \le 2\pi$, $\cos(bt - c)$ goes through one cycle in the interval $0 \le bt - c \le 2\pi$, that is, $c/b \le t \le (c + 2\pi)/b$. The graph is a standard (or upside-down) cosine curve with period $2\pi/b$ and phase shift $c/b$.

3. Divide the interval from $c/b$ to $(c + 2\pi)/b$ into four equal subintervals and sketch one cycle of the curve. For positive $A$, the curve decreases from its maximum height to a height of $d$ in the first subinterval, and to its minimum height in the second, then increases to a height of $d$ in the third subinterval and to its maximum height in the fourth. For negative $A$, the curve increases from its minimum height to a height of $d$ in the first subinterval, and to its maximum height in the second, then decreases to a height of $d$ in the third subinterval and to its minimum height in the fourth.

4. Show the behavior of the curve in further cycles as desired.

**21.5.** Sketch a graph of $u = 6\sin \frac{1}{2}t$.

Amplitude $= 6$. The graph is a standard sine curve. Period $= 2\pi \div 1/2 = 4\pi$. Phase shift $= 0$; $d = 0$. Divide the interval from 0 to $4\pi$ into four equal subintervals and sketch the curve with maximum height 6 and minimum height $-6$. See Fig. 21-13.

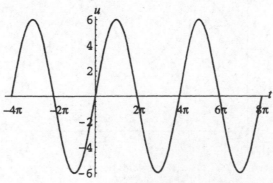

Figure 21-13

**21.6.** Sketch a graph of $u = 3\cos\pi t + 2$.

Amplitude = 3. The graph is a standard cosine curve. Period = $2\pi \div \pi = 2$. Phase shift = 0; $d = 2$. Divide the interval from 0 to 2 into four equal subintervals and sketch the curve with maximum height 5 and minimum height $-1$. See Fig. 21-14.

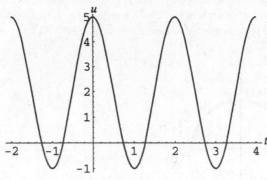

Figure 21-14

**21.7.** Sketch a graph of $u = 2\sin(5t - \pi)$.

Amplitude = 2. The graph is a standard sine curve. Period = $2\pi/5$. Phase shift = $\pi/5$; $d = 0$. Divide the interval from $\pi/5$ to $3\pi/5$ (= phase shift + one period) into four equal subintervals and sketch the curve with maximum height 2 and minimum height $-2$. See Fig. 21-15.

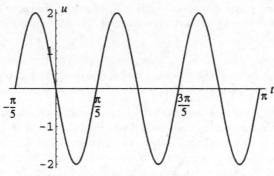

Figure 21-15

**21.8.** Sketch a graph of $u = -\dfrac{1}{2}\cos\left(3t + \dfrac{\pi}{4}\right) + \dfrac{3}{2}$.

Amplitude = $\dfrac{1}{2}$. The graph is an upside down cosine curve. Period = $\dfrac{2\pi}{3}$. Phase shift = $\left(-\dfrac{\pi}{4}\right) \div 3 = -\dfrac{\pi}{12}$. Divide the interval from $-\dfrac{\pi}{12}$ to $\dfrac{7\pi}{12}$ (= phase shift + one period) into four equal subintervals and sketch the curve with maximum height 2 and minimum height 1. See Fig. 21-16.

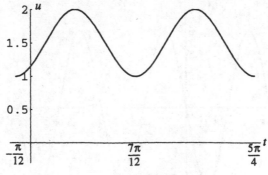

Figure 21-16

**21.9.** Sketch a graph of $u = |\sin t|$.

The graph is the same as the graph of $u = \sin t$ in the intervals for which $\sin t$ is positive, that is, $(0,\pi)$, $(2\pi,3\pi)$, $(-2\pi,-\pi)$, and so on. In the intervals for which $\sin t$ is negative, that is, $(\pi,2\pi)$, $(-\pi,0)$, and so on, since $|\sin t| = -\sin t$ in these intervals, the graph is the same as the graph of $u = -\sin t$, that is, the graph of $u = \sin t$ reflected with respect to the $t$ axis (Fig. 21-17).

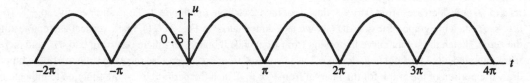

Figure 21-17

**21.10.** Explain the properties of the graph of the tangent function.

Recall that $\tan t$ is defined as the ratio $y/x$ of the coordinates of the point $P(t)$ obtained by proceeding a distance $|t|$ around the unit circle from the point $(1,0)$. (See Fig. 21-18.) As $t$ increases from 0 to $\pi/4$ this ratio increases from 0 to 1; as $t$ continues to increase from $\pi/4$ toward $\pi/2$, the ratio continues to increase beyond all bounds, as $t \to \pi/2^-$ (approaches from the left), $\tan t \to \infty$. Thus, the line $t = \pi/2$ is a vertical asymptote for the graph. Since tangent is an odd function, the graph has origin symmetry, the line $t = -\pi/2$ is also a vertical asymptote, and the curve is as shown in Fig. 21-19 for the interval $(-\pi/2,\pi/2)$. Since the tangent function has period $\pi$, the graph repeats this cycle for the intervals $(\pi/2,3\pi/2)$, $(3\pi/2,5\pi/2)$, $(-3\pi/2,-\pi/2)$, and so on.

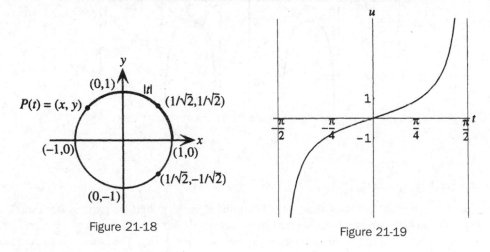

Figure 21-18

Figure 21-19

**21.11.** Sketch a graph of $u = \tan(t - \pi/3)$.

The graph is the same as the graph of $u = \tan t$ shifted $\pi/3$ units to the right, and has period $\pi$. Since $\tan T$ goes through one cycle in the interval $-\pi/2 < T < \pi/2$, $\tan(t - \pi/3)$ goes through one cycle in the interval $-\pi/2 < t - \pi/3 < \pi/2$, that is, $-\pi/6 < t < 5\pi/6$. Sketch the graph in this interval and repeat the cycle with period $\pi$.

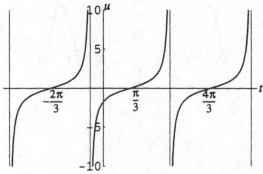

Figure 21-20

**21.12.** Explain the properties and sketch the graph of the secant function.

Since $\sec t$ is the reciprocal of $\cos t$, it is convenient to understand the graph of the secant function in terms of the graph of the cosine function: the secant function is even, has period $2\pi$, and has vertical asymptotes at the zeros of the cosine function, that is, at $t = \pi/2 + 2\pi n$ or $3\pi/2 + 2\pi n$, $n$ any integer. Where $\cos t = 1$, $\sec t = 1$, that is, for $t = 0 + 2\pi n$, $n$ any integer. Where $\cos t = -1$, $\sec t = -1$, that is, for $t = \pi + 2\pi n$, $n$ any integer. As $t$ increases from 0 to $\pi/2$, $\cos t$ decreases from 1 to 0; thus, $\sec t$ increases from 1 beyond all bounds; as $t$ increases from $\pi/2$ to $\pi$, $\cos t$ decreases from 0 to $-1$, thus, $\sec t$ increases from unboundedly large and negative to $-1$. To graph $u = \sec t$, sketch a graph of $u = \cos t$ (shown as a dotted curve in Fig. 21-21), mark vertical asymptotes through the zeros, sketch the secant curve increasing from 1 beyond all bounds at $t$ increases from 0 to $\pi/2$ and increasing from unboundedly large and negative to $-1$ as $t$ increases from $\pi/2$ to $\pi$, Use the even property of the function to draw the portion of the graph for the interval from $-\pi$ to 0, then the periodicity of the function to indicate further portions of the graph.

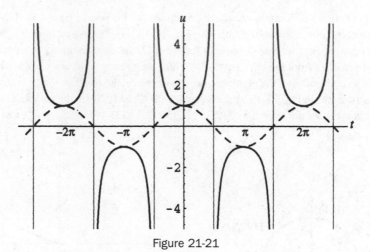

Figure 21-21

**21.13.** Sketch a graph of $u = t \sin t$.

Since $|\sin t| \le 1$, $0 \le |t| \, |\sin t| \le |t|$, thus $-|t| \le |t| \, |\sin t| \le |t|$, for all $t$. Thus the graph of $u = t \sin t$ lies between the lines $u = t$ and $u = -t$. Moreover, since $t \sin t = 0$ at $t = n\pi$ and $t \sin t = \pm t$ at $t = n\pi + \pi/2$, the graph of $u = t \sin t$ has $t$ intercepts at $t = n\pi$ and touches the lines at $t = n\pi + \pi/2$. The function is an even function; the graph is as shown in Fig. 21-22.

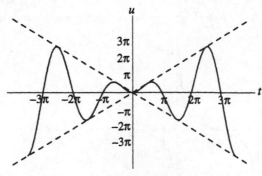

Figure 21-22

## SUPPLEMENTARY PROBLEMS

**21.14.** State the amplitude and period of (a) $u = \sin \pi t$; (b) $u = 2 \cos t - 4$.

   *Ans.* (a) amplitude = 1, period = 2; (b) amplitude = 2, period = $2\pi$.

**21.15.** Sketch a graph of (a) $u = \sin \pi t$; (b) $u = 2 \cos t - 4$.

   *Ans.* (a) Fig. 21-23; (b) Fig. 21-24.

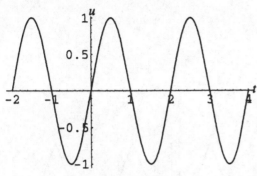

Figure 21-23

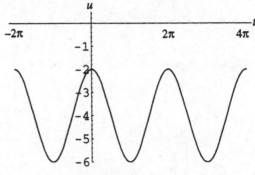

Figure 21-24

**21.16.** State the amplitude, period, and phase shift of (a) $u = \frac{1}{3}\cos 2t$; (b) $u = -2 \sin\left(\frac{1}{3}t - \pi\right) + 4$.

   *Ans.* (a) amplitude = $\frac{1}{3}$, period = $\pi$, phase shift = 0; (b) amplitude = 2, period = $6\pi$, phase shift = $3\pi$.

**21.17.** Sketch a graph of (a) $u = \frac{1}{3}\cos 2t$; (b) $u = -2 \sin\left(\frac{1}{3}t - \pi\right) + 4$.

   *Ans.* (a) Fig. 21-25; (b) Fig. 21-26.

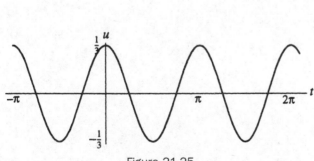

Figure 21-25

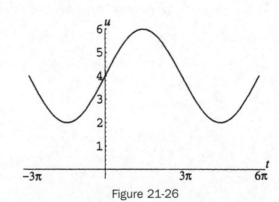

Figure 21-26

**21.18.** State the period of (a) $u = \tan\frac{1}{2}t$; (b) $u = -\sec 2t$.

   *Ans.* (a) $2\pi$; (b) $\pi$

**21.19.** Sketch a graph of (a) $u = \tan\frac{1}{2}t$; (b) $u = -\sec 2t$.

   *Ans.* (a) Fig. 21-27; (b) Fig. 21-28.

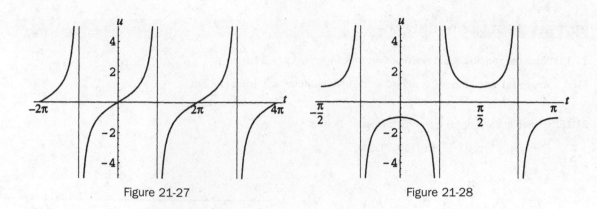

Figure 21-27                    Figure 21-28

**21.20.** Sketch a graph of (a) $u = e^{-t} \cos 2t$; (b) $u = 2 - |\cos t|$.

    *Ans.*   (a) Fig. 21-29; (b) Fig. 21-30.

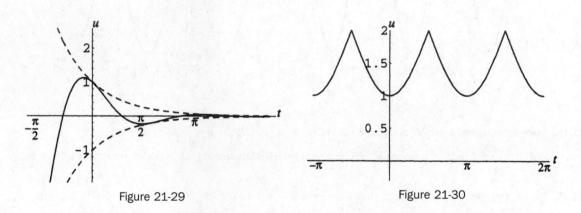

Figure 21-29                    Figure 21-30

**21.21.** Explain the properties of the graphs of the cotangent and cosecant functions.

# Angles

## Trigonometric Angles

A trigonometric angle is determined by rotating a ray about its endpoint, called the *vertex* of the angle. The starting position of the ray is called the *initial side* and the ending position is the *terminal side*. (See Fig. 22-1.)

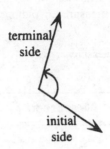

Figure 22-1

If the displacement of the ray from its starting position is in the counterclockwise direction, the angle is assigned a positive measure, if in the clockwise direction, a negative measure. A zero angle corresponds to zero displacement; the initial and terminal sides of a zero angle are coincident.

## Angles in Standard Position

An angle is in standard position in a Cartesian coordinate system if its vertex is at the origin and its initial side is the positive x-axis. Angles in standard position are categorized by their terminal sides: If the terminal side falls along an axis, the angle is called a quadrantal angle; if the terminal side is in quadrant *n*, the angle is referred to as a quadrant *n* angle (see Figs. 22-2 to 22-5).

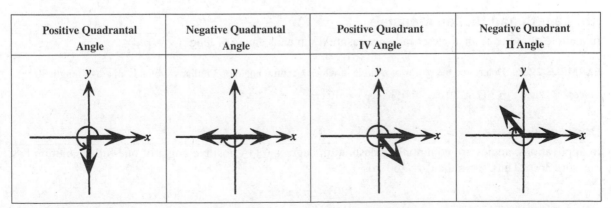

| Positive Quadrantal Angle | Negative Quadrantal Angle | Positive Quadrant IV Angle | Negative Quadrant II Angle |
|---|---|---|---|
| Figure 22-2 | Figure 22-3 | Figure 22-4 | Figure 22-5 |

## Radian Measurement of Angles

In calculus, angles are normally measured in radian measure. One radian is defined as the measure of an angle that, if placed with vertex at the center of a circle, subtends (intersects) an arc of length equal to the radius of the circle. In Fig. 22-6, angle $\theta$ has measure 1 radian.

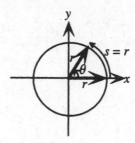

Figure 22-6

Since the circumference of a circle of radius $r$ has length $2\pi r$, a positive angle of one full revolution corresponds to an arc length of $2\pi r$ and thus has measure $2\pi$ radians.

**EXAMPLE 22.1** Draw examples of angles of measures $\pi$, $\dfrac{\pi}{2}$, and $\dfrac{3\pi}{2}$ radians.

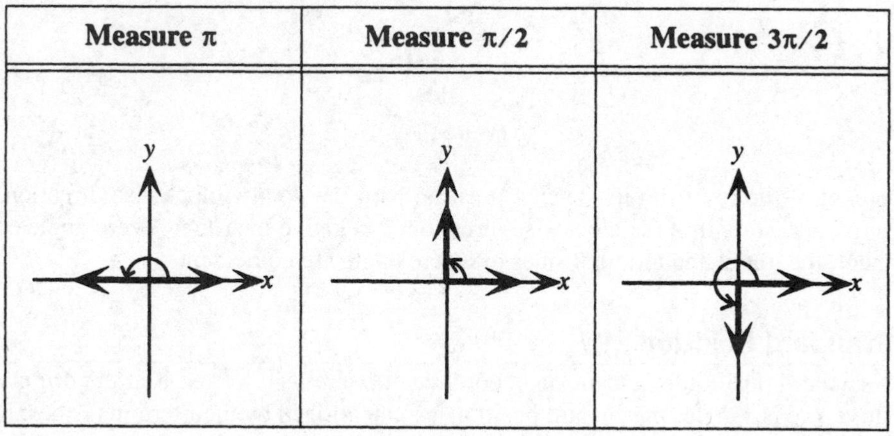

| **Measure π** | **Measure π/2** | **Measure 3π/2** |
|---|---|---|
| | | |

Figure 22-7

## Arc Length and Radian Measure

In a circle of radius $r$, an angle of radian measure $\theta$ subtends an arc of length $s = r\theta$.

**EXAMPLE 22.2** Determine the radius of a circle in which a central angle of 3 radians subtends an arc of length 30 cm.

Since $\theta = 3$ and $s = 30$ cm, $30$ cm $= 3r$; hence $r = 10$ cm.

## Degree Measure

In applications, angles are commonly measured in degrees (°). A positive angle of one full revolution has measure 360°. Thus, $2\pi$ radians = 360°, or

$$180° = \pi \text{ radians}$$

To transform radian measure into degrees, use this relation in the form $180°/\pi = 1$ radian and multiply the radian measure by $180°/\pi$. To transform degree measure into radians, use the relation in the form $1° = \pi/180$ radians and multiply the degree measure by $\pi/180°$. The following table summarizes the measure of common angles:

| Degree Measure | 0° | 30° | 45° | 60° | 90° | 120° | 135° | 150° | 180° | 270° | 360° |
|---|---|---|---|---|---|---|---|---|---|---|---|
| Radian Measure | 0 | $\dfrac{\pi}{6}$ | $\dfrac{\pi}{4}$ | $\dfrac{\pi}{3}$ | $\dfrac{\pi}{2}$ | $\dfrac{2\pi}{3}$ | $\dfrac{3\pi}{4}$ | $\dfrac{5\pi}{6}$ | $\pi$ | $\dfrac{3\pi}{2}$ | $2\pi$ |

**EXAMPLE 22.3** (a) Transform 210° into radians. (b) Transform $6\pi$ radians into degrees.

(a) $210° = 210° \cdot \dfrac{\pi}{180°}$ radians $= \dfrac{7\pi}{6}$ radians; (b) $6\pi$ radians $= 6\pi \cdot \dfrac{180°}{\pi} = 1080°$

## Degrees, Minutes, and Seconds

If measurements smaller than a degree are required, the degree may be subdivided into decimal fractions. Alternatively, a degree is subdivided into minutes (′) and seconds (″). Thus, $1° = 60′$ and $1′ = 60″$; hence, $1° = 3600″$.

**EXAMPLE 22.4** Transform $35°24′36″$ into decimal degrees.

$$35°24′36″ = \left(35 + \frac{24}{60} + \frac{36}{3600}\right)° = 35.41°$$

## Terminology for Special Angles

An angle of measure between 0 and $\pi/2$ radians (between 0° and 90°) is called an *acute* angle. An angle of measure $\pi/2$ radians (90°) is called a *right* angle. An angle of measure between $\pi/2$ and $\pi$ radians (between 90° and 180°) is called an *obtuse* angle. An angle of measure $\pi$ radians (180°) is called a *straight* angle. An angle is normally referred to by giving its measure; thus $\theta = 30°$ means that $\theta$ has a measure of 30°.

## Complementary and Supplementary Angles

If $\alpha$ and $\beta$ are two angles such that $\alpha + \beta = \pi/2$, $\alpha$ and $\beta$ are called complementary angles. If $\alpha$ and $\beta$ are two angles such that $\alpha + \beta = \pi$, $\alpha$ and $\beta$ are called supplementary angles.

**EXAMPLE 22.5** Find an angle complementary to $\theta$ if (a) $\theta = \pi/3$; (b) $\theta = 37°15′$.

(a) The complementary angle to $\theta$ is $\dfrac{\pi}{2} - \theta = \dfrac{\pi}{2} - \dfrac{\pi}{3} = \dfrac{\pi}{6}$.
(b) The complementary angle to $\theta$ is $90° - \theta = 90° - 37°15′ = 89°60′ - 37°15′ = 52°45′$.

## Coterminal Angles

Two angles in standard position are coterminal if they have the same terminal side. There are an infinite number of angles coterminal with a given angle. To find an angle coterminal with a given angle, add or subtract $2\pi$ (if the angle is measured in radians) or 360° (if the angle is measured in degrees).

**EXAMPLE 22.6** Find two angles coterminal with (a) 2 radians; (b) $-60°$.

(a) Coterminal with 2 radians are $2 + 2\pi$ and $2 - 2\pi$ radians, as well as many other angles.
(b) Coterminal with $-60°$ are $-60° + 360° = 300°$ and $-60° - 360° = -420°$, as well as many other angles.

## Trigonometric Functions of Angles

If $\theta$ is an angle with radian measure $t$, then the value of each trigonometric function of $\theta$ is its value at the real number $t$.

**EXAMPLE 22.7** Find (a) cos 90°; (b) tan 135°.

(a) $\cos 90° = \cos\dfrac{\pi}{2} = 0$; (b) $\tan 135° = \tan\left(135° \cdot \dfrac{\pi}{180°}\right) = \tan\dfrac{3\pi}{4} = -1$

## Trigonometric Functions of Angles as Ratios

Let $\theta$ be an angle in standard position, and $P(x, y)$ be any point except the origin on the terminal side of $\theta$. If $r = \sqrt{x^2 + y^2}$ is the distance from $P$ to the origin, then the six trigonometric functions of $\theta$ are given by:

$$\sin\theta = \frac{y}{r} \qquad\qquad \csc\theta = \frac{r}{y} \quad (\text{if } y \neq 0)$$

$$\cos\theta = \frac{x}{r} \qquad\qquad \sec\theta = \frac{r}{x} \quad (\text{if } x \neq 0)$$

$$\tan\theta = \frac{y}{x} \quad (\text{if } x \neq 0) \qquad \cot\theta = \frac{x}{y} \quad (\text{if } y \neq 0)$$

**EXAMPLE 22.8** Let $\theta$ be an angle in standard position with $P(-3, 4)$ a point on the terminal side of $\theta$ (see Fig. 22-8). Find the six trigonometric functions of $\theta$.

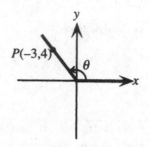

$P(-3,4)$

Figure 22-8

$x = -3, y = 4, r = \sqrt{x^2 + y} = \sqrt{(-3)^2 + 4^2} = 5$; hence

$$\sin\theta = \frac{y}{r} = \frac{4}{5} \qquad \cos\theta = \frac{x}{r} = \frac{-3}{5} = -\frac{3}{5} \qquad \tan\theta = \frac{y}{x} = \frac{4}{-3} = -\frac{4}{3}$$

$$\csc\theta = \frac{r}{y} = \frac{5}{4} \qquad \sec\theta = \frac{r}{x} = \frac{5}{-3} = -\frac{5}{3} \qquad \cot\theta = \frac{x}{y} = \frac{-3}{4} = -\frac{3}{4}$$

## Trigonometric Functions of Acute Angles

If $\theta$ is an acute angle, it can be regarded as an angle of a right triangle. If $\theta$ is placed in standard position, and the sides of the right triangle are named as hypotenuse (hyp), opposite (opp), and adjacent (adj), the lengths of the adjacent and opposite sides are the $x$- and $y$-coordinates, respectively, of a point on the terminal side of the angle. The length of the hypotenuse is $r = \sqrt{x^2 + y^2}$. (See Fig. 22-9.)

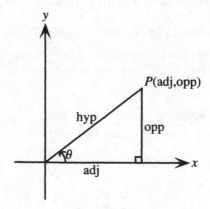

Figure 22-9

For an acute angle $\theta$, the trigonometric functions of $\theta$ are then as follows:

$$\sin\theta = \frac{y}{r} = \frac{\text{opp}}{\text{hyp}} \qquad \csc\theta = \frac{r}{y} = \frac{\text{hyp}}{\text{opp}}$$

$$\cos\theta = \frac{x}{r} = \frac{\text{adj}}{\text{hyp}} \qquad \sec\theta = \frac{r}{x} = \frac{\text{hyp}}{\text{adj}}$$

$$\tan\theta = \frac{y}{x} = \frac{\text{opp}}{\text{adj}} \qquad \cot\theta = \frac{x}{y} = \frac{\text{adj}}{\text{opp}}$$

**EXAMPLE 22.9**   Find the six trigonometric functions of $\theta$ as shown in Fig. 22-10.

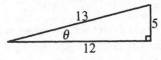

Figure 22-10

For $\theta$ as shown, opp = 5, adj = 12, hyp = 13, hence

$$\sin\theta = \frac{\text{opp}}{\text{hyp}} = \frac{5}{13} \qquad \cos\theta = \frac{\text{adj}}{\text{hyp}} = \frac{12}{13} \qquad \tan\theta = \frac{\text{opp}}{\text{adj}} = \frac{5}{12}$$

$$\csc\theta = \frac{\text{hyp}}{\text{opp}} = \frac{13}{5} \qquad \sec\theta = \frac{\text{hyp}}{\text{adj}} = \frac{13}{12} \qquad \cot\theta = \frac{\text{adj}}{\text{opp}} = \frac{12}{5}$$

## Reference Angles

The reference angle for $\theta$, a nonquadrantal angle in standard position, is the acute angle $\theta_R$ between the $x$-axis and the terminal side of $\theta$. Fig. 22-11 shows angles and reference angles for cases $0 < \theta < 2\pi$. To find reference angles for other nonquadrantal angles, first add or subtract multiples of $2\pi$ to obtain an angle coterminal with $\theta$ that satisfies $0 < \theta < 2\pi$.

| **Quadrant I** | **Quadrant II** | **Quadrant III** | **Quadrant IV** |
|:---:|:---:|:---:|:---:|
| $\theta_R = \theta$ | $\theta_R = \pi - \theta$ $= 180° - \theta$ | $\theta_R = \theta - \pi$ $= \theta - 180°$ | $\theta_R = 2\pi - \theta$ $= 360° - \theta$ |

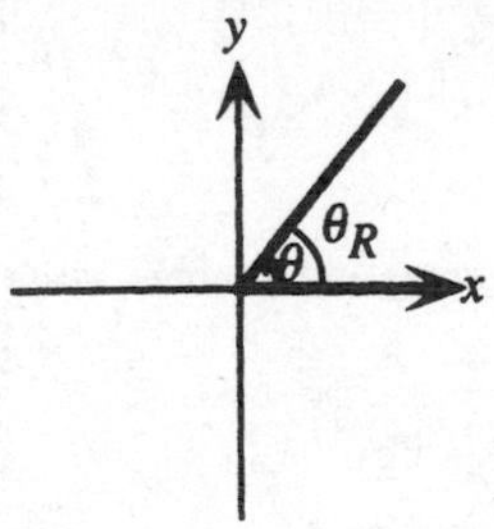

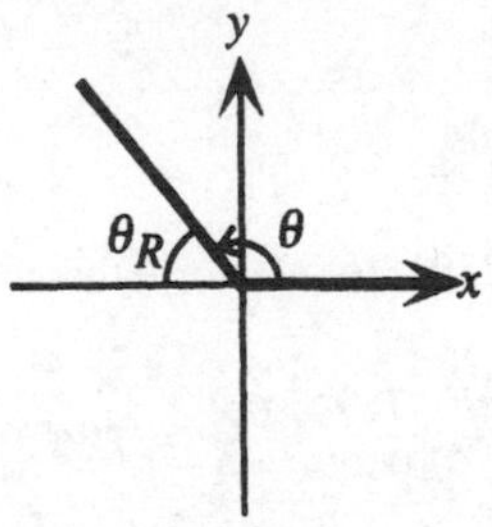

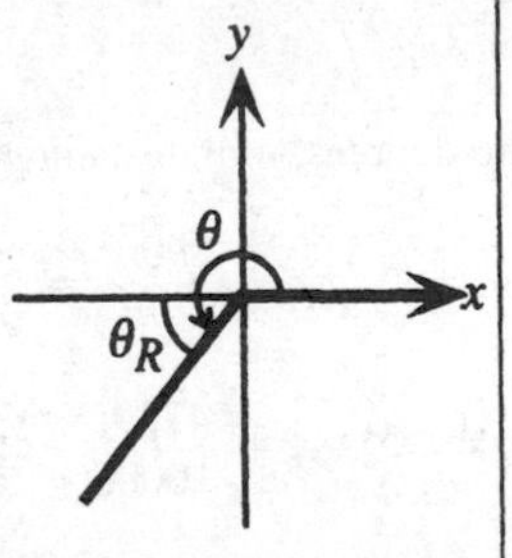

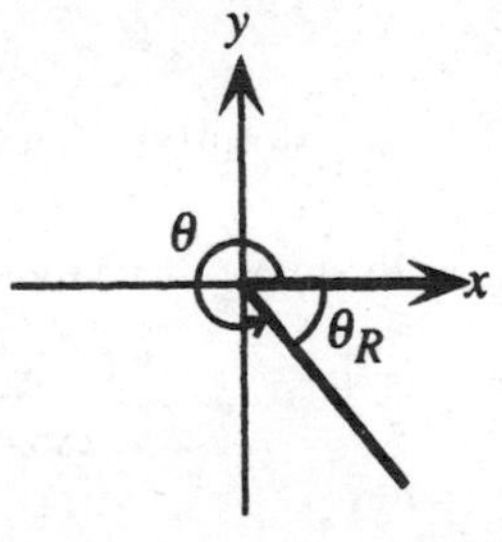

Figure 22-11

## Trigonometric Functions of Angles in Terms of Reference Angles

For any nonquadrantal angle $\theta$, each trigonometric function of $\theta$ has the same absolute value as the same trigonometric function of $\theta_R$. To find a trigonometric function of $\theta$, find the function of $\theta_R$, then apply the correct sign for the quadrant of $\theta$.

**EXAMPLE 22.10**   Find $\cos\dfrac{3\pi}{4}$.

The reference angle for $\dfrac{3\pi}{4}$, a second quadrant angle, is $\pi - \dfrac{3\pi}{4} = \dfrac{\pi}{4}$. In quadrant II, the sign of the cosine function is negative. Hence, $\cos\dfrac{3\pi}{4} = -\cos\dfrac{\pi}{4} = -\dfrac{1}{\sqrt{2}}$.

## SOLVED PROBLEMS

**22.1.** List all angles coterminal with (a) $40°$; (b) $\dfrac{2\pi}{3}$ radians.

    (a) To find angles coterminal with $40°$, add or subtract any integer multiple of $360°$. Thus, $400°$ and $-320°$ are examples of angles coterminal with $40°$, and all angles coterminal with $40°$ can be expressed as $40° + n360°$, where $n$ is any integer.

    (b) To find angles coterminal with $2\pi/3$, add or subtract any integer multiple of $2\pi$. Thus, $8\pi/3$ and $-4\pi/3$ are examples of angles coterminal with $2\pi/3$ and all angles coterminal with $2\pi/3$ can be expressed as $2\pi/3 + 2\pi n$, where $n$ is any integer.

**22.2.** Find the trigonometric functions of (a) $180°$; (b) $-360°$.

    (a) $180° = \pi$ radians; hence, $\sin 180° = \sin \pi = 0$, $\cos 180° = \cos \pi = -1$, $\tan 180° = \tan \pi = 0$, $\cot 180° = \cot \pi$ is undefined, $\sec 180° = \sec \pi = -1$, $\csc 180° = \csc \pi$ is undefined.

    (b) $-360° = -2\pi$ radians, hence, $\sin(-360°) = \sin(-2\pi) = 0$, $\cos(-360°) = \cos(-2\pi) = 1$, $\tan(-360°) = \tan(-2\pi) = 0$, $\cot(-360°) = \cot(-2\pi)$ is undefined, $\sec(-360°) = \sec(-2\pi) = 1$, $\csc(-360°) = \csc(-2\pi)$ is undefined.

**22.3.** Find an angle supplementary to $\theta$ if (a) $\theta = \pi/3$; (b) $\theta = 37°15'$.

    (a) Supplementary to $\dfrac{\pi}{3}$ is $\pi - \dfrac{\pi}{3} = \dfrac{2\pi}{3}$.

    (b) Supplementary to $37°15'$ is $180° - 37°15' = 179°60' - 37°15' = 142°45'$.

**22.4.** Transform 5 radians into degrees, minutes, and seconds.

First note that 5 radians $= 5 \cdot \dfrac{180°}{\pi} = \dfrac{900°}{\pi} \approx 286.4789°$. To transform this into degrees and minutes, write

$$286.4789° = 286° + \frac{4789°}{10000} = 286° + \frac{4789°}{10000} \cdot \frac{60'}{1°} = 286° + 28.734'$$

To transform this into degrees, minutes, and seconds, write

$$286° + 28.734' = 286° + 28' + \frac{734'}{1000} = 286° + 28' + \frac{734'}{1000} \cdot \frac{60''}{1'} = 286°28'44.04''$$

**22.5.** Transform $424°34'24''$ into radians.

First note that $424°34'24'' = \left(424 + \dfrac{34}{60} + \dfrac{24}{3600}\right)° \approx 424.57333°$. To transform this into radians, write

$$424.57333° = 424.57333° \cdot \frac{\pi}{180°} \approx 7.41 \text{ radians.}$$

**22.6.** (a) Derive the relationship $s = r\theta$. (b) Find the angle in radians subtended by an arc of length 5 cm on a circle of radius 3 cm. (c) Find the linear distance traveled by a point on the rim of a bicycle wheel of radius 26 in as the wheel makes 10 rotations.

(a) Draw two circles of radius $r$, as shown in Fig. 22-12.

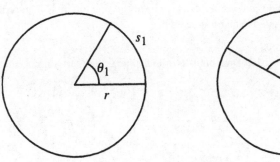

Figure 22-12

From plane geometry it is known that the ratio of the arc lengths equals the ratio of the angles.

Thus

$$\frac{s}{s_1} = \frac{\theta}{\theta_1}$$

Take $\theta_1 = 1$ radian, then $s_1 = r$; hence, $\frac{s}{r} = \frac{\theta}{1}$, that is, $s = r\theta$.

(b) Use $s = r\theta$ with $s = 5$ cm, $r = 3$ cm, then $5 = 3\theta$; thus $\theta = \frac{5}{3}$ radians.

(c) First note that 10 rotations represents an angle of $10 \cdot 2\pi = 20\pi$ radians. Hence,

$$s = r\theta = 26 \text{ in} \cdot 20\pi \text{ radians} = 520\pi \text{ in} \approx 136 \text{ ft.}$$

**22.7.** Show that the definitions of the trigonometric functions as ratios are consistent with the definitions of the trigonometric functions of angles.

Let $\theta$ be a nonquadrantal angle in standard position. Choose an arbitrary point $Q(x, y)$ on the terminal side of $\theta$. (See Fig. 22-13).

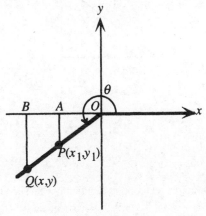

Figure 22-13

Then $r = \sqrt{x^2 + y^2}$. Let $P(x_1, y_1)$ be a point on the terminal side of $\theta$ with $\sqrt{x_1^2 + y_1^2} = 1$. Then $P$ lies on the unit circle and $\sin\theta = y_1$. Drop perpendicular lines from $P$ and $Q$ to the $x$ axis at $A$ and $B$, respectively. Then triangles $OAP$ and $OBQ$ are similar, and hence ratios of corresponding sides are equal; thus

$$\frac{|y|}{r} = \frac{|y_1|}{1}$$

Since $y$ and $y_1$ have the same sign, it follows that

$$y_1 = \frac{y}{r}$$

Thus $\sin\theta = y_1 = \frac{y}{r}$ and the two definitions are consistent for the sine function. The proof is easily extended to the other trigonometric functions and to quadrantal angles.

**22.8.** If $\theta$ is in standard position and $(-20, 21)$ lies on its terminal side, find the trigonometric functions of $\theta$.

$x = -20$ and $y = 21$ hence $r = \sqrt{x^2 + y^2} = \sqrt{(-20)^2 + 21^2} = 29$. Therefore,

$$\sin\theta = \frac{y}{r} = \frac{21}{29} \qquad \cos\theta = \frac{x}{r} = \frac{-20}{29} = -\frac{20}{29} \qquad \tan\theta = \frac{y}{x} = \frac{21}{-20} = -\frac{21}{20}$$

$$\csc\theta = \frac{r}{y} = \frac{29}{21} \qquad \sec\theta = \frac{r}{x} = \frac{29}{-20} = -\frac{29}{20} \qquad \cot\theta = \frac{x}{y} = \frac{-20}{21} = -\frac{20}{21}$$

**22.9.** If $\theta$ is in standard position and its terminal side lies in quadrant I on the line $y = 2x$, find the trigonometric functions of $\theta$.

To find the trigonometric functions of $\theta$, any point on the terminal side of $\theta$ may be chosen; let $x = 1$, then $y = 2$ and $r = \sqrt{x^2 + y^2} = \sqrt{1^2 + 2^2} = \sqrt{5}$. Therefore

$$\sin\theta = \frac{y}{r} = \frac{2}{\sqrt{5}} \qquad \cos\theta = \frac{x}{r} = \frac{1}{\sqrt{5}} \qquad \tan\theta = \frac{y}{x} = \frac{2}{1} = 2$$

$$\csc\theta = \frac{r}{y} = \frac{\sqrt{5}}{2} \qquad \sec\theta = \frac{r}{x} = \frac{\sqrt{5}}{1} = \sqrt{5} \qquad \cot\theta = \frac{x}{y} = \frac{1}{2}$$

**22.10.** If $\theta$ is an acute angle, find the other trigonometric functions of $\theta$, given

(a) $\sin\theta = \frac{3}{5}$; (b) $\tan\theta = \frac{2}{3}$.

(a) Draw a figure. In the right triangle, take opp $= 3$ and hyp $= 5$. Then the third side is found from the Pythagorean theorem: adj $= \sqrt{5^2 - 3^2} = 4$. See Fig. 22-14.

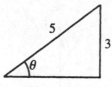

Figure 22-14

Hence,

$$\sin\theta = \frac{\text{opp}}{\text{hyp}} = \frac{3}{5} \qquad \cos\theta = \frac{\text{adj}}{\text{hyp}} = \frac{4}{5} \qquad \tan\theta = \frac{\text{opp}}{\text{adj}} = \frac{3}{4}$$

$$\csc\theta = \frac{\text{hyp}}{\text{opp}} = \frac{5}{3} \qquad \sec\theta = \frac{\text{hyp}}{\text{adj}} = \frac{5}{4} \qquad \cot\theta = \frac{\text{adj}}{\text{opp}} = \frac{4}{3}$$

(b) Draw a figure. In the right triangle, take opp $= 2$ and adj $= 3$. Then the third side is found from the Pythagorean theorem: hyp $= \sqrt{2^2 + 3^2} = \sqrt{13}$. See Fig. 22-15.

Figure 22-15

Hence,

$$\sin\theta = \frac{\text{opp}}{\text{hyp}} = \frac{2}{\sqrt{13}} \qquad \cos\theta = \frac{\text{adj}}{\text{hyp}} = \frac{3}{\sqrt{13}} \qquad \tan\theta = \frac{\text{opp}}{\text{adj}} = \frac{2}{3}$$

$$\csc\theta = \frac{\text{hyp}}{\text{opp}} = \frac{\sqrt{13}}{2} \qquad \sec\theta = \frac{\text{hyp}}{\text{adj}} = \frac{\sqrt{13}}{3} \qquad \cot\theta = \frac{\text{adj}}{\text{opp}} = \frac{3}{2}$$

**22.11.** Find the trigonometric functions of 30°, 45°, and 60°.

To find the trigonometric functions of 30° and 60°, draw a 30−60° right triangle (Fig. 22-16). Since the side opposite the 30° angle is one-half the hypotenuse, for 30° take opp = 1, hyp = 2. Then from the Pythagorean theorem, adj = $\sqrt{3}$.

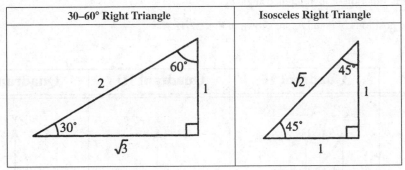

| 30–60° Right Triangle | Isosceles Right Triangle |

Figure 22-16          Figure 22-17

Hence

$$\sin 30° = \frac{\text{opp}}{\text{hyp}} = \frac{1}{2} \qquad \cos 30° = \frac{\text{adj}}{\text{hyp}} = \frac{\sqrt{3}}{2} \qquad \tan 30° = \frac{\text{opp}}{\text{adj}} = \frac{1}{\sqrt{3}}$$

$$\csc 30° = \frac{\text{hyp}}{\text{opp}} = \frac{2}{1} = 2 \qquad \sec 30° = \frac{\text{hyp}}{\text{adj}} = \frac{2}{\sqrt{3}} \qquad \cot 30° = \frac{\text{adj}}{\text{opp}} = \frac{\sqrt{3}}{1} = \sqrt{3}$$

Fig. 22-16 can also be used to determine the trigonometric functions of 60°, even though the 60° angle is not in standard position. Take opp = $\sqrt{3}$, adj = 1, and hyp = 2, then

$$\sin 60° = \frac{\text{opp}}{\text{hyp}} = \frac{\sqrt{3}}{2} \qquad \cos 60° = \frac{\text{adj}}{\text{hyp}} = \frac{1}{2} \qquad \tan 60° = \frac{\text{opp}}{\text{adj}} = \frac{\sqrt{3}}{1} = \sqrt{3}$$

$$\csc 60° = \frac{\text{hyp}}{\text{opp}} = \frac{2}{\sqrt{3}} \qquad \sec 60° = \frac{\text{hyp}}{\text{adj}} = \frac{2}{1} = 2 \qquad \cot 60° = \frac{\text{adj}}{\text{opp}} = \frac{1}{\sqrt{3}}$$

To find the trigonometric functions of 45°, draw an isosceles right triangle (Fig. 22-17). Take opp = 1, adj = 1, and hyp = $\sqrt{2}$, then

$$\sin 45° = \frac{\text{opp}}{\text{hyp}} = \frac{1}{\sqrt{2}} \qquad \cos 45° = \frac{\text{adj}}{\text{hyp}} = \frac{1}{\sqrt{2}} \qquad \tan 45° = \frac{\text{opp}}{\text{adj}} = \frac{1}{1} = 1$$

$$\csc 45° = \frac{\text{hyp}}{\text{opp}} = \frac{\sqrt{2}}{1} = \sqrt{2} \qquad \sec 45° = \frac{\text{hyp}}{\text{adj}} = \frac{\sqrt{2}}{1} = \sqrt{2} \qquad \cot 45° = \frac{\text{adj}}{\text{opp}} = \frac{1}{1} = 1$$

**22.12.** Form a table of the trigonometric functions of 0, $\frac{\pi}{6}$, $\frac{\pi}{4}$, $\frac{\pi}{3}$, and $\frac{\pi}{2}$ radians.

The trigonometric functions of 0 and $\pi/2$ radians are the same as the functions of the real numbers 0 and $\pi/2$, respectively, calculated in Problems 20.7 and 20.8. The trigonometric functions of $\pi/6$, $\pi/4$, and $\pi/3$ are the same as the functions of 30°, 45°, and 60°, calculated in Problem 22.11. Summarizing yields the following table (U stands for undefined):

| $\theta$ (radians) | $\theta$ (degrees) | $\sin\theta$ | $\cos\theta$ | $\tan\theta$ | $\cot\theta$ | $\sec\theta$ | $\csc\theta$ |
|---|---|---|---|---|---|---|---|
| 0 | 0° | 0 | 1 | 0 | U | 1 | U |
| $\pi/6$ | 30° | 1/2 | $\sqrt{3}/2$ | $1/\sqrt{3}$ | $\sqrt{3}$ | $2/\sqrt{3}$ | 2 |
| $\pi/4$ | 45° | $1/\sqrt{2}$ | $1/\sqrt{2}$ | 1 | 1 | $\sqrt{2}$ | $\sqrt{2}$ |
| $\pi/3$ | 60° | $\sqrt{3}/2$ | 1/2 | $\sqrt{3}$ | $1/\sqrt{3}$ | 2 | $2/\sqrt{3}$ |
| $\pi/2$ | 90° | 1 | 0 | U | 0 | U | 1 |

**22.13.** Show that for any nonquadrantal angle $\theta$, each trigonometric function of $\theta$ has the same absolute value as the same trigonometric function of its reference angle $\theta_R$.

The four possible positions of $\theta$ and $\theta_R$ are shown in Fig. 22-18.

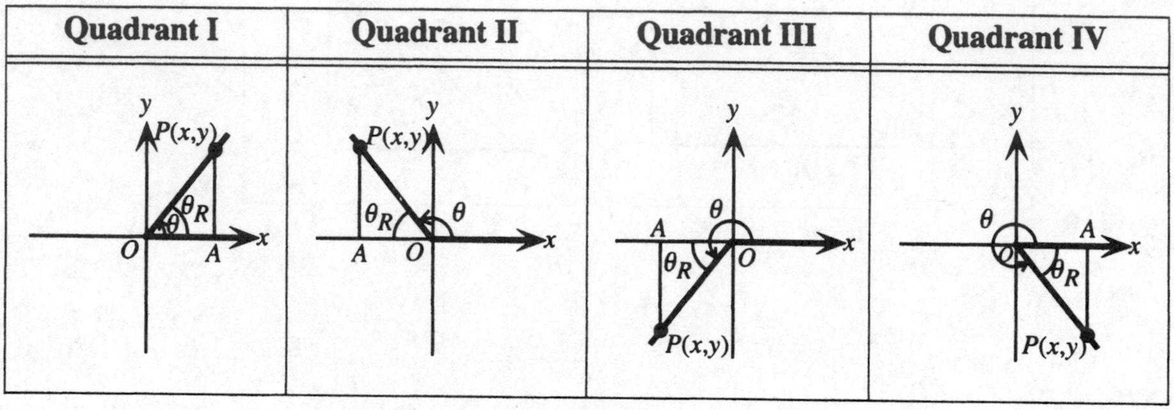

Figure 22-18

In each case, let $P(x, y)$ be a point on the terminal side of $\theta$ and draw a line from $P$ perpendicular to the $x$-axis at $A$. In triangle $OAP$, $\theta_R$ is an acute angle with opp $= |y|$, adj $= |x|$, and hyp $= \sqrt{x^2 + y^2} = r$.

Therefore

$$|\sin\theta| = \left|\frac{y}{r}\right| = \frac{|y|}{r} = \sin\theta_R \qquad |\cos\theta| = \left|\frac{x}{r}\right| = \frac{|x|}{r} = \cos\theta_R \qquad |\tan\theta| = \left|\frac{y}{x}\right| = \frac{|y|}{|x|} = \tan\theta_R$$

and similarly for the other trigonometric functions.

**22.14.** Find the reference angle for (a) 480°; (b) $-\dfrac{3\pi}{4}$ radians.

(a) First note that $480° - 360° = 120°$ is an angle between 0° and 360° coterminal with 480°. Since $90° < 120° < 180°$, 120° is a second quadrant angle. Hence the reference angle for 120° and therefore for 480° is $180° - 120° = 60°$.

(b) First note that $-\dfrac{3\pi}{4} + 2\pi = \dfrac{5\pi}{4}$ is an angle between 0 and $2\pi$ radians coterminal with $-\dfrac{3\pi}{4}$. Since $\pi < \dfrac{5\pi}{4} < \dfrac{3\pi}{2}$, $\dfrac{5\pi}{4}$ is a third quadrant angle. Hence the reference angle for $\dfrac{5\pi}{4}$ and therefore for $-\dfrac{3\pi}{4}$ is $\dfrac{5\pi}{4} - \pi = \dfrac{\pi}{4}$.

**22.15.** Find the trigonometric functions for (a) $480°$; (b) $-\dfrac{3\pi}{4}$ radians.

(a) To find the trigonometric functions of an angle, find the functions of its reference angle and attach the correct sign for the quadrant. $480°$ is a second quadrant angle. In quadrant II, sine and cosecant are positive, and the other trigonometric functions are negative. Using the reference angle found in the previous problem yields:

$$\sin 480° = \sin 60° = \frac{\sqrt{3}}{2} \qquad \cos 480° = -\cos 60° = -\frac{1}{2} \qquad \tan 480° = -\tan 60° = -\sqrt{3}$$

$$\csc 480° = \csc 60° = \frac{2}{\sqrt{3}} \qquad \sec 480° = -\sec 60° = -2 \qquad \cot 480° = -\cot 60° = -\frac{1}{\sqrt{3}}$$

(b) $-3\pi/4$ is a third quadrant angle. In quadrant III, tangent and cotangent are positive, and the other trigonometric functions are negative. Using the reference angle found in the previous problem yields:

$$\sin\left(-\frac{3\pi}{4}\right) = -\sin\frac{\pi}{4} = -\frac{1}{\sqrt{2}} \qquad \cos\left(-\frac{3\pi}{4}\right) = -\cos\frac{\pi}{4} = -\frac{1}{\sqrt{2}} \qquad \tan\left(-\frac{3\pi}{4}\right) = \tan\frac{\pi}{4} = 1$$

$$\csc\left(-\frac{3\pi}{4}\right) = -\csc\frac{\pi}{4} = -\sqrt{2} \qquad \sec\left(-\frac{3\pi}{4}\right) = -\sec\frac{\pi}{4} = -\sqrt{2} \qquad \cot\left(-\frac{3\pi}{4}\right) = \cot\frac{\pi}{4} = 1$$

**22.16.** Find all angles $\theta$, $0 \le \theta < 2\pi$, such that (a) $\sin\theta = \dfrac{1}{2}$; (b) $\sin\theta = -\dfrac{\sqrt{3}}{2}$.

The sine function is increasing on the interval from $0$ to $\pi/2$; thus it is a one-to-one function on this interval. For values of $a$ in the interval $0 \le a \le 1$, the notation $t = \sin^{-1} a$ is used to denote the unique value $t$ in the interval $0 \le t \le \pi/2$ such that $\sin t = a$. (See Chapter 25 for a fuller discussion of inverse trigonometric functions.)

(a) From the table in Problem 22.12, $\sin\dfrac{\pi}{6} = \dfrac{1}{2}$; thus, $\dfrac{\pi}{6} = \sin^{-1}\dfrac{1}{2}$. Since the sine function is positive in quadrants I and II, there is also an angle $\theta$ in quadrant II with reference angle $\dfrac{\pi}{6}$ and $\sin\theta = \dfrac{1}{2}$. This angle must be $\pi - \dfrac{\pi}{6} = \dfrac{5\pi}{6}$. Then $\dfrac{\pi}{6}$ and $\dfrac{5\pi}{6}$ are the two required angles.

(b) From the table in Problem 22.12, $\sin\dfrac{\pi}{3} = \dfrac{\sqrt{3}}{2}$, thus, $\dfrac{\pi}{3} = \sin^{-1}\dfrac{\sqrt{3}}{2}$. Since the sine function is negative in quadrants III and IV, the required angles are the angles $\theta_1$ and $\theta_2$ in these quadrants with reference angle $\dfrac{\pi}{3}$ and $\sin\theta_1 = \sin\theta_2 = -\dfrac{\sqrt{3}}{2}$. In quadrant III, $\theta_1 - \pi = \dfrac{\pi}{3}$; thus, $\theta_1 = \dfrac{4\pi}{3}$. In quadrant IV, $2\pi - \theta_2 = \dfrac{\pi}{3}$; thus, $\theta_2 = \dfrac{5\pi}{3}$.

**22.17.** Find all angles $\theta$, $0 \le \theta < 2\pi$, such that (a) $\cos\theta = \dfrac{1}{\sqrt{2}}$; (b) $\cos\theta = -\dfrac{1}{2}$.

The cosine function is decreasing on the interval from $0$ to $\pi/2$; thus it is a one-to-one function on this interval. For values of $a$ in the interval $0 \le a \le 1$, the notation $t = \cos^{-1} a$ is used to denote the unique value $t$ in the interval $0 \le t \le \pi/2$ such that $\cos t = a$.

(a) From the table in Problem 22.12, $\cos\dfrac{\pi}{4} = \dfrac{1}{\sqrt{2}}$, thus, $\dfrac{\pi}{4} = \cos^{-1}\dfrac{1}{\sqrt{2}}$. Since the cosine function is positive in quadrants I and IV, there is also an angle $\theta$ in quadrant IV with reference angle $\dfrac{\pi}{4}$ and $\cos\theta = \dfrac{1}{\sqrt{2}}$. This angle must satisfy $2\pi - \theta = \dfrac{\pi}{4}$; thus, $\theta = \dfrac{7\pi}{4}$. Thus $\dfrac{\pi}{4}$ and $\dfrac{7\pi}{4}$ are the two required angles.

(b) From the table in Problem 22.12, $\cos\dfrac{\pi}{3} = \dfrac{1}{2}$; thus $\dfrac{\pi}{3} = \cos^{-1}\dfrac{1}{2}$. Since the cosine function is negative in quadrants II and III, the required angles are the angles $\theta_1$ and $\theta_2$ in these quadrants with reference angle $\dfrac{\pi}{3}$ and $\cos\theta_1 = \cos\theta_2 = -\dfrac{1}{2}$. In quadrant II, $\pi - \theta_1 = \dfrac{\pi}{3}$; thus $\theta_1 = \dfrac{2\pi}{3}$. In quadrant III, $\theta_2 - \pi = \dfrac{\pi}{3}$; thus, $\theta_2 = \dfrac{4\pi}{3}$.

**22.18.** Find all angles $\theta$, $0° \leq \theta < 360°$, such that (a) $\tan\theta = \sqrt{3}$; (b) $\tan\theta = -1$.

The tangent function is increasing on the interval from 0 to $\pi/2$, thus it is a one-to-one function on this interval. For nonnegative values of $a$, the notation $t = \tan^{-1} a$ is used to denote the unique nonnegative value $t$ such that $\tan t = a$.

(a) From the table in Problem 22.12, $\tan\dfrac{\pi}{3} = \sqrt{3}$, thus $\dfrac{\pi}{3} = \tan^{-1}\sqrt{3}$. Hence 60° is one required angle. Since the tangent function is positive in quadrants I and III, there is also an angle $\theta$ in quadrant III with reference angle 60° and $\tan\theta = \sqrt{3}$. In quadrant III, $\theta - 180° = 60°$; thus $\theta = 240°$. The required angles are 60° and 240°.

(b) From the table in Problem 22.12, $\tan\dfrac{\pi}{4} = 1$; thus $\dfrac{\pi}{4} = \tan^{-1}1$. Since the tangent function is negative in quadrants II and IV, the required angles are the angle $\theta_1$ and $\theta_2$ in these quadrants with reference angle $\dfrac{\pi}{4} = 45°$. In quadrant II, $180° - \theta_1 = 45°$; thus $\theta_1 = 135°$. In quadrant IV, $360° - \theta_2 = 45°$; thus $\theta_2 = 315°$.

**22.19.** Use a scientific calculator to find approximate values for (a) $\sin 42°$; (b) $\cos 238°$; (c) $\tan(-61.5°)$; (d) $\sec 341°25'$.

In using a scientific calculator for trigonometric calculations, it is crucial to make certain that the correct mode (degree mode or radian mode) is selected. Refer to the calculator manual for instructions for choosing the mode. In this problem, put the calculator in degree mode. (a) $\sin 42° = 0.6691$; (b) $\cos 238° = -0.5299$; (c) $\tan(-61.5°) = -1.8418$; (d) secant cannot be calculated directly on a calculator; use a trigonometric identity:

$$\sec(341°25') = \frac{1}{\cos(341°25')} = \frac{1}{\cos(341 + 25/60)°} = 1.055$$

**22.20.** Use a scientific calculator to find approximate values for (a) $\sin 3$; (b) $\cos(-5.3)$; (c) $\tan(2.356)$; (d) $\cot(12.3)$.

See comments in previous problem. In this problem, put the calculator in radian mode. (a) $\sin 3 = 0.1411$; (b) $\cos(-5.3) = 0.5544$; (c) $\tan(2.356) = -1.0004$; (d) cotangent cannot be calculated directly on a calculator; use a trigonometric identity:

$$\cot(12.3) = \frac{1}{\tan(12.3)} = -3.6650$$

**22.21.** Use a scientific calculator to find approximate values for all angles $\theta$, $0 \leq \theta < 2\pi$, such that

(a) $\sin\theta = 0.7543$ (b) $\tan\theta = -4.412$.

Put the calculator into radian mode.

(a) First find $\sin^{-1} 0.7543 = 0.8546$. Since the sine function is positive in quadrants I and II, there is also an angle $\theta$ in quadrant II with reference angle 0.8546 and $\sin\theta = 0.7543$. This angle must be $\pi - 0.8546 = 2.2870$. Then 0.8546 and 2.2870 are the two required angles.

(b) First find $\tan^{-1} 4.412 = 1.3479$. Since the tangent function is negative in quadrants II and IV, the required angles are the angles $\theta_1$ and $\theta_2$ in these quadrants with reference angle 1.3479. In quadrant II, $\pi - \theta_1 = 1.3479$; thus $\theta_1 = 1.7937$. In quadrant IV, $2\pi - \theta_2 = 1.3479$, thus $\theta_2 = 4.9353$.

**22.22.** Use a scientific calculator to find approximate values for all angles $\theta$, $0° \leq \theta < 360°$, such that

(a) $\cos\theta = 0.8455$; (b) $\csc\theta = -3$; (c) $\sec\theta = 0.333$.

Put the calculator into degree mode.

(a) First find $\cos^{-1}0.8455 = 0.5633 = 32.27°$. Since the cosine function is positive in quadrants I and IV, there is also an angle $\theta$ in quadrant IV with reference angle 32.27° and $\cos\theta = 0.8455$. This angle must satisfy $360° - \theta = 32.27°$; thus $\theta = 327.73°$. Then 32.27° and 327.73° are the two required angles.

(b) Cosecant cannot be calculated directly on a calculator; use a trigonometric identity. $\csc \theta = -3$ is equivalent to $1/(\sin \theta) = -3$, thus, $\sin \theta = -\frac{1}{3}$.

First find $\sin^{-1} \frac{1}{3} = 0.3398 = 19.47°$. Since the sine function is negative in quadrants III and IV, the required angles are the angles $\theta_1$ and $\theta_2$ in these quadrants with reference angle $19.47°$ and $\sin \theta_1 = \sin \theta_2 = -\frac{1}{3}$. In quadrant III, $\theta_1 - 180° = 19.47°$; thus, $\theta_1 = 199.47°$. In quadrant IV, $360° - \theta_2 = 19.47°$; thus $\theta_2 = 340.53°$.

(c) There is no angle that satisfies $\sec \theta = 0.333$ since 0.333 is not in the range of the secant function. A calculator will return an error message.

## SUPPLEMENTARY PROBLEMS

**22.23.** List all angles coterminal with (a) $\theta$ radians; (b) $\theta$ degrees.

*Ans.* (a) $\theta + 2\pi n$, $n$ any integer; (b) $\theta + n360°$, $n$ any integer

**22.24.** Find the trigonometric functions of $270°$.

*Ans.* $\sin 270° = -1$, $\cos 270° = 0$, $\tan 270°$ is undefined, $\cot 270° = 0$, $\sec 270°$ is undefined, $\csc 270° = -1$

**22.25.** Complete the proof in Problem 22.7 that the definitions of the trigonometric functions as ratios are consistent with the definitions of the trigonometric functions of angles.

**22.26.** Find (a) $\sin 120°$; (b) $\cos \dfrac{5\pi}{6}$; (c) $\tan(-45°)$; (d) $\cot \dfrac{7\pi}{6}$; (e) $\sec 240°$; (f) $\csc \dfrac{2\pi}{3}$.

*Ans.* (a) $\dfrac{\sqrt{3}}{2}$; (b) $-\dfrac{\sqrt{3}}{2}$; (c) $-1$; (d) $\sqrt{3}$; (e) $-2$; (f) $\dfrac{2}{\sqrt{3}}$

**22.27.** Find (a) $\sin \dfrac{7\pi}{4}$; (b) $\cos 450°$; (c) $\tan \dfrac{8\pi}{3}$; (d) $\cot(-720°)$; (e) $\sec \dfrac{17\pi}{6}$; (f) $\csc(-510°)$.

*Ans.* (a) $-\dfrac{1}{\sqrt{2}}$; (b) 0; (c) $-\sqrt{3}$; (d) undefined; (e) $-\dfrac{2}{\sqrt{3}}$; (f) $-2$

**22.28.** If $\theta$ is in standard position and $(-1, -4)$ lies on its terminal side, find the trigonometric functions of $\theta$.

*Ans.* $\sin \theta = -\dfrac{4}{\sqrt{17}}$, $\cos \theta = -\dfrac{1}{\sqrt{17}}$, $\tan \theta = 4$, $\cot \theta = \dfrac{1}{4}$, $\sec \theta = -\sqrt{17}$, $\csc \theta = -\dfrac{\sqrt{17}}{4}$

**22.29.** If $\theta$ is an acute angle, find the other trigonometric functions of $\theta$, given

(a) $\sin \theta = \dfrac{12}{13}$; (b) $\cos \theta = \dfrac{5}{7}$; (c) $\tan \theta = \dfrac{1}{\sqrt{2}}$.

*Ans.* (a) $\cos \theta = \dfrac{5}{13}$, $\tan \theta = \dfrac{12}{5}$, $\cot \theta = \dfrac{5}{12}$, $\sec \theta = \dfrac{13}{5}$, $\csc \theta = \dfrac{13}{12}$

(b) $\sin \theta = \dfrac{\sqrt{24}}{7}$, $\tan \theta = \dfrac{\sqrt{24}}{5}$, $\cot \theta = \dfrac{5}{\sqrt{24}}$, $\sec \theta = \dfrac{7}{5}$, $\csc \theta = \dfrac{7}{\sqrt{24}}$

(c) $\sin \theta = \dfrac{1}{\sqrt{3}}$, $\cos \theta = \sqrt{\dfrac{2}{3}}$, $\cot \theta = \sqrt{2}$, $\sec \theta = \sqrt{\dfrac{3}{2}}$, $\csc \theta = \sqrt{3}$

**22.30.** If $\theta$ is in standard position and its terminal side lies in quadrant II on the line $x + 3y = 0$, find the trigonometric functions of $\theta$.

*Ans.* $\sin\theta = \dfrac{1}{\sqrt{10}}$, $\cos\theta = -\dfrac{3}{\sqrt{10}}$, $\tan\theta = -\dfrac{1}{3}$, $\cot\theta = -3$, $\sec\theta = -\dfrac{\sqrt{10}}{3}$, $\csc\theta = \sqrt{10}$

**22.31.** Find approximate values for all angles $\theta$, $0 \le \theta < 2\pi$, such that

(a) $\sin\theta = 0.1188$; (b) $\tan\theta = 8.7601$; (c) $\sec\theta = -2.3$.

*Ans.* (a) 0.1191, 3.0225; (b) 1.4571, 4.5987; (c) 2.0206, 4.2626

**22.32.** Find approximate values for all angles $\theta$, $0° \le \theta < 360°$, such that

(a) $\cos\theta = 0.0507$; (b) $\cot\theta = 62$; (c) $\csc\theta = -5.2$.

*Ans.* (a) 87.09°, 272.91°; (b) 0.92°, 180.92°; (c) 191.09°, 348.91°

**22.33.** The angular speed $\omega$ of a point moving in a circle is defined as the quotient $\theta/t$, where $\theta$ is the angle in radians through which the point travels in time $t$.

(a) Find the angular speed of a point that moves through an angle of 4 radians in 6 seconds.

(b) Find the angular speed of a point on the rim of a wheel that travels at 60 rpm (revolutions per minute).

(c) Show that the linear speed $v$ of a point moving in a circle is related to the angular speed by the formula $v = r\omega$.

(d) A car is moving at the rate of 60 miles per hour, and the diameter of each wheel is 2.5 feet. Find the angular speed of the wheels.

*Ans.* (a) $\frac{2}{3}$ rad/sec  (b) $120\pi$ rad/min  (d) 4224 rad/min

# CHAPTER 23

# Trigonometric Identities and Equations

## Definition of Identity

An identity is a statement that two quantities are equal that is true for all values of the variables for which the statement is meaningful.

**EXAMPLE 23.1**   Which of the following statements is an identity?

(a) $x + 3 = 3 + x$; (b) $x + 3 = 5$; (c) $x \cdot \dfrac{1}{x} = 1$.

(a) is an identity since it is always true; (b) is not an identity since it is true only if $x = 2$; (c) is an identity since it is true unless $x = 0$, in which case it is not meaningful.

## Basic Trigonometric Identities

Basic trigonometric identities are repeated below for reference:

1. **PYTHAGOREAN IDENTITIES.** For all $t$ for which both sides are defined:

$$\cos^2 t + \sin^2 t = 1 \qquad 1 + \tan^2 t = \sec^2 t \qquad \cot^2 t + 1 = \csc^2 t$$

$$\cos^2 t = 1 - \sin^2 t \qquad \tan^2 t = \sec^2 t - 1 \qquad \cot^2 t = \csc^2 t - 1$$

$$\sin^2 t = 1 - \cos^2 t \qquad 1 = \sec^2 t - \tan^2 t \qquad 1 = \csc^2 t - \cot^2 t$$

2. **RECIPROCAL IDENTITIES.** For all $t$ for which both sides are defined:

$$\sin t = \frac{1}{\csc t} \qquad \cos t = \frac{1}{\sec t} \qquad \tan t = \frac{1}{\cot t}$$

$$\csc t = \frac{1}{\sin t} \qquad \sec t = \frac{1}{\cos t} \qquad \cot t = \frac{1}{\tan t}$$

3. **QUOTIENT IDENTITIES.** For all $t$ for which both sides are defined:

$$\tan t = \frac{\sin t}{\cos t} \qquad \cot t = \frac{\cos t}{\sin t}$$

4. **IDENTITIES FOR NEGATIVES.** For all $t$ for which both sides are defined:

$$\sin(-t) = -\sin t \qquad \cos(-t) = \cos t \qquad \tan(-t) = -\tan t$$

$$\csc(-t) = -\csc t \qquad \sec(-t) = \sec t \qquad \cot(-t) = -\cot t$$

## Simplifying Trigonometric Expressions

The basic trigonometric identities are used to reduce trigonometric expressions to simpler form:

**EXAMPLE 23.2**   Simplify: $\dfrac{1 - \cos^2\alpha}{\sin\alpha}$

From the Pythagorean identity, $1 - \cos^2\alpha = \sin^2\alpha$. Hence, $\dfrac{1 - \cos^2\alpha}{\sin\alpha} = \dfrac{\sin^2\alpha}{\sin\alpha} = \sin\alpha$.

## Verifying Trigonometric Identities

To verify that a given statement is an identity, show that one side can be transformed into the other by using algebraic techniques, including simplification and substitution, and trigonometric techniques, frequently including reducing other functions to sines and cosines.

**EXAMPLE 23.3**   Verify that $(1 - \cos\theta)(1 + \cos\theta) = \sin^2\theta$ is an identity.

Starting with the left side, an obvious first step is to perform algebraic operations:

$$(1 - \cos\theta)(1 + \cos\theta) = 1 - \cos^2\theta \quad \text{Algebra}$$

$$= \sin^2\theta \quad \text{Pythagorean identity}$$

**EXAMPLE 23.4**   Verify that $\dfrac{\sin t\cos t}{\tan t} = \cos^2 t$ is an identity.

Starting with the left side, an obvious first step is to reduce to sines and cosines:

$$\frac{\sin t\cos t}{\tan t} = \frac{\sin t\cos t}{\sin t/\cos t} \quad \text{Quotient identity}$$

$$= \sin t\cos t \div \frac{\sin t}{\cos t} \quad \text{Algebra}$$

$$= \sin t\cos t \cdot \frac{\cos t}{\sin t} \quad \text{Algebra}$$

$$= \cos^2 t \quad \text{Algebra}$$

## Nonidentity Statements

If a statement is meaningful yet not true for even one value of the variable or variables, it is not an identity. To show that it is not an identity, it is sufficient to find one value of the variable or variables that would make it false.

**EXAMPLE 23.5**   Show that $\sin t + \cos t = 1$ is not an identity.

Although this statement is true for some values of $t$, for example $t = 0$, it is not an identity. For example, choose $t = \pi/4$. Then

$$\sin\frac{\pi}{4} + \cos\frac{\pi}{4} = \frac{1}{\sqrt{2}} + \frac{1}{\sqrt{2}} = \frac{2}{\sqrt{2}} = \sqrt{2} \neq 1$$

## Inverses of Trigonometric Functions

The trigonometric functions are periodic, and therefore are not one-to-one. However, in the first quadrant, sine and tangent are increasing functions and cosine is decreasing; hence, in this region the functions are one-to-one and thus have inverses. For present purposes, the following notation is used:

$t = \sin^{-1}a$ (read: inverse sine of $a$) if $0 \leq t \leq \pi/2$ and $\sin t = a$.
$t = \cos^{-1}a$ (read: inverse cosine of $a$) if $0 \leq t \leq \pi/2$ and $\cos t = a$.
$t = \tan^{-1}a$ (read: inverse tangent of $a$) if $0 \leq t < \pi/2$ and $\tan t = a$.

A complete treatment of inverse trigonometric functions is given in Chapter 25.

**EXAMPLE 23.6**   Find (a) $\sin^{-1}\dfrac{\sqrt{3}}{2}$; (b) $\cos^{-1}0$ (c) $\tan^{-1}1$.

(a) There is exactly one value of $t$ such that $\sin t = \sqrt{3}/2$ and $0 \le t \le \pi/2$, that is, $\pi/3$. Hence, $\sin^{-1} \dfrac{\sqrt{3}}{2} = \dfrac{\pi}{3}$.

(b) There is exactly one value of $t$ such that $\cos t = 0$ and $0 \le t \le \pi/2$, that is, $\pi/2$. Hence, $\cos^{-1} 0 = \dfrac{\pi}{2}$.

(c) There is exactly one value of $t$ such that $\tan t = 1$ and $0 \le t < \pi/2$, that is, $\pi/4$. Hence, $\tan^{-1} 1 = \dfrac{\pi}{4}$.

## Trigonometric Equations

Trigonometric equations can be solved by a mixture of algebraic and trigonometric techniques, including reducing other functions to sines and cosines, substitution from known trigonometric identities, algebraic simplification, and so on.

1. **BASIC TRIGONOMETRIC EQUATIONS** are equations of the form $\sin t = a$, $\cos t = b$, $\tan t = c$. These are solved by using inverses of trigonometric functions to express all solutions in the interval $[0, 2\pi)$ and then extending to the entire set of solutions. Some problems, however, specify that only solutions in the interval $[0, 2\pi)$ are to be found.
2. **OTHER TRIGONOMETRIC EQUATIONS** are solved by reducing to basic equations using algebraic and trigonometric techniques.

**EXAMPLE 23.7**  Find all solutions of $\cos t = \frac{1}{2}$.

First find all solutions in the interval $[0, 2\pi)$: Start with

$$t = \cos^{-1} \frac{1}{2} = \frac{\pi}{3}$$

Since cosine is positive in quadrant I and IV, there is also a solution in quadrant IV with reference angle $\pi/3$, namely $2\pi - \pi/3 = 5\pi/3$.

Extending to the entire real line, since cosine is periodic with period $2\pi$, all solutions can be written as $\pi/3 + 2\pi n, 5\pi/3 + 2\pi n, n$ any integer.

**EXAMPLE 23.8**  Find all solutions in the interval $[0, 2\pi)$ for $5 \tan t = 3 \tan t - 2$.

First reduce this to a basic trigonometric equation by isolating the quantity $\tan t$.

$$2 \tan t = -2$$

$$\tan t = -1$$

Now find all solutions of this equation in the interval $[0, 2\pi)$. Start with $\tan^{-1} 1 = \pi/4$. Since tangent is negative in quadrants II and IV, the solutions are the angles in these quadrants with reference angle $\pi/4$. These are $\pi - \pi/4 = 3\pi/4$ and $2\pi - \pi/4 = 7\pi/4$.

## SOLVED PROBLEMS

**23.1.**  Verify that $\csc t - \sin t = \cot t \cos t$ is an identity.

Starting with the left side, an obvious first step is to reduce to sines and cosines:

$$\csc t - \sin t = \frac{1}{\sin t} - \sin t \qquad \text{Reciprocal identity}$$

$$\csc t - \sin t = \frac{1 - \sin^2 t}{\sin t} \qquad \text{Algebra}$$

$$= \frac{\cos^2 t}{\sin t} \qquad \text{Pythagorean identity}$$

$$= \frac{\cos t}{\sin t} \cdot \cos t \qquad \text{Algebra}$$

$$= \cot t \cos t \qquad \text{Quotient identity}$$

**23.2.** Verify that $\sin^4\theta - \cos^4\theta = \sin^2\theta - \cos^2\theta$ is an identity.

Starting with the left side, an obvious first step is to express the fourth powers in terms of squares:

$$
\begin{aligned}
\sin^4\theta - \cos^4\theta &= (\sin^2\theta)^2 - (\cos^2\theta)^2 && \text{Algebra} \\
&= (\sin^2\theta - \cos^2\theta)(\sin^2\theta + \cos^2\theta) && \text{Algebra} \\
&= (\sin^2\theta - \cos^2\theta)(1) && \text{Pythagorean identity} \\
&= \sin^2\theta - \cos^2\theta && \text{Algebra}
\end{aligned}
$$

**23.3.** Verify that $\dfrac{1}{1 - \cos x} + \dfrac{1}{1 + \cos x} = 2\csc^2 x$ is an identity.

Starting with the left side, an obvious first step is to combine the two fractional expressions into one:

$$
\begin{aligned}
\frac{1}{1 - \cos x} + \frac{1}{1 + \cos x} &= \frac{(1 + \cos x) + (1 - \cos x)}{(1 - \cos x)(1 + \cos x)} \\
&= \frac{2}{1 - \cos^2 x}
\end{aligned}
$$

Now apply a Pythagorean identity:

$$
\begin{aligned}
\frac{2}{1 - \cos^2 x} &= \frac{2}{\sin^2 x} && \text{Pythagorean identity} \\
&= 2\csc^2 x && \text{Reciprocal identity}
\end{aligned}
$$

**23.4.** Verify that $\dfrac{1 - \cos\theta}{\sin\theta} = \dfrac{\sin\theta}{1 + \cos\theta}$ is an identity.

Often in this context squares of sines and cosines are easier to work with than the functions themselves. Starting with the right side, it is effective to multiply numerator and denominator by the expression $1 - \cos\theta$. (This is analogous to the operations in rationalizing the denominator.)

$$
\begin{aligned}
\frac{\sin\theta}{1 + \cos\theta} &= \frac{\sin\theta(1 - \cos\theta)}{(1 + \cos\theta)(1 - \cos\theta)} && \text{Algebra} \\
&= \frac{\sin\theta(1 - \cos\theta)}{1 - \cos^2\theta} && \text{Algebra} \\
&= \frac{\sin\theta(1 - \cos\theta)}{\sin^2\theta} && \text{Pythagorean identity} \\
&= \frac{1 - \cos\theta}{\sin\theta} && \text{Algebra}
\end{aligned}
$$

**23.5.** Verify that $\dfrac{\tan x + \tan y}{1 - \tan x \tan y} = \dfrac{\sin x \cos y + \cos x \sin y}{\cos x \cos y - \sin x \sin y}$ is an identity.

Starting with the left side, reduce to sines and cosines, then simplify the complex fraction that results by multiplying numerator and denominator by $\cos x \cos y$, the LCD of the internal fractions:

$$
\begin{aligned}
\frac{\tan x + \tan y}{1 - \tan x \tan y} &= \frac{\dfrac{\sin x}{\cos x} + \dfrac{\sin y}{\cos y}}{1 - \dfrac{\sin x \sin y}{\cos x \cos y}} && \text{Quotient identity} \\[2em]
&= \frac{\dfrac{\sin x}{\cos x} + \dfrac{\sin y}{\cos y}}{1 - \dfrac{\sin x \sin y}{\cos x \cos y}} \cdot \frac{\cos x \cos y}{\cos x \cos y} && \text{Algebra} \\[2em]
&= \frac{\sin x \cos y + \cos x \sin y}{\cos x \cos y - \sin x \sin y} && \text{Algebra}
\end{aligned}
$$

**23.6.** Show that $\sqrt{1 - \cos^2 t} = \sin t$ is not an identity.

From the Pythagorean identity, the statement is generally true if the left and right sides have the same sign. To show that it is not an identity, choose a value of $t$ for which $\sin t$ is negative, for example, $t = 3\pi/2$. Then $\sqrt{1 - \cos^2 3\pi/2} = \sqrt{1 - 0^2} = 1$ but $\sin 3\pi/2 = -1$.

**23.7.** Show that $(\sin\theta + \cos\theta)^2 = \sin^2\theta + \cos^2\theta$ is not an identity.

This statement arises from the very common algebraic error of confusing $(a + b)^2$ with $a^2 + b^2$. To show that it is not an identity, choose any value of $\theta$ for which neither $\sin\theta$ nor $\cos\theta$ is zero, for example, $\theta = \pi/6$.

Then $\left(\sin\dfrac{\pi}{6} + \cos\dfrac{\pi}{6}\right)^2 = \left(\dfrac{1}{2} + \dfrac{\sqrt{3}}{2}\right)^2 = \dfrac{4 + 2\sqrt{3}}{4}$ but $\sin^2\dfrac{\pi}{6} + \cos^2\dfrac{\pi}{6} = 1$ (from the Pythagorean identity).

**23.8.** Simplify the expression $\sqrt{25 - x^2}$ by making the substitution $x = 5\sin u$, $-\dfrac{\pi}{2} \le u \le \dfrac{\pi}{2}$.

Making the substitution and factoring the expression under the radical yields an expression that can be simplified by applying a Pythagorean identity:

$$\sqrt{25 - x^2} = \sqrt{25 - (5\sin u)^2} = \sqrt{25 - 25\sin^2 u} = \sqrt{25(1 - \sin^2 u)} = \sqrt{25\cos^2 u} = 5|\cos u|$$

The last expression can be further simplified by observing that the restriction $-\dfrac{\pi}{2} \le u \le \dfrac{\pi}{2}$ confines $u$ to quadrants I and IV in which $\cos\theta$ is never negative. In this region, $5|\cos u| = 5\cos u$.

**23.9.** Simplify the expression $\dfrac{1}{\sqrt{16 + x^2}}$ by making the substitution $x = 4\tan u$, $-\dfrac{\pi}{2} < u < \dfrac{\pi}{2}$.

Proceed as in the previous problem:

$$\frac{1}{\sqrt{16 + x^2}} = \frac{1}{\sqrt{16 + (4\tan u)^2}} = \frac{1}{\sqrt{16 + 16\tan^2 u}} = \frac{1}{\sqrt{16(1 + \tan^2 u)}} = \frac{1}{\sqrt{16\sec^2 u}} = \frac{1}{4|\sec u|}$$

The last expression can be further simplified by observing that the restriction $-\dfrac{\pi}{2} < u < \dfrac{\pi}{2}$ confines $u$ to quadrants I and IV, in which $\sec u$ is never negative. In this region, $\dfrac{1}{4|\sec u|} = \dfrac{1}{4\sec u} = \dfrac{\cos u}{4}$.

**23.10.** Find all solutions for $\sin t = \dfrac{\sqrt{3}}{2}$.

For this basic trigonometric equation, begin by finding all solutions in the interval $[0, 2\pi)$. Start with

$$t = \sin^{-1}\frac{\sqrt{3}}{2} = \frac{\pi}{3}$$

Since sine is positive in quadrants I and II, there is also a solution in quadrant II with reference angle $\pi/3$, namely, $\pi - \pi/3 = 2\pi/3$.
Extending to the entire real line, since sine is periodic with period $2\pi$, all solutions can be written as $\pi/3 + 2\pi n$, $2\pi/3 + 2\pi n$, $n$ any integer.

**23.11.** Find all solutions for $3 - 4\cos^2\theta = 0$.

First reduce this to a basic trigonometric equation by isolating the quantity $\cos\theta$.

$$\cos^2\theta = \frac{3}{4}$$

$$\cos\theta = \pm\frac{\sqrt{3}}{2}$$

Since $\cos^{-1}\dfrac{\sqrt{3}}{2} = \dfrac{\pi}{6}$, there are four solutions in the interval $[0, 2\pi)$, namely, $\dfrac{\pi}{6}$ (positive cosine), $\pi - \dfrac{\pi}{6} = \dfrac{5\pi}{6}$ (negative cosine), $\pi + \dfrac{\pi}{6} = \dfrac{7\pi}{6}$ (negative cosine), and $2\pi - \dfrac{\pi}{6} = \dfrac{11\pi}{6}$ (positive cosine). Extending to the entire real line, since cosine is periodic with period $2\pi$, all solutions can be written as $\pi/6 + 2\pi n$, $5\pi/6 + 2\pi n$, $7\pi/6 + 2\pi n$, $11\pi/6 + 2\pi n$, $n$ any integer.

**23.12.** Find all solutions for $2\cos 2x - 1 = 0$.

Reducing to a basic trigonometric equation yields $\cos 2x = \dfrac{1}{2}$. To solve this, begin with $2x = \cos^{-1}\dfrac{1}{2}$. Thus, in the interval $[0, 2\pi)$, $2x = \dfrac{\pi}{3}$ and $2x = 2\pi - \dfrac{\pi}{3} = \dfrac{5\pi}{3}$; extending to the entire real line yields $2x = \dfrac{\pi}{3} + 2\pi n$ and $2x = \dfrac{5\pi}{3} + 2\pi n$. Hence, isolating $x$, all solutions are given by $x = \dfrac{\pi}{6} + \pi n, \dfrac{5\pi}{6} + \pi n$, $n$ any integer.

**23.13.** Find all solutions on the interval $[0, 2\pi)$ for $2\sin^2 u + \sin u = 0$.

This is an equation in quadratic form in the quantity $\sin u$. It is most efficiently solved by factoring (alternatively: make the substitution $v = \sin u$):

$$\sin u(2\sin u + 1) = 0$$
$$\sin u = 0 \quad \text{or} \quad 2\sin u + 1 = 0$$
$$\sin u = -\dfrac{1}{2}$$

$\sin u = 0$ has solutions $0$ and $\pi$ on the interval $[0, 2\pi)$. $\sin u = -\dfrac{1}{2}$ has solutions $\dfrac{7x}{6}$ and $\dfrac{11\pi}{6}$ on the interval.

Solutions: $0, \pi, \dfrac{7\pi}{6}, \dfrac{11\pi}{6}$

**23.14.** Find all angles in the interval $[0°, 360°)$ that satisfy $2\sin^2\theta = 1 - \cos\theta$.

First, use a Pythagorean identity to reduce to one trigonometric function:

$$2(1 - \cos^2\theta) = 1 - \cos\theta$$

This is an equation in quadratic form in $\cos\theta$. Reduce to standard form:

$$2 - 2\cos^2\theta = 1 - \cos\theta$$
$$2\cos^2\theta - \cos\theta - 1 = 0$$

This is most efficiently solved by factoring:

$$(2\cos\theta + 1)(\cos\theta - 1) = 0$$
$$2\cos\theta + 1 = 0 \quad \text{or} \quad \cos\theta - 1 = 0$$
$$\cos\theta = -\dfrac{1}{2} \qquad\qquad \cos\theta = 1$$

$\cos\theta = -\dfrac{1}{2}$ has solutions in quadrants II and III. Since $\cos^{-1}\dfrac{1}{2} = \dfrac{\pi}{3} = 60°$, in the interval $[0°, 360°)$ the required angles are $180° - 60° = 120°$ and $180° + 60° = 240°$. The only solution of $\cos\theta = 1$ in the interval is $0°$.

Solutions: $0°, 120°, 240°$

**23.15.** Find all solutions on the interval $[0, 2\pi)$ for $\sin x + \cos x = 1$.

As remarked in Problem 23.4, squares of sines and cosines are often easier to work with than the functions themselves. Isolate $\sin x$ and square both sides. Recall (Chapter 5) that raising both sides of an equation to an even power is permissible if all solutions to the resulting equation are checked to see whether they are solutions of the original equation.

$$\sin x = 1 - \cos x$$
$$\sin^2 x = (1 - \cos x)^2$$
$$= 1 - 2\cos x + \cos^2 x$$

Now use a Pythagorean identity to reduce to one trigonometric function:

$$1 - \cos^2 x = 1 - 2\cos x + \cos^2 x$$

This is an equation in quadratic form in $\cos x$. Reduce to standard form:

$$0 = 2\cos^2 x - 2\cos x$$

Solve by factoring:

$$2\cos x\,(\cos x - 1) = 0$$
$$2\cos x = 0 \quad \text{or} \quad \cos x - 1 = 0$$
$$\cos x = 0 \qquad\qquad \cos x = 1$$

On the interval $[0, 2\pi)$, $\cos x = 0$ has solutions $\dfrac{\pi}{2}$ and $\dfrac{3\pi}{2}$; $\cos x = 1$ has solution 0. It is necessary to check each of these solutions in the original equation:

Check: $x = 0$: $\sin 0 + \cos 0 = 1$? $\qquad x = \dfrac{\pi}{2}$: $\sin\dfrac{\pi}{2} + \cos\dfrac{\pi}{2} = 1$? $\qquad x = \dfrac{3\pi}{2}$: $\sin\dfrac{3\pi}{2} + \cos\dfrac{3\pi}{2} = 1$?

$\qquad\qquad 0 + 1 = 1 \qquad\qquad\qquad\qquad 1 + 0 = 0 \qquad\qquad\qquad\qquad -1 + 0 \neq 1$

$\qquad\qquad$ A solution $\qquad\qquad\qquad\qquad$ A solution $\qquad\qquad\qquad\qquad$ Not a solution

Solutions: $0, \dfrac{\pi}{2}$

**23.16.** Find approximate values for all solutions to $\tan^2 t - \tan t - 6 = 0$.

This is an equation in quadratic form in the quantity $\tan t$. It is most efficiently solved by factoring:

$$\tan^2 t - \tan t - 6 = 0$$
$$(\tan t - 3)(\tan t + 2) = 0$$
$$\tan t - 3 = 0 \quad \text{or} \quad \tan t + 2 = 0$$
$$\tan t - 3 \qquad\qquad \tan t = -2$$

Use the calculator to find approximate values for solutions of these equations: $\tan t = 3$ has solutions in quadrants I and III; since $\tan^{-1} 3 = 1.2490$, the solutions in the interval $[0, 2\pi)$ are 1.2490 and $\pi + 1.2490 = 4.3906$. $\tan t = -2$ has solutions in quadrants II and IV; since $\tan^{-1} 2 = 1.1071$, these solutions are $\pi - 1.1071 = 2.0344$ and $2\pi - 1.1071 = 5.1761$. Extending to the entire real line, all solutions are given by $1.2490 + 2\pi n$, $2.0344 + 2\pi n$, $4.3906 + 2\pi n$, $5.1761 + 2\pi n$, $n$ any integer. More compactly, since tangent has period $\pi$, all solutions can be written as $1.2490 + \pi n$, $2.0344 + \pi n$, $n$ any integer.

**23.17.** Find approximate values for all solutions on the interval $[0, 2\pi)$ for $3\sin^2 x - 5\sin x = 2$.

This is an equation in quadratic form in the quantity $\sin x$. It is most efficiently solved by factoring:

$$3\sin^2 x - 5\sin x - 2 = 0$$
$$(3\sin x + 1)(\sin x - 2) = 0$$
$$3\sin x + 1 = 0 \quad \text{or} \quad \sin x - 2 = 0$$
$$\sin x = -\tfrac{1}{3} \qquad\qquad \sin x = 2$$

Use the calculator to find approximate values for solutions of these equations: $\sin x = -\frac{1}{3}$ has solutions in quadrants III and IV; since $\sin^{-1}\frac{1}{3} = 0.339$, the solutions in the interval $[0, 2\pi)$ are $\pi + 0.3398 = 3.4814$ and $2\pi - 0.3398 = 5.9434$. $\sin x = 2$ has no solutions, since 2 is not in the range of the sine function; a calculator will return an error message.

Solutions: 3.4814, 5.9434.

**23.18.** Find approximate values for all angles in the interval $[0°, 360°)$ that satisfy $3\cos^2 A + 5\cos A - 1 = 0$.

This is an equation in quadratic form in the quantity $\cos A$. Since it is not factorable in the integers, use the quadratic formula, with $a = 3, b = 5, c = -1$.

$$\cos A = \frac{-5 \pm \sqrt{5^2 - 4(3)(-1)}}{2\cdot 3}$$
$$\cos A = \frac{-5 \pm \sqrt{37}}{6}$$

Using a calculator to approximate these values yields $\cos A = 0.1805$ and $\cos A = -1.8471$. The first of these has solutions in quadrants I and IV; since $\cos^{-1}\left(\frac{-5 + \sqrt{37}}{6}\right) = 1.3893 = 79.6°$, the solutions are 79.6° and $360° - 79.6° = 280.4°$. $\cos A = -1.8471$ has no solutions, since $-1.8471$ is not in the range of the cosine function; a calculator will return an error message.

Solutions: 79.6°, 280.4°.

## SUPPLEMENTARY PROBLEMS

**23.19.** Simplify　(a) $\sin^2 x\cot^2 x$　　(b) $\cos t(1 + \tan^2 t)$

(c) $(\cot\theta + \csc\theta)(\cot\theta - \csc\theta)$　(d) $\dfrac{\cos\theta}{1 - \sin\theta} - \dfrac{\cos\theta}{1 + \sin\theta}$

*Ans.*　(a) $\cos^2 x$; (b) $\sec t$; (c) $-1$; (d) $2\tan\theta$

**23.20.** Simplify　(a) $\csc x\tan x$;　(b) $1 - \dfrac{\sin^2 x}{1 + \cos x}$;

(c) $\dfrac{\sin^4 u - \cos^4 u}{\sin u + \cos u}$;　(d) $\dfrac{\sec x}{\csc x} + \dfrac{\sin x}{\cos x}$

Ans.　(a) $\sec x$; (b) $\cos x$; (c) $\sin u - \cos u$; (d) $2\tan x$

**23.21.** Verify that the following are identities:

(a) $\dfrac{1}{\sin x} - \sin x = \dfrac{\cos^2 x}{\sin x}$　(b) $\dfrac{1 + \sin x}{\cos x} + \dfrac{\cos x}{1 + \sin x} = 2\sec x$　(c) $(\sec\beta + \tan\beta)^2 = \dfrac{1 + \sin\beta}{1 - \sin\beta}$

**23.22.** Verify that the following are identities:

(a) $\dfrac{\cos t}{\csc t - \sin t} = \tan t$; (b) $\sec^2 x - (1 + \tan x)^2 = -2\tan x$; (c) $\tan t + \dfrac{1}{\tan t} = \dfrac{1}{\sin t\cos t}$

**23.23.** Verify that the following are identities:

(a) $\sin x(1 - 2\cos^2 x + \cos^4 x) = \sin^5 x$; (b) $\dfrac{\sin^3 t - \cos^3 t}{\sin t - \cos t} = 1 + \sin t \cos t$;

(c) $\sin^2 u - \cos^2 u = \dfrac{\tan u - \cot u}{\tan u + \cot u}$

**23.24.** Verify that $\ln(\csc x) = -\ln(\sin x)$ is an identity.

**23.25.** Simplify the following algebraic expressions by making the indicated substitution:

(a) $\dfrac{1}{x\sqrt{4 - x^2}}$, substitute $x = 2\sin u, -\dfrac{\pi}{2} \le u \le \dfrac{\pi}{2}$;

(b) $\dfrac{(4x^2 + 9)^{3/2}}{x}$, substitute $x = \dfrac{3}{2}\tan u, -\dfrac{\pi}{2} < u < \dfrac{\pi}{2}$;

(c) $\dfrac{\sqrt{x^2 - a^2}}{x}$, substitute $x = a, \sec u, a > 0, 0 \le u < \dfrac{\pi}{2}$

*Ans.* (a) $\dfrac{1}{4}\sec u \csc u$; (b) $18\sec^2 u \csc u$; (c) $\sin u$

**23.26.** Show that the following are not identities:

(a) $\sec\theta = \sqrt{\tan^2\theta + 1}$; (b) $\cos 2\theta = 2\cos\theta$

**23.27.** Find all solutions:

(a) $4\sin x + 2\sqrt{3} = 0$ (b) $\tan 3t = 1$ (c) $2\cos^2 u = \cos u$

(d) $4 - \sin^2\theta = 1$ (e) $\ln\sin x = 0$

*Ans.* (a) $x = 4\pi/3 + 2\pi n, 5\pi/3 + 2\pi n$; (b) $t = \pi/12 + n\pi/3$;

(c) $u = \pi/2 + 2\pi n, 3\pi/2 + 2\pi n, \pi/3 + 2\pi n, 5\pi/3 + 2\pi n$; (d) no solution; (e) $x = \pi/2 + 2\pi n$

**23.28.** Find all solutions on the interval $[0, 2\pi)$.

(a) $2\cos^2 4\theta = 1$; (b) $\dfrac{1 + \sin x}{\cos x} + \dfrac{\cos x}{1 + \sin x} = 4$; (c) $2\cos^2 x + 3\sin x = 3$; (d) $\tan x - \sec x = 1$

*Ans.* (a) $\theta = \dfrac{\pi}{16}, \dfrac{3\pi}{16}, \dfrac{5\pi}{16}, \dfrac{7\pi}{16}, \dfrac{9\pi}{16}, \dfrac{11\pi}{16}, \dfrac{13\pi}{16}, \dfrac{15\pi}{16}$; (b) $x = \dfrac{\pi}{3}, \dfrac{5\pi}{3}$; (c) $x = \dfrac{\pi}{6}, \dfrac{5\pi}{6}, \dfrac{\pi}{2}$; (d) $x = \pi$

**23.29.** Find approximate values for all solutions on the interval $[0°, 360°)$.

(a) $4\sin^2 A - 4\sin A - 1 = 0$; (b) $2\cos^2 2A + 3\cos 2A - 1 = 0$.

*Ans.* (a) $191.95°, 348.05°$; (b) $36.85°, 143.15°, 216.85°, 323.15°$

**CHAPTER 24**

# Sum, Difference, Multiple, and Half-Angle Formulas

## Sum and Difference Formulas

Sum and difference formulas for sines, cosines, and tangents: Let $u$ and $v$ be any real numbers; then

$$\sin(u + v) = \sin u \cos v + \cos u \sin v \qquad\qquad \sin(u - v) = \sin u \cos v - \cos u \sin v$$

$$\cos(u + v) = \cos u \cos v - \sin u \sin v \qquad\qquad \cos(u - v) = \cos u \cos v + \sin u \sin v$$

$$\tan(u + v) = \frac{\tan u + \tan v}{1 - \tan u \tan v} \qquad\qquad \tan(u - v) = \frac{\tan u - \tan v}{1 + \tan u \tan v}$$

**EXAMPLE 24.1**  Calculate an exact value for $\sin\dfrac{\pi}{12}$.

Noting that $\dfrac{\pi}{12} = \dfrac{\pi}{3} - \dfrac{\pi}{4}$, apply the difference formula for sines:

$$\begin{aligned}
\sin\frac{\pi}{12} &= \sin\left(\frac{\pi}{3} - \frac{\pi}{4}\right) \\
&= \sin\frac{\pi}{3}\cos\frac{\pi}{4} - \cos\frac{\pi}{3}\sin\frac{\pi}{4} \\
&= \frac{\sqrt{3}}{2}\cdot\frac{1}{\sqrt{2}} - \frac{1}{2}\cdot\frac{1}{\sqrt{2}} \\
&= \frac{\sqrt{3}-1}{2\sqrt{2}} \text{ or } \frac{\sqrt{6}-\sqrt{2}}{4}
\end{aligned}$$

## Cofunction Formulas

Cofunction formulas for the trigonometric functions: Let $\theta$ be any real number; then

$$\sin\left(\frac{\pi}{2} - \theta\right) = \cos\theta \qquad \cos\left(\frac{\pi}{2} - \theta\right) = \sin\theta \qquad \tan\left(\frac{\pi}{2} - \theta\right) = \cot\theta$$

$$\csc\left(\frac{\pi}{2} - \theta\right) = \sec\theta \qquad \sec\left(\frac{\pi}{2} - \theta\right) = \csc\theta \qquad \cot\left(\frac{\pi}{2} - \theta\right) = \tan\theta$$

## Double-Angle Formulas

Double-angle formulas for sines, cosines, and tangents: Let $\theta$ be any real number; then

$$\sin 2\theta = 2\sin\theta\cos\theta \qquad \cos 2\theta = \cos^2\theta - \sin^2\theta \qquad \tan 2\theta = \frac{2\tan\theta}{1 - \tan^2\theta}$$

Also,

$$\cos 2\theta = 2\cos^2\theta - 1 = 1 - 2\sin^2\theta$$

**EXAMPLE 24.2** Given $\cos\theta = \frac{2}{3}$, find $\cos 2\theta$.

Use a double-angle formula for cosine: $\cos 2\theta = 2\cos^2\theta - 1 = 2\left(\frac{2}{3}\right)^2 - 1 = -\frac{1}{9}$

## Half-Angle Identities

Half-angle identities for sine and cosine: Let $u$ be any real number; then

$$\sin^2 u = \frac{1 - \cos 2u}{2} \qquad \cos^2 u = \frac{1 + \cos 2u}{2}$$

## Half-Angle Formulas

Half-angle formulas for sine, cosine, and tangent: Let $A$ be any real number; then

$$\sin\frac{A}{2} = (\pm)\sqrt{\frac{1 - \cos A}{2}} \qquad \cos\frac{A}{2} = (\pm)\sqrt{\frac{1 + \cos A}{2}} \qquad \tan\frac{A}{2} = (\pm)\sqrt{\frac{1 - \cos A}{1 + \cos A}}$$

$$= \frac{1 - \cos A}{\sin A}$$

$$= \frac{\sin A}{1 + \cos A}$$

The sign of the square root in these formulas cannot be specified in general; in any particular case, it is determined by the quadrant in which $A/2$ lies.

**EXAMPLE 24.3** Given $\cos\theta = \frac{2}{3}$, $\frac{3\pi}{2} < \theta < 2\pi$, find $\sin\frac{\theta}{2}$ and $\cos\frac{\theta}{2}$.

Use the half-angle formulas for sine and cosine. Since $\frac{3\pi}{2} < \theta < 2\pi$, dividing all sides of this inequality by 2 yields $\frac{3\pi}{4} < \frac{\theta}{2} < \pi$. Therefore, $\frac{\theta}{2}$ lies in quadrant II and the sign of $\sin\frac{\theta}{2}$ is to be chosen positive, while the sign of $\cos\frac{\theta}{2}$ is to be chosen negative.

$$\sin\frac{\theta}{2} = +\sqrt{\frac{1 - \frac{2}{3}}{2}} = \sqrt{\frac{1}{6}} \qquad \cos\frac{\theta}{2} = -\sqrt{\frac{1 + \frac{2}{3}}{2}} = -\sqrt{\frac{5}{6}}$$

## Product-To-Sum Formulas

Let $u$ and $v$ be any real numbers:

$$\sin u \cos v = \frac{1}{2}[\sin(u + v) + \sin(u - v)] \qquad \cos u \sin v = \frac{1}{2}[\sin(u + v) - \sin(u - v)]$$

$$\cos u \cos v = \frac{1}{2}[\cos(u + v) + \cos(u - v)] \qquad \sin u \sin v = \frac{1}{2}[\cos(u - v) - \cos(u + v)]$$

## Sum-To-Product Formulas

Let $a$ and $b$ be any real numbers:

$$\sin a + \sin b = 2\sin\frac{a + b}{2}\cos\frac{a - b}{2} \qquad \cos a + \cos b = 2\cos\frac{a + b}{2}\cos\frac{a - b}{2}$$

$$\sin a - \sin b = 2\cos\frac{a + b}{2}\sin\frac{a - b}{2} \qquad \cos a - \cos b = -2\sin\frac{a + b}{2}\sin\frac{a - b}{2}$$

**EXAMPLE 24.4** Express $\sin 10x - \sin 6x$ as a product.

Use the formula for $\sin a - \sin b$ with $a = 10x$ and $b = 6x$.

$$\sin 10x - \sin 6x = 2\cos\frac{10x + 6x}{2}\sin\frac{10x - 6x}{2} = 2\cos 8x \sin 2x$$

## SOLVED PROBLEMS

**24.1.** Derive the difference formula for cosines.

Let $u$ and $v$ be any two real numbers. Shown in Fig. 24-1 is a case for which $u$, $v$, and $u - v$ are positive.

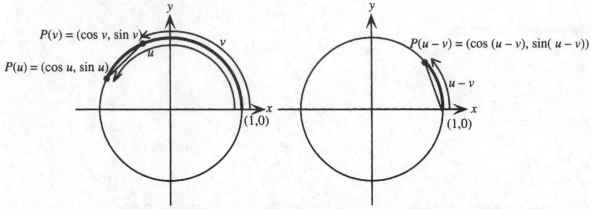

Figure 24-1

The arc with endpoints $P(u)$ and $P(v)$, shown on the left-hand unit circle, has length $u - v$. The arc with endpoints $(1, 0)$ and $P(u - v)$, shown on the right-hand unit circle, has the same length. Since congruent arcs on congruent circles have congruent chords, the distance from $P(v)$ to $P(u)$ must equal the distance from $(1, 0)$ to $P(u - v)$. Hence, from the distance formula,

$$\sqrt{(\cos u - \cos v)^2 + (\sin u - \sin v)^2} = \sqrt{(\cos(u - v) - 1)^2 + (\sin(u - v) - 0)^2}$$

Squaring both sides and expanding the squared expressions yields:

$$\cos^2 u - 2\cos u \cos v + \cos^2 v + \sin^2 u - 2\sin u \sin v + \sin^2 v = \cos^2(u - v) - 2\cos(u - v) + 1 + \sin^2(u - v)$$

Since $\cos^2 u + \sin^2 u = 1$, $\cos^2 v + \sin^2 v = 1$, and $\cos^2(u - v) + \sin^2(u - v) = 1$ (applying the Pythagorean identity three times), this simplifies to:

$$2 - 2\cos u \cos v - 2\sin u \sin v = 2 - 2\cos(u - v)$$

Subtracting 2 from each side and dividing by $-2$ yields:

$$\cos u \cos v + \sin u \sin v = \cos(u - v)$$

as required. The proof can be extended to cover all cases, $u$ and $v$ any real numbers.

**24.2.** Derive

(a) the sum formula for cosines; (b) the cofunction formulas for sines and cosines.

(a) Start with the difference formula for cosines and replace $v$ with $-v$.

$$\cos(u - v) = \cos u \cos v + \sin u \sin v$$

$$\cos[u - (-v)] = \cos u \cos(-v) + \sin u \sin(-v)$$

Now apply the identities for negatives, $\cos(-v) = \cos v$ and $\sin(-v) = -\sin v$, and simplify:

$$\cos(u + v) = \cos u \cos v + \sin u(-\sin v)$$

$$\cos(u + v) = \cos u \cos v - \sin u \sin v$$

(b) Use the difference formula for cosines, with $u = \dfrac{\pi}{2}$ and $v = \theta$:

$$\cos\left(\frac{\pi}{2} - \theta\right) = \cos\frac{\pi}{2}\cos\theta + \sin\frac{\pi}{2}\sin\theta$$

Since $\cos(\pi/2) = 0$ and $\sin(\pi/2) = 1$, it follows that

$$\cos\left(\frac{\pi}{2} - \theta\right) = 0\cos\theta + 1\sin\theta = \sin\theta.$$

Now replace $\theta$ with $\pi/2 - \theta$ :

$$\cos\left[\frac{\pi}{2} - \left(\frac{\pi}{2} - \theta\right)\right] = \sin\left(\frac{\pi}{2} - \theta\right)$$

Simplifying yields

$$\cos\theta = \sin\left(\frac{\pi}{2} - \theta\right)$$

**24.3.** Derive the difference formula for sines.

Start with a cofunction formula, for example, $\sin\theta = \cos\left(\dfrac{\pi}{2} - \theta\right)$ and replace $\theta$ with $u - v$:

$$\sin(u - v) = \cos\left[\frac{\pi}{2} - (u - v)\right] = \cos\left(\frac{\pi}{2} - u + v\right) = \cos\left[\left(\frac{\pi}{2} - u\right) + v\right]$$

In the last expression, apply the sum formula for cosines with $u$ replaced by $\pi/2 - u$.

$$\sin(u - v) = \cos\left[\left(\frac{\pi}{2} - u\right) + v\right] = \cos\left(\frac{\pi}{2} - u\right)\cos v - \sin\left(\frac{\pi}{2} - u\right)\sin v = \sin u \cos v - \cos u \sin v$$

using the cofunction formulas again at the last step.

**24.4.** Derive the difference formula for tangents.

Start with the quotient identity for $\tan(u - v)$.

$$\tan(u - v) = \frac{\sin(u - v)}{\cos(u - v)} = \frac{\sin u \cos v - \cos u \sin v}{\cos u \cos v + \sin u \sin v}.$$

To obtain the required expression in terms of tangents, divide numerator and denominator of the last expression by the quantity $\cos u \cos v$ and apply the quotient identity again:

$$\tan(u - v) = \frac{\sin u \cos v - \cos u \sin v}{\cos u \cos v + \sin u \sin v} = \frac{\dfrac{\sin u \cos v}{\cos u \cos v} - \dfrac{\cos u \sin v}{\cos u \cos v}}{\dfrac{\cos u \cos v}{\cos u \cos v} + \dfrac{\sin u \sin v}{\cos u \cos v}} = \frac{\dfrac{\sin u}{\cos u} - \dfrac{\sin v}{\cos v}}{1 + \dfrac{\sin u}{\cos u}\dfrac{\sin v}{\cos v}} = \frac{\tan u - \tan v}{1 + \tan u \tan v}$$

**24.5.** Given $\sin x = \frac{3}{5}$, $x$ in quadrant I, and $\cos y = \frac{2}{3}$, $y$ in quadrant IV, find (a) $\cos(x + y)$; (b) $\tan(x + y)$; (c) the quadrant in which $x + y$ must lie.

(a) From the sum formula for cosines, $\cos(x + y) = \cos x \cos y - \sin x \sin y$. $\sin x$ and $\cos y$ are given; $\cos x$ and $\sin y$ must be determined.

Since $\sin x = 3/5$ and $x$ is in quadrant I, $\cos x = +\sqrt{1 - (3/5)^2} = 4/5$. Since $\cos y = 2/3$ and $y$ is in quadrant IV, $\sin y = -\sqrt{1 - (2/3)^2} = -(\sqrt{5})/3$. Hence,

$$\cos(x + y) = \cos x \cos y - \sin x \sin y = \frac{4}{5}\cdot\frac{2}{3} - \frac{3}{5}\left(-\frac{\sqrt{5}}{3}\right) = \frac{8 + 3\sqrt{5}}{15}$$

(b) $\tan(x + y)$ can be found from the sum formula for tangents, using the given quantities and noting that

$$\tan x = \frac{\sin x}{\cos x} = \frac{3/5}{4/5} = \frac{3}{4} \text{ and } \tan y = \frac{\sin y}{\cos y} = \frac{-(\sqrt{5})/3}{2/3} = -\frac{\sqrt{5}}{2}. \text{ Hence,}$$

$$\tan(x + y) = \frac{\tan x + \tan y}{1 - \tan x \tan y} = \frac{\frac{3}{4} + \left(-\frac{\sqrt{5}}{2}\right)}{1 - \frac{3}{4}\left(-\frac{\sqrt{5}}{2}\right)} = \frac{6 - 4\sqrt{5}}{8 + 3\sqrt{5}}$$

(c) Since $\cos(x + y)$ is positive and $\tan(x + y)$ is negative, $x + y$ must lie in quadrant IV.

**24.6.** Derive the double-angle formulas for sine and cosine.

Start with the sum formulas for sines and let $u = v = \theta$. Then

$$\sin 2\theta = \sin(\theta + \theta) = \sin\theta \cos\theta + \cos\theta \sin\theta = 2\sin\theta \cos\theta$$

Similarly for cosines:

$$\cos 2\theta = \cos(\theta + \theta) = \cos\theta \cos\theta - \sin\theta \sin\theta = \cos^2\theta - \sin^2\theta$$

To derive the other forms of the double-angle formula for cosine, apply Pythagorean identities.

$$\cos 2\theta = \cos^2\theta - \sin^2\theta = 1 - \sin^2\theta - \sin^2\theta = 1 - 2\sin^2\theta$$

Also

$$\cos 2\theta = \cos^2\theta - \sin^2\theta = \cos^2\theta - (1 - \cos^2\theta) = 2\cos^2\theta - 1$$

**24.7.** Given $\tan t = \frac{7}{24}$, $t$ in quadrant III, find $\sin 2t$ and $\cos 2t$.

From the given information, it follows that $\sec t = -\sqrt{1 + \tan^2 t} = -\sqrt{1 + \left(\frac{7}{24}\right)^2} = -\frac{25}{24}$. Hence,
$\cos t = -\frac{24}{25}$ and $\sin t = -\frac{7}{25}$. Now apply the double-angle formulas for sine and cosine:

$$\sin 2t = 2\sin t \cos t = 2\left(-\frac{7}{25}\right)\left(-\frac{24}{25}\right) = \frac{336}{625}$$

$$\cos 2t = \cos^2 t - \sin^2 t = \left(-\frac{24}{25}\right)^2 - \left(-\frac{7}{25}\right)^2 = \frac{527}{625}$$

To check this, note that $\sin^2 2t + \cos^2 2t = \left(\frac{336}{625}\right)^2 + \left(\frac{527}{625}\right)^2 = 1$ as expected.

**24.8.** Derive the half-angle identities.

Start with the double-angle formulas for cosine, and solve for $\cos^2\theta$ and $\sin^2\theta$. Then

$$2\cos^2\theta - 1 = \cos 2\theta \qquad\qquad\qquad 1 - 2\sin^2\theta = \cos 2\theta$$
$$2\cos^2\theta = 1 + \cos 2\theta \qquad\qquad\qquad -2\sin^2\theta = \cos 2\theta - 1$$
$$\cos^2\theta = \frac{1 + \cos 2\theta}{2} \qquad\qquad\qquad 2\sin^2\theta = 1 - \cos 2\theta$$
$$\sin^2\theta = \frac{1 - \cos 2\theta}{2}$$

**24.9.** Use a double-angle identity to derive an expression for $\cos 4t$ in terms of $\cos t$.

$$\cos 4t = \cos 2(2t) = 2\cos^2 2t - 1 = 2(2\cos^2 t - 1)^2 - 1$$
$$= 2(4\cos^4 t - 4\cos^2 t + 1) - 1$$
$$= 8\cos^4 t - 8\cos^2 t + 1$$

**24.10.** Use a half-angle identity to derive an expression for $\cos^4 t$ in terms of cosines with exponent 1.

Applying a half-angle identity once yields:

$$\cos^4 t = (\cos^2 t)^2 = \left(\frac{1 + \cos 2t}{2}\right)^2 = \frac{1 + 2\cos 2t + \cos^2 2t}{4}$$

Applying the identity again, this time to $\cos^2 2t$, yields:

$$\cos^4 t = \frac{1 + 2\cos 2t + \cos^2 2t}{4} = \frac{1 + 2\cos 2t + \dfrac{1 + \cos 2(2t)}{2}}{4}$$

$$= \frac{2 + 4\cos 2t + 1 + \cos 4t}{8}$$

$$= \frac{3 + 4\cos 2t + \cos 4t}{8}$$

**24.11.** Derive the half-angle formulas.

In each case, start with the half-angle identities and set $A = 2\theta$, thus, $\theta = \frac{A}{2}$. Then

$$\cos^2\frac{A}{2} = \frac{1 + \cos A}{2} \qquad\qquad \sin^2\frac{A}{2} = \frac{1 - \cos A}{2}$$

$$\cos\frac{A}{2} = \pm\sqrt{\frac{1 + \cos A}{2}} \qquad\qquad \sin\frac{A}{2} = \pm\sqrt{\frac{1 - \cos A}{2}}$$

For tangent, the derivation is more complicated. First derive the first form of the formula by starting with the quotient identity:

$$\tan\frac{A}{2} = \frac{\sin\dfrac{A}{2}}{\cos\dfrac{A}{2}} = \frac{\pm\sqrt{\dfrac{1 - \cos A}{2}}}{\pm\sqrt{\dfrac{1 + \cos A}{2}}} = \pm\sqrt{\frac{1 - \cos A}{1 + \cos A}}$$

Note that for all three formulas the sign cannot be determined in general, but depends on the quadrant in which $A/2$ lies. To derive the second form of the half-angle formula for tangents and eliminate the sign ambiguity in this single case, eliminate the fraction under the radical symbol:

$$\tan\frac{A}{2} = \pm\sqrt{\frac{1 - \cos A}{1 + \cos A}} = \pm\sqrt{\frac{1 - \cos A}{1 + \cos A} \cdot \frac{1 - \cos A}{1 - \cos A}} = \pm\sqrt{\frac{(1 - \cos A)^2}{1 - \cos^2 A}}$$

$$= \pm\sqrt{\frac{(1 - \cos A)^2}{\sin^2 A}} = \pm\left|\frac{1 - \cos A}{\sin A}\right|$$

Thus $\tan\dfrac{A}{2}$ and $\dfrac{1 - \cos A}{\sin A}$ are always equal in absolute value. To show that in fact these quantities are always equal, it is sufficient to show that they have the same sign for any value of $A$ between 0 and $2\pi$, which can be done as follows: First note that $1 - \cos A$ is never negative, so the sign of the fractional expression depends only on the sign of $\sin A$. If $0 \le A \le \pi$, both $\sin A$ and $\tan(A/2)$ are nonnegative; if $\pi \le A \le 2\pi$, both are nonpositive. Summarizing,

$$\tan\frac{A}{2} = \frac{1 - \cos A}{\sin A}$$

The derivation of the third form is left to the student (Problem 24.26).

**24.12.** Given $\tan u = 4$, $\pi < u < \dfrac{3\pi}{2}$, find $\tan \dfrac{u}{2}$.

Use the half-angle formula for tangent. Since $\pi < u < \dfrac{3\pi}{2}$, $u$ lies in quadrant III and the signs of $\sin u$ and $\cos u$ are to be chosen negative. To find $\sin u$ and $\cos u$, use a Pythagorean identity.

$$\sec u = -\sqrt{1 + \tan^2 u} = -\sqrt{1 + 4^2} = -\sqrt{17}$$

Hence

$$\cos u = \frac{1}{\sec u} = -\frac{1}{\sqrt{17}} \quad \text{and} \quad \sin u = \cos u \tan u = -\frac{1}{\sqrt{17}} \cdot 4 = -\frac{4}{\sqrt{17}}.$$

From the half-angle formula for tangent,

$$\tan \frac{u}{2} = \frac{1 - \cos u}{\sin u} = \frac{1 - (-1/\sqrt{17})}{-4/(\sqrt{17})} = \frac{\sqrt{17} + 1}{4}$$

**24.13.** Derive the product-to-sum formulas.

Start with the sum and difference formulas for sine, and add left sides and right sides:

$$\begin{aligned}
\sin(u + v) &= \sin u \cos v + \cos u \sin v \\
\sin(u - v) &= \sin u \cos v - \cos u \sin v \\
\hline
\sin(u + v) + \sin(u - v) &= 2 \sin u \cos v
\end{aligned}$$

Dividing both sides by 2 yields

$$\sin u \cos v = \frac{1}{2}[\sin(u + v) + \sin(u - v)]$$

as required.

Now start with the difference and sum formulas for cosine, and add left sides and right sides:

$$\begin{aligned}
\cos(u - v) &= \cos u \cos v + \sin u \sin v \\
\cos(u + v) &= \cos u \cos v - \sin u \sin v \\
\hline
\cos(u - v) + \cos(u + v) &= 2 \cos u \cos v
\end{aligned}$$

Dividing both sides by 2 yields

$$\cos u \cos v = \frac{1}{2}[\cos(u - v) + \cos(u + v)]$$

as required.

The other two formulas are obtained similarly, except that the two sides are subtracted instead of being added.

**24.14.** Use a product-to-sum formula to rewrite $\cos 5x \cos x$ as a sum.

Use $\cos u \cos v = \dfrac{1}{2}[\cos(u - v) + \cos(u + v)]$ with $u = 5x$ and $v = x$. Then

$$\cos 5x \cos x = \frac{1}{2}[\cos(5x - x) + \cos(5x + x)] = \frac{1}{2}(\cos 4x + \cos 6x)$$

**24.15.** Derive the sum-to-product formulas.

Start with the product-to-sum formula $\sin u \cos v = \frac{1}{2}[\sin(u + v) + \sin(u - v)]$. Make the substitution $a = u + v$ and $b = u - v$. Then $a + b = 2u$ and $a - b = 2v$, hence $u = \dfrac{a + b}{2}$ and $v = \dfrac{a - b}{2}$.

Thus $\sin\dfrac{a+b}{2}\cos\dfrac{a-b}{2} = \dfrac{1}{2}[\sin a + \sin b]$. Multiplying by 2 yields $\sin a + \sin b = 2\sin\dfrac{a+b}{2}\cos\dfrac{a-b}{2}$ as required.

The other three sum-to-product formulas are derived by performing the same substitutions in the other three product-to-sum formulas.

**24.16.** Use a sum-to-product formula to rewrite $\sin 5u + \sin 3u$ as a product.

Use $\sin a + \sin b = 2\sin\dfrac{a+b}{2}\cos\dfrac{a-b}{2}$ with $a = 5u$ and $b = 3u$. Then

$$\sin 5u + \sin 3u = 2\sin\frac{5u+3u}{2}\cos\frac{5u-3u}{2} = 2\sin 4u\cos u$$

**24.17.** Verify the identity $\sin 3\theta = 3\sin\theta - 4\sin^3\theta$.

Starting with the left side, use the sum formula for sines to get an expression in terms of $\sin\theta$.

$$
\begin{array}{lll}
\sin 3\theta &= \sin(\theta + 2\theta) & \text{Algebra} \\
&= \sin\theta\cos 2\theta + \cos\theta\sin 2\theta & \text{Sum formula for sines} \\
&= \sin\theta(1 - 2\sin^2\theta) + \cos\theta \cdot 2\sin\theta\cos\theta & \text{Double-angle formulas} \\
&= \sin\theta - 2\sin^3\theta + 2\sin\theta\cos^2\theta & \text{Algebra} \\
&= \sin\theta - 2\sin^3\theta + 2\sin\theta(1 - \sin^2\theta) & \text{Pythagorean identity} \\
&= \sin\theta - 2\sin^3\theta + 2\sin\theta - 2\sin^3\theta & \text{Algebra} \\
&= 3\sin\theta - 4\sin^3\theta & \text{Algebra}
\end{array}
$$

**24.18.** If $f(x) = \sin x$, show that the difference quotient for $f(x)$ can be written as $\sin x\left(\dfrac{\cos h - 1}{h}\right) + \cos x\dfrac{\sin h}{h}$. (See Chapter 9.)

$$
\begin{array}{ll}
\dfrac{f(x+h) - f(x)}{h} = \dfrac{\sin(x+h) - \sin x}{h} & \text{Substitution} \\[2mm]
= \dfrac{\sin x\cos h + \cos x\sin h - \sin x}{h} & \text{Sum formula for sine} \\[2mm]
= \dfrac{\sin x(\cos h - 1) + \cos x\sin h}{h} & \text{Algebra} \\[2mm]
= \sin x\left(\dfrac{\cos h - 1}{h}\right) + \cos x\dfrac{\sin h}{h} & \text{Algebra}
\end{array}
$$

**24.19.** Verify the reduction formulas: (a) $\sin(\theta + \pi) = -\sin\theta$; (b) $\tan\left(\theta + \dfrac{\pi}{2}\right) = -\cot\theta$.

(a) Apply the sum formula for sines, then substitute known values:

$$\sin(\theta + \pi) = \sin\theta\cos\pi + \cos\theta\sin\pi = \sin\theta(-1) + \cos\theta(0) = -\sin\theta$$

(b) Proceeding directly as in (a) fails because $\tan(\pi/2)$ is undefined. However, first applying a quotient identity eliminates this difficulty.

$$\tan\left(\theta + \frac{\pi}{2}\right) = \frac{\sin\left(\theta + \dfrac{\pi}{2}\right)}{\cos\left(\theta + \dfrac{\pi}{2}\right)} = \frac{\sin\theta\cos\dfrac{\pi}{2} + \cos\theta\sin\dfrac{\pi}{2}}{\cos\theta\cos\dfrac{\pi}{2} - \sin\theta\sin\dfrac{\pi}{2}} = \frac{\sin\theta(0) + \cos\theta(1)}{\cos\theta(0) - \sin\theta(1)} = \frac{\cos\theta}{-\sin\theta} = -\cot\theta$$

**24.20.** Find all solutions on the interval $[0, 2\pi)$ for $\cos t - \sin 2t = 0$.

First use the double-angle formula for sines to obtain an equation involving only functions of $t$, then solve by factoring:

$$\cos t - \sin 2t = 0$$
$$\cos t - 2\sin t \cos t = 0$$
$$\cos t(1 - 2\sin t) = 0$$
$$\cos t = 0 \quad \text{or} \quad 1 - 2\sin t = 0$$

The solutions of $\cos t = 0$ on the interval $[0, 2\pi)$ are $\pi/2$ and $3\pi/2$. The solutions of $1 - 2\sin t = 0$, that is, $\sin t = 1/2$, on this interval, are $\pi/6$ and $5\pi/6$.

Solutions: $\dfrac{\pi}{6}, \dfrac{\pi}{2}, \dfrac{5\pi}{6}, \dfrac{3\pi}{2}$

**24.21.** Find all solutions on the interval $[0, 2\pi)$ for $\cos 5x - \cos 3x = 0$.

First use a sum-to-product formula to put the equation into the form $ab = 0$.

$$\cos 5x - \cos 3x = 0$$
$$-2\sin\left(\frac{5x + 3x}{2}\right)\sin\left(\frac{5x - 3x}{2}\right) = 0$$
$$-2\sin 4x \sin x = 0$$
$$\sin 4x \sin x = 0$$
$$\sin 4x = 0 \quad \text{or} \quad \sin x = 0$$

The solutions of $\sin 4x = 0$ on the interval $[0, 2\pi)$ are $0, \pi/4, \pi/2, 3\pi/4, \pi, 5\pi/4, 3\pi/2$, and $7\pi/4$. The solutions of $\sin x = 0$ on this interval are $0$ and $\pi$, which have already been listed.

## SUPPLEMENTARY PROBLEMS

**24.22.** Derive the sum formulas for sine and tangent.

**24.23.** Derive the cofunction formulas (a) for tangents and cotangents; (b) for secants and cosecants.

**24.24.** Use sum or difference formulas to find exact values for (a) $\sin\dfrac{5\pi}{12}$; (b) $\cos 105°$; (c) $\tan\left(-\dfrac{\pi}{12}\right)$.

*Ans.* (a) $\dfrac{1 + \sqrt{3}}{2\sqrt{2}}$ or $\dfrac{\sqrt{2} + \sqrt{6}}{4}$; (b) $\dfrac{1 - \sqrt{3}}{2\sqrt{2}}$ or $\dfrac{\sqrt{2} - \sqrt{6}}{4}$; (c) $\dfrac{1 - \sqrt{3}}{1 + \sqrt{3}}$ or $\sqrt{3} - 2$

**24.25.** Given $\sin u = -\frac{2}{5}$, $u$ in quadrant III, and $\cos v = \frac{3}{4}$, $v$ in quadrant IV, find (a) $\sin(u + v)$; (b) $\cos(u - v)$; (c) $\tan(v - u)$.

*Ans.* (a) $\dfrac{-6 + 7\sqrt{3}}{20}$; (b) $\dfrac{-3\sqrt{21} + 2\sqrt{7}}{20}$; (c) $\dfrac{-6 - 7\sqrt{3}}{3\sqrt{21} - 2\sqrt{7}}$ or $-\dfrac{32\sqrt{21} + 75\sqrt{7}}{161}$

**24.26.** Derive (a) the double-angle formula for tangents; (b) the third form of the half-angle formula for tangents.

**24.27.** Given $\sec t = -3$, $\dfrac{\pi}{2} < t < \pi$, find (a) $\sin 2t$; (b) $\tan 2t$; (c) $\cos\dfrac{t}{2}$; (d) $\tan\dfrac{t}{2}$.

*Ans.* (a) $-\dfrac{4\sqrt{2}}{9}$; (b) $\dfrac{4\sqrt{2}}{7}$; (c) $\dfrac{1}{\sqrt{3}}$; (d) $\sqrt{2}$.

**24.28.** Use a double-angle identity to derive an expression for (a) $\sin 4x$ in terms of $\sin x$ and $\cos x$; (b) $\cos 6u$ in terms of $\cos u$.

   *Ans.*   (a) $\sin 4x = 4 \sin x \cos x (1 - 2 \sin^2 x)$; (b) $\cos 6u = 32 \cos^6 u - 48 \cos^4 u + 18 \cos^2 u + 1$

**24.29.** Use a half-angle identity to derive an expression in terms of cosines with exponent 1 for (a) $\sin^2 2t \cos^2 2t$; (b) $\sin^4 \dfrac{x}{2}$.

   *Ans.*   (a) $\dfrac{1 - \cos 8t}{8}$; (b) $\dfrac{3 - 4\cos x + \cos 2x}{8}$

**24.30.** Complete the derivations of the product-to-sum and sum-to-product formulas (see Problems 24.13 and 24.14).

**24.31.** (a) Write $\sin 120\pi t + \sin 110\pi t$ as a product. (b) Write $\sin \dfrac{\pi n}{L} x \cos \dfrac{k\pi n}{L} t$ as a sum.

   *Ans.*   (a) $2 \sin 115\pi t \cos 5\pi t$; (b) $\dfrac{1}{2} \sin \dfrac{\pi n}{L}(x + kt) + \dfrac{1}{2} \sin \dfrac{\pi n}{L}(x - kt)$

**24.32.** Verify that the following are identities: (a) $\dfrac{1 + \sin 2x - \cos 2x}{1 + \sin 2x + \cos 2x} = \tan x$; (b) $\tan \dfrac{u}{2} = \csc u - \cot u$;

   (c) $1 + \tan \alpha \tan \dfrac{\alpha}{2} = \sec \alpha$; (d) $\dfrac{\cos a - \cos b}{\sin a - \sin b} = -\tan\left(\dfrac{a + b}{2}\right)$

**24.33.** Verify the reduction formulas:

   (a) $\sin(n\pi + \theta) = (-1)^n \sin \theta$, for $n$ any integer; (b) $\cos(n\pi + \theta) = (-1)^n \cos \theta$, for $n$ any integer.

**24.34.** If $f(x) = \cos x$, show that the difference quotient for $f(x)$ can be written as

$$\cos x\left(\frac{\cos h - 1}{h}\right) - \sin x \frac{\sin h}{h}.$$

**24.35.** Find all solutions on the interval $[0, 2\pi)$ for the following equations:

   (a) $\sin 2\theta - \sin \theta = 0$; (b) $\cos x + \cos 3x = \cos 2x$.

   *Ans.*   (a) $0, \dfrac{\pi}{3}, \pi, \dfrac{5\pi}{3}$; (b) $\dfrac{\pi}{4}, \dfrac{\pi}{3}, \dfrac{3\pi}{4}, \dfrac{5\pi}{4}, \dfrac{5\pi}{3}, \dfrac{7\pi}{4}$

**24.36.** Find approximate values for all solutions on the interval $[0°, 360°)$ for $\cos x = 2\cos 2x$.

   *Ans.*   $32.53°, 126.38°, 233.62°, 327.47°$

# CHAPTER 25

# Inverse Trigonometric Functions

## Periodicity and Inverses

The trigonometric functions are periodic; hence, they are not one-to-one, and no inverses can be defined for the entire domain of a basic trigonometric function. By redefining each trigonometric function on a carefully chosen subset of its domain, the new function can be specified one-to-one and therefore has an inverse function.

## Redefined Trigonometric Functions

The table shows domains chosen on which each trigonometric function is one-to-one:

| Function $f(x) =$ | Domain | Range | Function $f(x) =$ | Domain | Range |
|---|---|---|---|---|---|
| $\sin x$ | $\left[-\dfrac{\pi}{2}, \dfrac{\pi}{2}\right]$ | $[-1, 1]$ | $\csc x$ | $\left(-\pi, -\dfrac{\pi}{2}\right] \cup \left(0, \dfrac{\pi}{2}\right]$ | $(-\infty, -1] \cup [1, \infty)$ |
| $\cos x$ | $[0, \pi]$ | $[-1, 1]$ | $\sec x$ | $\left[0, \dfrac{\pi}{2}\right) \cup \left[\pi, \dfrac{3\pi}{2}\right)$ | $(-\infty, -1] \cup [1, \infty)$ |
| $\tan x$ | $\left(-\dfrac{\pi}{2}, \dfrac{\pi}{2}\right)$ | $R$ | $\cot x$ | $[0, \pi]$ | $R$ |

Note that in each case, although the domain has been restricted, the entire range of the original function is retained.

Note also that in each case the restricted domain (sometimes called the *principal* domain) is the result of a choice. Other choices might be possible, and in the case of the secant and cosecant functions, *no universal agreement exists*. The choice used here is the one most commonly made in elementary calculus texts.

## Definitions of Inverse Trigonometric Functions

1. **INVERSE SINE** $f(x) = \sin^{-1}x$ is defined by $y = \sin^{-1}x$ if and only if $x = \sin y$ with $-1 \le x \le 1$ and $-\dfrac{\pi}{2} \le y \le \dfrac{\pi}{2}$. The values the function takes on lie in quadrants I and IV.

2. **INVERSE COSINE** $f(x) = \cos^{-1}x$ is defined by $y = \cos^{-1}x$ if and only if $x = \cos y$ with $-1 \le x \le 1$ and $0 \le y \le \pi$. The values the function takes on lie in quadrants I and II.

3. **INVERSE TANGENT** $f(x) = \tan^{-1}x$ is defined by $y = \tan^{-1}x$ is and only if $x = \tan y$ with $x \in R$ and $-\dfrac{\pi}{2} < y < \dfrac{\pi}{2}$. The values the function takes on lie in quadrants I and IV.

4. **INVERSE COTANGENT** $f(x) = \cot^{-1}x$ is defined by $y = \cot^{-1}x$ if and only if $x = \cot y$ with $x \in R$ and $0 < y < \pi$. The values the function takes on lie in quadrants I and II.

5. **INVERSE SECANT** $f(x) = \sec^{-1}x$ is defined by $y = \sec^{-1}x$ if and only if $x = \sec y$ with either $x \geq 1$ and $0 \leq y < \dfrac{\pi}{2}$ or $x \leq -1$ and $\pi \leq y < \dfrac{3\pi}{2}$. The values the function takes on lie in quadrants I and III.

6. **INVERSE COSECANT** $f(x) = \csc^{-1}x$ is defined by $y = \csc^{-1}x$ if and only if $x = \csc y$ with either $x \geq 1$ and $0 < y \leq \dfrac{\pi}{2}$ or $x \leq -1$ and $-\pi < y \leq -\dfrac{\pi}{2}$. The values the function takes on lie in quadrants I and III.

**EXAMPLE 25.1**  Evaluate (a) $\sin^{-1}\dfrac{1}{2}$; (b) $\sin^{-1}\left(-\dfrac{1}{2}\right)$.

(a) $y = \sin^{-1}\dfrac{1}{2}$ is equivalent to $\sin y = \dfrac{1}{2}$, $-\dfrac{\pi}{2} \leq y \leq \dfrac{\pi}{2}$. The only solution of the equation on the interval is $\dfrac{\pi}{6}$; hence $\sin^{-1}\dfrac{1}{2} = \dfrac{\pi}{6}$.

(b) $y = \sin^{-1}\left(-\dfrac{1}{2}\right)$ is equivalent to $\sin y = -\dfrac{1}{2}$, $-\dfrac{\pi}{2} \leq y \leq \dfrac{\pi}{2}$. The only solution of the equation on the interval is $-\dfrac{\pi}{6}$; hence $\sin^{-1}\left(-\dfrac{1}{2}\right) = -\dfrac{\pi}{6}$. Note that this value is in quadrant IV.

**EXAMPLE 25.2**  Evaluate (a) $\cos^{-1}\dfrac{1}{2}$; (b) $\cos^{-1}\left(-\dfrac{1}{2}\right)$.

(a) $y = \cos^{-1}\dfrac{1}{2}$ is equivalent to $\cos y = \dfrac{1}{2}$, $0 \leq y \leq \pi$. The only solution of the equation on the interval is $\dfrac{\pi}{3}$; hence $\cos^{-1}\dfrac{1}{2} = \dfrac{\pi}{3}$.

(b) $y = \cos^{-1}\left(-\dfrac{1}{2}\right)$ is equivalent to $\cos y = -\dfrac{1}{2}$, $0 \leq y \leq \pi$. The only solution of the equation on the interval is $\dfrac{2\pi}{3}$; hence $\cos^{-1}\left(-\dfrac{1}{2}\right) = \dfrac{2\pi}{3}$. Note that this value is in quadrant II.

## Alternative Notation

The inverse trigonometric functions are also referred to as the arc functions. In this notation:

$$\sin^{-1}x = \arcsin x \qquad \cos^{-1}x = \arccos x \qquad \tan^{-1}x = \arctan x$$
$$\csc^{-1}x = \operatorname{arccsc} x \qquad \sec^{-1}x = \operatorname{arcsec} x \qquad \cot^{-1}x = \operatorname{arccot} x$$

**EXAMPLE 25.3**  Evaluate $\arctan 1$.

$y = \arctan 1 = \tan^{-1}1$ is equivalent to $\tan y = 1$, $-\dfrac{\pi}{2} < y < \dfrac{\pi}{2}$. The only solution of the equation on the interval is $\dfrac{\pi}{4}$; hence $\arctan 1 = \dfrac{\pi}{4}$.

## Phase-Shift Identity

Let $A$ be any positive real number and $B$ and $x$ be any real numbers. Then

$$A\cos bx + B\sin bx = C\cos(bx - d)$$

where $C = \sqrt{A^2 + B^2}$ and $d = \tan^{-1}\dfrac{B}{A}$.

**EXAMPLE 25.4**  Write $\sin x + \cos x$ in the form $C\cos(bx - d)$.

Here $A = B = 1$; hence $C = \sqrt{A^2 + B^2} = \sqrt{1^2 + 1^2} = \sqrt{2}$ and $d = \tan^{-1}\dfrac{1}{1} = \dfrac{\pi}{4}$. It follows from the phase-shift identity that $\sin x + \cos x = \sqrt{2}\cos\left(x - \dfrac{\pi}{4}\right)$.

**SOLVED PROBLEMS**

**25.1.** Sketch a graph of the sine function showing the interval of redefinition.

The redefined sine function is restricted to the domain $[-\pi/2, \pi/2]$. Draw a graph of the basic sine function, with the portion in this interval emphasized (see Fig. 25-1).

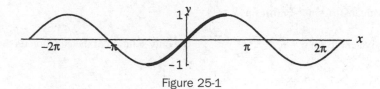

Figure 25-1

**25.2.** Sketch a graph of the cosine function showing the interval of redefinition.

The redefined cosine function is restricted to the domain $[0, \pi]$. Draw a graph of the basic cosine function, with the portion in this interval emphasized (see Fig. 25-2).

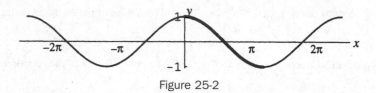

Figure 25-2

**25.3.** Sketch a graph of the inverse sine function.

The domain of the inverse sine function is the range of the (redefined) sine function: $[-1, 1]$. The range of the inverse sine function is the domain of the redefined sine function: $[-\pi/2, \pi/2]$. The graph of the inverse sine function is the graph of the redefined sine function, reflected in the line $y = x$. Form a table of values (see the table of values for the trigonometric functions in Problem 22.12) and sketch the graph (Fig. 25-3).

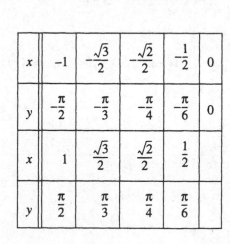

| $x$ | $-1$ | $-\dfrac{\sqrt{3}}{2}$ | $-\dfrac{\sqrt{2}}{2}$ | $-\dfrac{1}{2}$ | $0$ |
|---|---|---|---|---|---|
| $y$ | $-\dfrac{\pi}{2}$ | $-\dfrac{\pi}{3}$ | $-\dfrac{\pi}{4}$ | $-\dfrac{\pi}{6}$ | $0$ |
| $x$ | $1$ | $\dfrac{\sqrt{3}}{2}$ | $\dfrac{\sqrt{2}}{2}$ | $\dfrac{1}{2}$ | |
| $y$ | $\dfrac{\pi}{2}$ | $\dfrac{\pi}{3}$ | $\dfrac{\pi}{4}$ | $\dfrac{\pi}{6}$ | |

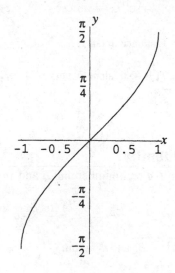

Figure 25-3

**25.4.** Sketch a graph of the inverse cosine function.

The domain of the inverse cosine function is the range of the (redefined) cosine function: $[-1, 1]$. The range of the inverse cosine function is the domain of the redefined cosine function: $[0, \pi]$. The graph of the inverse cosine function is the graph of the redefined cosine function, reflected in the line $y = x$. Form a table of values (see the table of values for the trigonometric functions in Problem 22.12) and sketch the graph (Fig. 25-4).

| x | −1 | $-\dfrac{\sqrt{3}}{2}$ | $-\dfrac{\sqrt{2}}{2}$ | $-\dfrac{1}{2}$ | 0 |
|---|----|----|----|----|---|
| y | π | $\dfrac{5\pi}{6}$ | $\dfrac{3\pi}{4}$ | $\dfrac{2\pi}{3}$ | $\dfrac{\pi}{2}$ |
| x | 1 | $\dfrac{\sqrt{3}}{2}$ | $\dfrac{\sqrt{2}}{2}$ | $\dfrac{1}{2}$ | |
| y | 0 | $\dfrac{\pi}{6}$ | $\dfrac{\pi}{4}$ | $\dfrac{\pi}{3}$ | |

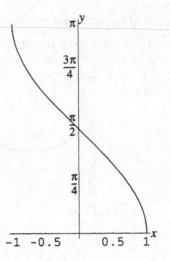

Figure 25-4

**25.5.** Sketch a graph of the tangent function showing the interval of redefinition.

The redefined tangent function is restricted to the domain $(-\pi/2, \pi/2)$. Draw a graph of the basic tangent function, with the portion in this interval emphasized (see Fig. 25-5).

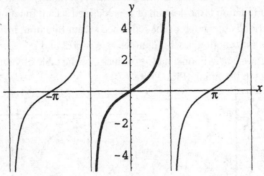

Figure 25-5

**25.6.** Sketch a graph of the inverse tangent function.

The domain of the inverse tangent function is the range of the (redefined) tangent function: $\boldsymbol{R}$. The range of the inverse tangent function is the domain of the redefined tangent function: $(-\pi/2, \pi/2)$. The graph of the inverse tangent function is the graph of the redefined tangent function, reflected in the line $y = x$. Since the graph of the redefined tangent function has asymptotes at $x = \pm\pi/2$, the reflected graph will have asymptotes at $y = \pm\pi/2$. Form a table of values (see the table of values for the trigonometric functions in Problem 22.12) and sketch the graph (Fig. 25-6).

| x | $-\sqrt{3}$ | −1 | $-\dfrac{1}{\sqrt{3}}$ | 0 |
|---|----|----|----|---|
| y | $-\dfrac{\pi}{3}$ | $-\dfrac{\pi}{4}$ | $-\dfrac{\pi}{6}$ | 0 |
| x | $\sqrt{3}$ | 1 | $\dfrac{1}{\sqrt{3}}$ | |
| y | $\dfrac{\pi}{3}$ | $\dfrac{\pi}{4}$ | $\dfrac{\pi}{6}$ | |

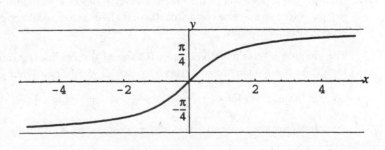

Figure 25-6

**25.7.** Sketch a graph of the secant function showing the interval of redefinition.

The redefined secant function is restricted to the domain $[0, \pi/2) \cup [\pi, 3\pi/2)$. Draw a graph of the basic secant function, with the portion in this interval emphasized.

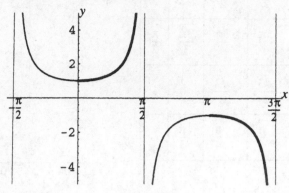

Figure 25-7

**25.8.** Sketch a graph of the inverse secant function.

The domain of the inverse secant function is the range of the (redefined) secant function: $(-\infty, -1] \cup [1, \infty)$. The range of the inverse secant function is the domain of the redefined secant function $[0, \pi/2) \cup [\pi, 3\pi/2)$. The graph of the inverse secant function is the graph of the redefined secant function, reflected in the line $y = x$. Since the graph of the redefined secant function has asymptotes at $x = \pi/2$ and $x = 3\pi/2$, the reflected graph will have asymptotes at $y = \pi/2$ and $y = 3\pi/2$. Form a table of values (see the table of values for the trigonometric functions in Problem 22.12) and sketch the graph (Fig. 25-8).

| $x$ | $-2$ | $-\sqrt{2}$ | $-\dfrac{2}{\sqrt{3}}$ | $-1$ |
|---|---|---|---|---|
| $y$ | $\dfrac{4\pi}{3}$ | $\dfrac{5\pi}{4}$ | $\dfrac{7\pi}{6}$ | $\pi$ |
| $x$ | $2$ | $\sqrt{2}$ | $\dfrac{2}{\sqrt{3}}$ | $1$ |
| $y$ | $\dfrac{\pi}{3}$ | $\dfrac{\pi}{4}$ | $\dfrac{\pi}{6}$ | $0$ |

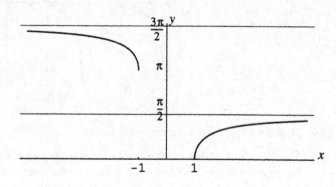

Figure 25-8

**25.9.** Analyze the application of the function-inverse function relation to (a) the sine and inverse sine functions, (b) the cosine and inverse cosine functions, (c) the tangent and inverse tangent functions, and (d) the secant and inverse secant functions.

The function-inverse function relation (Chapter 13) states that if $g$ is the inverse function of $f$, then $g(f(x)) = x$ for all $x$ in the domain of $f$, and $f(g(y)) = y$ for all $y$ in the domain of $g$. Hence

(a)  $\sin^{-1}(\sin x) = x$ for all $x, -\dfrac{\pi}{2} \le x \le \dfrac{\pi}{2}$;  $\sin(\sin^{-1}y) = y$ for all $y, -1 \le y \le 1$.

(b)  $\cos^{-1}(\cos x) = x$ for all $x, 0 \le x \le \pi$;  $\cos(\cos^{-1}y) = y$ for all $y, -1 \le y \le 1$.

(c)  $\tan^{-1}(\tan x) = x$ for all $x, -\dfrac{\pi}{2} < x < \dfrac{\pi}{2}$;  $\tan(\tan^{-1}y) = y$ for all $y \in R$.

(d)  $\sec^{-1}(\sec x) = x$ for all $x, 0 \le x < \dfrac{\pi}{2}$ or $\pi < x \le \dfrac{3\pi}{2}$;  $\sec(\sec^{-1}y) = y$ for all $y \ge 1$ or $y \le -1$.

**25.10.** Simplify: (a) $\sin\left(\sin^{-1}\frac{\sqrt{3}}{2}\right)$ (b) $\sin\left(\sin^{-1}\frac{1}{3}\right)$; (c) $\sin(\sin^{-1}2)$.

(a) Since $-1 \le \sqrt{3}/2 \le 1$, $\sqrt{3}/2$ is in the domain of the inverse sine function. Hence, applying the function-inverse-function relation, $\sin\left(\sin^{-1}\frac{\sqrt{3}}{2}\right) = \frac{\sqrt{3}}{2}$. Alternatively, note that $\sin^{-1}\frac{\sqrt{3}}{2} = \frac{\pi}{3}$, hence

$$\sin\left(\sin^{-1}\frac{\sqrt{3}}{2}\right) = \sin\frac{\pi}{3} = \frac{\sqrt{3}}{2}.$$

(b) Since $-1 \le \frac{1}{3} \le 1$, $\frac{1}{3}$ is in the domain of the inverse sine function. Hence, applying the function-inverse-function relation, $\sin\left(\sin^{-1}\frac{1}{3}\right) = \frac{1}{3}$.

(c) Since $2 > 1$, 2 is not in the domain of the inverse sine function. Hence, $\sin(\sin^{-1}2)$ is undefined.

**25.11.** Simplify: (a) $\tan^{-1}\left(\tan\frac{\pi}{6}\right)$; (b) $\tan^{-1}\left(\tan\left(-\frac{1}{4}\right)\right)$; (c) $\tan^{-1}\left(\tan\frac{2\pi}{3}\right)$

(a) Since $-\pi/2 < \pi/6 < \pi/2$, $\pi/6$ is in the domain of the restricted tangent function. Hence, applying the function-inverse-function relationship, $\tan^{-1}\left(\tan\frac{\pi}{6}\right) = \frac{\pi}{6}$.

Alternatively, note that $\tan\frac{\pi}{6} = \frac{1}{\sqrt{3}}$, hence $\tan^{-1}\left(\tan\frac{\pi}{6}\right) = \tan^{-1}\frac{1}{\sqrt{3}} = \frac{\pi}{6}$.

(b) Since $-\pi/2 < -1/4 < \pi/2$, $-1/4$ is in the domain of the restricted tangent function. Hence, applying the function-inverse-function relationship, $\tan^{-1}\left(\tan\left(-\frac{1}{4}\right)\right) = -\frac{1}{4}$.

(c) Since $2\pi/3 > \pi/2$, $2\pi/3$ is not in the domain of the restricted tangent function, and the function-inverse-function relation cannot be used. However, $2\pi/3$ is in the domain of the general tangent function, thus:

$$\tan^{-1}\left(\tan\frac{2\pi}{3}\right) = \tan^{-1}(-\sqrt{3}) = -\frac{\pi}{3}$$

**25.12.** Simplify: (a) $\cos\left(\sin^{-1}\frac{3}{5}\right)$; (b) $\sin\left(\cos^{-1}\left(-\frac{2}{3}\right)\right)$; (c) $\tan\left(\sec^{-1}\left(-\frac{5}{2}\right)\right)$; (d) $\cot(\cos^{-1}3)$

(a) Since $\frac{3}{5}$ is in the domain of the inverse sine function, let $u = \sin^{-1}\frac{3}{5}$. Then, by the definition of the inverse sine, $\sin u = \frac{3}{5}$, $-\frac{\pi}{2} \le u \le \frac{\pi}{2}$. It follows from the Pythagorean identities that

$$\cos\left(\sin^{-1}\frac{3}{5}\right) = \cos u = \sqrt{1 - \sin^2 u} = \sqrt{1 - \left(\frac{3}{5}\right)^2} = \frac{4}{5}$$

Note that the positive sign is taken on the square root since $u$ must be in quadrant I or IV, where the sign of the cosine is positive.

(b) Since $-\frac{2}{3}$ is the domain of the inverse cosine function, let $u = \cos^{-1}\left(-\frac{2}{3}\right)$. Then, by the definition of the inverse cosine, $\cos u = -\frac{2}{3}$, $0 \le u \le \pi$. It follows from the Pythagorean identities that

$$\sin\left(\cos^{-1}\left(-\frac{2}{3}\right)\right) = \sin u = \sqrt{1 - \cos^2 u} = \sqrt{1 - \left(-\frac{2}{3}\right)^2} = \frac{\sqrt{5}}{3}$$

Note that the positive sign is taken on the square root since $u$ must be in quadrant I or II, where the sign of the sine is positive.

(c) Since $-\frac{5}{2}$ is in the domain of the inverse secant function, let $u = \sec^{-1}\left(-\frac{5}{2}\right)$. Then, by the definition of the inverse secant, $\sec u = -\frac{5}{2}$, $\pi \le u < \frac{3\pi}{2}$ (since $\sec u$ is negative). If follows from the Pythagorean identities that

$$\tan\left(\sec^{-1}\left(-\frac{5}{2}\right)\right) = \tan u = \sqrt{\sec^2 u - 1} = \sqrt{\left(-\frac{5}{2}\right)^2 - 1} = \frac{\sqrt{21}}{2}$$

Note that the positive sign is taken on the square root since $u$ must be in quadrant III, where the sign of the tangent is positive.

(d)  Since 3 is not in the domain of the inverse cosine in function, $\cot(\cos^{-1} 3)$ is undefined.

**25.13.**  Simplify: (a) $\sin\left(\sin^{-1}\dfrac{1}{3} + \sin^{-1}\dfrac{2}{3}\right)$; (b) $\tan\left(\cos^{-1}\dfrac{3}{5} - \sin^{-1}\dfrac{5}{6}\right)$

(a)  Let $u = \sin^{-1}\dfrac{1}{3}$ and $v = \sin^{-1}\dfrac{2}{3}$. Then $\sin u = \dfrac{1}{3}$ and $\sin v = \dfrac{2}{3}$, $-\dfrac{\pi}{2} \le u, v \le \dfrac{\pi}{2}$. From the sum formula for sines, it follows that

$$\sin\left(\sin^{-1}\frac{1}{3} + \sin^{-1}\frac{2}{3}\right) = \sin(u + v) = \sin u \cos v + \cos u \sin v$$

Now $\sin u$ and $\sin v$ are given. Proceeding as in the previous problem,

$$\cos u = \sqrt{1 - \sin^2 u} = \sqrt{1 - \left(\frac{1}{3}\right)^2} = \frac{2\sqrt{2}}{3} \qquad \cos v = \sqrt{1 - \sin^2 v} = \sqrt{1 - \left(\frac{2}{3}\right)^2} = \frac{\sqrt{5}}{3}$$

Hence

$$\sin\left(\sin^{-1}\frac{1}{3} + \sin^{-1}\frac{2}{3}\right) = \sin u \cos v + \cos u \sin v = \frac{1}{3} \cdot \frac{\sqrt{5}}{3} + \frac{2\sqrt{2}}{3} \cdot \frac{2}{3} = \frac{\sqrt{5} + 4\sqrt{2}}{9}$$

(b)  Let $u = \cos^{-1}\dfrac{3}{5}$ and $v = \sin^{-1}\dfrac{5}{6}$. Then $\cos u = \dfrac{3}{5}$, $0 \le u \le \pi$, and $\sin v = \dfrac{5}{6}$, $-\dfrac{\pi}{2} \le v \le \dfrac{\pi}{2}$. From the difference formula for tangents, if follows that

$$\tan\left(\cos^{-1}\frac{3}{5} - \sin^{-1}\frac{5}{6}\right) = \tan(u - v) = \frac{\tan u - \tan v}{1 + \tan u \tan v}$$

From the Pythagorean and quotient identities

$$\tan u = \frac{\sin u}{\cos u} = \frac{\sqrt{1 - \cos^2 u}}{\cos u} = \frac{\sqrt{1 - (3/5)^2}}{3/5} = \frac{4}{3}$$

$$\tan v = \frac{\sin v}{\cos v} = \frac{\sin v}{\sqrt{1 - \sin^2 v}} = \frac{5/6}{\sqrt{1 - (5/6)^2}} = \frac{5}{\sqrt{11}}$$

Hence

$$\tan\left(\cos^{-1}\frac{3}{5} - \sin^{-1}\frac{5}{6}\right) = \frac{\tan u - \tan v}{1 + \tan u \tan v} = \frac{\dfrac{4}{3} - \dfrac{5}{\sqrt{11}}}{1 + \left(\dfrac{4}{3}\right)\left(\dfrac{5}{\sqrt{11}}\right)} = \frac{4\sqrt{11} - 15}{3\sqrt{11} + 20} = \frac{125\sqrt{11} - 432}{301}$$

**25.14.**  Simplify (a) $\cos\left(2\cos^{-1}\dfrac{5}{13}\right)$; (b) $\sin\left(\dfrac{1}{2}\sin^{-1}\left(-\dfrac{7}{25}\right)\right)$.

(a)  Let $u = \cos^{-1}\dfrac{5}{13}$. Then $\cos u = \dfrac{5}{13}$, $0 \le u \le \pi$. From the double-angle formula for cosines, it follows that

$$\cos\left(2\cos^{-1}\frac{5}{13}\right) = \cos 2u = 2\cos^2 u - 1 = 2\left(\frac{5}{13}\right)^2 - 1 = -\frac{119}{169}$$

(b)  Let $u = \sin^{-1}\left(-\dfrac{7}{25}\right)$. Then $\sin u = -\dfrac{7}{25}$, $-\dfrac{\pi}{2} \le u \le 0$ (since $\sin u$ is negative). From the half-angle formula for sines, it follows that

$$\sin\left(\frac{1}{2}\sin^{-1}\left(-\frac{7}{25}\right)\right) = \sin\left(\frac{1}{2}u\right) = -\sqrt{\frac{1 - \cos u}{2}}$$

where the negative sign is taken on the square root since $-\dfrac{\pi}{4} \le \dfrac{u}{2} \le 0$. From the Pythagorean identity, $\cos u = \dfrac{24}{25}$, hence

$$\sin\left(\frac{1}{2}\sin^{-1}\left(-\frac{7}{25}\right)\right) = -\sqrt{\frac{1 - \cos u}{2}} = -\sqrt{\frac{1 - \frac{24}{25}}{2}} = -\frac{1}{\sqrt{50}}$$

**25.15.** Find an algebraic expression for $\sin(\cos^{-1}x)$.

Let $u = \cos^{-1}x$. Then $\cos u = x$, $0 \le u \le \pi$. It follows from the Pythagorean identities that

$$\sin(\cos^{-1}x) = \sin u = \sqrt{1 - \cos^2 u} = \sqrt{1 - x^2}$$

where the positive sign is taken on the square root, since $u$ must be in quadrant I or II, where the sign of the sine is positive.

**25.16.** Derive the phase-shift formula.

Given a quantity of form $A\cos bx + B\sin bx$ with $A$ any positive real number and $B$ and $x$ any real numbers, let $C = \sqrt{A^2 + B^2}$ and $d = \tan^{-1}\dfrac{B}{A}$. From the Pythagorean and quotient identities, $\cos d = \dfrac{A}{\sqrt{A^2 + B^2}}$ and $\sin d = \dfrac{B}{\sqrt{A^2 + B^2}}$. From algebra, it follows that:

$$A\cos bx + B\sin bx = \frac{\sqrt{A^2 + B^2}}{\sqrt{A^2 + B^2}}(A\cos bx + B\sin bx) = \sqrt{A^2 + B^2}\left(\frac{A}{\sqrt{A^2 + B^2}}\cos bx + \frac{B}{\sqrt{A^2 + B^2}}\sin bx\right)$$

Hence,

$$A\cos bx + B\sin bx = \sqrt{A^2 + B^2}(\cos d \cos bx + \sin d \sin bx)$$

By the difference formula for cosine, the quantity in parentheses must equal $\cos(bx - d)$, hence,

$$A\cos bx + B\sin bx = \sqrt{A^2 + B^2}\cos(bx - d) = C\cos(bx - d)$$

**25.17.** (a) Use the phase-shift formula to rewrite $\sqrt{2}\cos 3x + \sqrt{2}\sin 3x$.

(b) Draw a graph of $f(x) = \sqrt{2}\cos 3x + \sqrt{2}\sin 3x$ using the result of part (a).

(a) Let $A = B = \sqrt{2}$, then $\sqrt{A^2 + B^2} = \sqrt{(\sqrt{2})^2 + (\sqrt{2})^2} = 2$ and

$$\tan^{-1}\frac{B}{A} = \tan^{-1}\frac{\sqrt{2}}{\sqrt{2}} = \tan^{-1}1 = \frac{\pi}{4}.$$ It follows that

$$\sqrt{2}\cos 3x + \sqrt{2}\sin 3x = 2\cos\left(3x - \frac{\pi}{4}\right)$$

(b) To sketch $y = f(x) = 2\cos\left(3x - \dfrac{\pi}{4}\right)$, note amplitude $= 2$. The graph (Fig. 25-9) is a basic cosine curve. Period $= \dfrac{2\pi}{3}$. Phase shift $= \dfrac{\pi}{4} \div 3 = \dfrac{\pi}{12}$. Divide the interval from $\dfrac{\pi}{12}$ to $\dfrac{3\pi}{4}$ ($=$ phase shift $+$ one period) into four equal subintervals and sketch the curve with maximum height 2 and minimum height $-2$.

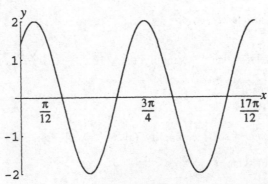

Figure 25-9

## SUPPLEMENTARY PROBLEMS

**25.18.** Evaluate: (a) $\sin^{-1}\left(-\dfrac{\sqrt{3}}{2}\right)$; (b) $\cos^{-1}\left(-\dfrac{\sqrt{3}}{2}\right)$; (c) $\tan^{-1}\left(-\dfrac{1}{\sqrt{3}}\right)$; (d) $\sec^{-1}\left(-\dfrac{2}{\sqrt{3}}\right)$

*Ans.*    (a) $-\dfrac{\pi}{3}$; (b) $\dfrac{5\pi}{6}$; (c) $-\dfrac{\pi}{6}$; (d) $\dfrac{7\pi}{6}$

**25.19.** Evaluate: (a) $\sin\left(\sin^{-1}\dfrac{1}{3}\right)$; (b) $\cos\left(\cos^{-1}\left(-\dfrac{3}{4}\right)\right)$; (c) $\tan(\tan^{-1}0)$; (d) $\sec\left(\sec^{-1}\dfrac{1}{2}\right)$

*Ans.*    (a) $\dfrac{1}{3}$; (b) $-\dfrac{3}{4}$; (c) 0; (d) not defined

**25.20.** Evaluate: (a) $\sin^{-1}\left(\sin\dfrac{\pi}{3}\right)$; (b) $\cos^{-1}(\cos 1)$; (c) $\tan^{-1}\left(\tan\dfrac{5\pi}{3}\right)$; (d) $\sec^{-1}\left(\sec\left(-\dfrac{\pi}{4}\right)\right)$

*Ans.*    (a) $\dfrac{\pi}{3}$; (b) 1; (c) $-\dfrac{\pi}{3}$; (d) $\dfrac{5\pi}{4}$

**25.21.** Evaluate: (a) $\cos\left(\sin^{-1}\dfrac{2}{5}\right)$; (b) $\sin(\tan^{-1}2)$; (c) $\tan(\sec^{-1}(-3))$; (d) $\cos\left(\tan^{-1}\left(-\dfrac{5}{12}\right)\right)$

*Ans.*    (a) $\dfrac{\sqrt{21}}{5}$; (b) $\dfrac{2}{\sqrt{5}}$; (c) $\sqrt{8}$; (d) $\dfrac{12}{13}$

**25.22.** Evaluate: (a) $\sin\left(\sin^{-1}\dfrac{1}{3}+\cos^{-1}\dfrac{2}{3}\right)$; (b) $\cos\left(\cos^{-1}\dfrac{3}{5}-\sin^{-1}\dfrac{12}{13}\right)$; (c) $\tan\left(\tan^{-1}\dfrac{3}{4}+\sin^{-1}\dfrac{7}{25}\right)$

*Ans.*    (a) $\dfrac{2+2\sqrt{10}}{9}$; (b) $\dfrac{63}{65}$; (c) $\dfrac{4}{3}$

**25.23.** Evaluate: (a) $\cos\left(2\sin^{-1}\dfrac{2}{3}\right)$; (b) $\sec\left(2\tan^{-1}\dfrac{1}{2}\right)$; (c) $\sin\left(\dfrac{1}{2}\cos^{-1}\dfrac{4}{5}\right)$; (d) $\sec\left(\dfrac{1}{2}\sin^{-1}\dfrac{2}{3}\right)$

*Ans.*    (a) $\dfrac{1}{9}$; (b) $\dfrac{5}{3}$; (c) $\dfrac{1}{\sqrt{10}}$; (d) $\dfrac{\sqrt{18-6\sqrt{5}}}{2}$

**25.24.** Simplify: (a) $\sin(\cos^{-1}x)$; (b) $\cos(\tan^{-1}x)$; (c) $\tan(2\cos^{-1}x)$; (d) $\cos\left(\dfrac{1}{2}\sin^{-1}x\right)$

*Ans.*    (a) $\sqrt{1-x^2}$; (b) $\dfrac{1}{\sqrt{1+x^2}}$; (c) $\dfrac{2x\sqrt{1-x^2}}{2x^2-1}$; (d) $\sqrt{\dfrac{1+\sqrt{1-x^2}}{2}}$

**25.25.** (a) Show that for $-1 \le x \le 1$, $-\dfrac{\pi}{2} \le \sin^{-1}x + \cos^{-1}x \le \dfrac{\pi}{2}$. (b) Show that $\sin(\sin^{-1}x + \cos^{-1}x) = 1$.
(c) From (a) and (b), deduce that for $-1 \le x \le 1$, $\sin^{-1}x + \cos^{-1}x = \dfrac{\pi}{2}$.

**25.26.** By making the substitution $u = \sin^{-1}\dfrac{x}{3}$, simplify: (a) $\sqrt{9-x^2}$; (b) $\dfrac{x^2}{\sqrt{9-x^2}}$.

*Ans.*    (a) $3\cos u$; (b) $3\sin u \tan u$

**25.27.** By making the substitution $u = \tan^{-1}\dfrac{x}{4}$, simplify; (a) $\sqrt{16+x^2}$; (b) $\dfrac{\sqrt{16+x^2}}{x^3}$.

*Ans.*    (a) $4\sec u$; (b) $\dfrac{\cos^2 u}{16\sin^3 u}$

**25.28.** By making the substitution $u = \sec^{-1}\dfrac{x}{2}$, simplify: (a) $x = x\sqrt{x^2-4}$; (b) $(x^2-4)^{3/2}$

*Ans.*    (a) $4\tan u \sec u$; (b) $8\tan^3 u$

**25.29.** Given $y = 3 \sin^{-1}(x - 5)$, (a) state the possible values of $x$ and $y$; (b) solve for $x$ in terms of $y$.

   *Ans*   (a) $4 \leq x \leq 6$, $-3\pi/2 \leq y \leq 3\pi/2$; (b) $x = 5 + \sin(y/3)$

**25.30.** Use the phase-shift formula to rewrite:

   (a) $6 \cos 3x - 6 \sin 3x$; (b) $3 \cos 4x + \sqrt{3} \sin 4x$; (c) $3 \cos \frac{1}{2}x + 4 \sin \frac{1}{2}x$

   *Ans.*   (a) $6\sqrt{2} \cos\left(3x + \frac{\pi}{4}\right)$; (b) $2\sqrt{3} \cos\left(4x - \frac{\pi}{6}\right)$; (c) $5 \cos\left(\frac{1}{2}x - \tan^{-1} \frac{4}{3}\right)$

# CHAPTER 26

# Triangles

## Conventional Notation for a Triangle

The conventional notation for a triangle $ABC$ is shown in Fig. 26-1.

| Right Triangle | Acute Triangle | Obtuse Triangle |
|:---:|:---:|:---:|
| | | |

Figure 26-1

A triangle that contains no right angle is called an *oblique* triangle. The six *parts* of the triangle $ABC$ are the three sides $a$, $b$, and $c$, together with the three angles $\alpha$, $\beta$, and $\gamma$.

## Solving a Triangle

Solving a triangle is the process of determining all the parts of the triangle. In general, given three parts of a triangle, including at least one side, the other parts can be determined. (Exceptions are cases where two possible triangles are determined or where no triangle can be shown to be consistent with the given data.)

## Right Triangles

Here one part is known from the outset to be an angle of 90°. Given either two sides, or one side and one of the acute angles, the other parts can be determined using the definitions of the trigonometric functions for acute angles, the Pythagorean theorem, and the fact that the sum of the three angles in a plane triangle is 180°.

**EXAMPLE 26.1**   Given a right triangle $ABC$ with $c = 20$ and $\alpha = 30°$, solve the triangle.

Here it is assumed that $\gamma = 90°$.

Solve for $\beta$:

Since $\alpha + \beta + \gamma = 180°$, $\beta = 180° - \alpha - \gamma = 180° - 30° - 90° = 60°$.

Solve for $a$:

In the right triangle $ABC$, $\sin \alpha = \frac{a}{c}$, hence $a = c \sin \alpha = 20 \sin 30° = 10$.

Solve for $b$:

From the Pythagorean theorem, $c^2 = a^2 + b^2$; hence $b = \sqrt{c^2 - a^2} = \sqrt{20^2 - 10^2} = \sqrt{300} = 10\sqrt{3}$

## Oblique Triangles

Oblique triangles are solved using the *law of sines* and the *law of cosines*. Normally five cases are recognized on the basis of which parts are given: **AAS** (two angles and a nonincluded side are given), **ASA** (two angles and an included side), **SSA** (two sides and a nonincluded angle), **SAS** (two sides and an included angle), and **SSS** (three sides).

## Law of Sines

In any triangle, the ratio of each side to the sine of the angle opposite that side is the same for all three sides:

$$\frac{a}{\sin\alpha} = \frac{b}{\sin\beta} \qquad \frac{a}{\sin\alpha} = \frac{c}{\sin\gamma} \qquad \frac{b}{\sin\beta} = \frac{c}{\sin\gamma}$$

## Law of Cosines

In any triangle, the square of any side is equal to the sum of the squares of the other two sides, diminished by twice the product of the other two sides and the cosine of the angle included between them:

$$a^2 = b^2 + c^2 - 2bc\,\cos\alpha$$
$$b^2 = a^2 + c^2 - 2ac\,\cos\beta$$
$$c^2 = a^2 + b^2 - 2ab\,\cos\gamma$$

## Accuracy in Computations

In working with approximate data, the number of significant digits in a result cannot be greater than the number of significant digits in the given data. In interpreting calculator results for angles, the following table is useful:

| NUMBER OF SIGNIFICANT DIGITS FOR SIDES | DEGREE MEASURE OF ANGLES TO THE NEAREST |
|:---:|:---:|
| 2 | 1° |
| 3 | 0.1° or 10′ |
| 4 | 0.01° or 1′ |

## Bearing and Heading

In applications involving navigation and aviation, as well as some other situations, angles are normally specified with reference to a north-south axis:

1. **BEARING:** A direction is specified in terms of an angle measured east or west of a north-south axis. Thus Fig. 26-2 shows bearings of N30°E and S70°W.

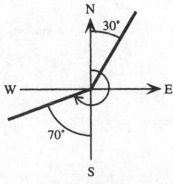

Figure 26-2

2. **HEADING:** A direction is specified in terms of an angle measured clockwise from north. Thus the same figure shows headings of 30° and 250° (that is, $180° + 70°$).

## Angles of Elevation and Depression

1. **ANGLE OF ELEVATION** is the angle from the horizontal measured upward to the line of sight of the observer.
2. **ANGLE OF DEPRESSION** is the angle from the horizontal measured downward to the line of sight of the observer.

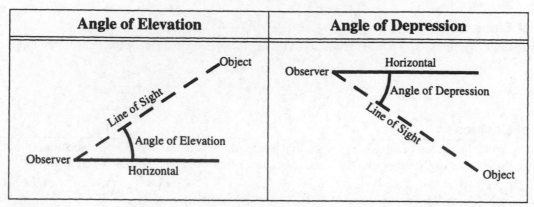

Figure 26-3

---

## SOLVED PROBLEMS

**26.1.** Given a right triangle $ABC$ with $\alpha = 42.7°$ and $a = 68.2$, solve the triangle. (See Fig. 26-1.)

Here it is assumed that $\gamma = 90°$. In any right triangle, the acute angles are complementary, since $\alpha + \beta + 90° = 180°$ implies $\alpha + \beta = 90°$.

Solve for $\beta$:

$$\beta = 90° - \alpha = 90° - 42.7° = 47.3°$$

Solve for $b$:

In the right triangle $ABC$, $\tan \alpha = \dfrac{a}{b}$; hence $b = \dfrac{a}{\tan \alpha} = \dfrac{68.2}{\tan 42.7°} = 73.9$.

Solve for $c$:

In the right triangle $ABC$, $\sin \alpha = \dfrac{a}{c}$; hence $c = \dfrac{a}{\sin \alpha} = \dfrac{68.2}{\sin 42.7°} = 100.6$.

Alternatively, use the Pythagorean theorem to find $c$, since $a$ and $b$ are known. However, it is preferable to use given data rather than calculated data wherever possible, since errors in calculations accumulate.

**26.2.** Given a right triangle $ABC$ with $c = 5.07$ and $a = 3.34$, solve the triangle. Express angles in degrees and minutes. (See Fig. 26-1.)

Here it is assumed that $\gamma = 90°$.

Solve for $\alpha$:

In the right triangle $ABC$, $\sin \alpha = \dfrac{a}{c}$; hence $\alpha = \sin^{-1} \dfrac{a}{c} = \sin^{-1} \dfrac{3.34}{5.07} = 41°12'$.

Solve for $\beta$:

In the right triangle $ABC$, $\alpha + \beta = 90°$; hence $\beta = 90° - \alpha = 90° - 41°12' = 48°48'$.

Solve for $b$:

From the Pythagorean theorem, $c^2 = a^2 + b^2$; hence $b = \sqrt{c^2 - a^2} = \sqrt{5.07^2 - 3.34^2} = 3.81$.

**26.3.** When the angle of elevation of the sun is 27°, a pole casts a shadow 14 meters long on level ground. Find the height of the pole.

Sketch a figure (see Fig. 26-4).

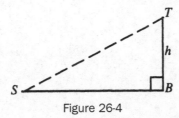

Figure 26-4

In right triangle *STB*, let $h$ = height of pole. Given that $SB = 14$ and $\angle S = 27°$, then $\tan S = \dfrac{h}{SB}$, thus

$$h = SB \tan S = 14 \tan 27° = 7.1 \text{ meters}$$

**26.4.** An airplane leaves an airport and travels at an average speed of 450 kilometers per hour on a heading of 250°. After three hours, how far south and how far west is it from its original position?

Sketch a figure (see Fig. 26-5).

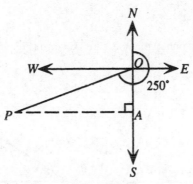

Figure 26-5

In the figure, the original position is *O* and the final position is *P*. Thus $OP = (450 \text{ km/hr})(3 \text{ hr}) = 1350 \text{ km}$. Since the heading is 250°, $\angle AOP$ must be $250° - 180° = 70°$.
Hence, in right triangle *AOP*,

$$\frac{OA}{OP} = \cos AOP, \text{ or } OA = OP \cos AOP = 1350 \cos 70° = 462 \text{ km south}$$

and

$$\frac{AP}{OP} = \sin AOP, \text{ or } AP = OP \sin AOP = 1350 \sin 70° = 1269 \text{ km west}$$

**26.5.** From a point on level ground the angle of elevation of the top of a building is 37.3°. From a point 50 yards closer, the angle of elevation is 56.2°. Find the height of the building.

Sketch a figure (Fig. 26-6).

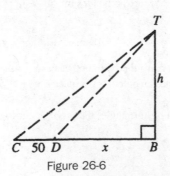

Figure 26-6

Introduce the auxiliary variable $x$. The cotangent function is chosen, as it leads to the simplest algebra in eliminating $x$. In right triangle $DBT$,

$$\cot TDB = \frac{x}{h}$$

In right triangle $CBT$,

$$\cot TCB = \frac{50 + x}{h}$$

Therefore,

$$\cot TCB - \cot TCB = \frac{50 + x}{h} - \frac{x}{h} = \frac{50}{h}$$

Hence,

$$h = \frac{50}{\cot TCB - \cot TDB} = \frac{50}{\cot 37.3° - \cot 56.2°} = 78 \text{ yd}$$

*Note*: The accuracy of the result is determined by the least accurate input measurement. Also note that in calculating cotangent on a scientific calculator, the identity $\cot u = 1/(\tan u)$ is used.

**26.6.** Derive the law of sines.

Two typical situations (acute and obtuse triangles) are sketched (Fig. 26-7).

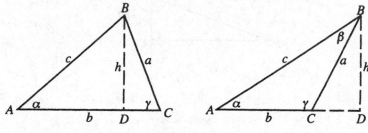

Figure 26-7

$h$ represents an altitude of the triangle, drawn perpendicular from one vertex (shown as $B$) to the opposite side. For the obtuse triangle, the altitude lies outside the triangle. In general, however, triangles $ADB$ and $CDB$ are right triangles. In triangle $ADB$,

$$\sin \alpha = \frac{h}{c} \qquad \text{so } h = c \sin \alpha$$

In triangle $CDB$, $\angle BCD$ is either $\gamma$ or $180° - \gamma$. In either case, $\sin BCD = \sin \gamma = \sin(180° - \gamma)$; hence

$$\sin BCD = \sin \gamma = \frac{h}{a} \qquad \text{so } h = a \sin \gamma$$

Therefore, $a \sin \gamma = c \sin \alpha$, or, dividing both sides by $\sin \alpha \sin \gamma$,

$$\frac{a}{\sin \alpha} = \frac{c}{\sin \gamma}$$

*Note*: Since the letters are assigned arbitrarily, the other cases of the law of sines can be immediately derived by replacement (sometimes called rotation of letters): replace $a$ with $b$, $b$ with $c$, $c$ with $a$, and also $\alpha$ with $\beta$, $\beta$ with $\gamma$, $\gamma$ with $\alpha$.

The law of sines also applies to a right triangle; the proof is left to the student.

**26.7.** Analyze the AAS and ASA cases of solving an oblique triangle.

In either case, two angles are known; hence the third can be found immediately since the sum of the angles of a triangle is 180°. With all three angles known, and one side given, there is enough information present to substitute into the law of sines to find the second and third sides. For example, given *a*, then *b* can be found, since

$$\frac{a}{\sin \alpha} = \frac{b}{\sin \beta}, \qquad \text{so } b = \frac{a \sin \beta}{\sin \alpha}$$

**26.8.** Solve triangle *ABC*, given $\alpha = 23.9°$, $\beta = 114°$, and $c = 82.8$.

Since two angles and the included side are given, this is the ASA case.

Solve for $\gamma$.

Since $\alpha + \beta + \gamma = 180°$, $\gamma = 180° - \alpha - \beta = 180° - 23.9° - 114° = 42.1°$.

Solve for *a*:

From the law of sines, $\dfrac{a}{\sin \alpha} = \dfrac{c}{\sin \gamma}$; hence

$$a = \frac{c \sin \alpha}{\sin \gamma} = \frac{82.8 \sin 23.9°}{\sin 42.1°} = 50.0$$

Solve for *b*:

Applying the law of sines again, $\dfrac{b}{\sin \beta} = \dfrac{c}{\sin \gamma}$, hence

$$b = \frac{c \sin \beta}{\sin \gamma} = \frac{82.8 \sin 114°}{\sin 42.1°} = 113$$

**26.9.** Fire station *B* is located 11.0 kilometers due east of fire station *A*. Smoke is spotted at a bearing of S23°40′E from station *A* and at S68°40′W from station *B*. How far is the fire from each fire station?

Sketch a figure (Fig. 26-8).

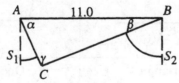

Figure 26-8

Given side $AB = c = 11.0$, $\angle S_1AC = 23°40′$, and $\angle S_2BC = 68°40′$, it follows that

$$\alpha = 90° - 23°40′ = 66°20' \text{ and } \beta = 90° - 68°40′ = 21°20'$$

Thus two angles and the included side are given and this is the ASA case.

Solve for $\gamma$:

$$\gamma = 180° - \alpha - \beta = 180° - 66°20' - 21°20' = 92°20'$$

Solve for *a*:

From the law of sines, $\dfrac{a}{\sin \alpha} = \dfrac{c}{\sin \gamma}$; hence

$$a = \frac{c \sin \alpha}{\sin \gamma} = \frac{11.0 \sin 66°20'}{\sin 92°20'} = 10.1 \text{ km from } B$$

Solve for $b$:

Applying the law of sines again, $\dfrac{b}{\sin\beta} = \dfrac{c}{\sin\gamma}$; hence

$$b = \frac{c\sin\beta}{\sin\gamma} = \frac{11.0\sin 21°20'}{\sin 92°20'} = 4.01 \text{ km from } A$$

**26.10.** Analyze the SSA case of solving an oblique triangle.

There are several possibilities. Assume for consistency that $a$, $b$, and $\alpha$ are given. Draw a line segment of unspecified length to represent $c$, then draw angle $\alpha$ and side $b$. Then the following cases can be distinguished:

$\alpha$ acute (see Fig. 26-9):

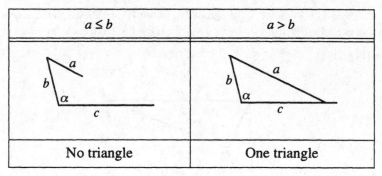

Figure 26-9

$\alpha$ obtuse (see Fig. 26-10):

| $a \le b$ | $a > b$ |
|:---:|:---:|
| No triangle | One triangle |

Figure 26-10

In every case, start by calculating the possible values of $\beta$, using the law of sines:

Since $\dfrac{a}{\sin\alpha} = \dfrac{b}{\sin\beta}$, it follows that $\sin\beta = \dfrac{b\sin\alpha}{a}$.

If the value of $\sin\beta$ calculated in this way is greater than 1, there is no solution to this equation and no triangle is possible. If this value of $\sin\beta = 1$, there is one (right) triangle possible. If this value of $\sin\beta < 1$, there are two solutions for $\beta$:

$$\beta = \sin^{-1}\frac{b\sin\alpha}{a} \quad \text{and} \quad \beta' = 180° - \sin^{-1}\frac{b\sin\alpha}{a}.$$

If both of these solutions, substituted into $\alpha + \beta + \gamma = 180°$, yield a positive value for $\gamma$, then two triangles are possible; if not, then only the first solution leads to a possible triangle and there is only one triangle.

The SSA case is sometimes referred to as the ambiguous case, both because there are so many possibilities and because there may be two triangles determined by the given information.

**26.11.** Solve triangle $ABC$, given $\alpha = 23.9°$, $a = 43.7$, and $b = 35.1$.

Sketch a figure, starting with a line segment of unspecified length to represent $c$: Since two sides and a nonincluded angle are given, this is the SSA case.

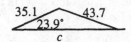

Figure 26-11

From Fig. 26-11, and the fact that $a > b$, only one triangle is determined by the given data.

Solve for $\beta$:

$$\sin\beta = \frac{b\sin\alpha}{a} = \frac{35.1\sin 23.9°}{43.7} = 0.3254$$
$$\beta = \sin^{-1}0.3254 = 19.0°$$

Solve for $\gamma$:

$$\gamma = 180° - \alpha - \beta = 180° - 23.9° - 19.0° = 137.1°$$

Note that the second solution of $\sin\beta = 0.3254$, namely, $\beta = 180° - \sin^{-1}0.3254 = 161°$, is too large to fit into the same triangle as $\alpha = 23.9°$; this possibility must be discarded.

Solve for $c$:

From the law of sines $\dfrac{a}{\sin\alpha} = \dfrac{c}{\sin\gamma}$, hence

$$c = \frac{a\sin\gamma}{\sin\alpha} = \frac{43.7\sin 137.1°}{\sin 23.9°} = 73.4$$

**26.12.** Derive the law of cosines.

Two typical situations (acute and obtuse angle $\gamma$) are sketched (Fig. 26-12).

In either case, since $A$ is a point $b$ units from the origin, on the terminal side of $\gamma$ in standard position, the coordinates of $A$ are given by $(b\cos\gamma, b\sin\gamma)$. The coordinates of $B$ are given by $(a, 0)$. Then the distance from $A$ to $B$, labeled $c$, is given by the distance formula as:

$$c = d(A,B) = \sqrt{(a - b\cos\gamma)^2 + (0 - b\sin\gamma)^2}$$

Squaring yields

$$c^2 = (a - b\cos\gamma)^2 + (b\sin\gamma)^2$$

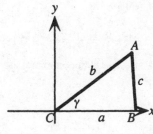
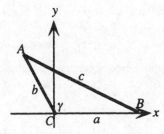

Fig 26-12

Simplifying yields:

$$c^2 = a^2 - 2ab\cos\gamma + b^2\cos^2\gamma + b^2\sin^2\gamma$$
$$= a^2 - 2ab\cos\gamma + b^2(\cos^2\gamma + \sin^2\gamma)$$

Hence by the Pythagorean identity,

$$c^2 = a^2 + b^2 - 2ab\cos\gamma$$

*Note*: Since the letters are assigned arbitrarily, the other cases of the law of cosines can be immediately derived by replacement (sometimes called rotation of letters): replace $a$ with $b$, $b$ with $c$, $c$ with $a$, and also $\alpha$ with $\beta$, $\beta$ with $\gamma$, $\gamma$ with $\alpha$.

The law of cosines also applies to a right triangle; for the side opposite the right angle, it reduces to the Pythagorean theorem; the proof is left to the student.

**26.13.** Analyze the SAS and SSS cases of solving an oblique triangle.

In the SAS case, two sides and the angle included between them are given. Label them $a$, $b$, and $\gamma$. Then none of the three ratios in the law of sines is known at the outset, so no information can be derived from this law. From the law of cosines, however, the third side, $c$, can be determined. Then with three sides and an angle known, the law of sines can be used to determine a second angle. If the smaller of the two unknown angles (smaller because it is opposite a smaller side) is chosen, this angle must be acute, hence there is only one possibility. The third angle follows immediately since the sum of the three angles must be 180°.

In the SSS case, again, none of the three ratios in the law of sines is known at the outset, so no information can be derived from this law. However, the law of cosines can be solved for the cosine of any unknown angle to yield:

$$\cos\alpha = \frac{b^2 + c^2 - a^2}{2bc} \qquad \cos\beta = \frac{a^2 + c^2 - b^2}{2ac} \qquad \cos\gamma = \frac{a^2 + b^2 - c^2}{2ab}$$

If the angle opposite the largest side is calculated first, then if the cosine of this angle is negative, the angle is obtuse; otherwise, the angle is acute. In either case, the other two angles cannot be obtuse and must be acute. Therefore, the second angle can be found from the law of sines without ambiguity, since this angle must be acute. The third angle follows immediately since the sum of the three angles must be 180°.

**26.14.** Solve triangle $ABC$, given $a = 3.562$, $c = 8.026$, and $\beta = 14°23'$.

Since two sides and the angle included between them are given, this is the SAS case.

Solve for $b$:

From the law of cosines:

$$b^2 = a^2 + c^2 - 2ac\cos\beta = (3.562)^2 + (8.026)^2 - 2(3.562)(8.026)\cos 14°23' = 21.7194$$

Hence

$$b = \sqrt{21.7194} = 4.660$$

Solve for the smaller of the two unknown angles, which must be $\alpha$:

From the law of sines, $\dfrac{a}{\sin\alpha} = \dfrac{b}{\sin\beta}$ hence $\sin\alpha = \dfrac{a\sin\beta}{b} = \dfrac{3.562\sin 14°23'}{4.660} = 0.18986$. The only acceptable solution of this equation must be an acute angle; hence $\alpha = \sin^{-1} 0.18986$, or, expressed in degrees and minutes, $\alpha = 10°56'$.

Solve for $\gamma$:

$$\gamma = 180° - \alpha - \beta = 180° - 10°56' - 14°23' = 154°41'$$

**26.15.** Solve triangle $ABC$, given $a = 29.4$, $b = 47.5$, and $c = 22.0$.

Since three sides are given, this is the SSS case. Start by solving for the largest angle, $\beta$, largest because it is opposite the largest side, $b$.

Solve for $\beta$:

From the law of cosines,

$$\cos\beta = \frac{a^2 + c^2 - b^2}{2ac} = \frac{(29.4)^2 + (22.0)^2 - (47.5)^2}{2(29.4)(22.0)} = -0.70183$$

The only acceptable solution of this equation must be an obtuse angle; hence $\beta = \cos^{-1}(-0.70183)$, or, expressed in degrees, $\beta = 134.6°$.

Solve for $\alpha$:

From the law of sines, $\dfrac{a}{\sin\alpha} = \dfrac{b}{\sin\beta}$; hence $\sin\alpha = \dfrac{a\sin\beta}{b} = \dfrac{29.4\sin[\cos^{-1}(-0.70183)]}{47.5} = 0.44090$. The

only acceptable solution of this equation must be an acute angle, hence $\alpha = \sin^{-1}0.44090$, or, expressed in degrees, $\alpha = 26.2°$.

Solve for $\gamma$:

$$\gamma = 180° - a - b = 180° - 26.2° - 134.6° = 19.2°$$

**26.16.** A car leaves an intersection traveling at an average speed of 56 miles per hour. Five minutes later, a second car leaves the same intersection and travels on a road making an angle of 112° with the first, at an average speed of 48 miles per hour. Assuming the roads are straight, how far apart are the cars 15 minutes after the first car has left?

Sketch a figure (see Fig. 26-13).

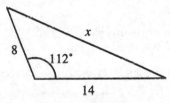

Figure 26-13

Let $x$ = the required distance. Since the first car travels 56 mph for $\frac{1}{4}$ hr, it goes a distance of $56\left(\frac{1}{4}\right) = 14$ miles. The second car travels 48 mph for $\frac{1}{6}$ hr, so it goes a distance of $48\left(\frac{1}{6}\right) = 8$ miles. In the triangle, two sides and the angle included between them are given, hence by the law of cosines,

$$x^2 = 8^2 + 14^2 - 2(8)(14)\cos 112° = 343.9$$

Hence $x = \sqrt{343.9} = 19$ miles to the accuracy of the input data.

**26.17.** A regular pentagon is inscribed in a circle of radius 10.0 units. Find the length of one side of the pentagon.

Sketch a figure (see Fig. 26-14).

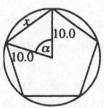

Figure 26-14

Let $x$ = the length of the side. Since the pentagon is regular, angle $\alpha = \frac{1}{5}$ of a full circle = 72°. Hence, from the law of cosines,

$$x^2 = (10.0)^2 + (10.0)^2 - 2(10.0)(10.0)\cos 72° = 138.2, \text{ so } x = \sqrt{138.2} = 11.8 \text{ units}$$

## SUPPLEMENTARY PROBLEMS

**26.18.** Solve a right triangle given $a = 350$ and $\alpha = 73°$.

*Ans.* $\beta = 17°, b = 107, c = 366$

**26.19.** Solve a right triangle given $b = 9.94$ and $c = 12.7$.

*Ans.* $a = 7.90, \beta = 51.5°, \alpha = 38.5°$

**26.20.** A rectangle is 173 meters long and 106 meters high. Find the angle between a diagonal and the longer side.

*Ans.* $31.5°$

**26.21.** From the top of a tower the angle of depression of a point on level ground is $56°30'$. If the height of the tower is 79.4 feet, how far is the point from the base of the tower?

*Ans.* 52.6 feet

**26.22.** A radio antenna is attached to the top of a building. From a point 12.5 meters from the base of the building, on level ground, the angle of elevation of the bottom of the antenna is $47.2°$ and the angle of elevation of the top is $51.8°$. Find the height of the antenna.

*Ans.* 2.39 meters

**26.23.** Show that the law of sines holds for a right triangle.

**26.24.** Show that the law of cosines holds for a right triangle, and reduces to the Pythagorean theorem for the side opposite the right angle.

**26.25.** Show that the area of a triangle can be expressed as one-half the product of any two sides times the sine of the angle included between them. $\left( A = \frac{1}{2}bc \sin\alpha \right)$

**26.26.** How many triangles are possible on the basis of the given data?

(a) $\alpha = 20°, b = 30, \gamma = 40°$; (b) $\alpha = 20°, b = 30, a = 5$; (c) $a = 30, c = 20, \gamma = 50°$;

(d) $a = 30, c = 30, \gamma = 100°$; (e) $\beta = 20°, b = 50, c = 30$

*Ans.* (a) 1; (b) 0; (c) 2; (d) 0; (e) 1

**26.27.** Solve triangle $ABC$ given (a) $\beta = 35.5°, \gamma = 82.6°, c = 7.88$; (b) $\alpha = 65°50', \beta = 78°20', c = 15.3$.

*Ans.* (a) $\alpha = 61.9°, b = 4.61, a = 7.01$; (b) $\gamma = 35°50', a = 23.8, b = 25.6$

**26.28.** Solve triangle $ABC$ given (a) $a = 12.3, b = 84.5, \alpha = 71.0°$; (b) $a = 84.5, b = 12.3, \alpha = 71.0°$;
(c) $a = 4.53, c = 6.47, \alpha = 39.3°$; (d) $a = 934, b = 1420, \beta = 108°$.

*Ans.* (a) No triangle can result from the given data; (b) $\beta = 7.91°, \gamma = 101°, c = 87.7$;
(c) Two triangles can result from the given data; triangle 1: $\gamma = 64.8°, \beta = 75.9°, b = 6.94$, triangle 2:
$\gamma' = 115.2°, \beta' = 25.5°, b' = 3.08$; (d) $\alpha = 38.7°, \gamma = 33.3°, c = 819$

**26.29.** How many triangles are possible on the basis of the given data?

(a) $a = 30, \beta = 40°, c = 50$; (b) $a = 80, b = 120, c = 30$; (c) $a = 40, b = 50, c = 35$;

(d) $\alpha = 75°, \beta = 35°, \gamma = 70°$; (e) $a = 40, b = 40, \gamma = 130°$

*Ans.*   (a) 1; (b) 0; (c) 1; (d) an infinite number; (e) 1

**26.30.** Solve triangle $ABC$ given (a) $b = 78, c = 150, \alpha = 83°$; (b) $a = 1260, b = 1440, c = 1710$.

*Ans.*   (a) $a = 160, \beta = 29°, \gamma = 68°$; (b) $\alpha = 46.2°, \beta = 55.5°, \gamma = 78.3°$

**26.31.** Points $A$ and $B$ are on opposite sides of a lake. To find the distance between them, a point $C$ is located 354 meters from $B$ and 286 meters from $A$. The angle between $AB$ and $AC$ is found to be $46°20'$. Find the distance between $A$ and $B$.

*Ans.*   485 meters

**26.32.** Two sides of a parallelogram are 9 and 15 units in length. The length of the shorter diagonal of the parallelogram is 14 units. Find the length of the long diagonal.

*Ans.*   $4\sqrt{26} \approx 20.4$ units

**26.33.** A plane travels 175 miles with heading 130° and then travels 85 miles with heading 255°. How far is the plane from its starting point?

*Ans.*   144 miles

**26.34.** (a) Use the law of sines to show that in any triangle $\dfrac{a + b}{c} = \dfrac{\sin \alpha + \sin \beta}{\sin \gamma}$.

(b) Use the result of part (a) to derive Mollweide's formula: $\dfrac{a + b}{c} = \dfrac{\cos \frac{1}{2}(\alpha - \beta)}{\sin(\gamma/2)}$.

**26.35.** Because it contains all six parts of a triangle, Mollweide's formula is sometimes used to check results in solving triangles. Use the formula to check the results in Problem 26.27a.

*Ans.*   $\dfrac{a + b}{c} = 1.4746$, $\dfrac{\cos \frac{1}{2}(\alpha - \beta)}{\sin(\gamma/2)} = 1.4751$; the two sides agree to the accuracy of the given data.

# CHAPTER 27

# Vectors

## Vectors and Vector Quantities

A quantity with both magnitude and direction is called a *vector quantity*. Examples include force, velocity, acceleration, and linear displacement. A vector quantity can be represented by a directed line segment, called a (geometric) *vector*. The length of the line segment represents the magnitude of the vector; the direction is indicated by the relative positions of the *initial point* and *terminal point* of the line segment. (See Fig. 27-1.)

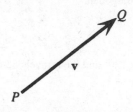

Figure 27-1

Vectors are indicated by boldface letters. In the figure, $P$ is the initial point of the vector $\mathbf{v}$ and $Q$ is the terminal point. Vector $\mathbf{v}$ would also be referred to as vector $\overrightarrow{PQ}$.

## Scalars and Scalar Quantities

A quantity with only magnitude is called a *scalar quantity*. Examples include mass, length, time, and temperature. The numbers used to measure scalar quantities are called *scalars*.

## Equivalent Vectors

Two vectors are called equivalent if they have the same magnitude and the same direction.

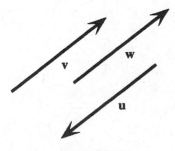

Figure 27-2

Normally, equivalence is indicated with the equality symbol. In Fig. 27-2, $\mathbf{v} = \mathbf{w}$ but $\mathbf{u} \neq \mathbf{v}$. Since there are an infinite number of line segments with a given magnitude and direction, there are an infinite number of vectors equivalent to a given vector (sometimes called copies of the vector).

## Zero Vector

A zero vector is defined as a vector with zero magnitude and denoted **0**. The initial and terminal points of a zero vector coincide; hence a zero vector may be thought of as a single point.

## Addition of Vectors

The sum of two vectors is defined in two equivalent ways, the triangle method and the parallelogram method.

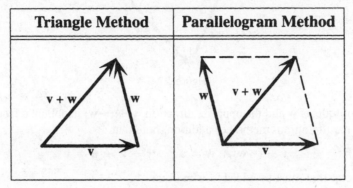

Figure 27-3

1. **TRIANGLE METHOD:** Given **v** and **w**, **v** + **w** is the vector formed as follows: place a copy of **w** with initial point coincident with the terminal point of **v**. Then **v** + **w** has the initial point of **v** and the terminal point of **w**.
2. **PARALLELOGRAM METHOD:** Given **v** and **w**, **v** + **w** is the vector formed as follows: place copies of **v** and **w** with the same initial point. Complete the parallelogram (assuming **v** and **w** are not parallel line segments). Then **v** + **w** is the diagonal of the parallelogram with this initial point.

## Multiplication of a Vector by a Scalar

Given a vector **v** and a scalar $c$, the product $c$**v** is defined as follows: If $c$ is positive, $c$**v** is a vector with the same direction as **v** and $c$ times the magnitude. If $c = 0$, then $c$**v** $= 0$**v** $= \mathbf{0}$. If $c$ is negative, $c$**v** is a vector with the opposite direction from **v** and $|c|$ times the magnitude.

**EXAMPLE 27.1**   Given **v** as shown, draw $2\mathbf{v}, \frac{1}{2}\mathbf{v}$, and $-2\mathbf{v}$.

The vector $2\mathbf{v}$ has the same direction as **v** and twice the magnitude. The vector $\frac{1}{2}\mathbf{v}$ has the same direction as **v** and one-half the magnitude. The vector $-2\mathbf{v}$ has the opposite direction from **v** and twice the magnitude (see Fig. 27-4).

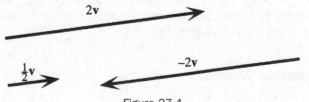

Figure 27-4

## Vector Subtraction

If **v** is a nonzero vector, $-\mathbf{v}$ is the vector with the same magnitude as **v** and the opposite direction. Then **v** $-$ **w** is defined as **v** $+ (-\mathbf{w})$.

**EXAMPLE 27.2** Illustrate the relations among **v**, **w**, −**w**, **v** − **w**, and **v** + (−**w**).

See Fig. 27-5.

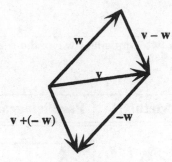

Figure 27-5

−**w** has the same magnitude as **w** and the opposite direction. **v** + (−**w**) is obtained from the triangle method of addition. From the parallelogram method of addition, note that

$$\mathbf{v} + (-\mathbf{w}) + \mathbf{w} = \mathbf{v} \qquad \mathbf{v} - \mathbf{w} + \mathbf{w} = \mathbf{v}$$

Thus, **v** − **w** is the vector that must be added to **w** to obtain **v**.

## Algebraic Vectors

If a vector **v** is placed in a Cartesian coordinate system such that $\mathbf{v} = \overrightarrow{P_1P_2}$, where $P_1$ has coordinates $(x_1,y_1)$ and $P_2$ has coordinates $(x_2,y_2)$, then the horizontal displacement from $P_1$ to $P_2$, $x_2 - x_1$, is called the *horizontal component* of **v**, and the vertical displacement $y_2 - y_1$ is called the *vertical component* of **v**. (See Fig. 27-6.)

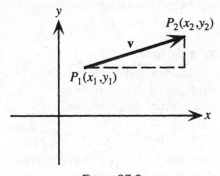

Figure 27-6

Given horizontal and vertical components $a$ and $b$, then **v** is completely determined by $a$ and $b$ and is written as the *algebraic vector* $\mathbf{v} = \langle a, b \rangle$. Then $\mathbf{v} = \overrightarrow{OP}$, where $O$ is the origin and $P$ has coordinates $(a, b)$. There is a one-to-one correspondence between algebraic and geometric vectors; any geometric vector corresponding to $\langle a, b \rangle$ is called a *geometric representative* of $\langle a, b \rangle$.

## Operations with Algebraic Vectors

Let $\mathbf{v} = \langle v_1, v_2 \rangle$ and $\mathbf{w} = \langle w_1, w_2 \rangle$. Then

$$\mathbf{v} + \mathbf{w} = \langle v_1 + w_1, v_2 + w_2 \rangle \qquad -\mathbf{w} = \langle -w_1, -w_2 \rangle$$

$$\mathbf{v} - \mathbf{w} = \langle v_1 - w_1, v_2 - w_2 \rangle \qquad c\mathbf{v} = \langle cv_1, cv_2 \rangle$$

**EXAMPLE 27.3** Given $\mathbf{a} = \langle 3, -8 \rangle$ and $\mathbf{b} = \langle 5, 2 \rangle$, find $\mathbf{a} + \mathbf{b}$.

$$\mathbf{a} + \mathbf{b} = \langle 3, -8 \rangle + \langle 5, 2 \rangle = \langle 3 + 5, -8 + 2 \rangle = \langle 8, -6 \rangle$$

## Magnitude of an Algebraic Vector

The magnitude of $\mathbf{v} = \langle v_1, v_2 \rangle$ is given by

$$|\mathbf{v}| = \sqrt{v_1^2 + v_2^2}$$

## Vector Algebra

Given vectors $\mathbf{u}$, $\mathbf{v}$, and $\mathbf{w}$, then

$\mathbf{v} + \mathbf{w} = \mathbf{w} + \mathbf{v}$ $\qquad$ $\mathbf{u} + (\mathbf{v} + \mathbf{w}) = (\mathbf{u} + \mathbf{v}) + \mathbf{w}$ $\qquad$ $\mathbf{v} + \mathbf{0} = \mathbf{v}$

$\mathbf{v} + (-\mathbf{v}) = \mathbf{0}$ $\qquad$ $c(\mathbf{v} + \mathbf{w}) = c\mathbf{v} + c\mathbf{w}$ $\qquad$ $(c + d)\mathbf{v} = c\mathbf{v} + d\mathbf{v}$

$(cd)\mathbf{v} = c(d\mathbf{v}) = d(c\mathbf{v})$ $\qquad$ $1\mathbf{v} = \mathbf{v}$ $\qquad$ $0\mathbf{v} = \mathbf{0}$

## Vector Multiplication

Given two vectors $\mathbf{v} = \langle v_1, v_2 \rangle$ and $\mathbf{w} = \langle w_1, w_2 \rangle$, the *dot product* of $\mathbf{v}$ and $\mathbf{w}$ is defined as $\mathbf{v} \cdot \mathbf{w} = v_1 w_1 + v_2 w_2$. Note that this is a scalar quantity.

**EXAMPLE 27.4** Given $\mathbf{a} = \langle 3, -8 \rangle$ and $\mathbf{b} = \langle 5, 2 \rangle$, find $\mathbf{a} \cdot \mathbf{b}$.

$$\mathbf{a} \cdot \mathbf{b} = 3 \cdot 5 + (-8)2 = -1$$

## Angle Between Two Vectors

If two nonzero vectors $\mathbf{v} = \langle v_1, v_2 \rangle$ and $\mathbf{w} = \langle w_1, w_2 \rangle$ have geometric representatives $\overrightarrow{OV}$ and $\overrightarrow{OW}$, then the angle between $\mathbf{v}$ and $\mathbf{w}$ is defined as angle $VOW$ (Fig. 27-7).

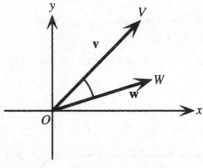

Figure 27-7

## Theorem on the Dot Product

If $\theta$ is the angle between two nonzero vectors $\mathbf{v}$ and $\mathbf{w}$, then $\mathbf{v} \cdot \mathbf{w} = |\mathbf{v}|\,|\mathbf{w}| \cos \theta$.

## Properties of the Dot Product

Given vectors $\mathbf{u}$, $\mathbf{v}$, and $\mathbf{w}$, and $a$ a real number, then

$\mathbf{u} \cdot \mathbf{v} = \mathbf{v} \cdot \mathbf{u}$ (commutative property) $\qquad$ $(a\mathbf{v}) \cdot \mathbf{w} = a(\mathbf{v} \cdot \mathbf{w})$ (associative property)

$\mathbf{u} \cdot (\mathbf{v} + \mathbf{w}) = \mathbf{u} \cdot \mathbf{v} + \mathbf{u} \cdot \mathbf{w}$ $\qquad\qquad\qquad\qquad$ (distributive property)

$\mathbf{v} \cdot \mathbf{v} \geq 0$ and $\mathbf{v} \cdot \mathbf{v} = 0$ if and only if $\mathbf{v} = \mathbf{0}$ $\qquad$ (nonzero property)

## SOLVED PROBLEMS

**27.1.** Given vectors **v** and **w** as shown, sketch $\mathbf{v} + \mathbf{w}$, $2\mathbf{v}$, and $2\mathbf{v} - \frac{1}{2}\mathbf{w}$.

To find $\mathbf{v} + \mathbf{w}$, place a copy of **w** with initial point coincident with the terminal point of **v**. Then $\mathbf{v} + \mathbf{w}$ has the initial point of **v** and the terminal point of **w**.
To find $2\mathbf{v}$, sketch a vector with the same direction as **v** and twice the magnitude.

To find $2\mathbf{v} - \frac{1}{2}\mathbf{w}$, sketch $-\frac{1}{2}\mathbf{w}$, a vector with the opposite direction to **w** and half the magnitude, with initial point coincident with the terminal point of $2\mathbf{v}$. Then $2\mathbf{v} - \frac{1}{2}\mathbf{w}$ has the initial point of $2\mathbf{v}$ and the terminal point of $-\frac{1}{2}\mathbf{w}$.

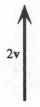
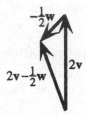

Figure 27-8

**27.2.** Given $\mathbf{v} = \langle -5, 3 \rangle$ and $\mathbf{w} = \langle 0, -4 \rangle$, find $\mathbf{v} + \mathbf{w}$, $4\mathbf{v}$, and $2\mathbf{v} - \frac{1}{2}\mathbf{w}$.

$$\mathbf{v} + \mathbf{w} = \langle -5, 3 \rangle + \langle 0, -4 \rangle = \langle -5 + 0, 3 + (-4) \rangle = \langle -5, -1 \rangle$$

$$4\mathbf{v} = 4\langle -5, 3 \rangle = \langle 4(-5), 4 \cdot 3 \rangle = \langle -20, 12 \rangle$$

$$2\mathbf{v} - \frac{1}{2}\mathbf{w} = 2\langle -5, 3 \rangle - \frac{1}{2}\langle 0, -4 \rangle = \langle -10, 6 \rangle - \langle 0, -2 \rangle = \langle -10, 8 \rangle$$

**27.3.** Given vector $\mathbf{v} = \langle v_1, v_2 \rangle$, (a) show that the magnitude of **v** is given by $|\mathbf{v}| = \sqrt{v_1^2 + v_2^2}$; (b) find the angle $\theta$ formed by vector $\mathbf{v} = \langle v_1, v_2 \rangle$ and the horizontal.

(a) Draw a copy of **v** with initial point at the origin; then the terminal point of **v** is $P(v_1, v_2)$.

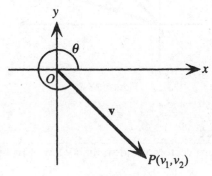

Figure 27-9

From the distance formula,

$$|\mathbf{v}| = d(O, P) = \sqrt{(v_1 - 0)^2 + (v_2 - 0)^2} = \sqrt{v_1^2 + v_2^2}$$

(b) From the definitions of the trigonometric functions as ratios (Chapter 22), since $P$ is a point on the terminal side of angle $\theta$,

$$\tan\theta = \frac{v_2}{v_1}$$

Thus, if $0 \le \theta < \pi/2$, then $\theta = \tan^{-1}\dfrac{v_2}{v_1}$. Otherwise, $\theta$ is an angle with this value as reference angle. (If $v_1 = 0$, then if $v_2 > 0$, $\theta$ may be taken as $\pi/2 + 2\pi n$; and if $v_2 < 0$, $\theta$ may be taken as $-\pi/2 + 2\pi n$.)

**27.4** Find $|\mathbf{v}|$ and the angle $\theta$ formed by vector $\mathbf{v} = \langle v_1, v_2 \rangle$ and the horizontal, given

(a) $\mathbf{v} = \langle 8, 5 \rangle$;  (b) $\mathbf{v} = \langle -6, -6 \rangle$.

(a) $|\mathbf{v}| = \sqrt{8^2 + 5^2} = \sqrt{89}$. $\tan\theta = \frac{5}{8}$. Since if the initial point of $\mathbf{v}$ is at the origin, the terminal point $(8, 5)$ is in quadrant I, $\theta$ may be taken as $\tan^{-1}\frac{5}{8}$.

(b) $|\mathbf{v}| = \sqrt{(-6)^2 + (-6)^2} = \sqrt{72} = 6\sqrt{2}$. $\tan\theta = \dfrac{-6}{-6} = 1$. Since if the initial point of $\mathbf{v}$ is at the origin, the terminal point $(-6, -6)$ is in quadrant III, $\theta$ may be taken as any solution of $\tan\theta = 1$ in this quadrant, for example, $5\pi/4$.

**27.5.** Resolve a vector $\mathbf{v}$ into horizontal and vertical components.

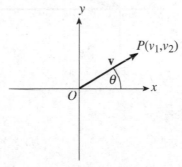

Figure 27-10

See Fig. 27-10. Vector $\mathbf{v} = \langle v_1, v_2 \rangle$. $v_1$ and $v_2$ are referred to, respectively, as the horizontal and vertical components of $\mathbf{v}$. Since the coordinates of $P$ are $(v_1, v_2)$,

$$\frac{v_1}{|\mathbf{v}|} = \cos\theta \text{ and } \frac{v_2}{|\mathbf{v}|} = \sin\theta,$$

hence $v_1 = |\mathbf{v}|\cos\theta$ and $v_2 = |\mathbf{v}|\sin\theta$ are the horizontal and vertical components of $\mathbf{v}$.

**27.6.** Show that for any two algebraic vectors $\mathbf{v}$ and $\mathbf{w}$, $\mathbf{v} + \mathbf{w} = \mathbf{w} + \mathbf{v}$ (vector addition is commutative).

Let $\mathbf{v} = \langle v_1, v_2 \rangle$ and $\mathbf{w} = \langle w_1, w_2 \rangle$. Then

$$\mathbf{v} + \mathbf{w} = \langle v_1, v_2 \rangle + \langle w_1, w_2 \rangle = \langle v_1 + w_1, v_2 + w_2 \rangle$$

and

$$\mathbf{w} + \mathbf{v} = \langle w_1, w_2 \rangle + \langle v_1, v_2 \rangle = \langle w_1 + v_1, w_2 + v_2 \rangle.$$

By the commutative law of addition for real numbers, $\langle v_1 + w_1, v_2 + w_2 \rangle = \langle w_1 + v_1, w_2 + v_2 \rangle$. Hence

$$\mathbf{v} + \mathbf{w} = \mathbf{w} + \mathbf{v}$$

**27.7.** Prove that if $\theta$ is the angle between two nonzero vectors $\mathbf{v}$ and $\mathbf{w}$, then $\mathbf{v} \cdot \mathbf{w} = |\mathbf{v}|\,|\mathbf{w}| \cos\theta$.

First consider the special case when $\mathbf{v}$ and $\mathbf{w}$ have the same direction. Then $\theta = 0$ and $\mathbf{w} = k\mathbf{v}$, where $k$ is positive. Hence

$$\mathbf{v} \cdot \mathbf{w} = \mathbf{v} \cdot k\mathbf{v} = \langle v_1, v_2 \rangle \cdot \langle kv_1, kv_2 \rangle = kv_1^2 + kv_2^2,$$

and

$$|\mathbf{v}|\,|\mathbf{w}| \cos\theta = \sqrt{v_1^2 + v_2^2}\,\sqrt{k^2v_1^2 + k^2v_2^2}\,\cos 0 = kv_1^2 + kv_2^2$$

Thus $\mathbf{v} \cdot \mathbf{w} = |\mathbf{v}|\,|\mathbf{w}| \cos\theta$ in this case.

A second special case occurs when $\mathbf{v}$ and $\mathbf{w}$ have opposite directions. This case is left to the student. Otherwise, take geometric vectors $\mathbf{v}$ and $\mathbf{w}$, each with initial point at the origin, and consider the triangle formed by $\mathbf{v}$, $\mathbf{w}$, and $\mathbf{v} - \mathbf{w}$. The terminal point of $\mathbf{v}$ is $V(v_1, v_2)$ and the terminal point of $\mathbf{w}$ is $W(w_1, w_2)$. $\mathbf{v} - \mathbf{w} = \overrightarrow{WV} = \langle v_1 - w_1, v_2 - w_2 \rangle$. Then, by the law of cosines applied to triangle $VOW$,

$$|\mathbf{v} - \mathbf{w}|^2 = |\mathbf{v}|^2 + |\mathbf{w}|^2 - 2|\mathbf{v}|\,|\mathbf{w}| \cos\theta$$

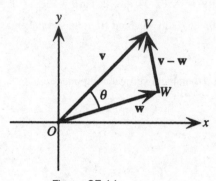

Figure 27-11

Or, writing in terms of components,

$$(v_1 - w_1)^2 + (v_2 - w_2)^2 = v_1^2 + v_2^2 + w_1^2 + w_2^2 - 2|\mathbf{v}|\,|\mathbf{w}| \cos\theta$$

Simplifying the left side and subtracting and dividing both sides by the same quantity yield, in turn,

$$v_1^2 - 2v_1w_1 + w_1^2 + v_2^2 - 2v_2w_2 + w_2^2 = v_1^2 + v_2^2 + w_1^2 + w_2^2 - 2|\mathbf{v}|\,|\mathbf{w}| \cos\theta$$

$$-2v_1w_1 - 2v_2w_2 = -2\,|\mathbf{v}|\,|\mathbf{w}| \cos\theta$$

$$v_1w_1 + v_2w_2 = |\mathbf{v}|\,|\mathbf{w}| \cos\theta$$

Since the left side, by definition, is $\mathbf{v} \cdot \mathbf{w}$, the proof is complete.

**27.8.** Find the angle $\theta$ between the vectors $\langle 5, 6 \rangle$ and $\langle 7, -8 \rangle$.

The formula in the previous problem is often written

$$\cos\theta = \frac{\mathbf{v} \cdot \mathbf{w}}{|\mathbf{v}||\mathbf{w}|}$$

In this case the formula is applied to obtain

$$\cos\theta = \frac{\langle 5, 6 \rangle \cdot \langle 7, -8 \rangle}{|\langle 5, 6 \rangle|\,|\langle 7, -8 \rangle|} = \frac{5 \cdot 7 + 6(-8)}{\sqrt{5^2 + 6^2}\sqrt{7^2 + (-8)^2}} = \frac{-13}{\sqrt{61}\,\sqrt{113}}$$

Thus $\theta = \cos^{-1}\dfrac{-13}{\sqrt{61}\,\sqrt{113}}$, or, expressed in degrees, $\theta \approx 99°$.

**27.9.** Prove the commutative property of the dot product.

Let $\mathbf{u} = \langle u_1, u_2 \rangle$ and $\mathbf{v} = \langle v_1, v_2 \rangle$. Then $\mathbf{u} \cdot \mathbf{v} = u_1 v_1 + u_2 v_2$ and $\mathbf{v} \cdot \mathbf{u} = v_1 u_1 + v_2 u_2$. By the commutative law of multiplication for real numbers, $u_1 v_1 = v_1 u_1$ and $u_2 v_2 = v_2 u_2$. Hence

$$\mathbf{u} \cdot \mathbf{v} = u_1 v_1 + u_2 v_2 = v_1 u_1 + v_2 u_2 = \mathbf{v} \cdot \mathbf{u}.$$

**27.10.** The vector sum of forces is generally called the *resultant* of the forces. Find the resultant of two forces, a force $\mathbf{F}_1$ of 55.0 pounds and a force $\mathbf{F}_2$ of 35.0 pounds acting at an angle of 120° to $\mathbf{F}_1$.

Denote the resultant force by $\mathbf{R}$. Sketch a figure (see Fig. 27-12).

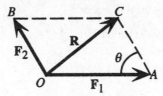

Figure 27-12

Since $\angle AOB$ is given as 120°, angle $\theta$ must measure 180° − 120° = 60°. From the law of cosines applied to triangle *OBC*,

$$|\mathbf{R}|^2 = |\mathbf{F}_1|^2 + |\mathbf{F}_2|^2 - 2|\mathbf{F}_1||\mathbf{F}_2|\cos\theta$$
$$= 55^2 + 35^2 - 2 \cdot 55 \cdot 35 \cos 60°$$
$$= 2325$$

Thus $|\mathbf{R}| = \sqrt{2325} = 48.2$ pounds. This determines the magnitude of the resultant force; since $\mathbf{R}$ is a vector, the direction of $\mathbf{R}$ must also be determined. From the law of sines applied to triangle *OBC*,

$$\frac{\sin AOC}{|\mathbf{F}_2|} = \frac{\sin\theta}{|\mathbf{R}|}$$

Hence

$$\angle AOC = \sin^{-1} \frac{|\mathbf{F}_2| \sin\theta}{|\mathbf{R}|} = \sin^{-1} \frac{35 \sin 60°}{48.2} = 38.9°$$

## SUPPLEMENTARY PROBLEMS

**27.11.** Let $\mathbf{v}$ be a vector with initial point (3,8) and terminal point (1,1). Let $\mathbf{w}$ be a vector with initial point (3,−4) and terminal point (0,0). (a) Express $\mathbf{v}$ and $\mathbf{w}$ in terms of components. (b) Find $\mathbf{v} + \mathbf{w}$, $\mathbf{v} - \mathbf{w}$, $3\mathbf{v} - 2\mathbf{w}$, and $\mathbf{v} \cdot \mathbf{w}$. (c) Find $|\mathbf{v}|$, $|\mathbf{w}|$, and the angle between $\mathbf{v}$ and $\mathbf{w}$.

*Ans.* (a) $\mathbf{v} = \langle -2, -7 \rangle$, $\mathbf{w} = \langle -3, 4 \rangle$;

(b) $\mathbf{v} + \mathbf{w} = \langle -5, -3 \rangle$, $\mathbf{v} - \mathbf{w} = \langle 1, -11 \rangle$, $3\mathbf{v} - 2\mathbf{w} = \langle 0, -29 \rangle$, $\mathbf{v} \cdot \mathbf{w} = -22$;

(c) $|\mathbf{v}| = \sqrt{53}$, $|\mathbf{w}| = 5$, angle $= \cos^{-1} \dfrac{-22}{5\sqrt{53}}$, or, expressed in degrees $\approx 127°$.

**27.12.** (a) Show that any vector $\mathbf{v}$ can be written as $\langle |\mathbf{v}| \cos\theta, |\mathbf{v}| \sin\theta \rangle$.

(b) Show that a vector parallel to a line with slope $m$ can be written as $a \langle 1, m \rangle$ for some value of $a$.

**27.13.** A unit vector is defined as a vector with magnitude 1. The unit vectors in the positive $x$ and $y$ directions are, respectively, referred to as $\mathbf{i}$ and $\mathbf{j}$.

(a) Show that any unit vector can be written as $\langle \cos\theta, \sin\theta \rangle$.

(b) Show that any vector $\mathbf{v} = \langle v_1, v_2 \rangle$ can be written as $v_1 \mathbf{i} + v_2 \mathbf{j}$.

**27.14** Two vectors that form an angle of $\pi/2$ are called *orthogonal*.

(a) Show that the dot product of two nonzero vectors is 0 if, and only if, the vectors are orthogonal.

(b) Show that $\langle 10, -6 \rangle$ and $\langle 9, 15 \rangle$ are orthogonal.

(c) Find a unit vector orthogonal to $\langle 2, -5 \rangle$ with horizontal component positive.

*Ans.* (c) $\langle 5/\sqrt{29}, 2/\sqrt{29} \rangle$

**27.15.** (a) Prove the associative property of the dot product;

(b) prove the distributive property of the dot product;

(c) prove the nonzero property of the dot product.

**27.16.** For $\mathbf{v}$ any vector, prove (a) $\mathbf{v} \cdot \mathbf{v} = |\mathbf{v}|^2$; (b) $\mathbf{0} \cdot \mathbf{v} = 0$.

**27.17.** A force of 46.3 pounds is applied at an angle of 34.8° to the horizontal. Resolve the force into horizontal and vertical components.

*Ans.* Horizontal: 38.0 pounds; vertical: 26.4 pounds

**27.18.** A weight of 75 pounds is resting on a surface inclined at an angle of 25° to the ground. Find the components of the weight parallel and perpendicular to the surface.

*Ans.* 32 pounds parallel to the surface, 68 pounds perpendicular to the surface

**27.19.** Find the resultant of two forces, one with magnitude 155 pounds and direction N50°W, and a second with magnitude 305 pounds and direction S55°W.

*Ans.* 376 pounds in the direction S78°W

# CHAPTER 28

# Polar Coordinates;
# Parametric Equations

## Polar Coordinate System

A polar coordinate system specifies points in the plane in terms of directed distances $r$ from a fixed point called the *pole* and angles $\theta$ measured from a fixed ray (with initial point the pole) called the *polar axis*. The polar axis is the positive half of a number line, drawn to the right of the pole. See Fig. 28-1.

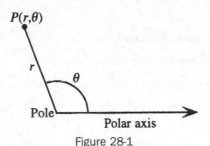

Figure 28-1

For any point $P$, $\theta$ is an angle formed by the polar axis and the ray connecting the pole to $P$, and $r$ is the distance measured along this ray from the pole to $P$. For any ordered pair $(r,\theta)$, if $r$ is positive, take $\theta$ as an angle with vertex the pole and initial side the polar axis, and measure $r$ units along the terminal side of $\theta$. If $r$ is negative, measure $|r|$ units along the ray directed *opposite* to the terminal side of $\theta$. Any pair with $r = 0$ represents the pole. In this manner, every ordered pair $(r,\theta)$ is represented by a unique point.

**EXAMPLE 28.1** Graph the points specified by $(3,\pi/3)$ and $(-3,\pi/3)$.

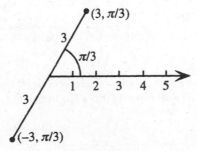

Figure 28-2

## Polar Coordinates of a Point Are not Unique

The polar coordinates of a point are not unique, however. Given point $P$, there is an infinite set of polar coordinates corresponding to $P$, since there are an infinite number of angles with terminal sides passing through $P$.

**EXAMPLE 28.2** List four alternative sets of polar coordinates corresponding to the point $P(3, \pi/3)$. Adding any multiple of $2\pi$ yields an angle coterminal with a given angle; hence $(3, 7\pi/3)$ and $(3, 13\pi/3)$ are two possible alternative polar coordinates. Since $\pi + \pi/3 = 4\pi/3$ has terminal side the ray opposite to $\pi/3$, the coordinates $(-3, 4\pi/3)$ and $(-3, 10\pi/3)$ are further alternative polar coordinates for $P$.

## Polar and Cartesian Coordinates

If a polar coordinate system is superimposed upon a Cartesian coordinate system, as in Fig. 28-3, the transformation relationships below hold between the two sets of coordinates.

If $P$ has polar coordinates $(r, \theta)$ and
Cartesian coordinates $(x, y)$, then
$$x = r\cos\theta \quad y = r\sin\theta$$
$$r^2 = x^2 + y^2 \quad \tan\theta = \frac{y}{x} \quad (x \neq 0)$$

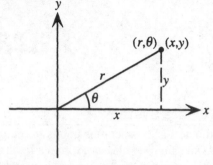

Figure 28-3

**EXAMPLE 28.3** Convert $(6, 2\pi/3)$ to Cartesian coordinates.

Since $r = 6$ and $\theta = 2\pi/3$, applying the transformation relationships yields

$$x = r\cos\theta = 6\cos 2\pi/3 = -3 \quad\quad y = r\sin\theta = 6\sin 2\pi/3 = 3\sqrt{3}$$

Thus the Cartesian coordinates are $(-3, 3\sqrt{3})$.

**EXAMPLE 28.4** Convert $(-5, -5)$ to polar coordinates with $r > 0$ and $0 \leq \theta \leq 2\pi$.

Since $x = -5$ and $y = -5$, applying the transformation relationships yields

$$r^2 = x^2 + y^2 = (-5)^2 + (-5)^2 = 50 \quad\quad \tan\theta = \frac{y}{x} = \frac{-5}{-5} = 1$$

Since $r$ is required positive, $r = \sqrt{50} = 5\sqrt{2}$. Since the point $(-5, -5)$ is in quadrant III, $\theta = 5\pi/4$. The polar coordinates that satisfy the given conditions are $(5\sqrt{2}, 5\pi/4)$.

## Equations in Polar Coordinates

Any equation in the variables $r$ and $\theta$ may be interpreted as a polar coordinate equation. Often $r$ is specified as a function of $\theta$.

**EXAMPLE 28.5** $r\theta = 1$ and $r^2 = 2\cos 2\theta$ are examples of polar coordinate equations. $r = 2\sin\theta$ and $r = 3 - 3\cos 2\theta$ are examples of polar coordinate equations with $r$ specified as a function of $\theta$.

## Parametric Equations

An equation for a curve may be given by specifying $x$ and $y$ separately as functions of a third variable, often $t$, called a *parameter*. These functions are called the *parametric equations* for the curve. Points on the curve may be found by assigning permissible values of $t$. Often, $t$ may be eliminated algebraically, but any restrictions placed on $t$ are needed to determine the portion of the curve that is specified by the parametric equations.

**EXAMPLE 28.6** Graph the curve specified by the parametric equations $x = 1 - t$, $y = 2t + 2$.

First note that $t$ can be eliminated by solving the equation specifying $x$ for $t$ to obtain $t = 1 - x$, then substituting into the equation specifying $y$ to obtain $y = 2(1 - x) + 2 = 4 - 2x$. Thus, for every value of $t$, the point $(x, y)$ lies on the graph of

$y = 4 - 2x$. Moreover, since there are no restrictions on $t$ and the functions $x(t)$ and $y(t)$ are one-to-one, it follows that $x$ and $y$ can take on any value and the graph is the entire line $y = 4 - 2x$. Form a table of values, then plot the points and connect them (Fig. 28-4).

| $t$ | 0 | 1 | 2 |
|---|---|---|---|
| $x$ | 1 | 0 | −1 |
| $y$ | 2 | 4 | 6 |

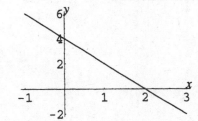

Figure 28-4

**EXAMPLE 28.7**  Graph the curve specified by the parametric equations $x = \cos^2 t$, $y = \sin^2 t$.

First note that $t$ can be eliminated by adding the equations specifying $x$ and $y$ to obtain $x + y = 1$. However, both variables are restricted by these equations to the interval $[0, 1]$. In fact, since both $x$ and $y$ are periodic with period $\pi$, the graph is the portion of the line $x + y = 1$ on the interval $0 \le x \le 1$, and is traced out repeatedly as $t$ varies through all possible real values. Form a table of values, then plot the points and connect them (Fig. 28-5).

| $t$ | 0 | $\pi/4$ | $\pi/2$ | $3\pi/4$ | $\pi$ |
|---|---|---|---|---|---|
| $x$ | 1 | 1/2 | 0 | 1/2 | 1 |
| $y$ | 0 | 1/2 | 1 | 1/2 | 0 |

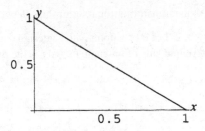

Figure 28-5

## Polar Coordinates and Parametric Equations

According to the transformation relationships, the Cartesian coordinates of a point are given in terms of its polar coordinates by the equations $x = r\cos\theta$ and $y = r\sin\theta$. Hence any polar coordinate equation specifying $r = f(\theta)$ can be regarded as giving parametric equations for $x$ and $y$ of the form $x = f(\theta)\cos\theta$, $y = f(\theta)\sin\theta$, with $\theta$ as the parameter.

**EXAMPLE 28.8**  Write the parametric equations for $x$ and $y$ specified by $r = 1 + \sin\theta$.

$$x = (1 + \sin\theta)\cos\theta \qquad y = (1 + \sin\theta)\sin\theta$$

## SOLVED PROBLEMS

**28.1.** Graph the points with the following polar coordinates:

(a) $A(4, \pi/6)$, $B(6, -\pi/4)$; (b) $C(-2, 5\pi/3)$, $D(-5, \pi)$.

(a) $(4, \pi/6)$ is located 4 units along the ray at an angle of $\pi/6$ to the polar axis. $(6, -\pi/4)$ is located 6 units along the ray at an angle of $-\pi/4$ to the polar axis (Fig. 28-6).

(b) $(-2, 5\pi/3)$ is located $|-2| = 2$ units along the ray directed opposite to $5\pi/3$. $(-5, \pi)$ is located $|-5| = 5$ units along the ray directed opposite to $\pi$ (Fig. 28-7).

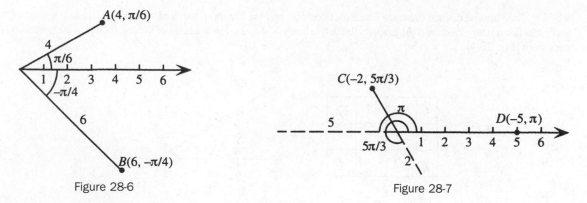

Figure 28-6                              Figure 28-7

**28.2.** Give all possible polar coordinates that can describe a point $P(r,\theta)$.

Since any angle of measure $\theta + 2\pi$, where $n$ is an integer, is coterminal with $\theta$, any point with coordinates $(r,\theta + 2\pi n)$ is coincident with $(r,\theta)$. Also, since the ray forming an angle of $\theta + \pi$ with the polar axis is directed opposite to the ray forming angle $\theta$, the coordinates $(r,\theta)$ and $(-r, \theta + \pi)$ name the same point. Finally, any point with coordinates $(-r, \theta + \pi + 2\pi n)$ is coincident with $(-r, \theta + \pi)$. Summarizing, the coordinates $(r, \theta + 2\pi n)$ and $(-r, \theta + (2n + 1)\pi)$ describe the same point as $(r,\theta)$.

**28.3.** Establish the transformation relationships between polar and Cartesian coordinates.

See Fig. 28-3. Let $P$ be a point with Cartesian coordinates $(x,y)$ and polar coordinates $(r,\theta)$. Since $P(x, y)$ is a point on the terminal side of an angle $\theta$, and $r$ is the distance of $P$ from the origin, it follows that

$$\tan \theta = \frac{y}{x} (x \neq 0) \quad \cos \theta = \frac{x}{r} \quad \sin \theta = \frac{y}{r}$$

Hence $x = r\cos\theta$ and $y = r\sin\theta$. From these relationships, it follows that

$$x^2 + y^2 = r^2\cos^2\theta + r^2\sin^2\theta = r^2(\cos^2\theta + \sin^2\theta) = r^2$$

The last relationship also follows immediately from the distance formula.

**28.4.** Convert to Cartesian coordinates: (a) $\left(4\sqrt{3}, \frac{4\pi}{3}\right)$; (b) $\left(-5, -\frac{\pi}{2}\right)$.

(a) Since $r = 4\sqrt{3}$ and $\theta = 4\pi/3$, it follows from the transformation relationships that

$$x = r\cos\theta = 4\sqrt{3} \cos(4\pi/3) = 4\sqrt{3}(-1/2) = -2\sqrt{3}$$

and

$$y = r\sin\theta = 4\sqrt{3} \sin(4\pi/3) = 4\sqrt{3}(-\sqrt{3}/2) = -6.$$

The Cartesian coordinates are $(-2\sqrt{3}, -6)$.

(b) Since $r = -5$ and $\theta = -\pi/2$, it follows from the transformation relationships that

$$x = r\cos\theta = -5\cos(-\pi/2) = 0 \quad \text{and} \quad y = r\sin\theta = -5\sin(-\pi/2) = 5.$$

The Cartesian coordinates are $(0,5)$.

**28.5.** Convert $(-8\sqrt{2}, 8\sqrt{2})$ to polar coordinates with $r > 0$ and $0 \leq \theta \leq 2\pi$.

Since $x = -8\sqrt{2}$ and $y = 8\sqrt{2}$, applying the transformation relationships yields

$$r^2 = x^2 + y^2 = (-8\sqrt{2})^2 + (8\sqrt{2})^2 = 256 \quad \text{and} \quad \tan \theta = \frac{y}{x} = \frac{8\sqrt{2}}{-8\sqrt{2}} = -1$$

Since $r$ is required positive, $r = \sqrt{256} = 16$. Since the point $(-8\sqrt{2}, 8\sqrt{2})$ is in quadrant II, $\theta = 3\pi/4$. The polar coordinates that satisfy the given conditions are $(16, 3\pi/4)$.

**28.6.** Transform the following polar coordinate equations to Cartesian coordinates:

(a) $r = 4$; (b) $r = 4\cos\theta$; (c) $r^2\sin 2\theta = 4$.

(a) Since the polar coordinate equation specifies all points that are at a distance of 4 units from the origin, this is the equation of a circle with radius 4 and center at the origin. Hence the Cartesian coordinate equation is $x^2 + y^2 = 16$.

(b) Multiply both sides by $r$ to obtain an equation that is easier to work with: $r^2 = 4r\cos\theta$. This operation adds the pole to the graph ($r = 0$). But the pole was already part of the graph (choose $\theta = \pi/2$), so nothing has been changed. Now apply the transformation relationships $r^2 = x^2 + y^2$ and $x = r\cos\theta$ to obtain $x^2 + y^2 = 4x$. This equation can be rewritten as $(x - 2)^2 + y^2 = 4$; thus, it is the equation of a circle with center at $(2,0)$ and radius 2.

(c) Rewrite $r^2\sin 2\theta = 4$ as follows:

$$
\begin{aligned}
r^2(2\sin\theta\cos\theta) &= 4 \qquad &\text{Double angle identity} \\
2r\cos\theta\, r\sin\theta &= 4 \qquad &\text{Algebra} \\
2xy &= 4 \qquad &\text{Transformation relationships} \\
xy &= 2 \qquad &\text{Algebra}
\end{aligned}
$$

**28.7.** Transform the following Cartesian coordinate equations to polar coordinates:

(a) $x + y = 3$; (b) $x^2 + y^2 = 3y$; (c) $y^2 = 4x$.

(a) Apply the transformation equations $x = r\cos\theta$ and $y = r\sin\theta$ to obtain $r\cos\theta + r\sin\theta = 3$.

(b) Apply the transformation equations $x^2 + y^2 = r^2$ and $y = r\sin\theta$ to obtain $r^2 = 3r\sin\theta$. This can be further simplified as follows:

$$
\begin{aligned}
r^2 - 3r\sin\theta &= 0 \\
r(r - 3\sin\theta) &= 0 \\
r = 0 \quad \text{or} \quad r - 3\sin\theta &= 0 \\
r &= 3\sin\theta
\end{aligned}
$$

The graph of $r = 0$ consists only of the pole. Since the pole is included in the graph of $r = 3\sin\theta$ (choose $\theta = 0$), it is sufficient to consider only $r = 3\sin\theta$ as the transformed equation.

(c) Apply the transformation equations $x = r\cos\theta$ and $y = r\sin\theta$ to obtain $r^2\sin^2\theta = 4r\cos\theta$. Proceeding as in part (b), this can be simplified to $r\sin^2\theta = 4\cos\theta$, which can be further rewritten as follows:

$$
\begin{aligned}
r &= \frac{4\cos\theta}{\sin^2\theta} \\
&= 4\frac{\cos\theta}{\sin\theta}\frac{1}{\sin\theta} \\
&= 4\cot\theta\,\csc\theta
\end{aligned}
$$

**28.8.** Sketch a graph of $r = 1 + \cos\theta$.

Before making a table of values it is helpful to consider the general behavior of the function $r(\theta) = 1 + \cos\theta$. From knowledge of the behavior of the cosine function:

| As $\theta$ increases | $\cos\theta$ | $1 + \cos\theta$ |
|---|---|---|
| from 0 to $\pi/2$ | decreases from 1 to 0 | decreases from 2 to 1 |
| from $\pi/2$ to $\pi$ | decreases from 0 to $-1$ | decreases from 1 to 0 |
| from $\pi$ to $3\pi/2$ | increases from $-1$ to 0 | increases from 0 to 1 |
| from $3\pi/2$ to $2\pi$ | increases from 0 to 1 | increases from 1 to 2 |

Since the cosine function is periodic with period $2\pi$, this shows the behavior of $1 + \cos\theta$ for all $\theta$. Now form a table of values and sketch the graph (see Fig. 28-8).

| $\theta$ | 0 | $\pi/4$ | $\pi/2$ | $3\pi/4$ | $\pi$ |
|---|---|---|---|---|---|
| $r$ | 2 | 1.7 | 1 | 0.3 | 0 |
| $\theta$ | | $5\pi/4$ | $3\pi/2$ | $7\pi/4$ | $2\pi$ |
| $r$ | | 0.3 | 1 | 1.7 | 2 |

Figure 28-8

The curve is known as a cardioid because of its heart shape.

**28.9.** Sketch a graph of $r = \cos 2\theta$.

Before making a table of values, it is helpful to consider the general behavior of the function $r(\theta) = \cos 2\theta$. From knowledge of the behavior of the cosine function:

| As $2\theta$ increases | $\theta$ increases | $\cos 2\theta$ |
|---|---|---|
| from 0 to $\pi/2$ | from 0 to $\pi/4$ | decreases from 1 to 0 |
| from $\pi/2$ to $\pi$ | from $\pi/4$ to $\pi/2$ | decreases from 0 to $-1$ |
| from $\pi$ to $3\pi/2$ | from $\pi/2$ to $3\pi/4$ | increases from $-1$ to 0 |
| from $3\pi/2$ to $2\pi$ | from $3\pi/4$ to $\pi$ | increases from 0 to 1 |
| from $2\pi$ to $5\pi/2$ | from $\pi$ to $5\pi/4$ | decreases from 1 to 0 |
| from $5\pi/2$ to $3\pi$ | from $5\pi/4$ to $3\pi/2$ | decreases from 0 to $-1$ |
| from $3\pi$ to $7\pi/2$ | from $3\pi/2$ to $7\pi/4$ | increases from $-1$ to 0 |
| from $7\pi/2$ to $4\pi$ | from $7\pi/4$ to $2\pi$ | increases from 0 to 1 |

Since the cosine function is periodic with period $2\pi$, this shows the behavior of $\cos 2\theta$ for all $\theta$. Now form a table of values and sketch the graph (Fig. 28-9).

| $\theta$ | 0 | $\pi/8$ | $\pi/4$ | $3\pi/8$ | $\pi/2$ |
|---|---|---|---|---|---|
| $r$ | 1 | 0.7 | 0 | –0.7 | –1 |
| $\theta$ | | $5\pi/8$ | $3\pi/4$ | $7\pi/8$ | $\pi$ |
| $r$ | | –0.7 | 0 | 0.7 | 1 |
| $\theta$ | | $9\pi/8$ | $5\pi/4$ | $11\pi/8$ | $3\pi/2$ |
| $r$ | | 0.7 | 0 | –0.7 | –1 |
| $\theta$ | | $13\pi/8$ | $7\pi/4$ | $15\pi/8$ | $2\pi$ |
| $r$ | | –0.7 | 0 | 0.7 | 1 |

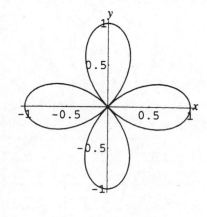

Figure 28-9

The curve is known as a four-leaved rose.

**28.10.** Graph the curve specified by the parametric equations $x = t^2$, $y = t^2 + 2$.

First note that $t$ can be eliminated by substituting $x$ for $t^2$ to obtain $y = x + 2$. However, both variables are restricted by these equations so that $x \geq 0$ and therefore $y \geq 2$. Hence the graph is the portion of the line $y = x + 2$ in the first quadrant, but is traced out twice, once for negative $t$ and once for positive $t$. Form a table of values, then plot the points and connect them (see Fig. 28-10).

| $t$ | $-2$ | $-1$ | $0$ | $1$ | $2$ |
|---|---|---|---|---|---|
| $x$ | $4$ | $1$ | $0$ | $1$ | $4$ |
| $y$ | $6$ | $3$ | $2$ | $3$ | $6$ |

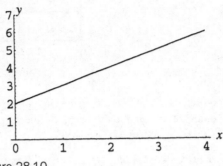

Figure 28-10

**28.11.** Graph the curve specified by the parametric equations $x = 2\cos t$, $y = 2\sin t$.

First note that $t$ can be eliminated by squaring the equations specifying $x$ and $y$ and adding to obtain $x^2 + y^2 = 4$. Thus the graph consists of the circle with center the origin and radius 2, and is traced out once each time $t$ increases by an amount $2\pi$. Form a table of values, then plot the points and draw the circle (Fig. 28-11).

| $t$ | $0$ | $\pi/4$ | $\pi/2$ | $3\pi/4$ | $\pi$ |
|---|---|---|---|---|---|
| $x$ | $2$ | $\sqrt{2}$ | $0$ | $-\sqrt{2}$ | $-2$ |
| $y$ | $0$ | $\sqrt{2}$ | $2$ | $\sqrt{2}$ | $0$ |
| $t$ | | $5\pi/4$ | $3\pi/2$ | $7\pi/4$ | $2\pi$ |
| $x$ | | $-\sqrt{2}$ | $0$ | $\sqrt{2}$ | $2$ |
| $y$ | | $-\sqrt{2}$ | $-2$ | $-\sqrt{2}$ | $0$ |

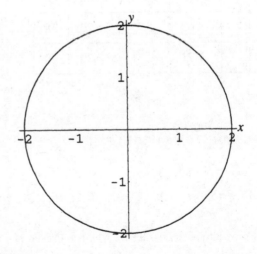

Figure 28-11

**28.12.** When a wheel of radius $a$ rolls without slipping on a horizontal surface, the curve traced out by a point on the rim of the wheel is called a cycloid. (a) Show that the parametric equations of a cycloid can be written as

$$x = a(\phi - \sin\phi)$$

$$y = a(1 - \cos\phi)$$

(b) Sketch a graph of a cycloid for $a = 1$.

(a) Draw a figure (see Fig. 28-12). The parameter $\phi$ is the angle through which the wheel has rotated.

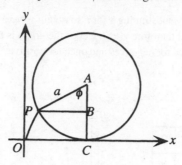

Figure 28-12

The coordinates of $P$, the point on the rim, are $(x, y)$. Because the wheel rotates without slipping, the length of arc $\overset{\frown}{PC}$ is equal to the length of line segment $\overline{OC}$. Hence $x = \overline{OC} - \overline{PB} = a\phi - a\sin\phi$ and $y = \overline{CB} = \overline{AC} - \overline{AB} = a - a\cos\phi$.

(b) In this case $x = \phi - \sin\phi$, $y = 1 - \cos\phi$. Form a table of values and connect the points. The curve (Fig. 28-13) is shown for $0 \le \phi \le 2\pi$; for other values of $\phi$ the arch shape is repeated, since $y$ is a periodic function of $\phi$.

| $\phi$ | 0 | $\pi/4$ | $\pi/2$ | $3\pi/4$ | $\pi$ |
|---|---|---|---|---|---|
| $x$ | 0 | 0.08 | 0.57 | 1.65 | $\pi$ |
| $y$ | 0 | 0.29 | 1 | 1.71 | 2 |
| $\phi$ | | $5\pi/4$ | $3\pi/2$ | $7\pi/4$ | $2\pi$ |
| $x$ | | 4.63 | 5.71 | 6.20 | $2\pi$ |
| $y$ | | 1.71 | 1 | 0.29 | 0 |

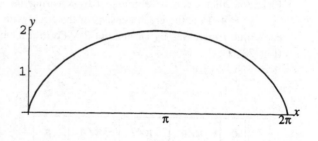

Figure 28-13

## SUPPLEMENTARY PROBLEMS

**28.13.** Convert to Cartesian coordinates: $(5,0)$, $(5,\pi)$, $(6, -\pi/3)$, $(-2\sqrt{2}, 3\pi/4)$, $(-20, -5\pi/2)$.

     *Ans.*    $(5,0)$, $(-5,0)$, $(3, -3\sqrt{3})$, $(2,-2)$, $(0,20)$

**28.14.** Convert to polar coordinates with $r > 0$ and $0 \le \theta \le 2\pi$: $(0,2)$, $(0,-3)$, $(-4,4)$, $(6, -6\sqrt{3})$.

     *Ans.*    $(2, \pi/2)$, $(3, 3\pi/2)$, $(4\sqrt{2}, 3\pi/4)$, $(12, 5\pi/3)$

**28.15.** Transform the following polar coordinate equations to Cartesian coordinates:

     (a) $r = 3\sin\theta$; (b) $\theta = \pi/4$; (c) $r = 2\tan\theta$ ; (d) $r = 1 + \cos\theta$.

     *Ans.*    (a) $x^2 + y^2 = 3y$; (b) $y = x$; (c) $x^4 + x^2y^2 = 4y^2$; (d) $x^4 + y^4 - 2x^3 - 2xy^2 + 2x^2y^2 - y^2 = 0$

**28.16.** Transform the following Cartesian coordinate equations to polar coordinates:

     (a) $y = 5$; (b) $xy = 4$; (c) $x^2 + y^2 = 16$; (d) $x^2 - y^2 = 16$.

     *Ans.*    (a) $r = 5\csc\theta$; (b) $r^2 \sin\theta\cos\theta = 4$; (c) $r = 4$; (d) $r^2\cos 2\theta = 16$

**28.17.** Sketch a graph of the following polar coordinate equations:

(a) $r = \theta$ $(0 \le \theta \le 4\pi)$; (b) $r = 1 + 2\sin\theta$

*Ans.*   (a) Fig. 28-14; (b) Fig. 28-15

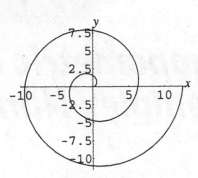

Figure 28-14

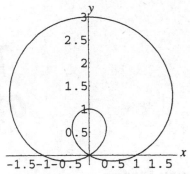

Figure 28-15

**28.18.** Eliminate the parameter $t$ and state any restrictions on the variables in the resulting equation:

(a) $x = 3t, y = 2t - 5$; (b) $x = \sqrt{t - 1}, y = t - 2$; (c) $x = e^t, y = e^{-t}$.

*Ans.*   (a) $2x - 3y = 15$; (b) $y = x^2 - 1, x \ge 0$; (c) $xy = 1, x, y > 0$

**28.19.** A projectile is fired at an angle of inclination $\alpha$ $(0 < \alpha < \pi/2)$ at an initial speed of $v_0$. Parametric equations for its path can be shown to be $x = v_0 t\cos\alpha, y = v_0 t\sin\alpha - (gt^2)/2$ ($t$ represents time).

(a) Eliminate the parameter $t$ and find the value of $t$ when the projectile hits the ground.

(b) Sketch the path of the projectile for the case $\alpha = \pi/6$, $v_0 = 32$ ft/sec, $g = 32$ ft/sec$^2$.

*Ans.*   (a) $y = x\tan\alpha - (gx^2\sec^2\alpha)/(2v_0^2)$; $y = 0$ when $t = \dfrac{2v_0\sin\alpha}{g}$; (b) Fig. 28-16

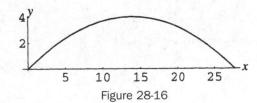

Figure 28-16

# CHAPTER 29

# Trigonometric Form of Complex Numbers

## The Complex Plane

Each complex number in standard form, $z = x + yi$, corresponds to an ordered pair of real numbers $(x, y)$ and thus to a point in a Cartesian coordinate system, referred to as the *complex plane*. The $x$-axis in this system is referred to as the *real axis*, and the $y$-axis as the *imaginary axis*.

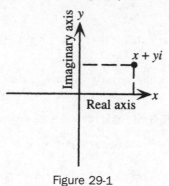

Figure 29-1

**EXAMPLE 29.1**    Show $4 + 2i$, $-2i$, and $-3 - i$ in a complex plane.

The points are represented geometrically by $(4,2)$, $(0,-2)$, and $(-3,-1)$.

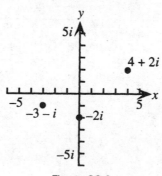

Figure 29-2

## Trigonometric Form of Complex Numbers

If a polar coordinate system is superimposed on the Cartesian coordinate system, then the relationships $x = r\cos\theta$ and $y = r\sin\theta$ hold. Thus every complex number $z$ can be written in *trigonometric form*:

$$z = r\cos\theta + ir\sin\theta$$

$$= r(\cos\theta + i\sin\theta)$$

This form is sometimes abbreviated as $z = r\operatorname{cis}\theta$. The standard form $z = x + yi$ is referred to as *rectangular form*. Since the polar coordinates of a point are not unique, there are an infinite number of equivalent trigonometric forms of a complex number. The relationships among $x$, $y$, $z$, $r$, and $\theta$ are shown in Fig. 29-3.

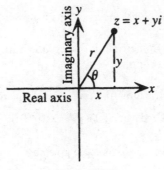

Figure 29-3

**EXAMPLE 29.2**  Write $5\left(\cos\dfrac{\pi}{2} + i\sin\dfrac{\pi}{2}\right)$ in rectangular form.

$$5\left(\cos\frac{\pi}{2} + i\sin\frac{\pi}{2}\right) = 5(0 + 1i) = 0 + 5i$$

## Modulus and Argument of a Complex Number

In writing a complex number in trigonometric form, the quantity $r$ is normally chosen positive. Then, since $r^2 = x^2 + y^2$, $r$ represents the distance of the complex number from the origin, and is referred to as the *modulus* (sometimes called the *absolute value*) of the complex number. The absolute value notation is used, thus:

$$|z| = r = \sqrt{x^2 + y^2}$$

The quantity $\theta$ is referred to as the *argument* of the complex number. Unless otherwise specified, $\theta$ is normally chosen so that $0 \le \theta < 2\pi$.

**EXAMPLE 29.3**  Write $z = -6 + 6i$ in trigonometric form and state the modulus and argument for $z$, choosing $0 \le \theta < 2\pi$.

$-6 + 6i$ corresponds to the geometric point $(-6, 6)$. Since $x = -6$ and $y = 6$, $r = |z| = \sqrt{(-6)^2 + 6^2} = \sqrt{72} = 6\sqrt{2}$ and $\tan\theta = \dfrac{6}{-6} = -1$. Since $(-6, 6)$ is in quadrant II, it follows that $\theta = \dfrac{3\pi}{4}$. Thus, in trigonometric form,

$z = 6\sqrt{2}\left(\cos\dfrac{3\pi}{4} + i\sin\dfrac{3\pi}{4}\right)$. The modulus of $z$ is $6\sqrt{2}$ and the argument of $z$ is $\dfrac{3\pi}{4}$. (Note that other, equally valid, arguments for $z$ can be obtained by adding integer multiples of $2\pi$ to the argument $3\pi/4$.)

## Products and Quotients of Complex Numbers

Let $z_1 = r_1(\cos\theta_1 + i\sin\theta_1)$ and $z_2 = r_2(\cos\theta_2 + i\sin\theta_2)$ be complex numbers in trigonometric form. Then (assuming $z_2 \ne 0$)

$$z_1 z_2 = r_1 r_2[\cos(\theta_1 + \theta_2) + i\sin(\theta_1 + \theta_2)] \quad \text{and} \quad \frac{z_1}{z_2} = \frac{r_1}{r_2}[\cos(\theta_1 - \theta_2) + i\sin(\theta_1 - \theta_2)]$$

## DeMoivre's Theorem

DeMoivre's Theorem on powers of complex numbers: Let $z = r(\cos\theta + i\sin\theta)$ be a complex number in trigonometric form. Then for any nonnegative integer $n$,

$$z^n = r^n(\cos n\theta + i\sin n\theta)$$

## Theorem on *n*th Roots of Complex Numbers

If $z = r(\cos\theta + i\sin\theta)$ is any nonzero complex number and if $n$ is any positive integer, then $z$ has exactly $n$ different $n$th roots $w_0, w_1, \ldots, w_{n-1}$. These roots are given by

$$w_k = \sqrt[n]{r}\left(\cos\frac{\theta + 2\pi k}{n} + i\sin\frac{\theta + 2\pi k}{n}\right)$$

for $k = 0, 1, \ldots, n-1$. The roots are symmetrically placed and equally spaced around a circle in the complex plane of radius $\sqrt[n]{r}$ and center the origin.

**EXAMPLE 29.4**　(a) Write $i$ in trigonometric form; (b) find the two square roots of $i$.

(a)　Since $i = 0 + 1i$ corresponds to the ordered pair $(0, 1)$, $r = \sqrt{0^2 + 1^2} = 1$ and $\theta = \pi/2$. Thus $i = 1[\cos(\pi/2) + i\sin(\pi/2)]$.

(b)　Since $n = 2$, $r = 1$, and $\theta = \pi/2$, the two square roots are given by

$$w_k = \sqrt{1}\left(\cos\frac{\pi/2 + 2\pi k}{2} + i\sin\frac{\pi/2 + 2\pi k}{2}\right)$$

for $k = 0, 1$. Thus

$$w_0 = 1\left(\cos\frac{\pi/2}{2} + i\sin\frac{\pi/2}{2}\right) = \cos\frac{\pi}{4} + i\sin\frac{\pi}{4} = \frac{\sqrt{2}}{2} + i\frac{\sqrt{2}}{2}$$

$$w_1 = 1\left(\cos\frac{\pi/2 + 2\pi}{2} + i\sin\frac{\pi/2 + 2\pi}{2}\right) = \cos\frac{5\pi}{4} + i\sin\frac{5\pi}{4} = -\frac{\sqrt{2}}{2} - i\frac{\sqrt{2}}{2}$$

## Polar Form of Complex Numbers

In advanced courses, it is shown that

$$e^{i\theta} = \cos\theta + i\sin\theta$$

Then any complex number can be written as

$$z = r(\cos\theta + i\sin\theta) = re^{i\theta}$$

Here, unless otherwise specified, $\theta$ is normally chosen between $-\pi$ and $\pi$. $e^{i\theta}$ obeys the standard properties for exponents, hence:

For $z = re^{i\theta}$, $z_1 = r_1 e^{i\theta_1}$, $z_2 = r_2 e^{i\theta_2}$, the previous formulas can be written:

$$z_1 z_2 = r_1 r_2 e^{i(\theta_1 + \theta_2)} \qquad \frac{z_1}{z_2} = \frac{r_1}{r_2} e^{i(\theta_1 - \theta_2)}$$

$$z^n = r^n e^{in\theta} \text{ (DeMoivre's theorem)}$$

The $n$ $n$th roots of $z = re^{i\theta}$ are given by $w_k = \sqrt[n]{r}e^{i(\theta + 2\pi k)/n}$ for $k = 0, 1, \ldots, n-1$ ($n$th roots theorem).

## SOLVED PROBLEMS

**29.1.**　Write in rectangular (standard) form:

(a)　$4(\cos 0 + i\sin 0)$; (b)　$3\left(\cos\frac{\pi}{6} + i\sin\frac{\pi}{6}\right)$; (c)　$10\left(\cos\frac{5\pi}{4} + i\sin\frac{5\pi}{4}\right)$;

(d)　$20\left[\cos\left(\tan^{-1}\frac{3}{4}\right) + i\sin\left(\tan^{-1}\frac{3}{4}\right)\right]$

(a)　$4(\cos 0 + i\sin 0) = 4(1 + 0i) = 4$

(b)　$3\left(\cos\frac{\pi}{6} + i\sin\frac{\pi}{6}\right) = 3\left(\frac{\sqrt{3}}{2} + i\left(\frac{1}{2}\right)\right) = \frac{3\sqrt{3} + 3i}{2}$

(c)　$10\left(\cos\frac{5\pi}{4} + i\sin\frac{5\pi}{4}\right) = 10\left(-\frac{\sqrt{2}}{2} - i\frac{\sqrt{2}}{2}\right) = -5\sqrt{2} - 5i\sqrt{2}$

(d) Let $u = \tan^{-1} \frac{3}{4}$. Then $\tan u = \frac{3}{4}$, $-\frac{\pi}{2} < u < \frac{\pi}{2}$. It follows that

$$\cos\left(\tan^{-1}\frac{3}{4}\right) = \cos u = \frac{4}{5} \quad \text{and} \quad \sin\left(\tan^{-1}\frac{3}{4}\right) = \sin u = \frac{3}{5}$$

Hence

$$20\left[\cos\left(\tan^{-1}\frac{3}{4}\right) + i\sin\left(\tan^{-1}\frac{3}{4}\right)\right] = 20\left[\frac{4}{5} + i\frac{3}{5}\right] = 16 + 12i$$

**29.2.** Write in trigonometric form: (a) $-8$; (b) $3i$; (c) $4 + 4i\sqrt{3}$; (d) $-3\sqrt{2} - 3i\sqrt{2}$; (e) $6 - 8i$.

(a) $-8 = -8 + 0i$ corresponds to the geometric point $(-8,0)$. Since $x = -8$ and $y = 0$,

$$r = \sqrt{x^2 + y^2} = \sqrt{(-8)^2 + 0^2} = 8 \quad \text{and} \quad \tan\theta = \frac{0}{-8} = 0.$$

Since $(-8, 0)$ is on the negative $x$-axis, $\theta = \pi$. It follows that $-8 = 8(\cos\pi + i\sin\pi)$.

(b) $3i = 0 + 3i$ corresponds to the geometric point $(0,3)$. Since $x = 0$ and $y = 3$,

$$r = \sqrt{x^2 + y^2} = \sqrt{0^2 + 3^2} = 3 \quad \text{and} \quad \tan\theta = \frac{3}{0} \text{ is undefined.}$$

Since $(0, 3)$ is on the positive $y$-axis, $\theta = \frac{\pi}{2}$. It follows that $3i = 3\left(\cos\frac{\pi}{2} + i\sin\frac{\pi}{2}\right)$.

(c) $4 + 4i\sqrt{3}$ corresponds to the geometric point $(4, 4\sqrt{3})$. Since $x = 4$ and $y = 4\sqrt{3}$,

$$r = \sqrt{x^2 + y^2} = \sqrt{4^2 + (4\sqrt{3})^2} = 8 \quad \text{and} \quad \tan\theta = \frac{4\sqrt{3}}{4} = \sqrt{3}.$$

Since $(4, 4\sqrt{3})$ is in quadrant I, $\theta = \frac{\pi}{3}$. It follows that $4 + 4i\sqrt{3} = 8\left(\cos\frac{\pi}{3} + i\sin\frac{\pi}{3}\right)$.

(d) $-3\sqrt{2} - 3i\sqrt{2}$ corresponds to the geometric point $(-3\sqrt{2}, -3\sqrt{2})$. Since $x = y = -3\sqrt{2}$,

$$r = \sqrt{x^2 + y^2} = \sqrt{(-3\sqrt{2})^2 + (-3\sqrt{2})^2} = 6 \quad \text{and} \quad \tan\theta = \frac{-3\sqrt{2}}{-3\sqrt{2}} = 1.$$

Since $(-3\sqrt{2}, -3\sqrt{2})$ is in quadrant III, $\theta = \frac{5\pi}{4}$. It follows that $-3\sqrt{2} - 3i\sqrt{2} = 6\left(\cos\frac{5\pi}{4} + i\sin\frac{5\pi}{4}\right)$.

(e) $6 - 8i$ corresponds to the geometric point $(6, -8)$. Since $x = 6$ and $y = -8$,

$$r = \sqrt{x^2 + y^2} = \sqrt{6^2 + (-8)^2} = 10 \quad \text{and} \quad \tan\theta = \frac{-8}{6} = -\frac{4}{3}.$$

Since $(6, -8)$ is in quadrant IV, $\theta$ may be chosen as $\tan^{-1}\left(-\frac{4}{3}\right)$. However, since this is a negative angle, the requirement that $0 \le \theta < 2\pi$ yields the alternative argument $\theta = 2\pi + \tan^{-1}\left(-\frac{4}{3}\right)$. With this argument, $6 - 8i = 10(\cos\theta + i\sin\theta)$.

**29.3.** Let $z_1 = r_1(\cos\theta_1 + i\sin\theta_1)$ and $z_2 = r_2(\cos\theta_2 + i\sin\theta_2)$ be complex numbers in trigonometric form. Assuming $z_2 \ne 0$, prove:

(a) $z_1 z_2 = r_1 r_2 [\cos(\theta_1 + \theta_2) + i\sin(\theta_1 + \theta_2)]$; (b) $\frac{z_1}{z_2} = \frac{r_1}{r_2}[\cos(\theta_1 - \theta_2) + i\sin(\theta_1 - \theta_2)]$.

(a) $z_1 z_2 = r_1(\cos\theta_1 + i\sin\theta_1)r_2(\cos\theta_2 + i\sin\theta_2)$

$= r_1 r_2(\cos\theta_1 + i\sin\theta_1)(\cos\theta_2 + i\sin\theta_2)$

$= r_1 r_2(\cos\theta_1\cos\theta_2 + i\sin\theta_2\cos\theta_1 + i\sin\theta_1\cos\theta_2 + i^2\sin\theta_1\sin\theta_2) \quad$ by FOIL

In this expression, use $i^2 = -1$ and combine real and imaginary terms:

$z_1 z_2 = r_1 r_2(\cos\theta_1\cos\theta_2 + i\sin\theta_2\cos\theta_1 + i\sin\theta_1\cos\theta_2 - \sin\theta_1\sin\theta_2)$

$= r_1 r_2[(\cos\theta_1\cos\theta_2 - \sin\theta_1\sin\theta_2) + i(\sin\theta_2\cos\theta_1 + \sin\theta_1\cos\theta_2)]$

The quantities in parentheses are recognized as $\cos(\theta_1 + \theta_2)$ and $\sin(\theta_1 + \theta_2)$, respectively, from the sum formulas for cosine and sine. Hence

$$z_1 z_2 = r_1 r_2 [\cos(\theta_1 + \theta_2) + i\sin(\theta_1 + \theta_2)]$$

(b) $\dfrac{z_1}{z_2} = \dfrac{r_1(\cos\theta_1 + i\sin\theta_1)}{r_2(\cos\theta_2 + i\sin\theta_2)} = \dfrac{r_1}{r_2}\dfrac{\cos\theta_1 + i\sin\theta_1}{\cos\theta_2 + i\sin\theta_2}$

In this expression, multiply numerator and denominator by $\cos\theta_2 - i\sin\theta_2$, the conjugate of the denominator, then use $i^2 = -1$ and combine real and imaginary terms:

$$\frac{z_1}{z_2} = \frac{r_1}{r_2}\frac{(\cos\theta_1 + i\sin\theta_1)(\cos\theta_2 - i\sin\theta_2)}{(\cos\theta_2 + i\sin\theta_2)(\cos\theta_2 - i\sin\theta_2)}$$

$$= \frac{r_1}{r_2}\frac{\cos\theta_1\cos\theta_2 - i\cos\theta_1\sin\theta_2 + i\sin\theta_1\cos\theta_2 - i^2\sin\theta_1\sin\theta_2}{\cos^2\theta_2 - i^2\sin^2\theta_2}$$

$$= \frac{r_1}{r_2}\frac{(\cos\theta_1\cos\theta_2 + \sin\theta_1\sin\theta_2) + i(\sin\theta_1\cos\theta_2 - \cos\theta_1\sin\theta_2)}{\cos^2\theta_2 + \sin^2\theta_2}$$

The quantities in parentheses are recognized as $\cos(\theta_1 - \theta_2)$ and $\sin(\theta_1 - \theta_2)$, respectively, from the difference formulas for cosine and sine, while $\cos^2\theta_2 + \sin^2\theta_2 = 1$ is from the Pythagorean identity. Hence

$$\frac{z_1}{z_2} = \frac{r_1}{r_2}[\cos(\theta_1 - \theta_2) + i\sin(\theta_1 - \theta_2)]$$

**29.4.** Let $z_1 = 40\left(\cos\dfrac{4\pi}{5} + i\sin\dfrac{4\pi}{5}\right)$ and $z_2 = 5\left(\cos\dfrac{3\pi}{5} + i\sin\dfrac{3\pi}{5}\right)$. Find $z_1 z_2$ and $\dfrac{z_1}{z_2}$.

$z_1 z_2 = 40\left(\cos\dfrac{4\pi}{5} + i\sin\dfrac{4\pi}{5}\right)5\left(\cos\dfrac{3\pi}{5} + i\sin\dfrac{3\pi}{5}\right)$      $\dfrac{z_1}{z_2} = \dfrac{40}{5}\left[\cos\left(\dfrac{4\pi}{5} - \dfrac{3\pi}{5}\right) + i\sin\left(\dfrac{4\pi}{5} - \dfrac{3\pi}{5}\right)\right]$

$\qquad = 40(5)\left[\cos\left(\dfrac{4\pi}{5} + \dfrac{3\pi}{5}\right) + i\sin\left(\dfrac{4\pi}{5} + \dfrac{3\pi}{5}\right)\right]$      $\qquad = 8\left(\cos\dfrac{\pi}{5} + i\sin\dfrac{\pi}{5}\right)$

$\qquad = 200\left(\cos\dfrac{7\pi}{5} + i\sin\dfrac{7\pi}{5}\right)$

**29.5.** Let $z_1 = 24i$ and $z_2 = 4\sqrt{3} - 4i$. Convert to trigonometric form and find $z_1 z_2$ and $\dfrac{z_1}{z_2}$ in trigonometric and in rectangular form.

In trigonometric form:

$$z_1 = 0 + 24i = 24\left(\cos\frac{\pi}{2} + i\sin\frac{\pi}{2}\right) \quad \text{and} \quad z_2 = 8\left(\cos\frac{11\pi}{6} + i\sin\frac{11\pi}{6}\right)$$

Hence

$z_1 z_2 = 24\left(\cos\dfrac{\pi}{2} + i\sin\dfrac{\pi}{2}\right)8\left(\cos\dfrac{11\pi}{6} + i\sin\dfrac{11\pi}{6}\right)$    $\dfrac{z_1}{z_2} = \dfrac{24}{8}\left[\cos\left(\dfrac{\pi}{2} - \dfrac{11\pi}{6}\right) + i\sin\left(\dfrac{\pi}{2} - \dfrac{11\pi}{6}\right)\right]$

$\qquad = 24(8)\left[\cos\left(\dfrac{\pi}{2} + \dfrac{11\pi}{6}\right) + i\sin\left(\dfrac{\pi}{2} + \dfrac{11\pi}{6}\right)\right]$    $\qquad = 3\left[\cos\left(-\dfrac{4\pi}{3}\right) + i\sin\left(-\dfrac{4\pi}{3}\right)\right]$

$\qquad = 192\left(\cos\dfrac{7\pi}{3} + i\sin\dfrac{7\pi}{3}\right)$

To satisfy the requirement that $0 \le \theta < 2\pi$, subtract $2\pi$ from the first argument and add $2\pi$ to the second. Thus

$$z_1 z_2 = 192\left[\cos\frac{\pi}{3} + i\sin\frac{\pi}{3}\right] \quad \text{and} \quad \frac{z_1}{z_2} = 3\left[\cos\frac{2\pi}{3} + i\sin\frac{2\pi}{3}\right]$$

In rectangular form:

$$z_1 z_2 = 192\left(\frac{1}{2} + i\frac{\sqrt{3}}{2}\right) = 96 + 96i\sqrt{3} \quad \text{and} \quad \frac{z_1}{z_2} = 3\left(-\frac{1}{2} + i\frac{\sqrt{3}}{2}\right) = -\frac{3}{2} + i\frac{3\sqrt{3}}{2}$$

**29.6.** Prove DeMoivre's theorem for $n = 2$ and $n = 3$.

Choose $z_1 = z_2 = z = r(\cos\theta + i\sin\theta)$. Then

$$z^2 = zz = r(\cos\theta + i\sin\theta)r(\cos\theta + i\sin\theta)$$
$$= r^2[\cos(\theta + \theta) + i\sin(\theta + \theta)]$$
$$= r^2(\cos 2\theta + i\sin 2\theta)$$

$$z^3 = z^2z = r^2(\cos 2\theta + i\sin 2\theta)r(\cos\theta + i\sin\theta)$$
$$= r^2r[\cos(2\theta + \theta) + i\sin(2\theta + \theta)]$$
$$= r^3(\cos 3\theta + i\sin 3\theta)$$

*Note*: Similar proofs can be given easily for $n = 4$, $n = 5$, and so on. These suggest the validity of DeMoivre's theorem for arbitrary integer $n$. A complete proof for arbitrary integral $n$ requires the principle of mathematical induction (Chapter 42).

**29.7.** Apply DeMoivre's theorem to find (a) $\left[2\left(\cos\dfrac{\pi}{9} + i\sin\dfrac{\pi}{9}\right)\right]^5$; (b) $(-1 + i)^6$

(a) $\left[2\left(\cos\dfrac{\pi}{9} + i\sin\dfrac{\pi}{9}\right)\right]^5 = 2^5\left(\cos\dfrac{5\pi}{9} + i\sin\dfrac{5\pi}{9}\right) = 32\left(\cos\dfrac{5\pi}{9} + i\sin\dfrac{5\pi}{9}\right)$

(b) First write $-1 + i$ in trigonometric form as $\sqrt{2}\left(\cos\dfrac{3\pi}{4} + i\sin\dfrac{3\pi}{4}\right)$. Then apply DeMoivre's theorem to obtain

$$(-1 + i)^6 = \left[\sqrt{2}\left(\cos\dfrac{3\pi}{4} + i\sin\dfrac{3\pi}{4}\right)\right]^6 = (\sqrt{2})^6\left(\cos\dfrac{9\pi}{2} + i\sin\dfrac{9\pi}{2}\right) = 8(0 + 1i) = 8i$$

**29.8.** Show that any complex number $w_k = \sqrt[n]{r}\left(\cos\dfrac{\theta + 2\pi k}{n} + i\sin\dfrac{\theta + 2\pi k}{n}\right)$, for nonnegative integral $k$, is an $n$th root of the complex number $z = r(\cos\theta + i\sin\theta)$.

Apply DeMoivre's theorem to $w_k$:

$$w_k^n = \left[\sqrt[n]{r}\left(\cos\dfrac{\theta + 2\pi k}{n} + i\sin\dfrac{\theta + 2\pi k}{n}\right)\right]^n = (\sqrt[n]{r})^n[\cos(\theta + 2\pi k) + i\sin(\theta + 2\pi k)]$$
$$= r(\cos\theta + i\sin\theta)$$

The last equality follows from the periodicity of the sine and cosine functions. Hence $w_k$ is an $n$th root of $z$.

**29.9.** Find the four fourth roots of $5(\cos 3 + i\sin 3)$.

Applying the theorem on $n$th roots with $n = 4$, $r = 5$, and $\theta = 3$, the four fourth roots are given by

$$w_k = \sqrt[4]{5}\left(\cos\dfrac{3 + 2\pi k}{4} + i\sin\dfrac{3 + 2\pi k}{4}\right)$$

for $k = 0, 1, 2, 3$. Thus

$$w_0 = \sqrt[4]{5}\left(\cos\dfrac{3}{4} + i\sin\dfrac{3}{4}\right) \qquad w_1 = \sqrt[4]{5}\left(\cos\dfrac{3 + 2\pi}{4} + i\sin\dfrac{3 + 2\pi}{4}\right)$$

$$w_2 = \sqrt[4]{5}\left(\cos\dfrac{3 + 4\pi}{4} + i\sin\dfrac{3 + 4\pi}{4}\right) \qquad w_3 = \sqrt[4]{5}\left(\cos\dfrac{3 + 6\pi}{4} + i\sin\dfrac{3 + 6\pi}{4}\right)$$

**29.10.** (a) Find the three cube roots of $-27i$; (b) sketch these numbers in a complex plane.

(a) First write $-27i$ in trigonometric form as $27\left(\cos\dfrac{3\pi}{2} + i\sin\dfrac{3\pi}{2}\right)$. Applying the theorem on $n$th roots with $n = 3$, $r = 27$, and $\theta = \dfrac{3\pi}{2}$, the three cube roots are given by

$$w_k = \sqrt[3]{27}\left(\cos\dfrac{3\pi/2 + 2\pi k}{3} + i\sin\dfrac{3\pi/2 + 2\pi k}{3}\right)$$

for $k = 0, 1, 2$. Thus

$$w_0 = \sqrt[3]{27}\left(\cos\dfrac{3\pi/2}{3} + i\sin\dfrac{3\pi/2}{3}\right) = 3\left(\cos\dfrac{\pi}{2} + i\sin\dfrac{\pi}{2}\right) = 3(0 + i1) = 3i$$

$$w_1 = \sqrt[3]{27}\left(\cos\dfrac{3\pi/2 + 2\pi}{3} + i\sin\dfrac{3\pi/2 + 2\pi}{3}\right) = 3\left(\cos\dfrac{7\pi}{6} + i\sin\dfrac{7\pi}{6}\right) = 3\left(-\dfrac{\sqrt{3}}{2} - i\dfrac{1}{2}\right) = -\dfrac{3\sqrt{3}}{2} - \dfrac{3}{2}i$$

$$w_2 = \sqrt[3]{27}\left(\cos\dfrac{3\pi/2 + 4\pi}{3} + i\sin\dfrac{3\pi/2 + 4\pi}{3}\right) = 3\left(\cos\dfrac{11\pi}{6} + i\sin\dfrac{11\pi}{6}\right) = 3\left(\dfrac{\sqrt{3}}{2} - i\dfrac{1}{2}\right) = \dfrac{3\sqrt{3}}{2} - \dfrac{3}{2}i$$

(b) All three cube roots have magnitude 3, and hence lie on the circle of radius 3 with center the origin (see Fig. 29-4).

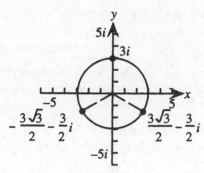

Figure 29-4

Note that since the arguments differ by $2\pi/3$, the three cube roots are symmetrically placed and equally spaced around the circle.

**29.11.** Find all complex solutions of $x^6 + 64 = 0$.

Since $x^6 + 64 = 0$ is equivalent to $x^6 = -64$, the solutions are the six complex sixth roots of $-64$.

Write $-64$ in trigonometric form as $64(\cos\pi + i\sin\pi)$. Applying the theorem on $n$th roots with $n = 6$, $r = 64$, and $\theta = \pi$, the six sixth roots are given by

$$w_k = \sqrt[6]{64}\left(\cos\frac{\pi + 2\pi k}{6} + i\sin\frac{\pi + 2\pi k}{6}\right)$$

for $k = 0, 1, 2, 3, 4, 5$. Thus the six complex solutions of $x^6 + 64 = 0$ are:

$$w_0 = \sqrt[6]{64}\left(\cos\frac{\pi}{6} + i\sin\frac{\pi}{6}\right) = 2\left(\frac{\sqrt{3}}{2} + i\frac{1}{2}\right) = \sqrt{3} + i$$

$$w_1 = \sqrt[6]{64}\left(\cos\frac{3\pi}{6} + i\sin\frac{3\pi}{6}\right) = 2(0 + i1) = 2i$$

$$w_2 = \sqrt[6]{64}\left(\cos\frac{5\pi}{6} + i\sin\frac{5\pi}{6}\right) = 2\left(-\frac{\sqrt{3}}{2} + i\frac{1}{2}\right) = -\sqrt{3} + i$$

$$w_3 = \sqrt[6]{64}\left(\cos\frac{7\pi}{6} + i\sin\frac{7\pi}{6}\right) = 2\left(-\frac{\sqrt{3}}{2} - i\frac{1}{2}\right) = -\sqrt{3} - i$$

$$w_4 = \sqrt[6]{64}\left(\cos\frac{9\pi}{6} + i\sin\frac{9\pi}{6}\right) = 2(0 - i1) = -2i$$

$$w_5 = \sqrt[6]{64}\left(\cos\frac{11\pi}{6} + i\sin\frac{11\pi}{6}\right) = 2\left(\frac{\sqrt{3}}{2} - i\frac{1}{2}\right) = \sqrt{3} - i$$

**29.12.** (a) Write $3e^{i(\pi/3)}$ in rectangular (standard) form; (b) write $6 - 6i$ in polar form.

(a) $3e^{i(\pi/3)} = 3\left(\cos\frac{\pi}{3} + i\sin\frac{\pi}{3}\right) = 3\left(\frac{1}{2} + \frac{\sqrt{3}}{2}i\right) = \frac{3}{2} + \frac{3\sqrt{3}}{2}i$

(b) $6 - 6i$ corresponds to the geometric point $(6, -6)$. Since $x = 6$ and $y = -6$,

$$r = \sqrt{x^2 + y^2} = \sqrt{6^2 + (-6)^2} = 6\sqrt{2} \quad \text{and} \quad \tan\theta = \frac{-6}{6} = -1$$

Since $(6, -6)$ is in quadrant IV, $\theta = -\frac{\pi}{4}$. It follows that $6 - 6i = 6\sqrt{2}e^{-i\pi/4}$.

**29.13.** For $z_1 = 12e^{i(5\pi/6)}$, $z_2 = 3e^{i\pi/3}$, find (a) $z_1z_2$; (b) $\dfrac{z_1}{z_2}$.

(a) $z_1z_2 = (12e^{i(5\pi/6)})(3e^{i\pi/3}) = 36e^{i(5\pi/6+\pi/3)} = 36e^{i(7\pi/6)}$. If $\theta$ is to be chosen between $-\pi$ and $\pi$, then write $36e^{i(7\pi/6)} = 36e^{-i(5\pi/6)}$.

(b) $\dfrac{z_1}{z_2} = \dfrac{12e^{i(5\pi/6)}}{3e^{i\pi/3}} = 4e^{i(5\pi/6-\pi/3)} = 4e^{i(\pi/2)}$

**29.14.** For $z = -1 + i\sqrt{3}$, find $z^3$ in (a) polar and (b) rectangular (standard) form.

(a) First write $-1 + i\sqrt{3}$ in polar form as $2e^{i(2\pi/3)}$. Then apply DeMoivre's theorem to obtain
$$(-1 + i\sqrt{3})^3 = [2e^{i(2\pi/3)}]^3 = 2^3 e^{2\pi i} = 8e^{2\pi i}$$

(b) In standard form $8e^{2\pi i} = 8(\cos 2\pi + i\sin 2\pi) = 8$.

**29.15.** Find the three cube roots of $64e^{i(5\pi/4)}$.

Applying the theorem on *n*th roots with $n = 3$, $r = 64$, and $\theta = 5\pi/4$, the three cube roots are given by $w_k = \sqrt[3]{64}e^{i(5\pi/4+2\pi k)/3}$, for $k = 0, 1, 2$. Thus

$$w_0 = \sqrt[3]{64}e^{i(5\pi/4)/3} = 4e^{i(5\pi/12)}$$

$$w_1 = \sqrt[3]{64}e^{i(5\pi/4+2\pi)/3} = 4e^{i(13\pi/4)/3} = 4e^{i(13\pi/12)}$$

$$w_2 = \sqrt[3]{64}e^{i(5\pi/4+4\pi)/3} = 4e^{i(21\pi/4)/3} = 4e^{i(7\pi/4)}$$

**29.16.** Show that $e^{i\pi} + 1 = 0$.

$$e^{i\pi} + 1 = \cos\pi + i\sin\pi + 1 = -1 + 0i + 1 = 0$$

## SUPPLEMENTARY PROBLEMS

**29.17.** Let $z_1 = 8\left(\cos\dfrac{4\pi}{9} + i\sin\dfrac{4\pi}{9}\right)$ and $z_2 = \cos\dfrac{2\pi}{9} + i\sin\dfrac{2\pi}{9}$. Find $z_1z_2$ and $\dfrac{z_1}{z_2}$.

*Ans.* $z_1z_2 = -4 + 4i\sqrt{3}$, $\dfrac{z_1}{z_2} = 8\left(\cos\dfrac{2\pi}{9} + i\sin\dfrac{2\pi}{9}\right)$

**29.18.** Write $-12$, $-8i$, $2 - 2i$, and $-\sqrt{3} + i$ in trigonometric form.

*Ans.* $12(\cos\pi + i\sin\pi)$, $8\left(\cos\dfrac{3\pi}{2} + i\sin\dfrac{3\pi}{2}\right)$, $2\sqrt{2}\left(\cos\dfrac{7\pi}{4} + i\sin\dfrac{7\pi}{4}\right)$, $2\left(\cos\dfrac{5\pi}{6} + i\sin\dfrac{5\pi}{6}\right)$

**29.19.** Use the results of the previous problem to find (a) $(-8i)(2 - 2i)$; (b) $\dfrac{-8i}{-\sqrt{3} + i}$; (c) $(2-2i)^3$.

*Ans.* (a) $-16 - 16i$; (b) $-2 + 2i\sqrt{3}$; (c) $-16 - 16i$

**29.20.** Prove DeMoivre's theorem for the cases $n = 0$, $n = 1$, and $n = 4$.

**29.21.** Show that every complex number $z = r(\cos\theta + i\sin\theta)$ has exactly $n$ different complex $n$th roots, for $n$ an integer greater than 1. [*Hint*: Set $w = s(\cos\alpha + i\sin\alpha)$ and consider the solutions of the equation $w^n = z$.]

**29.22.** Find the two square roots of $-1 + i\sqrt{3}$.

*Ans.* $\dfrac{\sqrt{2}}{2} + i\dfrac{\sqrt{6}}{2}$, $-\dfrac{\sqrt{2}}{2} - i\dfrac{\sqrt{6}}{2}$

**29.23.** (a) Find the three complex cube roots of 1. (b) Find the four complex fourth roots of $-1$.

> *Ans.*   (a) $1, -\dfrac{1}{2} + i\dfrac{\sqrt{3}}{2}, -\dfrac{1}{2} - i\dfrac{\sqrt{3}}{2}$; (b) $\dfrac{1+i}{\sqrt{2}}, \dfrac{-1+i}{\sqrt{2}}, \dfrac{-1-i}{\sqrt{2}}, \dfrac{1-i}{\sqrt{2}}$

**29.24.** (a) Write $12e^{i(3\pi/4)}$ in rectangular (standard) form; (b) Write $5i$ in polar form.

> *Ans.*   (a) $-6\sqrt{2} + 6i\sqrt{2}$; (b) $5e^{i\pi/2}$

**29.25.** For $z_1 = 20e^{i(4\pi/3)}$, $z_2 = 2e^{i\pi/2}$, find (a) $z_1 z_2$; (b) $\dfrac{z_1}{z_2}$.

> *Ans.*   (a) $40e^{i(11\pi/6)}$ or $40e^{-i\pi/6}$; (b) $10e^{i(5\pi/6)}$

**29.26.** For $z = 1 - i$, find $z^5$ in (a) polar and (b) rectangular (standard) form.

> *Ans.*   (a) $4\sqrt{2}e^{i(3\pi/4)}$; (b) $-4 + 4i$

**29.27.** Find the four fourth roots of $81e^{i(2\pi/3)}$.

> *Ans.*   $3e^{i\pi/6} = \dfrac{3\sqrt{3} + 3i}{2}, \; 3e^{i(2\pi/3)} = \dfrac{-3 + 3i\sqrt{3}}{2}, \; 3e^{i(7\pi/6)} = \dfrac{-3\sqrt{3} - 3i}{2}, \; 3e^{i(5\pi/3)} = \dfrac{3 - 3i\sqrt{3}}{2}$

# CHAPTER 30

# Systems of Linear Equations

## Systems of Equations

A system of equations consists of two or more equations, considered as simultaneous specifications on more than one variable. A *solution* to a system of equations is an ordered assignment of values of the variables that, when substituted, would make each of the equations into true statements. The process of finding the solutions of a system is called *solving* the system. The set of all solutions is called the *solution set* of the system. Systems with the same solution set are called *equivalent* systems.

**EXAMPLE 30.1**  Verify that $(x,y) = (-4,2)$ is a solution to the system

$$y^2 + x = 0 \qquad (1)$$
$$2x + 3y = -2 \qquad (2)$$

If $x = -4$ and $y = 2$, then equation (1) becomes $2^2 + (-4) = 0$ and equation (2) becomes $2(-4) + 3 \cdot 2 = -2$. Since these are both true statements, $(x, y) = (-4, 2)$ is a solution to the system.

## Systems of Linear Equations

A linear equation in several variables $x_1 x_2, \ldots, x_n$ is one that can be written in the form $a_1 x_1 + a_2 x_2 + \ldots + a_n x_n = b$, where the $a_i$ are constants. This is referred to as *standard form*. If all equations of a system are linear, the system is called a *linear system*; if all equations are in standard form, the system is also considered to be in standard form.

**EXAMPLE 30.2**  Rewrite the system

$$2x + 4y = 5x - 6y \qquad (1)$$
$$y + 5 = 3x + 5y \qquad (2)$$

in standard form.

An equation in standard form must have all variable terms on the left side and any constant terms on the right side. Here equation (1) violates the first of these conditions and equation (2) violates both. Hence, add $-5x + 6y$ to both sides of equation (1) to obtain $-3x + 10y = 0$, and add $-3x - 5y$ and $-5$ to both sides of equation (2) to obtain $-3x - 4y = -5$. The resulting equations are in standard form:

$$-3x + 10y = 0 \qquad (3)$$
$$-3x - 4y = -5 \qquad (4)$$

## Equivalent Systems

Equivalent systems of linear equations can be produced by the following *operations on equations*. (It is understood that "adding two equations" means adding left side to left side and right side to right side to produce a new equation and "multiple of an equation" means the result of multiplying left side and right side by the same constant.)

1. Interchanging two equations.
2. Replacing an equation by a nonzero multiple of itself.
3. Replacing an equation by the result of adding the equation to a multiple of another equation.

**EXAMPLE 30.3** For the system of the previous example, find an equivalent system in which one equation does not contain the variable $x$.

If both sides of equation (4) are multiplied by $-1$, the coefficient of $x$ will be the opposite of the coefficient of $x$ in equation (3). Hence, replacing equation (3) by itself added to $-1$ times equation (4) will achieve the required result:

$$14y = 5 \qquad (5)$$
$$-3x - 4y = -5 \qquad (4)$$

## Classification of Linear Systems

It is shown in advanced courses that systems of linear equations fall into one of three categories:

1. CONSISTENT AND INDEPENDENT. Such systems have exactly one solution.
2. INCONSISTENT. Such systems have no solutions.
3. DEPENDENT. Such systems have an infinite number of solutions.

## Solutions of Linear Systems in Two Variables

Solutions of linear systems in two variables are found by three methods:

1. GRAPHICAL METHOD. Graph each equation (each graph is a straight line). If the lines intersect in a single point, the coordinates of this point may be read from the graph. After checking by substitution in each equation, these coordinates are the solution of the system. If the lines coincide, the system is dependent, and there are an infinite number of solutions, with each solution to one equation being a solution of the others. If neither of these situations occurs, the system is inconsistent.
2. SUBSTITUTION METHOD. Solve one equation for one variable in terms of the other. Substitute this expression into the other equations to determine the value of the first variable (if possible). Then substitute this value to determine the value of the other variable.
3. ELIMINATION METHOD. Apply the operations on equations leading to equivalent systems to eliminate one variable from one equation, solve the resulting equation for this variable, and substitute this value to determine the value of the other variable.

In methods 2 and 3, the occurrence of an equation of the form $a = b$, where $a$ and $b$ are unequal constants, indicates an inconsistent system. If this does not occur, but all equations except one reduce to $0 = 0$, the system is dependent, and there are an infinite number of solutions, with each solution of one equation being a solution of the others.

## Solutions of Linear Systems in More than Two Variables

Solutions of linear systems in more than two variables are found by two methods:

1. SUBSTITUTION METHOD. Solve one equation for one variable in terms of the others. Substitute this expression into the other equations to obtain a system with one fewer variable. If this process can be continued until an equation in one variable is obtained, solve the resulting equation for this variable, and substitute this value to determine the value of the other variables.
2. ELIMINATION METHOD. Apply the operations on equations leading to equivalent systems to eliminate one variable from all equations except one. This leads to a system with one fewer variable. If this process can be continued until an equation in one variable is obtained, solve the resulting equation for this variable, and substitute this value to determine the value of the other variables.

Again, the occurrence of an equation of the form $a = b$, where $a$ and $b$ are unequal constants, indicates an inconsistent system. If this does not occur, but one or more equations reduce to $0 = 0$, leaving fewer nontrivial equations than there are variables, the system is dependent, and there are an infinite number of solutions, with each solution of one equation being a solution of the others.

**SOLVED PROBLEMS**

**30.1.** Solve the system $\begin{array}{l}2x + 3y = 6 \quad (1)\\ -3x - y = 5 \quad (2)\end{array}$ (a) graphically; (b) by substitution; (c) by elimination.

(a) Graph the two equations in the same Cartesian coordinate system (Fig. 30-1); the graphs are straight lines.

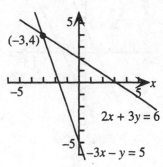

Figure 30-1

The two lines appear to intersect at $(-3,4)$. It is necessary to check this result: substituting $x = -3$ and $y = 4$ into equations (1) and (2) yields

$$2(-3) + 3 \cdot 4 = 6 \qquad\qquad -3(-3) - 4 = 5$$
$$\text{and}$$
$$6 = 6 \qquad\qquad 5 = 5$$

respectively. Thus $(-3,4)$ is the only solution of the system.

(b) It is correct to begin by solving either equation for either variable in terms of the other. The simplest choice seems to be to solve equation (2) for $y$ in terms of $x$ to obtain

$$y = -3x - 5$$

Substitute the expression $-3x - 5$ for $y$ into equation (1) to obtain

$$2x + 3(-3x - 5) = 6$$
$$-7x - 15 = 6$$
$$-7x = 21$$
$$x = -3$$

Substitute $-3$ for $x$ into equation (2) to obtain

$$-3(-3) - y = 5$$
$$9 - y = 5$$
$$y = 4$$

Again, $(-3,4)$ is the only solution of the system.

(c) If equation (2) is multiplied by 3, the coefficient of $y$ will "match" the coefficient of $y$ in equation (1); that is, it will be equal in absolute value and opposite in sign. Equation (2) then becomes

$$-9x - 3y = 15 \qquad (3)$$

If equation (1) is replaced by itself plus this multiple of equation (2), the following equivalent system results:

$$-7x = 21 \qquad (4)$$
$$-3x - y = 5 \qquad (2)$$

From equation (4), $x = -3$. Substituting into equation (2) yields $y = 4$, as before.

**30.2.** Solve the system $\begin{array}{ll} y = 2x + 2 & (1) \\ 4x - 2y = 8 & (2) \end{array}$ (a) graphically; (b) nongraphically.

(a) Graph the two equations in the same Cartesian coordinate system (Fig. 30-2); the graphs are straight lines.

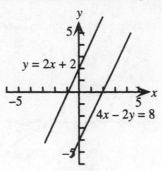

Figure 30-2

The lines appear to be parallel. In fact, since both have slope 2, but different $y$ intercepts, the lines are parallel; there is no point of intersection, and the system has no solution (inconsistent system).

(b) Solve by substitution: Substitute the expression $2x + 2$ for $y$ from equation (1) into equation (2).

$$4x - 2(2x + 2) = 8$$
$$4x - 4x - 4 = 8$$
$$-4 = 8$$

Thus there is no solution, and the system is inconsistent.

**30.3.** Solve the system $\begin{array}{ll} 4x + 2y = 6 & (1) \\ 6x + 3y = 9 & (2) \end{array}$ (a) graphically; (b) nongraphically.

(a) Graph the two equations in the same Cartesian coordinate system (Fig. 30-3); the graphs are straight lines.

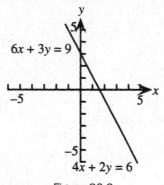

Figure 30-3

The lines appear to coincide. In fact, since both have slope $-2$ and $y$-intercept 3, they do coincide. The system is dependent; every solution of one equation is a solution of the other equation. All solutions can be summarized as follows:

(b) Let $y = c$, where $c$ is any real number. Then substituting $c$ for $y$ into one equation, say (1), and solving for $x$ yields:

$$4x + 2c = 6$$
$$4x = 6 - 2c$$
$$x = \frac{3 - c}{2}$$

Hence all solutions of the system can be written as $\left( \dfrac{3 - c}{2}, c \right)$, where $c$ is any real number.

**30.4.** Show in tabular form the algebraic and geometric interpretations of the types of systems of linear equations in two variables.

The systems are characterized as consistent and independent, inconsistent, or dependent. Algebraic interpretation means the number of solutions; geometric interpretation means the behavior of the graphs.

| Type of System | Number of Solutions | Behavior of Graphs |
|---|---|---|
| Consistent and independent | One | Lines intersect in one point |
| Inconsistent | None | Two lines parallel, more than two lines fail to intersect in one point |
| Dependent | Infinite | Lines coincide |

**30.5.** Solve the system

$$\begin{aligned} x - 3y + 2z &= 14 &&(1) \\ 2x + 5y - z &= -9 &&(2) \\ -3x - y + 2z &= 2 &&(3) \end{aligned}$$

(a) by substitution; (b) by elimination.

(a) Solve equation (1) for $x$ to obtain

$$x = 3y - 2z + 14 \qquad (4)$$

Substitute the expression $3y - 2z + 14$ for $x$ from equation (4) into equations (2) and (3).

$$2(3y - 2z + 14) + 5y - z = -9$$

$$-3(3y - 2z + 14) - y + 2z = 2$$

Simplifying yields:

$$11y - 5z = -37 \qquad (5)$$

$$-10y + 8z = 44 \qquad (6)$$

Solve equation (5) for $y$ to obtain

$$y = \frac{5z - 37}{11} \qquad (7)$$

Substitute the expression on the right for $y$ into equation (6).

$$-10\left(\frac{5z - 37}{11}\right) + 8z = 44$$

$$-50z + 370 + 88z = 484$$

$$38z = 114$$

$$z = 3$$

Substituting this value for $z$ into equation (7) yields $y = -2$. Substituting $y = -2$ and $z = 3$ into equation (4) yields $x = 2$. There is exactly one solution, written as an *ordered triple* $(2, -2, 3)$.

(b) Replacing equation (2) by itself plus $-2$ times equation (1) will eliminate $x$ from equation (2).

Thus:

$$\begin{aligned} 2x + 5y - z &= -9 &&(2) \\ \underline{-2x + 6y - 4z} &= \underline{-28} &&(-2) \cdot \text{Eq. (1)} \\ 11y - 5z &= -37 &&(5) \end{aligned}$$

Similarly, replacing equation (3) by itself plus 3 times equation (1) will eliminate $x$ from equation (3):

$$
\begin{array}{ll}
-3x - y + 2z = 2 & (3) \\
\underline{3x - 9y + 6z = 42} & 3 \cdot \text{Eq. (1)} \\
-10y + 8z = 44 & (6)
\end{array}
$$

Solving the system (5), (6) by elimination yields the same solution as above: $(2, -2, 3)$.

**30.6.** Solve the system
$$
\begin{array}{ll}
x - 4y - 5z = 8 & (1) \\
4x \quad\quad - 2z = 10 & (2). \\
5x - 4y - 7z = 3 & (3)
\end{array}
$$

Replace equation (2) by itself $-4$ times equation (1):

$$
\begin{array}{ll}
4x \quad\quad - 2z = 10 & (2) \\
\underline{-4x + 16y + 20z = -32} & (-4) \cdot \text{Eq. (1)} \\
16y + 18z = -22 & (4)
\end{array}
$$

Replace equation (3) by itself plus $-5$ times equation (1).

$$
\begin{array}{ll}
5x \quad - 4y \quad - 7z = 3 & (3) \\
\underline{-5x + 20y + 25z = -40} & (-5) \cdot \text{Eq. (1)} \\
16y + 18z = -37 & (5)
\end{array}
$$

The system (1), (4), (5) is clearly inconsistent, since adding $-1$ times equation (4) to equation (5) yields $0 = -15$. Thus there is no solution.

**30.7.** Solve the system
$$
\begin{array}{ll}
x + y + z = 1 & (1) \\
2x - 2y - 10z = -6 & (2). \\
-x + 3y + 11z = 7 & (3)
\end{array}
$$

Replace equation (2) by itself plus $-2$ times equation (1):

$$
\begin{array}{ll}
2x - 2y - 10z = -6 & (2) \\
\underline{-2x - 2y - 2z = -2} & (-2) \cdot \text{Eq. (1)} \\
-4y - 12z = -8 & (4)
\end{array}
$$

Replace equation (3) by itself plus equation (1):

$$
\begin{array}{ll}
-x + 3y + 11z = 7 & (3) \\
\underline{x + y + z = 1} & (1) \\
4y + 12z = 8 & (5)
\end{array}
$$

The system (1), (4), (5)

$$
\begin{array}{ll}
x + y + z = 1 & (1) \\
-4y - 12z = -8 & (4) \\
4y + 12z = 8 & (5)
\end{array}
$$

is clearly dependent, since replacing equation (5) by itself plus equation (4) yields

$$
\begin{array}{ll}
x + y + z = 1 & (1) \\
-4y - 12z = -8 & (4) \\
0 = 0 & (6)
\end{array}
$$

Thus there is an infinite number of solutions. To express them all, let $z = c$, $c$ any real number. Then solving $-4y - 12c = -8$ for $y$ yields $y = 2 - 3c$. Substituting $y = 2 - 3c$ and $z = c$ into equation (1) yields

$$x + 2 - 3c + c = 1$$

$$x = 2c - 1$$

Thus all solutions can be written as ordered triples $(2c - 1, 2 - 3c, c)$, $c$ any real number.

**30.8.** $8000 is to be invested, part at 6% interest, and part at 11% interest. How much should be invested at each rate if a total return of 9% is desired?

Use the formula $I = Prt$ with $t$ understood to be one year. Let $x$ = amount invested at 6% and $y$ = amount invested at 11%; a tabular arrangement is helpful:

| | **P: Amount Invested** | **r: Rate of Interest** | **I: Interest Earned** |
|---|---|---|---|
| First account | $x$ | 0.06 | $0.06x$ |
| Second account | $y$ | 0.11 | $0.11y$ |
| Total investment | 8000 | 0.09 | 0.09(8000) |

Since the amounts invested add up to the total investment,

$$x + y = 8000 \qquad (1)$$

Since the interest earned adds up to the total interest,

$$0.06x + 0.11y = 0.09(8000) \qquad (2)$$

The system (1), (2) can be solved by elimination. Replace equation (2) by itself plus $-0.06$ times equation (1):

$$
\begin{aligned}
0.06x + 0.11y &= 0.09(8000) & (2)\\
\underline{-0.06x - 0.06y} &= \underline{-0.06(8000)} & (-0.06) \cdot \text{Eq. (1)}\\
0.05y &= 0.03(8000) & (3)
\end{aligned}
$$

Hence $y = 4800$. Substituting into equation (1) yields $x = 3200$, hence $3200 should be invested at 6% and $4800 at 11%.

**30.9.** Find $a$, $b$, and $c$ so that the graph of the circle with equation $x^2 + y^2 + ax + by + c = 0$ passes through the points (1,5), (4,4), and (3,1).

If a point lies on the graph of an equation, the coordinates of the point satisfy the equation. Hence, substitute, in turn, $(x, y) = (1,5)$, $(x, y) = (4,4)$, and $(x, y) = (3,1)$ to obtain

$$
\begin{aligned}
1 + 25 + a1 + b5 + c &= 0 & & & a + 5b + c &= -26 & (1)\\
16 + 16 + a4 + b4 + c &= 0 & &\text{or, simplifying} & 4a + 4b + c &= -32 & (2)\\
9 + 1 + a3 + b1 + c &= 0 & & & 3a + b + c &= -10 & (3)
\end{aligned}
$$

To solve the system (1), (2), (3), eliminate $a$ from equations (2) and (3) as follows:

$$
\begin{aligned}
a + 5b + c &= -26 & (1)\\
-16b - 3c &= 72 & (4) = \text{Eq. (2)} + (-4) \cdot \text{Eq. (1)}\\
-14b - 2c &= 68 & (5) = \text{Eq. (3)} + (-3) \cdot \text{Eq. (1)}
\end{aligned}
$$

Now eliminate $b$ from equation (5) by replacing it with itself plus $-7/8$ times equation (4).

$$a + 5b + c = -26 \qquad (1)$$

$$-16b - 3c = 72 \qquad (4)$$

$$\frac{5}{8}c = 5 \qquad (6)$$

Finally, solve equation (6) to obtain $c = 8$ and substitute in turn into equations (4) and (1) to obtain $b = -6$ and $a = -4$.

The equation of the circle is $x^2 + y^2 - 4x - 6y + 8 = 0$.

## SUPPLEMENTARY PROBLEMS

**30.10.** Solve the systems (a) $\begin{array}{l} 2x - 3y = 4 \\ 3x + 2y = 19 \end{array}$ (b) $\begin{array}{l} 6x - 4y = 8 \\ 9x - 6y = 12 \end{array}$ (c) $\begin{array}{l} 2y = 3x + 4 \\ 9x - 6y = 4 \end{array}$

*Ans.* (a) $(5, 2)$; (b) $\left(\dfrac{2c + 4}{3}, c\right)$, $c$ any real number; (c) no solution

**30.11.** Solve the systems (a) $\begin{array}{l} 3x - 2y = 0 \\ x + 3y = 0 \\ 2x - y = 0 \end{array}$ (b) $\begin{array}{l} x - 3y = 0 \\ 2x + 3y = 2 \\ -x + y = 1 \end{array}$ (c) $\begin{array}{l} x + 2y = 2 \\ 2x - y = 3 \\ 3x + y = 5 \end{array}$

*Ans.* (a) $(0, 0)$; (b) no solution; (c) $\left(\dfrac{8}{5}, \dfrac{1}{5}\right)$

**30.12.** Solve the systems:

(a) $\begin{array}{l} x + y + z = 5 \\ x - 4y - 3z = 11 \\ -2x + 2y + 5z = -30 \end{array}$ (b) $\begin{array}{l} x + y - 2z = 4 \\ 2x - 5y + z = 7 \\ x + 8y - 7z = 2 \end{array}$ (c) $\begin{array}{l} -x + 2y + 2z = -13 \\ 5x + y - 8z = 0 \\ 3x - y = 12 \end{array}$

*Ans.* (a) $(7, 2, -4)$; (b) no solution; (c) $\left(2, -6, \dfrac{1}{2}\right)$

**30.13.** Solve the systems (a) $\begin{array}{l} 2x - y - z = 0 \\ x - y + z = 0 \\ 3x + 2y + z = 0 \end{array}$ (b) $\begin{array}{l} x + y - z = 5 \\ 3x - y + z = 3 \\ y - z = 3 \end{array}$ (c) $\begin{array}{l} 3x - 3y - 6z = -15 \\ -2x + 2y + 4z = 10 \end{array}$

*Ans.* (a) $(0, 0, 0)$; (b) $(2, 3 + c, c)$, $c$ any real number; (c) $(c + 2d - 5, c, d)$, $c$ and $d$ any real numbers

**30.14.** $16,500 was invested in three accounts, yielding an annual return of 5%, 8%, and 10%, respectively. The amount invested at 5% was equal to the amount invested at 8% plus twice the amount invested at 10%. How much was invested at each rate if the total return on the investment was $1085?

*Ans.* $9500 at 5%, $4500 at 8%, $2500 at 10%

**30.15.** Find $a$, $b$, and $c$ so that the equation of the parabola $y = ax^2 + bx + c$ passes through $(1,4)$, $(-1,6)$, and $(2,12)$.

*Ans.* $a = 3, b = -1, c = 2$

# CHAPTER 31

# Gaussian and Gauss-Jordan Elimination

## Matrix Notation

Elimination methods for solving systems of equations are carried out more efficiently by means of matrices. A *matrix* is a rectangular array of numbers, arranged in rows and columns and enclosed in brackets, thus:

$$\begin{bmatrix} a_{11} & a_{12} & a_{13} & a_{14} \\ a_{21} & a_{22} & a_{23} & a_{24} \\ a_{31} & a_{32} & a_{33} & a_{34} \end{bmatrix}$$

The numbers are called *elements* of the matrix. The above matrix would be said to have three rows (first row: $a_{11}$ $a_{12}$ $a_{13}$ $a_{14}$, and so on) and four columns, and would be called a matrix of *order* $3 \times 4$. The elements are referred to by two subscripts; thus, the element in row 2, column 3 is element $a_{23}$. A matrix may have any number of rows and any number of columns; a general matrix is said to have *order* $m \times n$, thus, $m$ rows and $n$ columns.

## Row-Equivalent Matrices

Two matrices are said to be row-equivalent if one can be transformed into the other by successive applications of the following *row operations* on matrices:

1. Interchange two rows. (Symbol: $R_i \leftrightarrow R_j$)
2. Replace a row by a nonzero multiple of itself. (Symbol: $kR_i \rightarrow R_i$)
3. Replace a row by itself plus a multiple of another row. (Symbol: $kR_i + R_j \rightarrow R_j$)

Note the exact correspondence to the operations that result in equivalent systems of equations (Chapter 30).

**EXAMPLE 31.1** Given the matrix $\begin{bmatrix} 5 & -2 \\ 2 & 6 \end{bmatrix}$, show the result of applying, in turn, (a) $R_1 \leftrightarrow R_2$;

(b) $\frac{1}{2}R_1 \rightarrow R_1$, (c) $-5R_1 + R_2 \rightarrow R_2$

(a) $R_1 \leftrightarrow R_2$ interchanges the two rows to yield $\begin{bmatrix} 2 & 6 \\ 5 & -2 \end{bmatrix}$.

(b) $\frac{1}{2}R_1 \rightarrow R_1$ replaces each element in the new first row 2 6 by one-half its value to yield $\begin{bmatrix} 1 & 3 \\ 5 & -2 \end{bmatrix}$.

(c) $-5R_1 + R_2 \rightarrow R_2$ replaces each element in the new second row by itself plus $-5$ times the corresponding element in the first row $-5$ $-15$ to yield $\begin{bmatrix} 1 & 3 \\ 0 & -17 \end{bmatrix}$.

## Matrices and Systems of Linear Equations

To every linear system of $m$ equations in $n$ variables in standard form there corresponds a matrix of order $m \times n + 1$ called the augmented matrix of the system. Thus to the system:

$$
\begin{aligned}
3x + 5y - 2z &= 4 \\
-2x - 3y \phantom{+ 5y} &= 6 \\
2x + 4y + z &= -3
\end{aligned}
\qquad \text{corresponds the augmented matrix} \qquad
\left[
\begin{array}{ccc|c}
3 & 5 & -2 & 4 \\
-2 & -3 & 0 & 6 \\
2 & 4 & 1 & -3
\end{array}
\right]
$$

The vertical bar has no mathematical significance and serves only to separate the coefficients of the variables from the constant terms.

## Row-Echelon Form of a Matrix

A matrix is in row-echelon form if it satisfies the following conditions:

1. The first nonzero number in each row is a 1.
2. The column containing the first nonzero number in each row is to the left of the column containing the first nonzero number in rows below it.
3. Any row containing only zeros appears below any row having any nonzero numbers.

**EXAMPLE 31.2**   Perform row operations to find a matrix in row-echelon form that is row-equivalent to the matrix $\left[\begin{array}{cc|c} 1 & -1 & 3 \\ 3 & 2 & -1 \end{array}\right]$.

$$
\left[\begin{array}{cc|c} 1 & -1 & 3 \\ 3 & 2 & -1 \end{array}\right]
\; -3R_1 + R_2 \to R_2 \;
\left[\begin{array}{cc|c} 1 & -1 & 3 \\ 0 & 5 & -10 \end{array}\right]
\; \tfrac{1}{5}R_2 \to R_2 \;
\left[\begin{array}{cc|c} 1 & -1 & 3 \\ 0 & 1 & -2 \end{array}\right]
$$

## Gaussian Elimination

Gaussian elimination (with back substitution) is the following process for solving systems of linear equations:

1. Write the system in standard form.
2. Write the augmented matrix of the system.
3. Apply row operations to this augmented matrix to obtain a row-equivalent matrix in row-echelon form.
4. Write the system of equations to which this matrix corresponds.
5. Find the solution of this system; it can be solved readily by substituting values from each equation into the one above it, starting with the last nonzero equation.

**EXAMPLE 31.3**   Solve the system $\begin{aligned} x - y &= 3 \\ 3x + 2y &= -1 \end{aligned}$ by Gaussian elimination.

The system is in standard form. The augmented matrix of the system is $\left[\begin{array}{cc|c} 1 & -1 & 3 \\ 3 & 2 & -1 \end{array}\right]$, considered in the previous example.

Reducing this to row-echelon form yields $\left[\begin{array}{cc|c} 1 & -1 & 3 \\ 0 & 1 & -2 \end{array}\right]$. This matrix corresponds to the system $\begin{aligned} x - y &= 3 \\ y &= -2 \end{aligned}$.

Thus $y = -2$. Substituting this into the first equation yields $x - (-2) = 3$ or $x = 1$. Thus the solution of the system is $(1, -2)$.

## Reduced Row-Echelon Form

A matrix is in *reduced* row-echelon form (often called just *reduced form*) if it satisfies the conditions for row-echelon form and, in addition, the entries *above* the first 1 in each row are all 0.

**EXAMPLE 31.4**   Find a matrix in reduced form that is row-equivalent to the matrix $\left[\begin{array}{cc|c} 1 & -1 & 3 \\ 0 & 1 & -2 \end{array}\right]$ from Example 31.2.

$$
\left[\begin{array}{cc|c} 1 & -1 & 3 \\ 0 & 1 & -2 \end{array}\right]
\; R_2 + R_1 \to R_1 \;
\left[\begin{array}{cc|c} 1 & 0 & 1 \\ 0 & 1 & -2 \end{array}\right]
$$

Note that the solution of the system can be read off immediately after writing the system that corresponds to this matrix ($x = 1$, $y = -2$).

## Gauss-Jordan Elimination

Gauss-Jordan elimination is the following process for solving systems of linear equations:

1. Write the system in standard form.
2. Write the augmented matrix of the system.
3. Apply row operations to this augmented matrix to obtain a row-equivalent matrix in reduced row-echelon form.
4. Write the system of equations to which this matrix corresponds.
5. Find the solution of this system. If there is a unique solution, it can be read off immediately. If there are infinite solutions, the system will be such that after assigning arbitrary real values to undetermined variables, the other variables are immediately expressed in terms of these.

The process of finding a matrix in row-echelon form or reduced row-echelon form that is row-equivalent to a given matrix thus plays a key role in solving systems of linear equations. This process is usually abbreviated as "Transform to row-echelon (or reduced row-echelon) form."

## SOLVED PROBLEMS

**31.1.** Show the result of applying (a) $R_1 \leftrightarrow R_3$; (b) $-\frac{1}{3}R_2 \to R_2$; (c) $2R_2 + R_1 \to R_1$ to the matrix

$$\begin{bmatrix} 5 & 3 & -2 & | & 3 \\ -3 & 6 & 12 & | & -3 \\ 1 & 0 & -4 & | & 5 \end{bmatrix}.$$

(a) $R_1 \leftrightarrow R_3$ interchanges rows 1 and 3, yielding $\begin{bmatrix} 1 & 0 & -4 & | & 5 \\ -3 & 6 & 12 & | & -3 \\ 5 & 3 & -2 & | & 3 \end{bmatrix}$.

(b) $-\frac{1}{3}R_2 \to R_2$ replaces row 2 by $-\frac{1}{3}$ times itself, yielding $\begin{bmatrix} 5 & 3 & -2 & | & 3 \\ 1 & -2 & -4 & | & 1 \\ 1 & 0 & -4 & | & 5 \end{bmatrix}$.

(c) $2R_2 + R_1 \to R_1$ adds row $-6 \ 12 \ 24 \ | -6$ to the existing row 1, yielding $\begin{bmatrix} -1 & 15 & 22 & | & -3 \\ -3 & 6 & 12 & | & -3 \\ 1 & 0 & -4 & | & 5 \end{bmatrix}$.

**31.2.** Transform the matrix $\begin{bmatrix} 1 & 2 & -2 & | & 3 \\ 2 & 5 & 0 & | & -7 \\ 3 & 7 & -2 & | & -4 \end{bmatrix}$ to row-echelon form.

The first element in row 1 is a 1. Use this to produce zeros in the first position in the lower rows:

$$\begin{bmatrix} 1 & 2 & -2 & | & 3 \\ 2 & 5 & 0 & | & -7 \\ 3 & 7 & -2 & | & -4 \end{bmatrix} \begin{matrix} R_2 + (-2)R_1 \to R_2 \\ R_3 + (-3)R_1 \to R_3 \end{matrix} \begin{bmatrix} 1 & 2 & -2 & | & 3 \\ 0 & 1 & 4 & | & -13 \\ 0 & 1 & 4 & | & -13 \end{bmatrix}$$

The first nonzero element in row 2 is now a 1. Use this to produce a zero in the corresponding position in the last row.

$$\begin{bmatrix} 1 & 2 & -2 & | & 3 \\ 0 & 1 & 4 & | & -13 \\ 0 & 1 & 4 & | & -13 \end{bmatrix} R_3 + (-1)R_2 \to R_3 \begin{bmatrix} 1 & 2 & -2 & | & 3 \\ 0 & 1 & 4 & | & -13 \\ 0 & 0 & 0 & | & 0 \end{bmatrix}$$

This matrix is in row-echelon form, and is row-equivalent to the original matrix.

**31.3.** Generalize the procedure of the previous problem to a general strategy for transforming the matrix of an arbitrary system to row-echelon form.

1. By interchanging rows if necessary, obtain a nonzero element in the first position in row 1. Replace row 1 by a multiple to make this element a 1.
2. Use this element to produce zeros in the first position in the lower rows.
3. If this produces rows that are zero to the left of the vertical bar, or all zeros, move these rows to the bottom. If there are no other rows, stop.
4. If there are nonzero elements in rows below the first, move the row with the leftmost nonzero element to row 2. Replace row 2 by a multiple to make this element a 1.
5. Use this element to produce zeros in the corresponding position in any rows below row 2 that are nonzero to the left of the vertical bar.
6. Proceed as in steps 3 to 5 for any remaining rows.

**31.4.** Solve the system
$$\begin{aligned} x + 2y - 2z &= 3 \\ 2x + 5y \phantom{{}- 2z} &= -7 \\ 3x + 7y - 2z &= -4 \end{aligned}$$
by Gaussian elimination.

The augmented matrix of the system is the matrix of Problem 31.2. Transforming to row-echelon form yields the matrix

$$\begin{bmatrix} 1 & 2 & -2 & | & 3 \\ 0 & 1 & 4 & | & -13 \\ 0 & 0 & 0 & | & 0 \end{bmatrix}$$ 
which corresponds to the system
$$\begin{aligned} x + 2y - 2z &= 3 & (1) \\ y + 4z &= -13 & (2) \\ 0 &= 0 & (3) \end{aligned}$$

This system has an infinite number of solutions. Let $z = r$, $r$ any real number. Then from equation (2), $y = -13 - 4r$. Substituting back into equation (1) yields:

$$x + 2(-13 - 4r) - 2r = 3$$
$$x = 10r + 29$$

Thus all solutions of the system can be written as $(10r + 29, -13 - 4r, r)$, $r$ any real number.

**31.5.** Transform the matrix $\begin{bmatrix} 1 & 0 & -1 & 1 & 2 & | & 2 \\ 1 & 1 & 0 & 2 & -3 & | & -4 \\ 2 & 0 & -1 & 1 & 3 & | & 3 \\ 0 & -2 & -1 & -3 & 9 & | & 11 \end{bmatrix}$ to reduced row-echelon form.

The first element in row 1 is a 1. Use this to produce zeros in the first position in the lower rows:

$$\begin{bmatrix} 1 & 0 & -1 & 1 & 2 & | & 2 \\ 1 & 1 & 0 & 2 & -3 & | & -4 \\ 2 & 0 & -1 & 1 & 3 & | & 3 \\ 0 & -2 & -1 & -3 & 9 & | & 11 \end{bmatrix} \begin{matrix} \\ R_2 + (-1)R_1 \to R_2 \\ R_3 + (-2)R_1 \to R_3 \\ \\ \end{matrix} \begin{bmatrix} 1 & 0 & -1 & 1 & 2 & | & 2 \\ 0 & 1 & 1 & 1 & -5 & | & -6 \\ 0 & 0 & 1 & -1 & -1 & | & -1 \\ 0 & -2 & -1 & -3 & 9 & | & 11 \end{bmatrix}$$

Now the first nonzero element in row 2 is a 1. Use this to produce zeros in the position below it in the lower rows (only row 4 lacks a zero).

$$\begin{bmatrix} 1 & 0 & -1 & 1 & 2 & | & 2 \\ 0 & 1 & 1 & 1 & -5 & | & -6 \\ 0 & 0 & 1 & -1 & -1 & | & -1 \\ 0 & -2 & -1 & -3 & 9 & | & 11 \end{bmatrix} R_4 + 2R_2 \to R_4 \begin{bmatrix} 1 & 0 & -1 & 1 & 2 & | & 2 \\ 0 & 1 & 1 & 1 & -5 & | & -6 \\ 0 & 0 & 1 & -1 & -1 & | & -1 \\ 0 & 0 & 1 & -1 & -1 & | & -1 \end{bmatrix}$$

Now the first nonzero element in row 3 is a 1. Use this to produce a zero in the position below it in row 4.

$$\begin{bmatrix} 1 & 0 & -1 & 1 & 2 & | & 2 \\ 0 & 1 & 1 & 1 & -5 & | & -6 \\ 0 & 0 & 1 & -1 & -1 & | & -1 \\ 0 & 0 & 1 & -1 & -1 & | & -1 \end{bmatrix} R_4 + (-1)R_3 \to R_4 \begin{bmatrix} 1 & 0 & -1 & 1 & 2 & | & 2 \\ 0 & 1 & 1 & 1 & -5 & | & -6 \\ 0 & 0 & 1 & -1 & -1 & | & -1 \\ 0 & 0 & 0 & 0 & 0 & | & 0 \end{bmatrix}$$

This matrix is in row-echelon form. To produce reduced row-echelon form, use the leading 1 in each row to produce zeros in the corresponding position in the rows above, starting from the bottom row.

$$\begin{bmatrix} 1 & 0 & -1 & 1 & 2 & | & 2 \\ 0 & 1 & 1 & 1 & -5 & | & -6 \\ 0 & 0 & 1 & -1 & -1 & | & -1 \\ 0 & 0 & 0 & 0 & 0 & | & 0 \end{bmatrix} \begin{matrix} R_1 + R_3 \to R_1 \\ R_2 + (-1)R_3 \to R_2 \end{matrix} \begin{bmatrix} 1 & 0 & 0 & 0 & 1 & | & 1 \\ 0 & 1 & 0 & 2 & -4 & | & -5 \\ 0 & 0 & 1 & -1 & -1 & | & -1 \\ 0 & 0 & 0 & 0 & 0 & | & 0 \end{bmatrix}$$

This matrix is in reduced row-echelon form.

**31.6.** Solve by Gauss-Jordan elimination:

$$x_1 \qquad - x_3 + x_4 + 2x_5 = 2$$
$$x_1 + x_2 \qquad + 2x_4 - 3x_5 = -4$$
$$2x_1 \qquad - x_3 + x_4 + 3x_5 = 3$$
$$-2x_2 - x_3 - 3x_4 + 9x_5 = 11$$

The augmented matrix of the system is the matrix of Problem 31.5. Transforming to reduced row-echelon form yields the matrix

$$\begin{bmatrix} 1 & 0 & 0 & 0 & 1 & | & 1 \\ 0 & 1 & 0 & 2 & -4 & | & -5 \\ 0 & 0 & 1 & -1 & -1 & | & -1 \\ 0 & 0 & 0 & 0 & 0 & | & 0 \end{bmatrix} \text{ which corresponds to the system } \begin{matrix} x_1 & & + x_5 = & 1 & (1) \\ & x_2 & + 2x_4 - 4x_5 = & -5 & (2) \\ & & x_3 - x_4 - x_5 = & -1 & (3) \end{matrix}$$

This system has an infinite number of solutions. Let $x_5 = r$, $x_4 = s$, $r$ and $s$ any real numbers. Then from equation (3), $x_3 = r + s - 1$; from equation (2), $x_2 = 4r - 2s - 5$; and from equation (1), $x_1 = 1 - r$. Thus all solutions can be written as $(1 - r, 4r - 2s - 5, r + s - 1, s, r)$, $r$ and $s$ any real numbers.

**31.7.** Pumps A, B, and C, working together, can fill a tank in 2 hours. If only A and C are used, it would take 4 hours. If only B and C are used, it would take 3 hours. How long would it take for each to fill the tank, working separately?

Let $t_1$, $t_2$, and $t_3$ be the times for pumps A, B, and C, respectively. Then the rate at which each pump works can be written as $r_1 = 1/t_1$, $r_2 = 1/t_2$, and $r_3 = 1/t_3$. Using quantity of work = (rate)(time), the following tabular arrangement can be made:

|        | RATE  | TIME | QUANTITY OF WORK |
|--------|-------|------|------------------|
| Pump A | $r_1$ | 2    | $2r_1$           |
| Pump B | $r_2$ | 2    | $2r_2$           |
| Pump C | $r_3$ | 2    | $2r_3$           |

Thus, if all three machines working together can fill the tank in 2 hours,

$$2r_1 + 2r_2 + 2r_3 = 1 \qquad (1)$$

Similarly,

$$4r_1 + 4r_3 = 1 \qquad (2)$$
$$3r_2 + 3r_3 = 1 \qquad (3)$$

The system (1), (2), (3) has the augmented matrix

$$\begin{bmatrix} 2 & 2 & 2 & | & 1 \\ 4 & 0 & 4 & | & 1 \\ 0 & 3 & 3 & | & 1 \end{bmatrix}$$ which transforms to the reduced row echelon form $$\begin{bmatrix} 1 & 0 & 0 & | & 1/6 \\ 0 & 1 & 0 & | & 1/4 \\ 0 & 0 & 1 & | & 1/12 \end{bmatrix}$$

Thus, $r_1 = 1/6$ job/hr, $r_2 = 1/4$ job/hr, and $r_3 = 1/12$ job/hr. Therefore, $t_1 = 6$ hr for pump A to fill the tank, $t_2 = 4$ hr for pump B to fill the tank, and $t_3 = 12$ hr for pump C to fill the tank, working alone.

**31.8.** An investor has \$800,000 that she wishes to divide among Certificates of Deposit (CDs) paying 6% interest, mutual funds paying 10% interest, growth stocks paying 12% interest, and venture capital paying 14% interest. Fox tax reasons, she wants to plan for an annual return of \$78,000, and she wants to have the total of all other investments three times the amount invested in CDs. How should she divide her investment?

Let $x_1$ = amount invested in CDs, $x_2$ = amount invested in mutual funds, $x_3$ = amount invested in growth stocks, and $x_4$ = amount invested as venture capital. Form a table:

| | AMOUNT INVESTED | RATE OF INTEREST | INTEREST EARNED |
|---|---|---|---|
| CDs | $x_1$ | 0.06 | $0.06x_1$ |
| Mutual funds | $x_2$ | 0.1 | $0.1x_2$ |
| Growth stocks | $x_3$ | 0.12 | $0.12x_3$ |
| Venture capital | $x_4$ | 0.14 | $0.14x_4$ |

Since the total investment is \$800,000, $x_1 + x_2 + x_3 + x_4 = 800,000$    (1)

Since the total income is \$78,000, $0.06x_1 + 0.1x_2 + 0.12x_3 + 0.14x_4 = 78,000$    (2)

Since the total of other investments is to equal three times the amount invested in CDs,

$$x_2 + x_3 + x_4 = 3x_1, \text{ or in standard form, } -3x_1 + x_2 + x_3 + x_4 = 0 \quad (3)$$

The system (1), (2), (3) has the following augmented matrix:

$$\begin{bmatrix} 1 & 1 & 1 & 1 & | & 800,000 \\ 0.06 & 0.1 & 0.12 & 0.14 & | & 78,000 \\ -3 & 1 & 1 & 1 & | & 0 \end{bmatrix}$$

Transforming this to reduced row-echelon form yields:

$$\begin{bmatrix} 1 & 0 & 0 & 0 & | & 200,000 \\ 0 & 1 & 0 & -1 & | & 300,000 \\ 0 & 0 & 1 & 2 & | & 300,000 \end{bmatrix}$$

This corresponds to the system of equations:

$$x_1 \qquad\qquad\qquad = 200,000$$
$$x_2 \qquad - x_4 = 300,000$$
$$x_3 + 2x_4 = 300,000$$

This has an infinite number of solutions. Let $x_4 = r$. Then all solutions can be written in the form $(200000, 300000 + r, 300000 - 2r, r)$. Thus the investor must put \$200,000 into CDs, but has a wide range of further options meeting the given conditions. An amount $r$ put into venture capital requires an amount \$300,000 more in mutual funds, and an amount \$300,000 − 2r in growth stocks. As long as these are all positive, the conditions of the problem are satisfied; for example, one solution would be to let $r = 100,000$, then $x_1 = $200,000$ in CDs, $x_2 = $400,000$ in mutual funds, $x_3 = $100,000$ in growth stocks, and $x_4 = $100,000$ in venture capital.

## SUPPLEMENTARY PROBLEMS

**31.9.** Transform to row-echelon form:

(a) $\begin{bmatrix} 2 & 5 & | & 3 \\ 4 & -2 & | & -6 \end{bmatrix}$
(b) $\begin{bmatrix} 2 & 5 & | & 3 \\ 4 & 10 & | & -6 \end{bmatrix}$
(c) $\begin{bmatrix} 2 & 5 & | & 3 \\ 4 & 10 & | & 6 \end{bmatrix}$

*Ans.* (a) $\begin{bmatrix} 1 & 5/2 & | & 3/2 \\ 0 & 1 & | & 1 \end{bmatrix}$; (b) $\begin{bmatrix} 1 & 5/2 & | & 3/2 \\ 0 & 0 & | & -12 \end{bmatrix}$; (c) $\begin{bmatrix} 1 & 5/2 & | & 3/2 \\ 0 & 0 & | & 0 \end{bmatrix}$

**31.10.** Solve, using the information from the previous problem:

(a) $\begin{array}{l} 2x + 5y = 3 \\ 4x - 2y = -6 \end{array}$
(b) $\begin{array}{l} 2x + 5y = 3 \\ 4x + 10y = -6 \end{array}$
(c) $\begin{array}{l} 2x + 5y = 3 \\ 4x + 10y = 6 \end{array}$

*Ans.* (a) $(-1, 1)$; (b) no solution: (c) $\left( \dfrac{3 - 5r}{2}, r \right)$, $r$ any real number.

**31.11.** Transform to reduced row-echelon form:

(a) $\begin{bmatrix} 2 & 3 & | & 8 \\ 3 & -1 & | & 12 \\ 5 & 2 & | & 20 \end{bmatrix}$
(b) $\begin{bmatrix} 2 & 3 & -4 & | & 8 \\ 3 & 4 & -5 & | & 6 \\ 1 & 1 & -1 & | & 2 \end{bmatrix}$
(c) $\begin{bmatrix} 1 & 3 & 4 & 5 & | & 1 \\ 3 & 5 & 2 & 6 & | & 7 \\ 4 & 8 & 6 & 11 & | & 8 \end{bmatrix}$

*Ans.* (a) $\begin{bmatrix} 1 & 0 & | & 4 \\ 0 & 1 & | & 0 \\ 0 & 0 & | & 0 \end{bmatrix}$; (b) $\begin{bmatrix} 1 & 0 & 1 & | & 2 \\ 0 & 1 & -2 & | & 0 \\ 0 & 0 & 0 & | & 4 \end{bmatrix}$; (c) $\begin{bmatrix} 1 & 0 & -7/2 & -7/4 & | & 4 \\ 0 & 1 & 5/2 & 9/4 & | & -1 \\ 0 & 0 & 0 & 0 & | & 0 \end{bmatrix}$

**32.12.** Solve, using the information from the previous problem:

(a) $\begin{array}{l} 2x + 3y = 8 \\ 3x - y = 12 \\ 5x + 2y = 20 \end{array}$
(b) $\begin{array}{l} 2x + 3y - 4z = 8 \\ 3x + 4y - 5z = 6 \\ x + y - z = 2 \end{array}$
(c) $\begin{array}{l} x_1 + 3x_2 + 4x_3 + 5x_4 = 1 \\ 3x_1 + 5x_2 + 2x_3 + 6x_4 = 7 \\ 4x_1 + 8x_2 + 6x_3 + 11x_4 = 8 \end{array}$

*Ans.* (a) $(4, 0)$; (b) no solution; (c) $\left( 4 + \dfrac{7s}{2} + \dfrac{7r}{4}, -1 - \dfrac{5s}{2} - \dfrac{9r}{4}, s, r \right)$, $r$ and $s$ any real numbers

**31.13.** A mixture of 140 pounds of nuts is to be made from almonds costing \$4 per pound, pecans costing \$6 per pound, and brazil nuts costing \$7.50 per pound. If the mixture will sell for \$5.50 per pound, what possible combinations of nuts can be made?

*Ans.* If $t =$ number of pounds of brazil nuts, then any combination of $105 - 1.75t$ pounds of pecans and $35 + 0.75t$ pounds of almonds for which all three are positive; thus, $0 < t < 60$, $105 > 105 - 1.75t > 0$, and $35 < 35 + 0.75t < 80$.

# CHAPTER 32

# Partial Fraction Decomposition

## Proper and Improper Rational Expressions

A rational expression is any quotient of form $\dfrac{f(x)}{g(x)}$, where $f$ and $g$ are polynomial expressions. (Here it is assumed that $f$ and $g$ have real coefficients.) If the degree of $f$ is less than the degree of $g$, the rational expression is called *proper*, otherwise *improper*. An improper rational expression can always be written, using the long division scheme (Chapter 14), as a polynomial plus a proper rational expression.

## Partial Fraction Decomposition

Any polynomial $g(x)$ can, theoretically, be written as the product of one or more linear and quadratic factors, where the quadratic factors have no real zeros (*irreducible* quadratic factors). It follows that any proper rational expression with denominator $g(x)$ can be written as a sum of one or more proper rational expressions, each having a denominator that is a power of a polynomial with degree less than or equal to 2. This sum is called the *partial fraction decomposition* of the rational expression.

**EXAMPLE 32.1** $\dfrac{x^2}{x+1}$ is an improper rational expression. It can be rewritten as the sum of a polynomial and a proper rational expression: $\dfrac{x^2}{x+1} = x - 1 + \dfrac{1}{x+1}$.

**EXAMPLE 32.2** $\dfrac{2x+1}{x^2+x}$ is a proper rational expression. Since its denominator factors as $x^2 + x = x(x+1)$, the partial fraction decomposition of $\dfrac{2x+1}{x^2+x}$ is $\dfrac{2x+1}{x^2+x} = \dfrac{1}{x} + \dfrac{1}{x+1}$, as can be verified by addition:

$$\frac{1}{x} + \frac{1}{x+1} = \frac{x+1}{x(x+1)} + \frac{x}{x(x+1)} = \frac{2x+1}{x^2+x}$$

**EXAMPLE 32.3** $\dfrac{x}{x^2+1}$ is already in partial fraction decomposed form, since the denominator is quadratic and has no real zeros.

## Procedure for Finding the Partial Fraction Decomposition

Procedure for finding the partial fraction decomposition of a rational expression:

1. If the expression is proper, go to step 2. If the expression is improper, divide to obtain a polynomial plus a proper rational expression and apply the following steps to the proper expression $f(x)/g(x)$.
2. Write the denominator as a product of powers of linear factors of form $(ax + b)^m$ and irreducible quadratic factors of form $(ax^2 + bx + c)^n$.
3. For each factor $(ax + b)^m$, write a partial fraction sum of form:

$$\frac{A_1}{ax+b} + \frac{A_2}{(ax+b)^2} + \cdots + \frac{A_m}{(ax+b)^m}$$

where the $A_i$ are as yet to be determined unknown coefficients.

4. For each factor $(ax^2 + bx + c)^n$, write a partial fraction sum of form:

$$\frac{B_1 x + C_1}{ax^2 + bx + c} + \frac{B_2 x + C_2}{(ax^2 + bx + c)^2} + \cdots + \frac{B_n x + C_n}{(ax^2 + bx + c)^n}$$

where the $B_j$ and $C_j$ are as yet to be determined unknown coefficients.
5. Set $f(x)/g(x)$ equal to the sum of the partial fractions from steps 4 and 5. Eliminate the denominator $g(x)$. by multiplying both sides to obtain the *basic equation* for the unknown coefficients.
6. Solve the basic equation for the unknown coefficients $A_i$, $B_j$, and $C_j$.

## General Method for Solving the Basic Equation

1. Expand both sides.
2. Collect terms in each power of $x$.
3. Equate coefficients of each power of $x$.
4. Solve the linear system in the unknowns $A_i$, $B_j$, and $C_j$ that results.

**EXAMPLE 32.4** Find the partial fraction decomposition of $\dfrac{4}{x^2 - 1}$.

This is a proper rational expression. The denominator $x^2 - 1$ factors as $(x - 1)(x + 1)$. Therefore, there are only two partial fraction sums, one with denominator $x - 1$ and the other with denominator $x + 1$. Then set

$$\frac{4}{x^2 - 1} = \frac{A_1}{x - 1} + \frac{A_2}{x + 1}$$

Multiply both sides by $x^2 - 1$ to obtain the basic equation

$$4 = A_1(x + 1) + A_2(x - 1)$$

Expanding yields

$$4 = A_1 x + A_1 + A_2 x - A_2$$

Collecting terms in each power of $x$ yields

$$0x + 4 = (A_1 + A_2)x + (A_1 - A_2)$$

For this to hold for all $x$, the coefficients of each power of $x$ on both sides of the equation must be equal; hence:

$$A_1 + A_2 = 0$$
$$A_1 - A_2 = 4$$

The system has one solution: $A_1 = 2$, $A_2 = -2$. Hence the partial fraction decomposition is

$$\frac{4}{x^2 - 1} = \frac{2}{x - 1} + \frac{-2}{x + 1}$$

## Alternate Method

Alternate method for solving the basic equation: Instead of expanding both sides of the basic equation, substitute values for $x$ into the equation. If, and only if, all partial fractions have distinct linear denominators, if the values chosen are the distinct zeros of these expressions, the values of the $A_i$ will be found immediately. In other situations, there will not be enough of these zeros to determine all the unknowns. Other values of $x$ may be chosen and the resulting system of equations solved, but in these situations the alternative method is not preferred.

**EXAMPLE 32.5** Use the alternative method for the previous example.

The basic equation is

$$4 = A_1(x + 1) + A_2(x - 1)$$

Substitute $x = 1$, then it follows that:

$$4 = A_1(1 + 1) + A_2(1 - 1)$$

$$4 = 2A_1$$

$$A_1 = 2$$

Now substitute $x = -1$, then it follows that:

$$4 = A_1(-1 + 1) + A_2(-1 - 1)$$

$$4 = -2A_2$$

$$A_2 = -2$$

This yields the same result as before.

## SOLVED PROBLEMS

**32.1.** Find the partial fraction decomposition of $\dfrac{x^2 + 7x - 2}{x^3 - x}$.

This is a proper rational expression. The denominator factors as follows:

$$x^3 - x = x(x^2 - 1) = x(x - 1)(x + 1)$$

Thus there are three partial fraction sums, one each with denominator $x$, $x - 1$, and $x + 1$. Set

$$\frac{x^2 + 7x - 2}{x^3 - x} = \frac{A_1}{x} + \frac{A_2}{x - 1} + \frac{A_3}{x + 1}$$

Multiplying both sides by $x^3 - x = x(x - 1)(x + 1)$ yields

$$(x^3 - x)\frac{x^2 + 7x - 2}{x^3 - x} = x(x - 1)(x + 1)\frac{A_1}{x} + x(x - 1)(x + 1)\frac{A_2}{x - 1} + x(x - 1)(x + 1)\frac{A_3}{x + 1}$$

$$x^2 + 7x - 2 = A_1(x - 1)(x + 1) + A_2 x(x + 1) + A_3 x(x - 1)$$

This is the basic equation. Since all the partial fractions have linear denominators, it is more efficient to apply the alternate method. Substitute for $x$, in turn, the zeros of the denominator $x(x - 1)(x + 1)$.

$x = 0$:

$$-2 = A_1(-1)(1) + A_2(0)(1) + A_3(0)(-1)$$

$$-2 = -A_1$$

$$A_1 = 2$$

$x = 1$:

$$1^2 + 7 \cdot 1 - 2 = A_1(1 - 1)(1 + 1) + A_2(1)(1 + 1) + A_3(1)(1 - 1)$$

$$6 = 2A_2$$

$$A_2 = 3$$

$x = -1$:

$$(-1)^2 + 7(-1) - 2 = A_1(-1 - 1)(-1 + 1) + A_2(-1)(-1 + 1) + A_3(-1)(-1 - 1)$$

$$-8 = 2A_3$$

$$A_3 = -4$$

Hence the partial fraction decomposition is

$$\frac{x^2 + 7x - 2}{x^3 - x} = \frac{2}{x} + \frac{3}{x - 1} + \frac{-4}{x + 1}$$

**32.2.** Find the partial fraction decomposition of $\dfrac{6x^3 + 5x^2 + 2x - 10}{6x^2 - x - 2}$.

This is an improper expression. Use the long division scheme to rewrite it as:

$$x + 1 + \frac{5x - 8}{6x^2 - x - 2}$$

The denominator factors as $(3x - 2)(2x + 1)$. Thus there are two partial fraction sums, one with denominator $3x - 2$ and the other with denominator $2x + 1$. Set

$$\frac{5x - 8}{6x^2 - x - 2} = \frac{A_1}{3x - 2} + \frac{A_2}{2x + 1}$$

Multiply both sides by $6x^2 - x - 2 = (3x - 2)(2x + 1)$ to obtain

$$5x - 8 = A_1(2x + 1) + A_2(3x - 2)$$

This is the basic equation. Since the zeros of the denominator involve fractions, the alternate method does not seem attractive. Expanding yields

$$5x - 8 = 2A_1 x + A_1 + 3A_2 x - 2A_2$$

Collecting terms in each power of $x$ yields

$$5x - 8 = (2A_1 + 3A_2)x + (A_1 - 2A_2)$$

For this to hold for all $x$, the coefficients of each power of $x$ on both sides of the equation must be equal; hence:

$$2A_1 + 3A_2 = 5$$
$$A_1 - 2A_2 = -8$$

The only solution of this system is $A_1 = -2$, $A_2 = 3$. Hence the partial fraction decomposition is

$$\frac{6x^3 + 5x^2 + 2x - 10}{6x^2 - x - 2} = x + 1 + \frac{-2}{3x - 2} + \frac{3}{2x + 1}$$

**32.3.** Find the partial fraction decomposition of $\dfrac{-x^5 - x^4 + 3x^3 + 5x^2 + 6x + 6}{x^4 + x^3}$.

This is an improper expression. Use the long division scheme to rewrite it as:

$$-x + \frac{3x^3 + 5x^2 + 6x + 6}{x^4 + x^3}$$

The denominator factors as $x^3(x + 1)$. The first factor is referred to as a *repeated linear factor;* one partial fraction sum must be considered for each power of $x$ from 1 to 3. Set

$$\frac{3x^3 + 5x^2 + 6x + 6}{x^4 + x^3} = \frac{A_1}{x} + \frac{A_2}{x^2} + \frac{A_3}{x^3} + \frac{A_4}{x + 1}$$

Multiply both sides by $x^4 + x^3 = x^3(x + 1)$ to obtain

$$3x^3 + 5x^2 + 6x + 6 = A_1 x^2(x + 1) + A_2 x(x + 1) + A_3(x + 1) + A_4 x^3$$

This is the basic equation. Expanding yields

$$3x^3 + 5x^2 + 6x + 6 = A_1 x^3 + A_1 x^2 + A_2 x^2 + A_2 x + A_3 x + A_3 + A_4 x^3$$

Collecting terms in each power of $x$ yields

$$3x^3 + 5x^2 + 6x + 6 = (A_1 + A_4)x^3 + (A_1 + A_2)x^2 + (A_2 + A_3)x + A_3$$

For this to hold for all $x$, the coefficients of each power of $x$ on both sides of the equation must be equal, hence:

$$A_1 + A_4 = 3$$

$$A_1 + A_2 = 5$$

$$A_2 + A_3 = 6$$

$$A_3 = 6$$

The only solution of this system is $A_1 = 5, A_2 = 0, A_3 = 6, A_4 = -2$. Hence the partial fraction decomposition is

$$\frac{-x^5 - x^4 + 3x^3 + 5x^2 + 6x + 6}{x^4 + x^3} = -x + \frac{5}{x} + \frac{6}{x^3} + \frac{-2}{x + 1}$$

**32.4.** Find the partial fraction decomposition of $\dfrac{x^3 - x^2 + 9x - 1}{x^4 - 1}$.

This is a proper rational expression. The denominator factors as follows:

$$x^4 - 1 = (x^2 - 1)(x^2 + 1) = (x - 1)(x + 1)(x^2 + 1)$$

Thus there are three partial fraction sums, one each with denominator $x - 1$, $x + 1$, and $x^2 + 1$. Note that the irreducible quadratic denominator $x^2 + 1$ requires a numerator of the form $B_1 x + C_1$, that is, a linear rather than a constant expression. Set

$$\frac{x^3 - x^2 + 9x - 1}{x^4 - 1} = \frac{A_1}{x - 1} + \frac{A_2}{x + 1} + \frac{B_1 x + C_1}{x^2 + 1}$$

Multiply both sides by $x^4 - 1 = (x - 1)(x + 1)(x^2 + 1)$ to obtain

$$x^3 - x^2 + 9x - 1 = A_1(x + 1)(x^2 + 1) + A_2(x - 1)(x^2 + 1) + (B_1 x + C_1)(x - 1)(x + 1)$$

This is the basic equation. Expanding yields

$$x^3 - x^2 + 9x - 1 = A_1 x^3 + A_1 x^2 + A_1 x + A_1 + A_2 x^3 - A_2 x^2 + A_2 x - A_2 + B_1 x^3 + C_1 x^2 - B_1 x - C_1$$

Collecting terms in each power of $x$ yields

$$x^3 - x^2 + 9x - 1 = (A_1 + A_2 + B_1)x^3 + (A_1 - A_2 + C_1)x^2 + (A_1 + A_2 - B_1)x + A_1 - A_2 - C_1$$

For this to hold for all $x$, the coefficients of each power of $x$ on both sides of the equation must be equal, hence:

$$A_1 + A_2 + B_1 = 1$$

$$A_1 - A_2 + C_1 = -1$$

$$A_1 + A_2 - B_1 = 9$$

$$A_1 - A_2 - C_1 = -1$$

The only solution of this system is $A_1 = 2, A_2 = 3, B_1 = -4, C_1 = 0$. Hence the partial fraction decomposition is

$$\frac{x^3 - x^2 + 9x - 1}{x^4 - 1} = \frac{2}{x - 1} + \frac{3}{x + 1} + \frac{-4x}{x^2 + 1}$$

**32.5.** Find the partial fraction decomposition of $\dfrac{5x^3 - 4x^2 + 21x - 28}{x^4 + 10x^2 + 9}$.

This is a proper rational expression. The denominator factors as $x^4 + 10x^2 + 9 = (x^2 + 1)(x^2 + 9)$. There are only two partial fractions, one each with denominator $x^2 + 1$ and $x^2 + 9$. Each irreducible quadratic denominator requires a linear, not a constant numerator. Set

$$\frac{5x^3 - 4x^2 + 21x - 28}{x^4 + 10x^2 + 9} = \frac{B_1 x + C_1}{x^2 + 1} + \frac{B_2 x + C_2}{x^2 + 9}$$

Multiply both sides by $x^4 + 10x^2 + 9 = (x^2 + 1)(x^2 + 9)$ to obtain

$$5x^3 - 4x^2 + 21x - 28 = (B_1x + C_1)(x^2 + 9) + (B_2x + C_2)(x^2 + 1)$$

This is the basic equation. Expanding yields

$$5x^3 - 4x^2 + 21x - 28 = B_1x^3 + C_1x^2 + 9B_1x + 9C_1 + B_2x^3 + C_2x^2 + B_2x + C_2$$

Collecting terms in each power of $x$ yields

$$5x^3 - 4x^2 + 21x - 28 = (B_1 + B_2)x^3 + (C_1 + C_2)x^2 + (9B_1 + B_2)x + 9C_1 + C_2$$

For this to hold for all $x$, the coefficients of each power of $x$ on both sides of the equation must be equal, hence:

$$B_1 + B_2 = 5$$
$$C_1 + C_2 = -4$$
$$9B_1 + B_2 = 21$$
$$9C_1 + C_2 = -28$$

The only solution of this system is $B_1 = 2$, $B_2 = 3$, $C_1 = -3$, $C_2 = -1$. Hence the partial fraction decomposition is

$$\frac{5x^3 - 4x^2 + 21x - 28}{x^4 + 10x^2 + 9} = \frac{2x - 3}{x^2 + 1} + \frac{3x - 1}{x^2 + 9}$$

**32.6.** A common error in setting up a partial fraction sum is to assign a constant numerator to a partial fraction with an irreducible quadratic denominator. Explain what would happen in the previous problem as a result of this error.

Assume the incorrect partial fraction sum

$$\frac{5x^3 - 4x^2 + 21x - 28}{x^4 + 10x^2 + 9} = \frac{A_1}{x^2 + 1} + \frac{A_2}{x^2 + 9}$$

is set up. Multiplying both sides by $x^4 + 10x^2 + 9 = (x^2 + 1)(x^2 + 9)$ would yield

$$5x^3 - 4x^2 + 21x - 28 = A_1(x^2 + 9) + A_2(x^2 + 1)$$

Expanding this incorrect basic equation would yield

$$5x^3 - 4x^2 + 21x - 28 = A_1x^2 + 9A_1 + A_2x^2 + A_2$$

Collecting terms in each power of $x$ would yield

$$5x^3 - 4x^2 + 21x - 28 = (A_1 + A_2)x^2 + 9A_1 + A_2$$

For this to hold for all $x$, the coefficients of each power of $x$ on both sides of the equation would have to be equal, but this is impossible; for example, the coefficient of $x^3$ on the left is 5, but on the right it is 0. So the problem has been tackled incorrectly.

**32.7.** Find the partial fraction decomposition of $\dfrac{3x^3 + 14x - 3}{x^4 + 8x^2 + 16}$.

This is a proper rational expression. The denominator factors as $x^4 + 8x^2 + 16 = (x^2 + 4)^2$. This is referred to as a repeated quadratic factor; one partial fraction sum must be considered for both $x^2 + 4$ and $(x^2 + 4)^2$. Each irreducible quadratic denominator requires a linear, not a constant, numerator. Set

$$\frac{3x^3 + 14x - 3}{x^4 + 8x^2 + 16} = \frac{B_1x + C_1}{x^2 + 4} + \frac{B_2x + C_2}{(x^2 + 4)^2}$$

Multiply both sides by $x^4 + 8x^2 + 16 = (x^2 + 4)^2$ to obtain

$$3x^3 + 14x - 3 = (B_1x + C_1)(x^2 + 4) + B_2x + C_2$$

This is the basic equation. Expanding yields

$$3x^3 + 14x - 3 = B_1x^3 + C_1x^2 + 4B_1x + 4C_1 + B_2x + C_2$$

Collecting terms in each power of $x$ yields

$$3x^3 + 14x - 3 = B_1x^3 + C_1x^2 + (4B_1 + B_2)x + 4C_1 + C_2$$

For this to hold for all $x$, the coefficients of each power of $x$ on both sides of the equation must be equal, hence:

$$B_1 = 3$$

$$C_1 = 0$$

$$4B_1 + B_2 = 14$$

$$4C_1 + C_2 = -3$$

The only solution of this system is $B_1 = 3$, $B_2 = 2$, $C_1 = 0$, $C_2 = -3$. Hence the partial fraction decomposition is

$$\frac{3x^3 + 14x - 3}{x^4 + 8x^2 + 16} = \frac{3x}{x^2 + 4} + \frac{2x - 3}{(x^2 + 4)^2}$$

## SUPPLEMENTARY PROBLEMS

**32.8.** Find the partial fraction decomposition of $\dfrac{11x - 10}{x^2 - 2x}$.

*Ans.*  $\dfrac{5}{x} + \dfrac{6}{x - 2}$

**32.9.** Find the partial fraction decomposition of $\dfrac{2x + 22}{x^2 + x - 12}$.

*Ans.*  $\dfrac{4}{x - 3} + \dfrac{-2}{x + 4}$

**32.10.** Find the partial fraction decomposition of $\dfrac{x^4 + 3x^3 - 2x^2 - 2x - 4}{x^2 - 1}$.

*Ans.*  $x^2 + 3x - 1 + \dfrac{-2}{x - 1} + \dfrac{3}{x + 1}$

**32.11.** Find the partial fraction decomposition of $\dfrac{4x^2 - 15x - 125}{x^3 - 25x}$.

*Ans.*  $\dfrac{5}{x} + \dfrac{-2}{x - 5} + \dfrac{1}{x + 5}$

**32.12.** Find the partial fraction decomposition of $\dfrac{-2x^2 + 46x - 3}{30x^3 + 39x^2 - 9x}$.

*Ans.*  $\dfrac{1}{3x} + \dfrac{3}{5x - 1} + \dfrac{-2}{2x + 3}$

**32.13.** Find the partial fraction decomposition of $\dfrac{x^2 - 4}{x^3 - 3x^2 + 3x - 1}$.

*Ans.*  $\dfrac{1}{x - 1} + \dfrac{2}{(x - 1)^2} + \dfrac{-3}{(x - 1)^3}$

**32.14.** Find the partial fraction decomposition of $\dfrac{x^6 - x^5 - 3x^3 + x^2 + 3x - 3}{x^4 - x^3}$.

*Ans.*  $x^2 + \dfrac{-1}{x} + \dfrac{3}{x^3} + \dfrac{-2}{x - 1}$

**32.15.** Find the partial fraction decomposition of $\dfrac{x^4 + 3x^2 - x - 8}{x^3 + 4x}$.

Ans. $x + \dfrac{-2}{x} + \dfrac{x - 1}{x^2 + 4}$

**32.16.** Find the partial fraction decomposition of $\dfrac{2x^3 - 4x}{x^4 + 2x^3 + 2x^2 + 2x + 1}$.

Ans. $\dfrac{2}{x + 1} + \dfrac{1}{(x + 1)^2} + \dfrac{-3}{x^2 + 1}$

**32.17.** Find the partial fraction decomposition of $\dfrac{2x^5 + 42x^3 + x^2 + 124x + 16}{x^4 + 20x^2 + 64}$.

Ans. $2x + \dfrac{1 - x}{x^2 + 4} + \dfrac{3x}{x^2 + 16}$

**32.18.** Find the partial fraction decomposition of $\dfrac{x^5 + x^4 + 2x^3 + 2x^2 + 4x - 1}{(x^2 + 1)^3}$.

Ans. $\dfrac{x + 1}{x^2 + 1} + \dfrac{3x - 2}{(x^2 + 1)^3}$

**32.19.** Find the partial fraction decomposition of $\dfrac{x^6 - x^5 + 2x^4 - x^3 - x^2 + 3x - 3}{x^4 + 2x^2 + 1}$.

Ans. $x^2 - x + \dfrac{x - 2}{x^2 + 1} + \dfrac{3x - 1}{(x^2 + 1)^2}$

**32.20.** Find the partial fraction decomposition of $\dfrac{5x^6 - x^5 + 33x^4 - 14x^3 + 51x^2 - 31x + 23}{(x^2 + 1)^2(x^2 + 4)^2}$.

Ans. $\dfrac{5}{x^2 + 4} + \dfrac{x + 3}{(x^2 + 4)^2} + \dfrac{-2x}{(x^2 + 1)^2}$

**32.21.** Find the partial fraction decomposition of $\dfrac{3x^2 - 6x + 6}{x^3 + 1}$.

Ans. $\dfrac{5}{x + 1} + \dfrac{1 - 2x}{x^2 - x + 1}$

**32.22.** Show that the partial fraction decomposition of $\dfrac{c}{x^2 - a^2}$ can be written as $\dfrac{c}{2a(x - a)} + \dfrac{-c}{2a(x + a)}$.

# CHAPTER 33

# Nonlinear Systems of Equations

## Definition of Nonlinear Systems of Equations

A system of equations in which any one equation is not linear is a nonlinear system. A nonlinear system may have no solutions, an infinite set of solutions, or any number of real or complex solutions.

## Solutions of Nonlinear Systems in Two Variables

Solutions of nonlinear systems in two variables can be found by three methods:

1. **GRAPHICAL METHOD.** Graph each equation. The coordinates of any points of intersection may be read from the graph. After checking by substitution in each equation, these coordinates are the real solutions of the system. Normally, only approximations to real solutions can be found by this method, but when the algebraic methods below fail, this method can still be used.
2. **SUBSTITUTION METHOD.** Solve one equation for one variable in terms of the other. Substitute this expression into the other equations to determine the value of the first variable (if possible). Then substitute this value to determine the value of the other variable.
3. **ELIMINATION METHOD.** Apply the operations on equations leading to equivalent systems to eliminate one variable from one equation, solve the resulting equation for this variable, and substitute this value to determine the value of the other variable.

**EXAMPLE 33.1**  Solve the system $\begin{aligned} y &= e^{-x} \\ y &= 1 + x \end{aligned}$ graphically.

The graph of $y = e^{-x}$ is an exponential decay curve; the graph of $y = 1 + x$ is a straight line.

Sketch the two graphs in the same coordinate system (see Fig. 33-1).

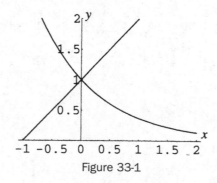

Figure 33-1

The graphs appear to intersect at $(0, 1)$. Substituting $x = 0$, $y = 1$ into $y = e^{-x}$ yields $1 = e^{-0}$ or $1 = 1$. Substituting into $y = 1 + x$ yields $1 = 1 + 0$. Thus, $(0, 1)$ is a solution of the system. The method does not rule out the possibility of other solutions, including nonreal complex solutions.

**EXAMPLE 33.2**  Solve the system $\begin{aligned} y &= x^2 - 2 \quad (1)\\ x + 2x &= 11 \quad (2)\end{aligned}$ by substitution.

Substitute the expression $x^2 - 2$ from equation (1) into equation (2) for $y$ to obtain

$$x + 2(x^2 - 2) = 11$$

Solving this quadratic equation in $x$ yields

$$2x^2 + x - 15 = 0$$

$$(2x - 5)(x + 3) = 0$$

$$2x - 5 = 0 \quad \text{or} \quad x + 3 = 0$$

$$x = \frac{5}{2} \qquad\qquad x = -3$$

Substituting these values for $x$ into equation (1) yields

$$x = \frac{5}{2}: y = \left(\frac{5}{2}\right)^2 - 2 = \frac{17}{4} \qquad\qquad x = -3: y = (-3)^2 - 2 = 7$$

Thus the solutions are $\left(\frac{5}{2}, \frac{17}{4}\right)$ and $(-3, 7)$.

**EXAMPLE 33.3**  Solve by elimination: $\begin{aligned} x^2 + y^2 &= 1 \quad (1)\\ x^2 - y^2 &= 7 \quad (2)\end{aligned}$

Replacing equation (2) by itself plus equation (1) yields the equivalent system:

$$x^2 + y^2 = 1 \quad (1)$$

$$2x^2 = 8 \quad (3)$$

Solving equation (3) for $x$ yields

$$x^2 = 4$$

$$x = 2 \quad \text{or} \quad x = -2$$

Substituting these values for $x$ into equation (1) yields

$$x = 2: \quad 2^2 + y^2 = 1 \qquad\qquad x = -2: \quad (-2)^2 + y^2 = 1$$

$$y^2 = -3 \qquad\qquad\qquad\qquad y^2 = -3$$

$$y = i\sqrt{3} \quad \text{or} \quad y = -i\sqrt{3} \qquad\qquad y = i\sqrt{3} \quad \text{or} \quad y = -i\sqrt{3}$$

Thus the solutions are $(2, i\sqrt{3}), (2, -i\sqrt{3}), (-2, i\sqrt{3}), (-2, -i\sqrt{3})$.

## No General Procedure Exists

There is no general procedure for solving nonlinear systems of equations. Sometimes a combination of the above methods is effective; frequently no algebraic method works and the graphical method can be used to find some approximate solutions, which can then be refined by advanced numerical methods.

## SOLVED PROBLEMS

**33.1.** Solve the system $\begin{aligned} y &= x^2 \quad (1)\\ x + y &= 2 \quad (2)\end{aligned}$ and illustrate graphically.

Solve by substitution: Substitute the expression $x^2$ from equation (1) into equation (2) for $y$ to obtain the quadratic equation

$$x + x^2 = 2$$

Solving yields

$$x^2 + x - 2 = 0$$

$$(x - 1)(x + 2) = 0$$

$$x = 1 \qquad \text{or} \qquad x = -2$$

Substituting these values for $x$ into equation (1) yields:

$$x = 1 : y = 1^2 = 1 \qquad x = -2 : y = (-2)^2 = 4$$

Thus the solutions are $(1, 1)$ and $(-2, 4)$.

The graph of $y = x^2$ is the basic parabola, opening up. The graph of $x + y = 2$ is a straight line with slope $-1$ and $y$ intercept 2. Sketch the two graphs in the same coordinate system (Fig. 33-2).

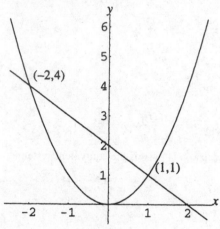

Figure 33-2

**33.2.** Solve the system $\begin{array}{ll} y = x^2 + 2 & (1) \\ y = 2x - 4 & (2) \end{array}$ and illustrate graphically.

Solve by substitution: Substitute the expression $x^2 + 2$ from equation (1) into equation (2) for $y$ to obtain the quadratic equation

$$x^2 + 2 = 2x - 4$$

Solving yields

$$x^2 - 2x + 6 = 0$$

$$x = \frac{-(-2) \pm \sqrt{(-2)^2 - 4(1)(6)}}{2(1)}$$

$$x = 1 \pm i\sqrt{5}$$

Substituting these values for $x$ into equation (1) yields:

$$x = 1 + i\sqrt{5} : y = (1 + i\sqrt{5})^2 + 2 = -2 + 2i\sqrt{5}$$

$$x = 1 - i\sqrt{5} : y = (1 - i\sqrt{5})^2 + 2 = -2 - 2i\sqrt{5}$$

Thus the solutions are $(1 + i\sqrt{5}, -2 + 2i\sqrt{5})$ and $(1 - i\sqrt{5}, -2 - 2i\sqrt{5})$.

The graph of $y = x^2 + 2$ is the basic parabola, opening up, shifted up 2 units. The graph of $y = 2x - 4$ is a straight line with slope 2 and $y$ intercept $-4$. Sketch the two graphs in the same coordinate system and note that the complex solutions correspond to the fact that the graphs do not intersect (Fig. 33-3).

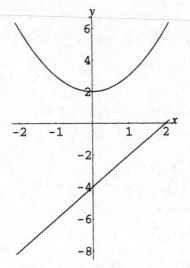

Figure 33-3

**33.3.** Solve the system $\begin{matrix} y^2 - 4x^2 = 4 & (1) \\ 9y^2 + 16x^2 = 140 & (2) \end{matrix}$.

The system is most efficiently solved by elimination. Replace equation (2) by itself plus four times equation (1):

$$
\begin{array}{ll}
4y^2 - 16x^2 = 16 & 4 \cdot \text{Eq. (1)} \\
\underline{9y^2 + 16x^2 = 140} & (2) \\
13y^2 \quad\quad = 156 & (3)
\end{array}
$$

Solving equation (3) yields

$$y^2 = 12$$

$$y = \pm 2\sqrt{3}$$

Substituting these values for $y$ into equation (1) yields

$$y = 2\sqrt{3}: \quad (2\sqrt{3})^2 - 4x^2 = 4 \qquad\qquad y = -2\sqrt{3}: \quad (-2\sqrt{3})^2 - 4x^2 = 4$$

$$x^2 = 2 \qquad\qquad\qquad\qquad\qquad x^2 = 2$$

$$x = \sqrt{2} \quad\text{or}\quad x = -\sqrt{2} \qquad\qquad\qquad x = \sqrt{2} \quad\text{or}\quad x = -\sqrt{2}$$

Thus the solutions are $(\sqrt{2}, 2\sqrt{3}), (\sqrt{2}, -2\sqrt{3}), (-\sqrt{2}, 2\sqrt{3}), (-\sqrt{2}, -2\sqrt{3})$.

**33.4.** Solve the system $\begin{matrix} x^2 + xy - 3y^2 = 3 & (1) \\ x^2 + 4xy + 3y^2 = 0 & (2) \end{matrix}$.

The system is most efficiently solved by substitution. Solve equation (2) for $x$ in terms of $y$:

$$(x + y)(x + 3y) = 0$$

$$x + y = 0 \quad\text{or}\quad x + 3y = 0$$

$$x = -y \qquad\qquad x = -3y$$

Now substitute these expressions for $x$ into equation (1):

$$\text{If } x = -y: (-y)^2 + (-y)\, y - 3y^2 = 3$$

$$-3y^2 = 3$$

$$y = i \quad\text{or}\quad y = -i$$

Since $x = -y$, when $y = i$, $x = -i$, and when $y = -i$, $x = i$.

$$\text{If } x = -3y: (-3y)^2 + (-3y)\,y - 3y^2 = 3$$

$$3y^2 = 3$$

$$y = 1 \quad \text{or} \quad y = -1$$

Since $x = -3y$, when $y = 1$, $x = -3$, and when $y = -1$, $x = 3$.

Thus the solutions are $(i, -i)$, $(-i, i)$, $(-3, 1)$, $(3, -1)$.

**33.5.** Solve the system $\begin{array}{ll} x^2 + xy - y^2 = -1 & (1) \\ x^2 + 2xy - y^2 = 1 & (2) \end{array}$.

This can be solved by a combination of elimination and substitution techniques. Replace equation (2) by itself plus $-1$ times equation (1):

$$\begin{array}{ll} -x^2 - xy + y^2 = 1 & (-1) \cdot \text{Eq. (1)} \\ \underline{x^2 + 2xy - y^2 = 1} & (2) \\ \quad xy \quad\quad = 2 & (3) \end{array}$$

Solving equation (3) for $y$ in terms of $x$ yields $y = \frac{2}{x}$. Substitute the expression $\frac{2}{x}$ into equation (1) for $y$ to obtain:

$$x^2 + x\left(\frac{2}{x}\right) - \left(\frac{2}{x}\right)^2 = -1$$

$$x^2 + 2 - \frac{4}{x^2} = -1$$

$$x^2 + 3 - \frac{4}{x^2} = 0$$

$$x^4 + 3x^2 - 4 = 0 \quad (x \neq 0)$$

$$(x - 1)(x + 1)(x - 2i)(x + 2i) = 0$$

$$x = 1 \quad \text{or} \quad x = -1 \quad \text{or} \quad x = 2i \quad \text{or} \quad x = -2i$$

Since $y = \frac{2}{x}$, when $x = 1$, $y = \frac{2}{1} = 2$ and when $x = -1$, $y = \frac{2}{-1} = -2$. Also, when $x = 2i$, $y = \frac{2}{2i} = -i$. and when $x = -2i$, $y = \frac{2}{-2i} = i$. Thus, the solutions are $(1, 2)$, $(-1, -2)$, $(2i, -i)$, and $(-2i, i)$.

**33.6.** An engineer wishes to design a rectangular television screen that is to have an area of 220 square inches and a diagonal of 21 inches. What dimensions should be used?

Let $x$ = width and $y$ = length of the screen. Sketch a figure (see Fig. 33-4).

Figure 33-4

Since the area of the rectangle is to be 220 square inches,

$$xy = 220 \quad (1)$$

Since the diagonal is to be 21 inches, from the Pythagorean theorem,

$$x^2 + y^2 = 21^2 \quad (2)$$

The system (1), (2) can be solved by substitution. Solving equation (1) for $y$ in terms of $x$ yields

$$y = \frac{220}{x}$$

Substitute the expression $\frac{220}{x}$ for $y$ into equation (2) to obtain:

$$x^2 + \left(\frac{220}{x}\right)^2 = 441$$

$$x^2 + \frac{48400}{x^2} = 441$$

$$x^4 + 48400 = 441x^2 \qquad (x \neq 0)$$

$$x^4 - 441x^2 + 48400 = 0$$

The last equation is quadratic in $x^2$, but not factorable. Use the quadratic formula to obtain:

$$x^2 = \frac{-(-441) \pm \sqrt{(-441)^2 - 4(1)(48400)}}{2(1)}$$

$$= \frac{441 \pm \sqrt{881}}{2}$$

$$x = \sqrt{\frac{441 \pm \sqrt{881}}{2}}$$

In the last step, only the positive square root is meaningful. Thus the two possible solutions are

$$x = \sqrt{\frac{441 + \sqrt{881}}{2}} \approx 15.34 \qquad \text{and} \qquad x = \sqrt{\frac{441 - \sqrt{881}}{2}} \approx 14.34$$

Since $y = 220/x$, if $x = 15.34$, $y = 14.34$, and conversely. Hence the only solution is for the dimensions of the screen to be $14.34 \times 15.34$ inches.

## SUPPLEMENTARY PROBLEMS

**33.7.** Solve the systems: (a) $\begin{array}{l} 2y = x^2 \\ 4y = x^3 \end{array}$; (b) $\begin{array}{l} x = y^2 \\ x^2 - y^2 = 2 \end{array}$

    *Ans.* (a) $(0,0)$, $(2,2)$; (b) $(2, \sqrt{2})$, $(2, -\sqrt{2})$, $(-1, i)$, $(-1, -i)$

**33.8.** Solve the systems (a) $\begin{array}{l} x^2 + y^2 = 16 \\ y^2 = 4 - x \end{array}$; (b) $\begin{array}{l} x^2 + y^2 = 8 \\ y - x = 4 \end{array}$

    *Ans.* (a) $(4,0)$, $(-3, \sqrt{7})$, $(-3, -\sqrt{7})$; (b) $(-2,2)$

**33.9.** Solve the systems: (a) $\begin{array}{l} x^2 + 4y^2 = 24 \\ x^2 - 4y = 0 \end{array}$; (b) $\begin{array}{l} x^2 - 8y^2 = 1 \\ x^2 + 4y^2 = 25 \end{array}$

    *Ans.* (a) $(\sqrt{8}, 2)$, $(-\sqrt{8}, 2)$, $(2i\sqrt{3}, -3)$, $(-2i\sqrt{3}, -3)$;

          (b) $(\sqrt{17}, \sqrt{2})$, $(\sqrt{17}, -\sqrt{2})$, $(-\sqrt{17}, \sqrt{2})$, $(-\sqrt{17}, -\sqrt{2})$

**33.10.** Solve and illustrate the solutions graphically: (a) $\begin{aligned} y &= x^2 - 1 \\ y &= 2x + 2 \end{aligned}$; (b) $\begin{aligned} y &= x^2 - 2 \\ y &= 2 - 2x - x^2 \end{aligned}$

    *Ans.*    (a) Solutions: $(-1,0)$, $(3,8)$; Fig. 33-5:          (b) Solutions: $(-2,2)$, $(1,-1)$; Fig. 33-6

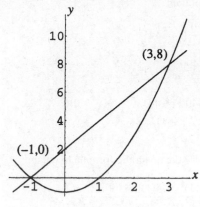
Figure 33-5

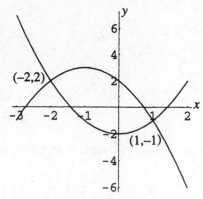
Figure 33-6

**33.11.** Solve the systems: (a) $\begin{aligned} 2x + 3y + xy &= 16 \\ xy - 5 &= 0 \end{aligned}$; (b) $\begin{aligned} 2x^2 - 5xy + 2y^2 &= 0 \\ 3x^2 + 2xy - y^2 &= 15 \end{aligned}$

    *Ans.*    (a) $\left(\dfrac{5}{2}, 2\right), \left(3, \dfrac{5}{3}\right)$; (b) $(2, 1)$, $(-2, -1)$, $(\sqrt{5}, 2\sqrt{5})$, $(-\sqrt{5}, -2\sqrt{5})$

**33.12.** A rectangle of perimeter 100 meters is to be constructed to have area 100 square meters. What dimensions are required?

    *Ans.*    $25 + \sqrt{525}$ by $25 - \sqrt{525}$, or approximately $47.91 \times 2.09$ meters

# CHAPTER 34

# Introduction to Matrix Algebra

## Definition of Matrix

A matrix is a rectangular arrangement of numbers in rows and columns, and enclosed in brackets, thus:

$$\begin{bmatrix} a_{11} & a_{12} & a_{13} & a_{14} \\ a_{21} & a_{22} & a_{23} & a_{24} \\ a_{31} & a_{32} & a_{33} & a_{34} \end{bmatrix}$$

The numbers are called *elements* of the matrix. The above matrix would be said to have three rows (first row: $a_{11}\ a_{12}\ a_{13}\ a_{14}$, and so on) and four columns, and would be called a matrix of *order* $3 \times 4$. The elements are referred to by two subscripts; thus, the element in row 2, column 3 is element $a_{23}$. A matrix may have any number of rows and any number of columns; a general matrix is said to have *order* $m \times n$, thus, $m$ rows and $n$ columns.

## Matrix Notation

Matrices are referred to by capital letters, thus: $A$, and by doubly subscripted lowercase letters enclosed in parentheses, thus: $(a_{ij})$. If necessary for clarity, the order of the matrix is specified as a subscript, thus: $A_{m \times n}$.

## Special Matrices

A matrix consisting of only one row is called a *row* matrix. A matrix consisting of only one column is called a *column* matrix. A matrix with equal numbers of rows and columns is called a *square* matrix. For a square matrix of order $n \times n$, the elements $a_{11}, a_{22}, \ldots, a_{nn}$ are called the *main diagonal* elements. A matrix with all elements equal to zero is called a *zero* matrix. A zero matrix of order $m \times n$ is denoted by $0_{m \times n}$, or, if the order is clear from the context, simply 0.

**EXAMPLE 34.1** $[5 \quad -2 \quad 0 \quad 9]$ is a row matrix (order $1 \times 4$). $\begin{bmatrix} 2 \\ -3 \end{bmatrix}$ is a column matrix (order $2 \times 1$).

Examples of square matrices are $[4]$, $\begin{bmatrix} 0 & 0 \\ 0 & 0 \end{bmatrix}$, and $\begin{bmatrix} -3 & 5 & -4 \\ 2 & 2 & -4 \\ 0 & 9 & -4 \end{bmatrix}$. $\begin{bmatrix} 0 & 0 \\ 0 & 0 \end{bmatrix}$ is also a zero matrix of order $2 \times 2$.

## Matrix Equality

Two matrices are equal if and only if they have the same order and corresponding elements are equal, thus, given $A = (a_{ij})$ and $B = (b_{ij})$, $A = B$ if and only if the matrices have the same order and $a_{ij} = b_{ij}$ for all $i$ and $j$.

## Matrix Addition

Given matrices of the same order $m \times n$, $A = (a_{ij})$ and $B = (b_{ij})$, the matrix sum $A + B$ is defined by $A + B = (a_{ij} + b_{ij})$, that is, $A + B$ is a matrix of order $m \times n$ with each element being the sum of the corresponding elements of $A$ and $B$. The sum of two matrices of different orders is not defined.

## Additive Inverse and Subtraction

The additive inverse, or negative, of an $m \times n$ matrix $A = (a_{ij})$ is the $m \times n$ matrix $-A = (-a_{ij})$. Subtraction of two matrices of the same order $m \times n$, $A = (a_{ij})$ and $B = (b_{ij})$, is defined by $A - B = (a_{ij} - b_{ij})$, that is, $A - B$ is a matrix of order $m \times n$ with each element being the difference of the corresponding elements of $A$ and $B$.

## Properties of Matrix Addition

Given $m \times n$ matrices $A$, $B$, $C$, and $O$, the following laws can be shown to hold ($O$ is a zero matrix):

1. COMMUTATIVE LAW: $A + B = B + A$
2. ASSOCIATIVE LAW: $A + (B + C) = (A + B) + C$
3. IDENTITY LAW: $A + O = A$
4. ADDITIVE INVERSE LAW: $A + (-A) = O$

## Product of a Matrix and a Scalar

The product of a matrix and a scalar is defined as follows: Given an $m \times n$ matrix $A = (a_{ij})$ and a scalar (real number) $c$, then $cA = (ca_{ij})$, that is, $cA$ is the $m \times n$ matrix formed by multiplying each element of $A$ by $c$. The following properties can be shown to hold ($A$ and $B$ both of order $m \times n$):

$$c(A + B) = cA + cB \qquad (c + d)A = cA + dA \qquad (cd)A = c(dA)$$

## SOLVED PROBLEMS

**34.1.** State the order of the following matrices: $A = \begin{bmatrix} 3 \\ 4 \end{bmatrix}$; $B = \begin{bmatrix} 4 & 3 \\ 5 & -2 \\ 6 & 4 \end{bmatrix}$; $C = \begin{bmatrix} 0 & 0 & -3 \\ 4 & 2 & 2 \end{bmatrix}$.

$A$ has 2 rows and 1 column; it is a $2 \times 1$ matrix.

$B$ has 3 rows and 2 columns; it is a $3 \times 2$ matrix.

$C$ has 2 rows and 3 columns; it is a $2 \times 3$ matrix.

**34.2.** Given the matrices $A = \begin{bmatrix} 5 & 0 \\ 2 & -3 \end{bmatrix}$, $B = \begin{bmatrix} 3 & -2 \\ -4 & 8 \end{bmatrix}$, $C = \begin{bmatrix} -3 & -2 & -3 \\ 4 & 0 & 2 \end{bmatrix}$, find

(a) $A + B$; (b) $-C$; (c) $B + C$; (d) $B - A$.

(a) $A + B = \begin{bmatrix} 5 & 0 \\ 2 & -3 \end{bmatrix} + \begin{bmatrix} 3 & -2 \\ -4 & 8 \end{bmatrix} = \begin{bmatrix} 5 & + 3 & 0 + (-2) \\ 2 + (-4) & (-3) + 8 \end{bmatrix} = \begin{bmatrix} 8 & -2 \\ -2 & 5 \end{bmatrix}$

(b) $-C = -\begin{bmatrix} -3 & -2 & -3 \\ 4 & 0 & 2 \end{bmatrix} = \begin{bmatrix} 3 & 2 & 3 \\ -4 & 0 & -2 \end{bmatrix}$

(c) Since $B$ is a $2 \times 2$ matrix and $C$ is a $2 \times 3$ matrix, $B + C$ is not defined.

(d) $B - A = \begin{bmatrix} 3 & -2 \\ -4 & 8 \end{bmatrix} - \begin{bmatrix} 5 & 0 \\ 2 & -3 \end{bmatrix} = \begin{bmatrix} 3 - 5 & (-2) - 0 \\ (-4) - 2 & 8 - (-3) \end{bmatrix} = \begin{bmatrix} -2 & -2 \\ -6 & 11 \end{bmatrix}$

**34.3.** Verify the commutative law for matrix addition: for any two $m \times n$ matrices $A$ and $B$, $A + B = B + A$.

Let $A = (a_{ij})$ and $B = (b_{ij})$. Since both $A$ and $B$ have order $m \times n$, both $A + B$ and $B + A$ are defined and have order $m \times n$. Then

$$A + B = (a_{ij}) + (b_{ij}) = (a_{ij} + b_{ij}) \qquad \text{and} \qquad B + A = (b_{ij}) + (a_{ij}) = (b_{ij} + a_{ij})$$

Since for all $i$ and $j$, $a_{ij} + b_{ij}$ and $b_{ij} + a_{ij}$ are real numbers, $a_{ij} + b_{ij} = b_{ij} + a_{ij}$, Hence $A + B = B + A$.

**34.4.** Verify the identity law for matrix addition: for any $m \times n$ matrix $A$, $A + 0_{m \times n} = A$.

Let $A = (a_{ij})$; by definition $0_{m \times n}$ is an $m \times n$ matrix with all entries equal to zero, that is, $0_{m \times n} = (0)$. Then $A + 0_{m \times n}$ is defined and has order $m \times n$, hence

$$A + 0_{m \times n} = (a_{ij}) + (0) = (a_{ij} + 0) = (a_{ij}) = A$$

**34.5.** Given the matrices $A = \begin{bmatrix} -2 & 6 & 2 \\ 0 & -3 & 4 \end{bmatrix}$, $B = \begin{bmatrix} 3 & -2 \\ -4 & 8 \end{bmatrix}$, $C = \begin{bmatrix} -3 & -2 & -3 \\ 4 & 0 & 2 \end{bmatrix}$, find

(a) $-2A$; (b) $0B$; (c) $5B + 3A$; (d) $-3C + 4A$.

(a) $-2A = -2\begin{bmatrix} -2 & 6 & 2 \\ 0 & -3 & 4 \end{bmatrix} = \begin{bmatrix} (-2)(-2) & (-2)6 & (-2)2 \\ (-2)0 & (-2)(-3) & (-2)4 \end{bmatrix} = \begin{bmatrix} 4 & -12 & -4 \\ 0 & 6 & -8 \end{bmatrix}$

(b) $0B = 0\begin{bmatrix} 3 & -2 \\ -4 & 8 \end{bmatrix} = \begin{bmatrix} 0(3) & 0(-2) \\ 0(-4) & 0(8) \end{bmatrix} = \begin{bmatrix} 0 & 0 \\ 0 & 0 \end{bmatrix}$

(c) Since $5B$ is a $2 \times 2$ matrix and $3A$ is a $2 \times 3$ matrix, $5B + 3A$ is not defined.

(d) $-3C + 4A = -3\begin{bmatrix} -3 & -2 & -3 \\ 4 & 0 & 2 \end{bmatrix} + 4\begin{bmatrix} -2 & 6 & 2 \\ 0 & -3 & 4 \end{bmatrix} = \begin{bmatrix} 9 & 6 & 9 \\ -12 & 0 & -6 \end{bmatrix} + \begin{bmatrix} -8 & 24 & 8 \\ 0 & -12 & 16 \end{bmatrix}$

$$= \begin{bmatrix} 1 & 30 & 17 \\ -12 & -12 & 10 \end{bmatrix}$$

**34.6.** Verify: If both $A$ and $B$ are $m \times n$ matrices, then for any scalar $c$, $c(A + B) = cA + cB$.

First note that $A + B$, $c(A + B)$, $cA$, $cB$, and hence $cA + cB$ are all defined and of order $m \times n$.

Let $A = (a_{ij})$ and $B = (b_{ij})$; then

$$c(A + B) = c((a_{ij}) + (b_{ij})) = c((a_{ij} + b_{ij})) = (c(a_{ij} + b_{ij}))$$

where the innermost multiplication is the product of two real numbers, and

$$cA + cB = c(a_{ij}) + c(b_{ij}) = (ca_{ij} + cb_{ij})$$

But by the distributive law for real numbers, $c(a_{ij} + b_{ij}) = ca_{ij} + cb_{ij}$ for any $i$ and $j$. Hence

$$c(A + B) = cA + cB$$

## SUPPLEMENTARY PROBLEMS

**34.7.** Given $A = \begin{bmatrix} 3 & 4 & -2 \\ 8 & 0 & 2 \\ 1 & 1 & -2 \end{bmatrix}$, $B = \begin{bmatrix} 4 & 2 \\ 4 & 2 \\ -4 & -2 \end{bmatrix}$, $C = \begin{bmatrix} 0 & 2 & 0 \\ -3 & -4 & 2 \\ 7 & 2 & -1 \end{bmatrix}$, find

(a) $A + B$; (b) $A + C$; (c) $B - B$; (d) $2C$.

*Ans.* (a) Not defined; (b) $\begin{bmatrix} 3 & 6 & -2 \\ 5 & -4 & 4 \\ 8 & 3 & -3 \end{bmatrix}$; (c) $\begin{bmatrix} 0 & 0 \\ 0 & 0 \\ 0 & 0 \end{bmatrix}$; (d) $\begin{bmatrix} 0 & 4 & 0 \\ -6 & -8 & 4 \\ 14 & 4 & -2 \end{bmatrix}$

**34.8.** Given $A$, $B$, and $C$ as in the previous problem, find (a) $3A + 2C$; (b) $\frac{1}{4}B$; (c) $-A - 2C$.

*Ans.* (a) $\begin{bmatrix} 9 & 16 & -6 \\ 18 & -8 & 10 \\ 17 & 7 & -8 \end{bmatrix}$; (b) $\begin{bmatrix} 1 & 1/2 \\ 1 & 1/2 \\ -1 & -1/2 \end{bmatrix}$; (c) $\begin{bmatrix} -3 & -8 & 2 \\ -2 & 8 & -6 \\ -15 & -5 & 4 \end{bmatrix}$

**34.9.** Verify the associative law for matrix addition: for any three $m \times n$ matrices $A$, $B$, and $C$, $A + (B + C) = (A + B) + C$.

**34.10.** Verify the additive inverse law for matrix addition: for any $m \times n$ matrix $A$, $A + (-A) = O_{m \times n}$.

**34.11.** Verify: for any two scalars $c$ and $d$ and any matrix $A$, $(c + d)A = cA + dA$.

**34.12.** Verify: for any two scalars $c$ and $d$ and any matrix $A$, $(cd)A = c(dA)$.

**34.13.** The *transpose* of an $m \times n$ matrix $A$ is a matrix $A^T$ formed by interchanging rows and columns of $A$, that is, an $n \times m$ matrix with the element in row $j$, column $i$ being $a_{ij}$. Find the transposes of matrices

(a) $A = \begin{bmatrix} 3 & 4 & -2 \\ 8 & 0 & 2 \\ 1 & 1 & -2 \end{bmatrix}$; (b) $B = \begin{bmatrix} 4 & 2 \\ 4 & 2 \\ -4 & -2 \end{bmatrix}$.

*Ans.* (a) $A^T = \begin{bmatrix} 3 & 8 & 1 \\ 4 & 0 & 1 \\ -2 & 2 & -2 \end{bmatrix}$; (b) $B^T = \begin{bmatrix} 4 & 4 & -4 \\ 2 & 2 & -2 \end{bmatrix}$

**34.14.** Verify: (a) $(A^T)^T = A$; (b) $(A + B)^T = A^T + B^T$; (c) $(cA)^T = cA^T$.

# CHAPTER 35

# Matrix Multiplication and Inverses

## Definition of Inner Product

The inner product of a row of matrix $A$ with a column of matrix $B$ is defined, if and only if the number of columns of matrix $A$ equals the number of rows of matrix $B$, as the following real number: multiply each element of the row of $A$ by the corresponding element of the column of $B$ and sum the results. Thus:

$$a_i \cdot b_j = a_{i1}a_{i2}\ldots a_{ip} \cdot \begin{bmatrix} b_{1j} \\ b_{2j} \\ \ldots \\ b_{pj} \end{bmatrix} \cdots = a_{i1}b_{1j} + a_{i2}b_{2j} + \cdots + a_{ip}b_{pj}$$

**EXAMPLE 35.1** Find the inner product of row 1 of $\begin{bmatrix} 3 & 4 \\ 6 & -2 \end{bmatrix}$ with column 2 of $\begin{bmatrix} 5 & 9 & 2 \\ 0 & 7 & 8 \end{bmatrix}$.

$$[3 \quad 4] \cdot \begin{bmatrix} 9 \\ 7 \end{bmatrix} = 3(9) + 4(7) = 55$$

## Multiplication of Matrices

The product of two matrices is defined, if and only if the number of columns of matrix $A$ equals the number of rows of matrix $B$, as the following matrix $AB$: Assuming $A$ is an $m \times p$ matrix and $B$ is a $p \times n$ matrix, then $C = AB$ is an $m \times n$ matrix with the element in row $i$, column $j$, being the inner product of row $i$ of matrix $A$ with column $j$ of matrix $B$.

**EXAMPLE 35.2** Let $A = \begin{bmatrix} 3 & 4 \\ 6 & -2 \end{bmatrix}$ and $B = \begin{bmatrix} 5 & 9 & 2 \\ 0 & 7 & 8 \end{bmatrix}$. Find $AB$.

First note that $A$ is a $2 \times 2$ matrix and $B$ is a $2 \times 3$ matrix, hence $AB$ is defined and is a $2 \times 3$ matrix. The element in row 1, column 1 of $AB$ is the inner product of row 1 of $A$ with column 1 of $B$, thus:

$$[3 \quad 4] \cdot \begin{bmatrix} 5 \\ 0 \end{bmatrix} = 3(5) + 4(0) = 15$$

Continuing in this manner, form

$$AB = \begin{bmatrix} 3(5) & + 4(0) & 3(9) & + 4(7) & 3(2) & + 4(8) \\ 6(5) + (-2)(0) & 6(9) + (-2)(7) & 6(2) + (-2)(8) \end{bmatrix} = \begin{bmatrix} 15 & 55 & 38 \\ 30 & 40 & -4 \end{bmatrix}$$

## Properties of Matrix Multiplication

In general, matrix multiplication is not commutative, that is, there is no guarantee that $AB$ should equal $BA$. In case the two results are equal, the matrices $A$ and $B$ are said to *commute*. The following properties can be proved for matrices $A$, $B$, and $C$ when all products are defined:

1. **ASSOCIATIVE LAW:** $A(BC) = (AB)C$
2. **LEFT DISTRIBUTIVE LAW:** $A(B + C) = AB + AC$
3. **RIGHT DISTRIBUTIVE LAW:** $(B + C)A = BA + CA$

## Identity Matrix

An $n \times n$ square matrix with all main diagonal elements equal to 1, and all other elements equal to 0, is called an identity matrix, and is denoted $I_n$, or, if the order is clear from the context, $I$. For any $n \times n$ square matrix $A$,

$$AI_n = I_nA = A$$

**EXAMPLE 35.3**   $I_2 = \begin{bmatrix} 1 & 0 \\ 0 & 1 \end{bmatrix}$. $I_3 = \begin{bmatrix} 1 & 0 & 0 \\ 0 & 1 & 0 \\ 0 & 0 & 1 \end{bmatrix}$.

## Inverses of Matrices

If $A$ is a square matrix, there may exist another square matrix of the same size, $B$, such that $AB = BA = I$. If this is the case, $B$ is called the (multiplicative) *inverse* of $A$; the notation $A^{-1}$ is used for $B$, thus,

$$AA^{-1} = A^{-1}A = I$$

Not every square matrix has an inverse; a matrix that has an inverse is called *nonsingular*; a matrix that has no inverse is called *singular*. If an inverse can be found for a matrix, this inverse is unique; any other inverse is equal to this one.

**EXAMPLE 35.4**   Show that $B = \begin{bmatrix} -5 & 2 \\ 3 & -1 \end{bmatrix}$ is an inverse for $A = \begin{bmatrix} 1 & 2 \\ 3 & 5 \end{bmatrix}$.

Multiply the matrices to find $AB$ and $BA$:

$$AB = \begin{bmatrix} 1 & 2 \\ 3 & 5 \end{bmatrix}\begin{bmatrix} -5 & 2 \\ 3 & -1 \end{bmatrix} = \begin{bmatrix} 1(-5) + 2(3) & 1(2) + 2(-1) \\ 3(-5) + 5(3) & 3(2) + 5(-1) \end{bmatrix} = \begin{bmatrix} 1 & 0 \\ 0 & 1 \end{bmatrix} = I$$

$$BA = \begin{bmatrix} -5 & 2 \\ 3 & -1 \end{bmatrix}\begin{bmatrix} 1 & 2 \\ 3 & 5 \end{bmatrix} = \begin{bmatrix} (-5)1 + 2(3) & (-5)2 + 2(5) \\ 3(1) + (-1)3 & 3(2) + (-1)5 \end{bmatrix} = \begin{bmatrix} 1 & 0 \\ 0 & 1 \end{bmatrix} = I$$

Since $AB = BA = I$, $B = A^{-1}$.

## Calculating the Inverse of a Given Matrix

To find the inverse of a given nonsingular square matrix $A$, perform the following operations:

1. Adjoin to $A$ the identity matrix of the same order to form a matrix schematically indicated by: $[A \mid I]$.
2. Perform row operations on this matrix as in Gauss-Jordan elimination until the portion to the left of the vertical bar has been reduced to $I$. (If this is not possible, a row of zeros will appear, and the original matrix $A$ was in fact singular.)
3. The entire matrix will now appear as $[I \mid A^{-1}]$ and the matrix $A^{-1}$ can be read at the right of the vertical bar.

**EXAMPLE 35.5** Find the inverse for the matrix $A = \begin{bmatrix} 1 & 0 \\ 1 & 1 \end{bmatrix}$.

First, form the matrix $\begin{bmatrix} 1 & 0 & | & 1 & 0 \\ 1 & 1 & | & 0 & 1 \end{bmatrix}$. One operation: $R_2 + (-1)R_1 \to R_2$, transforms the portion to the left of the bar

into $I_2$ and yields $\begin{bmatrix} 1 & 0 & | & 1 & 0 \\ 0 & 1 & | & -1 & 1 \end{bmatrix}$. Then $A^{-1} = \begin{bmatrix} 1 & 0 \\ -1 & 1 \end{bmatrix}$ is read from the right side of the bar. Checking, note that

$$AA^{-1} = \begin{bmatrix} 1 & 0 \\ 1 & 1 \end{bmatrix}\begin{bmatrix} 1 & 0 \\ -1 & 1 \end{bmatrix} = \begin{bmatrix} 1 & 0 \\ 0 & 1 \end{bmatrix} \text{ and } A^{-1}A = \begin{bmatrix} 1 & 0 \\ -1 & 1 \end{bmatrix}\begin{bmatrix} 1 & 0 \\ 1 & 1 \end{bmatrix} = \begin{bmatrix} 1 & 0 \\ 0 & 1 \end{bmatrix}.$$

## SOLVED PROBLEMS

**35.1.** Given $A = [3 \quad 8]$ and $B = \begin{bmatrix} 5 \\ 2 \end{bmatrix}$, find the inner product of each row of $A$ with each column of $B$.

There is only one row of $A$ and only one column of $B$. The inner product required is given by $3 \cdot 5 + 8 \cdot 2 = 31$.

**35.2.** Given $A = \begin{bmatrix} 2 & 1 & 4 \\ -3 & -1 & 6 \end{bmatrix}$ and $B = \begin{bmatrix} 9 \\ 5 \\ -2 \end{bmatrix}$, find the inner product of each row of $A$ with each column of $B$.
There is only one column of $B$.

The inner product of row 1 of $A$ with this column is given by $2 \cdot 9 + 1 \cdot 5 + 4(-2) = 15$. The inner product of row 2 of $A$ with the column is given by $(-3)9 + (-1)5 + 6(-2) = -44$.

**35.3.** Find the order of $AB$ and $BA$, given the following orders for $A$ and $B$:

(a) $A: 2 \times 3$, $B: 3 \times 2$     (b) $A: 2 \times 3$, $B: 3 \times 3$     (c) $A: 2 \times 4$, $B: 4 \times 3$

(d) $A: 3 \times 2$, $B: 3 \times 2$     (e) $A: 3 \times 3$, $B: 3 \times 3$     (f) $A: 1 \times 3$, $B: 2 \times 2$

(a) A $2 \times 3$ matrix multiplied times a $3 \times 2$ matrix yields a $2 \times 2$ matrix for $AB$. A $3 \times 2$ matrix multiplied times a $2 \times 3$ matrix yields a $3 \times 3$ matrix for $BA$.

(b) A $2 \times 3$ matrix multiplied times a $3 \times 3$ matrix yields a $2 \times 3$ matrix for $AB$. A $3 \times 3$ matrix can only be multiplied times a matrix with three rows, thus $BA$ is not defined.

(c) A $2 \times 4$ matrix multiplied times a $4 \times 3$ matrix yields a $2 \times 3$ matrix for $AB$. A $4 \times 3$ matrix can only be multiplied times a matrix with three rows, thus $BA$ is not defined.

(d) A $3 \times 2$ matrix can only be multiplied times a matrix with two rows, thus neither $AB$ nor $BA$ is defined.

(e) A $3 \times 3$ matrix multiplied times a $3 \times 3$ matrix yields a $3 \times 3$ matrix for both $AB$ and $BA$.

(f) A $1 \times 3$ matrix can only be multiplied times a matrix with three rows, thus $AB$ is not defined. A $2 \times 2$ matrix can only be multiplied times a matrix with two rows, thus $BA$ is not defined.

**35.4.** Given $A = \begin{bmatrix} 2 & 1 & 4 \\ -3 & -1 & 6 \end{bmatrix}$ and $B = \begin{bmatrix} 9 \\ 5 \\ -2 \end{bmatrix}$, find $AB$ and $BA$.

Since $A$ is a $2 \times 3$ matrix and $B$ is a $3 \times 1$ matrix, $AB$ is defined and is a $2 \times 1$ matrix. The element in row 1, column 1 of $AB$ is the inner product of row 1 of $A$ with column 1 of $B$. This was found in Problem 35.2 to be 15. The element in row 2, column 1 of $AB$ is the inner product of row 2 of $A$ with column 1 of $B$. This was found in Problem 35.2 to be $-44$. Hence

$$AB = \begin{bmatrix} 15 \\ -44 \end{bmatrix}$$

Since $B$ is a $3 \times 1$ matrix, it can only be multiplied times a matrix with 1 row, thus $BA$ is not defined.

**35.5.** Given $A = \begin{bmatrix} 5 & 2 \\ 3 & 1 \end{bmatrix}$ and $B = \begin{bmatrix} 1 & 6 \\ -8 & 4 \end{bmatrix}$, find $AB$ and $BA$.

Since $A$ is a $2 \times 2$ matrix and $B$ is a $2 \times 2$ matrix, $AB$ is defined and is a $2 \times 2$ matrix. Find the inner product of each row of $A$ with each column of $B$ and form $AB$:

$$AB = \begin{bmatrix} 5(1) + 2(-8) & 5(6) + 2(4) \\ 3(1) + 1(-8) & 3(6) + 1(4) \end{bmatrix} = \begin{bmatrix} -11 & 38 \\ -5 & 22 \end{bmatrix}$$

Since $B$ is a $2 \times 2$ matrix and $A$ is a $2 \times 2$ matrix, $BA$ is defined and is a $2 \times 2$ matrix. Find the inner product of each row of $B$ with each column of $A$ and form $BA$:

$$BA = \begin{bmatrix} 1(5) + 6(3) & 1(2) + 6(1) \\ (-8)5 + 4(3) & (-8)2 + 4(1) \end{bmatrix} = \begin{bmatrix} 23 & 8 \\ -28 & -12 \end{bmatrix}$$

Note that $AB \neq BA$.

**35.6.** Explain why there is no commutative law for matrix multiplication.

Given two matrices $A$ and $B$, there are a number of ways in which $AB$ can fail to equal $BA$. First, either $AB$ or $BA$ may fail to be defined (for example, if $A$ is a $2 \times 1$ matrix and $B$ is a $2 \times 2$ matrix, $AB$ is undefined, while $BA$ is defined). Second, both $AB$ and $BA$ may be defined, but be of different orders (for example, if $A$ is a $2 \times 3$ matrix and $B$ is a $3 \times 2$ matrix, $AB$ is a $2 \times 2$ matrix and $BA$ is a $3 \times 3$ matrix). Finally, both $AB$ and $BA$ may be defined and of the same order, as in the previous problem, but because $AB$ involves the inner product of the rows of $A$ with the columns of $B$, while $BA$ involves the inner product of the rows of $B$ with the columns of $A$, $AB \neq BA$.

**35.7.** Given $A = \begin{bmatrix} 2 & 1 & 0 \\ 3 & -2 & 5 \\ -2 & 5 & 0 \end{bmatrix}$ and $B = \begin{bmatrix} 4 & 4 & -1 \\ -3 & 0 & 2 \end{bmatrix}$, find $AB$ and $BA$.

Since $A$ is a $3 \times 3$ matrix and $B$ is a $2 \times 3$ matrix, $AB$ is not defined.

Since $B$ is a $2 \times 3$ matrix and $A$ is a $3 \times 3$ matrix, $BA$ is defined and has order $2 \times 3$. Find the inner product of each row of $B$ with each column of $A$ and form $BA$:

$$BA = \begin{bmatrix} 4(2) + 4(3) + (-1)(-2) & 4(1) + 4(-2) + (-1)5 & 4(0) + 4(5) + (-1)0 \\ (-3)(2) + 0(3) + 2(-2) & (-3)(1) + 0(-2) + 2(5) & (-3)0 + 0(5) + 2(0) \end{bmatrix} = \begin{bmatrix} 22 & -9 & 20 \\ -10 & 7 & 0 \end{bmatrix}$$

**35.8.** Given $A = \begin{bmatrix} 3 & 1 \\ 0 & -3 \end{bmatrix}$, $B = \begin{bmatrix} 8 & 3 \\ 3 & 8 \end{bmatrix}$, and $C = \begin{bmatrix} -5 & -1 \\ 4 & 2 \end{bmatrix}$, verify the associative law for matrix multiplication

$(AB)C = A(BC)$.

First find $AB$ and $BC$:

$$AB = \begin{bmatrix} 3 & 1 \\ 0 & -3 \end{bmatrix}\begin{bmatrix} 8 & 3 \\ 3 & 8 \end{bmatrix} = \begin{bmatrix} 27 & 17 \\ -9 & -24 \end{bmatrix} \quad \text{and} \quad BC = \begin{bmatrix} 8 & 3 \\ 3 & 8 \end{bmatrix}\begin{bmatrix} -5 & -1 \\ 4 & 2 \end{bmatrix} = \begin{bmatrix} -28 & -2 \\ 17 & 13 \end{bmatrix}$$

Hence

$$(AB)C = \begin{bmatrix} 27 & 17 \\ -9 & -24 \end{bmatrix}\begin{bmatrix} -5 & -1 \\ 4 & 2 \end{bmatrix} = \begin{bmatrix} -67 & 7 \\ -51 & -39 \end{bmatrix} \quad \text{and} \quad A(BC) = \begin{bmatrix} 3 & 1 \\ 0 & -3 \end{bmatrix}\begin{bmatrix} -28 & -2 \\ 17 & 13 \end{bmatrix} = \begin{bmatrix} -67 & 7 \\ -51 & -39 \end{bmatrix}.$$

Thus $(AB)C = A(BC)$.

**35.9.** Given $A = \begin{bmatrix} 3 & 1 \\ 0 & -3 \end{bmatrix}$, $B = \begin{bmatrix} 8 & 3 \\ 3 & 8 \end{bmatrix}$, and $C = \begin{bmatrix} -5 & -1 \\ 4 & 2 \end{bmatrix}$, verify the left distributive law for matrix multiplication $A(B + C) = AB + AC$.

First find $B + C$ and $AC$. ($AB$ was found in the previous problem.)

$$B + C = \begin{bmatrix} 8 & 3 \\ 3 & 8 \end{bmatrix} + \begin{bmatrix} -5 & -1 \\ 4 & 2 \end{bmatrix} = \begin{bmatrix} 3 & 2 \\ 7 & 10 \end{bmatrix} \text{ and } AC = \begin{bmatrix} 3 & 1 \\ 0 & -3 \end{bmatrix}\begin{bmatrix} -5 & -1 \\ 4 & 2 \end{bmatrix} = \begin{bmatrix} -11 & -1 \\ -12 & -6 \end{bmatrix}$$

Hence

$$A(B + C) = \begin{bmatrix} 3 & 1 \\ 0 & -3 \end{bmatrix}\begin{bmatrix} 3 & 2 \\ 7 & 10 \end{bmatrix} = \begin{bmatrix} 16 & 16 \\ -21 & -30 \end{bmatrix} \text{ and } AB + AC = \begin{bmatrix} 27 & 17 \\ -9 & -24 \end{bmatrix} + \begin{bmatrix} -11 & -1 \\ -12 & -6 \end{bmatrix} = \begin{bmatrix} 16 & 16 \\ -21 & -30 \end{bmatrix}.$$

Thus $A(B + C) = AB + AC$.

**35.10.** Verify that $I_3 A = A$ for any $3 \times 3$ matrix $A$.

Let $A = \begin{bmatrix} a_{11} & a_{12} & a_{13} \\ a_{21} & a_{22} & a_{23} \\ a_{31} & a_{32} & a_{33} \end{bmatrix}$. Then

$$I_3 A = \begin{bmatrix} 1 & 0 & 0 \\ 0 & 1 & 0 \\ 0 & 0 & 1 \end{bmatrix}\begin{bmatrix} a_{11} & a_{12} & a_{13} \\ a_{21} & a_{22} & a_{23} \\ a_{31} & a_{32} & a_{33} \end{bmatrix} = \begin{bmatrix} 1a_{11} + 0a_{21} + 0a_{31} & 1a_{12} + 0a_{22} + 0a_{32} & 1a_{13} + 0a_{23} + 0a_{33} \\ 0a_{11} + 1a_{21} + 0a_{31} & 0a_{12} + 1a_{22} + 0a_{32} & 0a_{13} + 1a_{23} + 0a_{33} \\ 0a_{11} + 0a_{21} + 1a_{31} & 0a_{12} + 0a_{22} + 1a_{32} & 0a_{13} + 0a_{23} + 1a_{33} \end{bmatrix}$$

$$= \begin{bmatrix} a_{11} & a_{12} & a_{13} \\ a_{21} & a_{22} & a_{23} \\ a_{31} & a_{32} & a_{33} \end{bmatrix}$$

Thus $I_3 A = A$.

**35.11.** Show that $I_n X = X$ for any $n \times 1$ matrix $X$.

Since $I_n A = A$ for any $n \times n$ matrix $A$, multiplying by $I$ must leave each column of $A$ unchanged. Since each column of $A$ can be viewed as an $n \times 1$ matrix, multiplying by $I_n$ must leave any $n \times 1$ matrix unchanged. Thus $I_n X = X$.

**35.12.** Show that if $A = [a_{11}]$ is a $1 \times 1$ matrix with $a_{11} \neq 0$, then $A^{-1} = [1/a_{11}]$.

Since $[a_{11}][1/a_{11}] = [a_{11}(1/a_{11})] = [1] = I_1$ and $[1/a_{11}][a_{11}] = [(1/a_{11})a_{11}] = [1] = I_1$, it follows that $[1/a_{11}] = [a_{11}]^{-1}$.

**35.13.** Find $A^{-1}$ given $A = \begin{bmatrix} 1 & 3 \\ 4 & 11 \end{bmatrix}$.

Form the matrix

$$[A|I] = \begin{bmatrix} 1 & 3 & 1 & 0 \\ 4 & 11 & 0 & 1 \end{bmatrix}$$

Apply row operations on this matrix until the portion to the left of the vertical bar has been reduced to $I$.

$$\begin{bmatrix} 1 & 3 & 1 & 0 \\ 4 & 11 & 0 & 1 \end{bmatrix} \quad -4R_1 + R_2 \to R_2 \quad \begin{bmatrix} 1 & 3 & 1 & 0 \\ 0 & -1 & -4 & 1 \end{bmatrix} \quad 3R_2 + R_1 \to R_1 \quad \begin{bmatrix} 1 & 0 & -11 & 3 \\ 0 & -1 & -4 & 1 \end{bmatrix}$$

$$-R_2 \to R_2 \begin{bmatrix} 1 & 0 & -11 & 3 \\ 0 & 1 & 4 & -1 \end{bmatrix} = [I|A^{-1}]$$

Thus $A^{-1} = \begin{bmatrix} -11 & 3 \\ 4 & -1 \end{bmatrix}$

**35.14.** Show that the matrix $A = \begin{bmatrix} 2 & 5 \\ 4 & 10 \end{bmatrix}$ has no multiplicative inverse.

Form the matrix

$$[A|I] = \begin{bmatrix} 2 & 5 & | & 1 & 0 \\ 4 & 10 & | & 0 & 1 \end{bmatrix}$$

Applying row operations on this matrix to attempt to reduce the portion to the left of the vertical bar to $I$ yields

$$\begin{bmatrix} 2 & 5 & | & 1 & 0 \\ 4 & 10 & | & 0 & 1 \end{bmatrix} -2R_1 + R_2 \rightarrow R_2 \begin{bmatrix} 2 & 5 & | & 1 & 0 \\ 0 & 0 & | & -2 & 1 \end{bmatrix}$$

There is no way to produce a 1 in row 2, column 2 without destroying the 0 in row 2, column 1. Thus the portion to the left of the vertical bar cannot be reduced to $I$ and there is no inverse for $A$.

**35.15.** Show that if an inverse $B$ exists for a given matrix $A$, this inverse is unique, that is, any other inverse $C$ is equal to $B$.

Assume that both $B$ and $C$ are inverses of $A$, then $BA = I$ and $CA = I$; hence $BA = CA$. Multiply both sides of this true statement by $B$, then

$$(BA)B = (CA)B$$

By the associative law for matrix multiplication,

$$B(AB) = C(AB)$$

But, since $B$ is an inverse for $A$, $AB = I$; hence $BI = CI$, thus, $B = C$.

**35.16.** Find $A^{-1}$ given $A = \begin{bmatrix} 5 & 3 & 4 \\ 2 & 2 & 3 \\ 2 & 0 & 0 \end{bmatrix}$.

Form the matrix

$$[A|I] = \begin{bmatrix} 5 & 3 & 4 & | & 1 & 0 & 0 \\ 2 & 2 & 3 & | & 0 & 1 & 0 \\ 2 & 0 & 0 & | & 0 & 0 & 1 \end{bmatrix}$$

Apply row operations on this matrix until the portion to the left of the vertical bar has been reduced to $I$.

$$\begin{bmatrix} 5 & 3 & 4 & | & 1 & 0 & 0 \\ 2 & 2 & 3 & | & 0 & 1 & 0 \\ 2 & 0 & 0 & | & 0 & 0 & 1 \end{bmatrix} R_1 \leftrightarrow R_3 \begin{bmatrix} 2 & 0 & 0 & | & 0 & 0 & 1 \\ 2 & 2 & 3 & | & 0 & 1 & 0 \\ 5 & 3 & 4 & | & 1 & 0 & 0 \end{bmatrix} \tfrac{1}{2}R_1 \rightarrow R_1 \begin{bmatrix} 1 & 0 & 0 & | & 0 & 0 & 1/2 \\ 2 & 2 & 3 & | & 0 & 1 & 0 \\ 5 & 3 & 4 & | & 1 & 0 & 0 \end{bmatrix}$$

$$\begin{matrix} R_2 + (-2)R_1 \rightarrow R_2 \\ R_3 + (-5)R_1 \rightarrow R_3 \end{matrix} \begin{bmatrix} 1 & 0 & 0 & | & 0 & 0 & 1/2 \\ 0 & 2 & 3 & | & 0 & 1 & -1 \\ 0 & 3 & 4 & | & 1 & 0 & -5/2 \end{bmatrix} \tfrac{1}{2}R_2 \rightarrow R_2 \begin{bmatrix} 1 & 0 & 0 & | & 0 & 0 & 1/2 \\ 0 & 1 & 3/2 & | & 0 & 1/2 & -1/2 \\ 0 & 3 & 4 & | & 1 & 0 & -5/2 \end{bmatrix}$$

$$R_3 + (-3)R_2 \rightarrow R_3 \begin{bmatrix} 1 & 0 & 0 & | & 0 & 0 & 1/2 \\ 0 & 1 & 3/2 & | & 0 & 1/2 & -1/2 \\ 0 & 0 & -1/2 & | & 1 & -3/2 & -1 \end{bmatrix} R_2 + 3R_3 \rightarrow R_2 \begin{bmatrix} 1 & 0 & 0 & | & 0 & 0 & 1/2 \\ 0 & 1 & 0 & | & 3 & -4 & -7/2 \\ 0 & 0 & -1/2 & | & 1 & -3/2 & -1 \end{bmatrix}$$

$$(-2)R_3 \rightarrow R_3 \begin{bmatrix} 1 & 0 & 0 & | & 0 & 0 & 1/2 \\ 0 & 1 & 0 & | & 3 & -4 & -7/2 \\ 0 & 0 & 1 & | & -2 & 3 & 2 \end{bmatrix} = [I|A^{-1}]$$

Thus $A^{-1} = \begin{bmatrix} 0 & 0 & 1/2 \\ 3 & -4 & -7/2 \\ -2 & 3 & 2 \end{bmatrix}$

**35.17.** Show that any system of $m$ linear equations in $n$ variables:

$$a_{11}x_1 + a_{12}x_2 + \cdots + a_{1n}x_n = b_1$$
$$a_{21}x_1 + a_{22}x_2 + \cdots + a_{2n}x_n = b_2$$
$$\cdots\cdots\cdots\cdots\cdots\cdots\cdots\cdots\cdots\cdots$$
$$a_{m1}x_1 + a_{m2}x_2 + \cdots + a_{mn}x_n = b_m$$

can be written as $AX = B$, where $A$ is called the *coefficient matrix* of the system, and $A$, $X$, and $B$ are given respectively by:

$$A = \begin{bmatrix} a_{11} & a_{12} & a_{13} & \cdots & a_{1n} \\ a_{21} & a_{22} & a_{23} & \cdots & a_{2n} \\ \cdots & \cdots & \cdots & \cdots & \cdots \\ a_{m1} & a_{m2} & a_{m3} & \cdots & a_{mn} \end{bmatrix} \qquad X = \begin{bmatrix} x_1 \\ x_2 \\ \cdots \\ x_n \end{bmatrix} \qquad B = \begin{bmatrix} b_1 \\ b_2 \\ \cdots \\ b_m \end{bmatrix}$$

With $A$ as the $m \times n$ matrix shown and $X$ as the $n \times 1$ matrix shown, the product $AX$ is the $m \times 1$ matrix:

$$\begin{bmatrix} a_{11}x_1 + a_{12}x_2 + \cdots + a_{1n}x_n \\ a_{21}x_1 + a_{22}x_2 + \cdots + a_{2n}x_n \\ \cdots\cdots\cdots\cdots\cdots\cdots\cdots\cdots\cdots \\ a_{m1}x_1 + a_{m2}x_2 + \cdots + a_{mn}x_n \end{bmatrix}$$

Thus, by the definition of matrix equality, the matrix equation $AX = B$ holds if and only if each entry of $AX$ is equal to the corresponding element of the $m \times 1$ matrix $B$, that is, if and only if the system of equations is satisfied. That is, the matrix equation is simply the system of equations written in matrix notation.

**35.18.** Show that if $A$ is a nonsingular square matrix, then the matrix $X$ that satisfies the matrix equation $AX = B$ is given by $X = A^{-1}B$, where

$$X = \begin{bmatrix} x_1 \\ x_2 \\ \cdots \\ x_n \end{bmatrix} \qquad \text{and} \qquad B = \begin{bmatrix} b_1 \\ b_2 \\ \cdots \\ b_n \end{bmatrix}$$

Let $AX = B$. Then, since $A$ is nonsingular, $A^{-1}$ exists; multiplying both sides of this equation by $A^{-1}$ yields:

$$A^{-1}AX = A^{-1}B$$

$$IX = A^{-1}B$$

$$X = A^{-1}B$$

**35.19.** Use the result of the previous problem to solve the system of equations

$$x_1 + x_2 + x_3 = b_1$$
$$x_1 + 2x_2 + 3x_3 = b_2$$
$$x_1 + x_2 + 2x_3 = b_3$$

given (a) $b_1 = 3, b_2 = 4, b_3 = 5$; (b) $b_1 = -7, b_2 = 9, b_3 = -6$.
The given system of equations can be written as $AX = B$, with

$$A = \begin{bmatrix} 1 & 1 & 1 \\ 1 & 2 & 3 \\ 1 & 1 & 2 \end{bmatrix} \qquad X = \begin{bmatrix} x_1 \\ x_2 \\ x_3 \end{bmatrix} \qquad B = \begin{bmatrix} b_1 \\ b_2 \\ b_3 \end{bmatrix}$$

To apply the result of the previous problem, first find $A^{-1}$. Start by forming the matrix

$$[A|I] = \begin{bmatrix} 1 & 1 & 1 & 1 & 0 & 0 \\ 1 & 2 & 3 & 0 & 1 & 0 \\ 1 & 1 & 2 & 0 & 0 & 1 \end{bmatrix}$$

Apply row operations on this matrix until the portion to the left of the vertical bar has been reduced to $I$.

$$\begin{bmatrix} 1 & 1 & 1 & 1 & 0 & 0 \\ 1 & 2 & 3 & 0 & 1 & 0 \\ 1 & 1 & 2 & 0 & 0 & 1 \end{bmatrix} \begin{matrix} R_2 + (-1)R_1 \to R_2 \\ R_3 + (-1)R_1 \to R_3 \end{matrix} \begin{bmatrix} 1 & 1 & 1 & 1 & 0 & 0 \\ 0 & 1 & 2 & -1 & 1 & 0 \\ 0 & 0 & 1 & -1 & 0 & 1 \end{bmatrix} \begin{matrix} R_1 + (-1)R_3 \to R_1 \\ R_2 + (-2)R_3 \to R_2 \end{matrix}$$

$$\begin{bmatrix} 1 & 1 & 0 & 2 & 0 & -1 \\ 0 & 1 & 0 & 1 & 1 & -2 \\ 0 & 0 & 1 & -1 & 0 & 1 \end{bmatrix} \; R_1 + (-1)R_2 \to R_1 \begin{bmatrix} 1 & 0 & 0 & 1 & -1 & 1 \\ 0 & 1 & 0 & 1 & 1 & -2 \\ 0 & 0 & 1 & -1 & 0 & 1 \end{bmatrix}$$

Thus $A^{-1} = \begin{bmatrix} 1 & -1 & 1 \\ 1 & 1 & -2 \\ -1 & 0 & 1 \end{bmatrix}$. Now the solutions of the given systems are given by $X = A^{-1}B$. Hence

(a) $\begin{bmatrix} x_1 \\ x_2 \\ x_3 \end{bmatrix} = \begin{bmatrix} 1 & -1 & 1 \\ 1 & 1 & -2 \\ -1 & 0 & 1 \end{bmatrix} \begin{bmatrix} b_1 \\ b_2 \\ b_3 \end{bmatrix} = \begin{bmatrix} 1 & -1 & 1 \\ 1 & 1 & -2 \\ -1 & 0 & 1 \end{bmatrix} \begin{bmatrix} 3 \\ 4 \\ 5 \end{bmatrix} = \begin{bmatrix} 4 \\ -3 \\ 2 \end{bmatrix}$, that is, $x_1 = 4, x_2 = -3, x_3 = 2$.

(b) $\begin{bmatrix} x_1 \\ x_2 \\ x_3 \end{bmatrix} = \begin{bmatrix} 1 & -1 & 1 \\ 1 & 1 & -2 \\ -1 & 0 & 1 \end{bmatrix} \begin{bmatrix} b_1 \\ b_2 \\ b_3 \end{bmatrix} = \begin{bmatrix} 1 & -1 & 1 \\ 1 & 1 & -2 \\ -1 & 0 & 1 \end{bmatrix} \begin{bmatrix} -7 \\ 9 \\ -6 \end{bmatrix} = \begin{bmatrix} -22 \\ 14 \\ 1 \end{bmatrix}$, that is, $x_1 = -22, x_2 = 14, x_3 = 1$.

Note that, in general, this method of solving systems of linear equations is not more efficient than elimination methods, since the calculation of the inverse matrix already requires all of the steps in a Gauss-Jordan elimination. However, the method is useful if, as in this problem, several systems with the same coefficient matrix and different right-hand sides are to be solved.

## SUPPLEMENTARY PROBLEMS

**35.20.** Given $A = \begin{bmatrix} 1 \\ 3 \end{bmatrix}$ and $B = [2 \quad 4]$, find $AB$ and $BA$.

*Ans.* $AB = \begin{bmatrix} 2 & 4 \\ 6 & 12 \end{bmatrix}$, $BA = [14]$

**35.21.** Given $A = \begin{bmatrix} 2 & 3 \\ -4 & 5 \end{bmatrix}$ and $B = \begin{bmatrix} 1 & -2 & 3 \\ 4 & 0 & 6 \end{bmatrix}$, find $AB$ and $BA$.

*Ans.* $AB = \begin{bmatrix} 14 & -4 & 24 \\ 16 & 8 & 18 \end{bmatrix}$, $BA$ is not defined

**35.22.** If $A$ is a square matrix, $A^2$ is defined as $AA$. Find $A^2$ if $A$ is given by:

(a) $\begin{bmatrix} 1 & -1 \\ -1 & 1 \end{bmatrix}$; (b) $\begin{bmatrix} 2 & 0 & 1 \\ 1 & 3 & 2 \\ -3 & -1 & 0 \end{bmatrix}$

*Ans.*  (a) $\begin{bmatrix} 2 & -2 \\ -2 & 2 \end{bmatrix}$; (b) $\begin{bmatrix} 1 & -1 & 2 \\ -1 & 7 & 7 \\ -7 & -3 & -5 \end{bmatrix}$

**35.23.** Given $A = \begin{bmatrix} 3 & 1 \\ 0 & -3 \end{bmatrix}$, $B = \begin{bmatrix} 8 & 3 \\ 3 & 8 \end{bmatrix}$, and $C = \begin{bmatrix} -5 & -1 \\ 4 & 2 \end{bmatrix}$, verify the right distributive law for matrix multiplication $(B + C)A = BA + CA$.

**35.24.** An orthonormal matrix is defined as a square matrix $A$ with its transpose equal to its inverse: $A^T = A^{-1}$. (See Problem 34.13.) Show that

$$\begin{bmatrix} 1/\sqrt{2} & -1/\sqrt{2} \\ 1/\sqrt{2} & 1/\sqrt{2} \end{bmatrix}$$

is an orthonormal matrix.

**35.25.** For square matrices $I, A, B$ of order $n \times n$, verify that (a) $I^{-1} = I$; (b) $(A^{-1})^{-1} = A$; (c) $(AB)^{-1} = B^{-1}A^{-1}$.

**35.26.** Find inverses for

(a) $\begin{bmatrix} 3 & 0 \\ 0 & 1/2 \end{bmatrix}$; (b) $\begin{bmatrix} 3 & 5 \\ -3 & -2 \end{bmatrix}$; (c) $\begin{bmatrix} 3 & 4 & 5 \\ 1 & 0 & 1 \\ 4 & 4 & 6 \end{bmatrix}$; (d) $\begin{bmatrix} 3 & 3 & 1 \\ 2 & -1 & 1 \\ -2 & -1 & -2 \end{bmatrix}$; (e) $\begin{bmatrix} 1 & 0 & 1 & 0 \\ 0 & 1 & 0 & 1 \\ -1 & 0 & 1 & 0 \\ 0 & -1 & 0 & 1 \end{bmatrix}$

*Ans.*  (a) $\begin{bmatrix} 1/3 & 0 \\ 0 & 2 \end{bmatrix}$; (b) $\frac{1}{9}\begin{bmatrix} -2 & -5 \\ 3 & 3 \end{bmatrix}$; (c) no inverse exists; (d) $\frac{1}{11}\begin{bmatrix} 3 & 5 & 4 \\ 2 & -4 & -1 \\ -4 & -3 & -9 \end{bmatrix}$;

(e) $\frac{1}{2}\begin{bmatrix} 1 & 0 & -1 & 0 \\ 0 & 1 & 0 & -1 \\ 1 & 0 & 1 & 0 \\ 0 & 1 & 0 & 1 \end{bmatrix}$

**35.27.** Use the result of Problem 35.26d to solve the system

$$3x + 3y + z = b_1$$

$$2x - y + z = b_2$$

$$-2x - y - 2z = b_3$$

for (a) $b_1 = -4, b_2 = 0, b_3 = 3$; (b) $b_1 = 11, b_2 = 22, b_3 = -11$; (c) $b_1 = 2, \ b_2 = -1, b_3 = 5$.

*Ans.*  (a) $x = 0, y = -1, z = -1$; (b) $x = 9, y = -5, z = -1$; (c) $x = \dfrac{21}{11}, y = \dfrac{3}{11}, z = -\dfrac{50}{11}$

# CHAPTER 36

# Determinants and Cramer's Rule

## Notation for the Determinant of a Matrix

Associated with every square matrix $A$ is a number called the *determinant* of the matrix, written $\det A$ or $|A|$. For a $1 \times 1$ matrix $A = [a_{11}]$, the determinant is written $|A|$ and its value is defined as $|A| = a_{11}$ (*Note*: the vertical bars do not denote absolute value).

## Determinant of a 2 × 2 Matrix

Let $A = \begin{bmatrix} a_{11} & a_{12} \\ a_{21} & a_{22} \end{bmatrix}$. Then the determinant of $A$ is written: $|A| = \begin{vmatrix} a_{11} & a_{12} \\ a_{21} & a_{22} \end{vmatrix}$;

its value is defined as $\begin{vmatrix} a_{11} & a_{12} \\ a_{21} & a_{22} \end{vmatrix} = a_{11}a_{22} - a_{21}a_{12}$.

The determinant of an $n \times n$ matrix is referred to as an $n \times n$ determinant.

**EXAMPLE 36.1** $\begin{vmatrix} 3 & 7 \\ 4 & 6 \end{vmatrix} = 3 \cdot 6 - 4 \cdot 7 = -10.$

## Minors and Cofactors

For any $n \times n$ matrix $(a_{ij})$ with $n > 1$, the following are defined:

1. The *minor* $M_{ij}$ of element $a_{ij}$ is the determinant of the $(n-1) \times (n-1)$ matrix found by deleting row $i$ and column $j$ from $(a_{ij})$.
2. The *cofactor* $A_{ij}$ of element $a_{ij}$ is $A_{ij} = (-1)^{i+j}M_{ij}$. A cofactor is sometimes referred to as a *signed minor*.

**EXAMPLE 36.2** Find $M_{12}$ and $A_{12}$ for the matrix $\begin{bmatrix} 8 & 2 \\ 3 & -5 \end{bmatrix}$.

Delete row 1 and column 2 to obtain $\begin{bmatrix} 8 & 2 \\ 3 & -5 \end{bmatrix}$

Then $M_{12} = 3$ and $A_{12} = (-1)^{1+2}M_{12} = (-1)^3(3) = -3$.

**EXAMPLE 36.3** Find $M_{23}$ and $A_{23}$ for the matrix $\begin{bmatrix} a_{11} & a_{12} & a_{13} \\ a_{21} & a_{22} & a_{23} \\ a_{31} & a_{32} & a_{33} \end{bmatrix}$.

Delete row 2 and column 3 to obtain $\begin{bmatrix} a_{11} & a_{12} & a_{13} \\ a_{21} & a_{22} & a_{23} \\ a_{31} & a_{32} & a_{33} \end{bmatrix}$. Hence:

$$M_{23} = \begin{vmatrix} a_{11} & a_{12} \\ a_{31} & a_{32} \end{vmatrix} = a_{11}a_{32} - a_{31}a_{12}$$

$$A_{23} = (-1)^{2+3}M_{23} = (-1)^5(a_{11}a_{32} - a_{31}a_{12}) = a_{31}a_{12} - a_{11}a_{32}$$

## Determinant of a 3 × 3 Matrix

The determinant of a 3 × 3 matrix is defined as follows:

$$|A| = \begin{vmatrix} a_{11} & a_{12} & a_{13} \\ a_{21} & a_{22} & a_{23} \\ a_{31} & a_{32} & a_{33} \end{vmatrix} = a_{11}A_{11} + a_{12}A_{12} + a_{13}A_{13}$$

That is, the value of the determinant is found by multiplying each element in row 1 by its cofactor, then adding these results. This definition is often referred to as *expanding by the first row*.

**EXAMPLE 36.4**   Evaluate $\begin{vmatrix} 3 & 1 & -2 \\ 2 & 4 & 1 \\ 3 & 6 & 5 \end{vmatrix}$.

$$\begin{vmatrix} 3 & 1 & -2 \\ 2 & 4 & 1 \\ 3 & 6 & 5 \end{vmatrix} = 3(-1)^{1+1}\begin{vmatrix} 4 & 1 \\ 6 & 5 \end{vmatrix} + 1(-1)^{1+2}\begin{vmatrix} 2 & 1 \\ 3 & 5 \end{vmatrix} + (-2)(-1)^{1+3}\begin{vmatrix} 2 & 4 \\ 3 & 6 \end{vmatrix}$$

$$= 3(4 \cdot 5 - 6 \cdot 1) - 1(2 \cdot 5 - 3 \cdot 1) - 2(2 \cdot 6 - 3 \cdot 4)$$

$$= 3 \cdot 14 - 1 \cdot 7 - 2 \cdot 0$$

$$= 35$$

## Determinant of an *n* × *n* Matrix

The determinant of an $n \times n$ matrix is defined as

$$|A| = a_{11}A_{11} + a_{12}A_{12} + \cdots + a_{1n}A_{1n}$$

Again, the value of the determinant is found by multiplying each element in row 1 by its cofactor, then adding these results.

## Properties of Determinants

The following can be proved in general for any $n \times n$ determinant.

1. The value of the determinant may be found by multiplying each element in any one row or any one column by its cofactor, then adding these results. (This is referred to as expanding by a particular row or column.)
2. The value of a determinant is unchanged if the matrix is replaced by its transpose, that is, each row is rewritten as a column. (This is referred to as interchanging rows and columns.)
3. If each element in any one row or any one column is multiplied by $c$, the value of the determinant is multiplied by $c$.
4. If a row operation $R_i \leftrightarrow R_j$ is performed on a determinant, that is, if any two rows are interchanged (or if any two columns are interchanged), the value of the determinant is multiplied by $-1$.
5. If two rows of a determinant are equal (that is, each element of row $i$ is equal to the corresponding element of row $j$), the value of the determinant is 0. If two columns of a determinant are equal, the value of the determinant is 0.
6. If any row or any column of a determinant consists entirely of zeros, the value of the determinant is 0.
7. If a row operation $R_i + kR_j \rightarrow R_i$ is performed on a determinant, that is, the elements of any row are replaced by their sum with a constant multiple of another row, the value of the determinant is unchanged. If an analogous column operation, written $C_i + kC_j \rightarrow C_i$, is performed, the value of the determinant is also unchanged.

## Cramer's Rule for Solving Systems of Equations

1. Let

$$a_{11}x + a_{12}y = b_1$$
$$a_{21}x + a_{22}y = b_2$$

be a 2 × 2 system of equations. Define the determinants

$$D = \begin{vmatrix} a_{11} & a_{12} \\ a_{21} & a_{22} \end{vmatrix} \quad D_1 = \begin{vmatrix} b_1 & a_{12} \\ b_2 & a_{22} \end{vmatrix} \quad D_2 = \begin{vmatrix} a_{11} & b_1 \\ a_{21} & b_2 \end{vmatrix}$$

$D$ is the determinant of the coefficient matrix of the system, and is referred to as the *determinant of the system*. $D_1$ and $D_2$ are the determinants found by replacing, respectively, the first and second columns of $D$ with the constants $b_j$. Cramer's rule states that if, and only if, $D \neq 0$, then the system has exactly one solution, given by

$$x = \frac{D_1}{D} \quad y = \frac{D_2}{D}$$

2. Let

$$a_{11}x_1 + a_{12}x_2 + a_{13}x_3 = b_1$$
$$a_{21}x_1 + a_{22}x_2 + a_{23}x_3 = b_2$$
$$a_{31}x_1 + a_{32}x_2 + a_{33}x_3 = b_3$$

be a 3 × 3 system of equations. Define the determinants

$$D = \begin{vmatrix} a_{11} & a_{12} & a_{13} \\ a_{21} & a_{22} & a_{23} \\ a_{31} & a_{32} & a_{33} \end{vmatrix} \quad D_1 = \begin{vmatrix} b_1 & a_{12} & a_{13} \\ b_2 & a_{22} & a_{23} \\ b_3 & a_{32} & a_{33} \end{vmatrix} \quad D_2 = \begin{vmatrix} a_{11} & b_1 & a_{13} \\ a_{21} & b_2 & a_{23} \\ a_{31} & b_3 & a_{33} \end{vmatrix} \quad D_3 = \begin{vmatrix} a_{11} & a_{12} & b_1 \\ a_{21} & a_{22} & b_2 \\ a_{31} & a_{32} & b_3 \end{vmatrix}$$

Again, $D$ is the determinant of the coefficient matrix of the system, and is referred to as the determinant of the system. $D_1$, $D_2$, and $D_3$ are the determinants found by replacing, respectively, the first, second, and third columns of $D$ with the constants $b_j$. Cramer's rule states that if, and only if, $D \neq 0$, then the system has exactly one solution, given by

$$x_1 = \frac{D_1}{D} \quad x_2 = \frac{D_2}{D} \quad x_3 = \frac{D_3}{D}$$

3. Cramer's rule can be extended to arbitrary systems of *n* equations in *n* variables. However, evaluation of large determinants is time-consuming; hence the rule is not a practical method of solving large systems (Gaussian or Gauss-Jordan elimination is generally more efficient); it is of theoretical importance, however.

## SOLVED PROBLEMS

**36.1.** Evaluate the determinants: (a) $\begin{vmatrix} 9 & 4 \\ 3 & 8 \end{vmatrix}$; (b) $\begin{vmatrix} 8 & 4 \\ 16 & 8 \end{vmatrix}$; (c) $\begin{vmatrix} 3 & 8 \\ 9 & 4 \end{vmatrix}$

(a) $\begin{vmatrix} 9 & 4 \\ 3 & 8 \end{vmatrix} = 9 \cdot 8 - 3 \cdot 4 = 60$; (b) $\begin{vmatrix} 8 & 4 \\ 16 & 8 \end{vmatrix} = 8 \cdot 8 - 16 \cdot 4 = 0$; (c) $\begin{vmatrix} 3 & 8 \\ 9 & 4 \end{vmatrix} = 3 \cdot 4 - 9 \cdot 8 = -60$

**36.2.** Evaluate the determinants: (a) $\begin{vmatrix} 5 & 2 & -2 \\ 3 & 4 & 0 \\ -4 & 2 & 6 \end{vmatrix}$; (b) $\begin{vmatrix} 5 & 2 & -2 \\ 3 & 4 & 0 \\ 8 & 6 & -2 \end{vmatrix}$

Use the definition of a $3 \times 3$ determinant (expanding by row 1):

(a) The value of the determinant is found by multiplying each element in row 1 by its cofactor, then adding the results:

$$\begin{vmatrix} 5 & 2 & -2 \\ 3 & 4 & 0 \\ -4 & 2 & 6 \end{vmatrix} = 5(-1)^{1+1}\begin{vmatrix} 4 & 0 \\ 2 & 6 \end{vmatrix} + 2(-1)^{1+2}\begin{vmatrix} 3 & 0 \\ -4 & 6 \end{vmatrix} + (-2)(-1)^{1+3}\begin{vmatrix} 3 & 4 \\ -4 & 2 \end{vmatrix}$$

$$= 5(4 \cdot 6 - 2 \cdot 0) - 2[3 \cdot 6 - (-4) \cdot 0] - 2[3 \cdot 2 - (-4) \cdot 4]$$

$$= 120 - 36 - 44$$

$$= 40$$

(b) Proceed as in (a):

$$\begin{vmatrix} 5 & 2 & -2 \\ 3 & 4 & 0 \\ 8 & 6 & -2 \end{vmatrix} = 5(-1)^{1+1}\begin{vmatrix} 4 & 0 \\ 6 & -2 \end{vmatrix} + 2(-1)^{1+2}\begin{vmatrix} 3 & 0 \\ 8 & -2 \end{vmatrix} + (-2)(-1)^{1+3}\begin{vmatrix} 3 & 4 \\ 8 & 6 \end{vmatrix}$$

$$= 5[4 \cdot (-2) - 6 \cdot 0] - 2[3 \cdot (-2) - 8 \cdot 0] - 2(3 \cdot 6 - 8 \cdot 4) = -40 + 12 + 28 = 0$$

**36.3.** Derive the following formula for a general $3 \times 3$ determinant:

$$\begin{vmatrix} a_{11} & a_{12} & a_{13} \\ a_{21} & a_{22} & a_{23} \\ a_{31} & a_{32} & a_{33} \end{vmatrix} = a_{11}a_{22}a_{33} - a_{11}a_{23}a_{32} - a_{12}a_{21}a_{33} + a_{12}a_{23}a_{31} + a_{13}a_{21}a_{32} - a_{13}a_{22}a_{31}$$

Expand the determinant by the first row:

$$\begin{vmatrix} a_{11} & a_{12} & a_{13} \\ a_{21} & a_{22} & a_{23} \\ a_{31} & a_{32} & a_{33} \end{vmatrix} = a_{11}(-1)^{1+1}\begin{vmatrix} a_{22} & a_{23} \\ a_{32} & a_{33} \end{vmatrix} + a_{12}(-1)^{1+2}\begin{vmatrix} a_{21} & a_{23} \\ a_{31} & a_{33} \end{vmatrix} + a_{13}(-1)^{1+3}\begin{vmatrix} a_{21} & a_{22} \\ a_{31} & a_{32} \end{vmatrix}$$

$$= a_{11}(a_{22}a_{33} - a_{32}a_{23}) - a_{12}(a_{21}a_{33} - a_{31}a_{23}) + a_{13}(a_{21}a_{32} - a_{31}a_{22})$$

$$= a_{11}a_{22}a_{33} - a_{11}a_{23}a_{32} - a_{12}a_{21}a_{33} + a_{12}a_{23}a_{31} + a_{13}a_{21}a_{32} - a_{13}a_{22}a_{31}$$

**36.4.** Property 1 of determinants states that the value of a determinant may be found by expanding by any row or column. Verify this for the above determinant for the case of expanding by the first column.

Multiply each element of the first column by its cofactor and add the results to obtain:

$$a_{11}A_{11} + a_{21}A_{21} + a_{31}A_{31} = a_{11}(-1)^{1+1}\begin{vmatrix} a_{22} & a_{23} \\ a_{32} & a_{33} \end{vmatrix} + a_{21}(-1)^{2+1}\begin{vmatrix} a_{12} & a_{13} \\ a_{32} & a_{33} \end{vmatrix} + a_{31}(-1)^{3+1}\begin{vmatrix} a_{12} & a_{13} \\ a_{22} & a_{23} \end{vmatrix}$$

$$= a_{11}(a_{22}a_{33} - a_{32}a_{23}) - a_{21}(a_{12}a_{33} - a_{32}a_{13}) + a_{31}(a_{12}a_{23} - a_{22}a_{13})$$

$$= a_{11}a_{22}a_{33} - a_{11}a_{32}a_{23} - a_{21}a_{12}a_{33} + a_{21}a_{32}a_{13} + a_{31}a_{12}a_{23} - a_{31}a_{22}a_{13}$$

$$= a_{11}a_{22}a_{33} - a_{11}a_{23}a_{32} - a_{12}a_{21}a_{33} + a_{12}a_{23}a_{31} + a_{13}a_{21}a_{32} - a_{13}a_{22}a_{31}$$

where the last equality follows by rearranging the order of factors and terms by the commutative and associative laws for multiplication and addition of real numbers. The last expression is precisely the quantity derived in Problem 36.3.

**36.5.** Find the value of $\begin{vmatrix} 5 & 2 & -3 \\ 4 & 0 & 1 \\ -2 & 0 & 3 \end{vmatrix}$

Use property 1 of determinants to expand by the second column. Then

$$\begin{vmatrix} 5 & 2 & -3 \\ 4 & 0 & 1 \\ -2 & 0 & 3 \end{vmatrix} = 2(-1)^{1+2}\begin{vmatrix} 4 & 1 \\ -2 & 3 \end{vmatrix} + 0(A_{22}) + 0(A_{32}) = -2[4 \cdot 3 - (-2)1] = -28$$

where the cofactors $A_{22}$ and $A_{32}$ need not be evaluated, since they are multiplied by 0.

**36.6.** Property 2 of determinants states that the value of a determinant is unchanged if the matrix is replaced by its transpose, that is, each row is rewritten as a column. Verify this for an arbitrary $2 \times 2$ determinant.

Consider the determinant

$$\begin{vmatrix} a_{11} & a_{12} \\ a_{21} & a_{22} \end{vmatrix} = a_{11}a_{22} - a_{21}a_{12} \quad \text{(by definition)}$$

The determinant of the transposed matrix is then

$$\begin{vmatrix} a_{11} & a_{21} \\ a_{12} & a_{22} \end{vmatrix}$$

But by the definition of the $2 \times 2$ determinant this must equal $a_{11}a_{22} - a_{12}a_{21}$, which is clearly the same as $a_{11}a_{22} - a_{21}a_{12}$. Thus the value of the determinant is unchanged by interchanging rows and columns.

**36.7.** Property 3 of determinants states that if each element in any one row or any one column is multiplied by $c$, the value of the determinant is multiplied by $c$. Verify this for the first row of a $2 \times 2$ determinant.

Consider

$$\begin{vmatrix} ca_{11} & ca_{12} \\ a_{21} & a_{22} \end{vmatrix} = ca_{11}a_{22} - ca_{21}a_{12} = c(a_{11}a_{22} - a_{21}a_{12}) = c\begin{vmatrix} a_{11} & a_{12} \\ a_{21} & a_{22} \end{vmatrix}$$

**36.8.** Property 4 of determinants states that if any two rows are interchanged (or if any two columns are interchanged), the value of the determinant is multiplied by $-1$. Verify this for the two rows of a $2 \times 2$ determinant.

Consider

$$\begin{vmatrix} a_{11} & a_{12} \\ a_{21} & a_{22} \end{vmatrix} = a_{11}a_{22} - a_{21}a_{12}$$

Now interchange the two rows to obtain

$$\begin{vmatrix} a_{21} & a_{22} \\ a_{11} & a_{12} \end{vmatrix}$$

By the definition of the $2 \times 2$ determinant, this must equal $a_{21}a_{12} - a_{11}a_{22} = -1(a_{11}a_{22} - a_{21}a_{12})$. Thus interchanging the two rows multiplies the value of the determinant by $-1$.

**36.9.** Property 7 of determinants states that if a row operation $R_i + kR_j \to R_i$ is performed on a determinant, that is, the elements of any row are replaced by their sum with a constant multiple of another row, the value of the determinant is unchanged. Verify this for the operation $R_1 + kR_2 \to R_1$ performed on a $2 \times 2$ determinant.

Consider

$$\begin{vmatrix} a_{11} & a_{12} \\ a_{21} & a_{22} \end{vmatrix} = a_{11}a_{22} - a_{21}a_{12}$$

Now perform the operation $R_1 + kR_2 \to R_1$ to obtain

$$\begin{vmatrix} a_{11} + ka_{21} & a_{12} + ka_{22} \\ a_{21} & a_{22} \end{vmatrix} = (a_{11} + ka_{21})a_{22} - a_{21}(a_{12} + ka_{22})$$

Simplifying the last expression yields

$$(a_{11} + ka_{21})a_{22} - a_{21}(a_{12} + ka_{22}) = a_{11}a_{22} + ka_{21}a_{22} - a_{21}a_{12} - ka_{21}a_{22} = a_{11}a_{22} - a_{21}a_{12},$$

that is, the value of the original determinant has not been changed.

**36.10.** Property 7 is used to evaluate large determinants by generating rows or columns in which many zeroes appear. Illustrate by applying property 7 to evaluate:

(a) $\begin{vmatrix} 5 & 6 & 7 \\ 5 & 7 & 9 \\ 10 & 9 & -1 \end{vmatrix}$; (b) $\begin{vmatrix} 1 & 2 & 3 & 4 \\ 0 & 3 & 0 & 2 \\ 2 & 4 & 5 & 6 \\ 3 & 7 & 8 & 2 \end{vmatrix}$

*Ans.* (a) $\begin{vmatrix} 5 & 6 & 7 \\ 5 & 7 & 9 \\ 10 & 9 & -1 \end{vmatrix} \begin{array}{l} R_2 + (-1)R_1 \to R_2 \\ R_3 + (-2)R_1 \to R_3 \end{array} \begin{vmatrix} 5 & 6 & 7 \\ 0 & 1 & 2 \\ 0 & -3 & -15 \end{vmatrix}$

The latter determinant can be efficiently evaluated by expanding by the first column:

$$\begin{vmatrix} 5 & 6 & 7 \\ 0 & 1 & 2 \\ 0 & -3 & -15 \end{vmatrix} = 5(-1)^{1+1} \begin{vmatrix} 1 & 2 \\ -3 & -15 \end{vmatrix} + 0(A_{21}) + 0(A_{31}) = 5[1(-15) - (-3)2] = -45$$

(b)

$\begin{vmatrix} 1 & 2 & 3 & 4 \\ 0 & 3 & 0 & 2 \\ 2 & 4 & 5 & 6 \\ 3 & 7 & 8 & 2 \end{vmatrix} \begin{array}{l} R_3 + (-2)R_1 \to R_3 \\ R_4 + (-3)R_1 \to R_4 \end{array} \begin{vmatrix} 1 & 2 & 3 & 4 \\ 0 & 3 & 0 & 2 \\ 0 & 0 & -1 & -2 \\ 0 & 1 & -1 & -10 \end{vmatrix} = 1(-1)^{1+1} \begin{vmatrix} 3 & 0 & 2 \\ 0 & -1 & -2 \\ 1 & -1 & -10 \end{vmatrix} = \begin{vmatrix} 3 & 0 & 2 \\ 0 & -1 & -2 \\ 1 & -1 & -10 \end{vmatrix}$

Apply property 7 to the latter determinant to produce a second zero in column 2:

$$\begin{vmatrix} 3 & 0 & 2 \\ 0 & -1 & -2 \\ 1 & -1 & -10 \end{vmatrix} R_3 + (-1)R_2 \to R_3 \begin{vmatrix} 3 & 0 & 2 \\ 0 & -1 & -2 \\ 1 & 0 & -8 \end{vmatrix}$$

This determinant can be efficiently evaluated by expanding by the second column:

$$\begin{vmatrix} 3 & 0 & 2 \\ 0 & -1 & -2 \\ 1 & 0 & -8 \end{vmatrix} = 0(A_{12}) + (-1)(-1)^{2+2} \begin{vmatrix} 3 & 2 \\ 1 & -8 \end{vmatrix} + 0(A_{32}) = (-1)[3(-8) - 1 \cdot 2] = 26$$

**36.11.** Show that the equation of the straight line through the points $(x_1, y_1)$ and $(x_2, y_2)$ can be expressed as:

$$\begin{vmatrix} x & y & 1 \\ x_1 & y_1 & 1 \\ x_2 & y_2 & 1 \end{vmatrix}$$

Expanding the determinant by the first row yields

$$xA_{11} + yA_{12} + 1A_{13} = 0,$$

where the three cofactors do not contain the variables $x$ and $y$; therefore, this is the equation of a straight line. Now set $x = x_1$ and $y = y_1$. Then the value of the determinant is 0, by property 5 of determinants, since two rows are equal. Therefore, the coordinates $(x_1, y_1)$ satisfy the equation of the line, hence the point is on the line. Similarly, setting $x = x_2$ and $y = y_2$ shows that the point $(x_2, y_2)$ is on the line. Hence the given equation is the equation of a straight line passing through the two given points.

**36.12.** Apply Cramer's rule for solving 2 × 2 system of equations to the systems:

(a) $\begin{array}{l} 3x + 4y = 5 \\ 4x + 3y = 16 \end{array}$; (b) $\begin{array}{l} 5x - 7y = 3 \\ 3x + 8y = 5 \end{array}$

(a) The determinant of the system is

$$D = \begin{vmatrix} 3 & 4 \\ 4 & 3 \end{vmatrix} = -7$$

Therefore the system has exactly one solution, given by

$$x = \frac{D_x}{D} = \frac{\begin{vmatrix} 5 & 4 \\ 16 & 3 \end{vmatrix}}{-7} = \frac{-49}{-7} = 7 \qquad y = \frac{D_y}{D} = \frac{\begin{vmatrix} 3 & 5 \\ 4 & 16 \end{vmatrix}}{-7} = \frac{28}{-7} = -4$$

(b) The determinant of the system is

$$D = \begin{vmatrix} 5 & -7 \\ 3 & 8 \end{vmatrix} = 61$$

Therefore the system has exactly one solution, given by

$$x = \frac{D_x}{D} = \frac{\begin{vmatrix} 3 & -7 \\ 5 & 8 \end{vmatrix}}{61} = \frac{59}{61} \qquad y = \frac{D_y}{D} = \frac{\begin{vmatrix} 5 & 3 \\ 3 & 5 \end{vmatrix}}{61} = \frac{16}{61}$$

**36.13.** Apply Cramer's rule for solving $3 \times 3$ systems of equations to the systems:

$$\text{(a)} \quad \begin{aligned} 3x_1 + 5x_2 - x_3 &= 4 \\ -x_1 + 4x_2 + 4x_3 &= 6 \\ 2x_1 \qquad\quad + 5x_3 &= -2 \end{aligned} \; ; \qquad \text{(b)} \quad \begin{aligned} 3x_1 + 5x_2 - x_3 &= 4 \\ -x_1 + 4x_2 + 4x_3 &= 6 \\ 2x_1 + 9x_2 + 3x_3 &= 10 \end{aligned}$$

(a) The determinant of the system is

$$D = \begin{vmatrix} 3 & 5 & -1 \\ -1 & 4 & 4 \\ 2 & 0 & 5 \end{vmatrix} = 133$$

Therefore, the system has exactly one solution, given by

$$x_1 = \frac{D_1}{D} = \frac{\begin{vmatrix} 4 & 5 & -1 \\ 6 & 4 & 4 \\ -2 & 0 & 5 \end{vmatrix}}{133} = -\frac{118}{133} \qquad x_2 = \frac{D_2}{D} = \frac{\begin{vmatrix} 3 & 4 & -1 \\ -1 & 6 & 4 \\ 2 & -2 & 5 \end{vmatrix}}{133} = \frac{176}{133}$$

$$x_3 = \frac{D_3}{D} = \frac{\begin{vmatrix} 3 & 5 & 4 \\ -1 & 4 & 6 \\ 2 & 0 & -2 \end{vmatrix}}{133} = -\frac{6}{133}$$

(b) The determinant of the system is

$$D = \begin{vmatrix} 3 & 5 & -1 \\ -1 & 4 & 4 \\ 2 & 9 & 3 \end{vmatrix} = 0$$

Therefore Cramer's rule cannot be used to solve the system. Gaussian elimination can be used to show that there are infinite solutions, given by $\left( \dfrac{24r - 14}{17}, \dfrac{22 - 11r}{17}, r \right)$, for $r$ any real number.

## SUPPLEMENTARY PROBLEMS

**36.14.** Evaluate the determinants: (a) $\begin{vmatrix} 11 & 12 \\ 13 & 14 \end{vmatrix}$; (b) $\begin{vmatrix} -5 & 8 \\ 25 & -40 \end{vmatrix}$; (c) $\begin{vmatrix} \cos t & -\sin t \\ \sin t & \cos t \end{vmatrix}$

*Ans.* (a) $-2$; (b) $0$; (c) $1$

**36.15.** Evaluate the determinants: (a) $\begin{vmatrix} 3 & -4 & -5 \\ 0 & -4 & 0 \\ 3 & 1 & 7 \end{vmatrix}$; (b) $\begin{vmatrix} 0 & -4 & -5 \\ -4 & 0 & 8 \\ -5 & 8 & 0 \end{vmatrix}$; (c) $\begin{vmatrix} 3 & -4 & -5 & 1 \\ 0 & 4 & 0 & 1 \\ 3 & 1 & 7 & 1 \\ 0 & 1 & 1 & 1 \end{vmatrix}$.

*Ans.* (a) $-144$; (b) 320; (c) 123

**36.16.** Verify property 5 of determinants: if two rows of a determinant are equal (or if two columns of a determinant are equal), the value of the determinant is 0. (Hint: consider what happens when the two rows or columns are interchanged.)

**36.17.** Verify property 6 of determinants: if one row of a determinant consists of all zeros, the value of the determinant is 0.

**36.18.** Evaluate the determinants (a) $\begin{vmatrix} i & j & k \\ 2 & 3 & 4 \\ 5 & -4 & 6 \end{vmatrix}$; (b) $\begin{vmatrix} i & j & k \\ 6 & -12 & 8 \\ -9 & 18 & -12 \end{vmatrix}$.

*Ans.* (a) $34i + 8j - 23k$; (b) 0

**36.19.** Use properties of determinants to verify: $\begin{vmatrix} 1 & 1 & 1 \\ a & b & c \\ a^2 & b^2 & c^2 \end{vmatrix} = (a - b)(b - c)(c - a)$.

**36.20.** Apply Cramer's rule to the solution of the systems

(a) $\begin{aligned} 5x - 6y &= 9 \\ 3x + 8y &= -5 \end{aligned}$; (b) $\begin{aligned} x_1 - 2x_2 - 5x_3 &= -28 \\ 2x_1 + 6x_2 + 5x_3 &= 44 \\ -3x_1 + 3x_2 - 4x_3 &= 25 \end{aligned}$; (c) $\begin{aligned} 2x_1 - 3x_2 + 4x_3 &= 0 \\ 4x_1 + x_2 - 3x_3 &= 3 \\ 10x_1 - x_2 - 2x_3 &= 5 \end{aligned}$

*Ans.* (a) $x = \dfrac{21}{29}, y = -\dfrac{26}{29}$; (b) $x_1 = -4, x_2 = 7, x_3 = 2$; (c) Since the determinant of the system is 0, Cramer's rule does not yield a solution; Gaussian elimination shows that the system has no solution.

# CHAPTER 37

# *Loci; Parabolas*

## Set of All Points

The set of all points that satisfy specified conditions is called the *locus* (plural: *loci*) of the point under the conditions.

**EXAMPLE 37.1**   The locus of a point with positive coordinates is the first quadrant ($x > 0$, $y > 0$).

**EXAMPLE 37.2**   The locus of points with distance 3 from the origin is the circle $x^2 + y^2 = 9$ with center at $(0, 0)$ and radius 3.

## Distance Formulas

Distance formulas are often used in finding loci.

1. **DISTANCE BETWEEN TWO POINTS** formula (derived in Chapter 8): The distance between two points $P_1(x_1, y_1)$ and $P_2(x_2, y_2)$ is given by

$$d(P_1, P_2) = \sqrt{(x_2 - x_1)^2 + (y_2 - y_1)^2}$$

2. **DISTANCE FROM A POINT TO A LINE** formula: The distance from a point $P_1(x_1, y_1)$ to a straight line $Ax + By + C = 0$ is given by

$$d = \frac{|Ax_1 + By_1 + C|}{\sqrt{A^2 + B^2}}$$

**EXAMPLE 37.3**   Find the locus of points $P(x, y)$ equidistant from $P_1(1,0)$ and $P_2(3,0)$.

Set $d(P, P_1) = d(P, P_2)$. Then $\sqrt{(x - 1)^2 + (y - 0)^2} = \sqrt{(x - 3)^2 + (y - 0)^2}$. Simplifying yields:

$$(x - 1)^2 + (y - 0)^2 = (x - 3)^2 + (y - 0)^2$$

$$x^2 - 2x + 1 + y^2 = x^2 - 6x + 9 + y^2$$

$$4x = 8$$

$$x = 2$$

The locus is a vertical line that forms the perpendicular bisector of $P_1P_2$.

## Parabola

A parabola is defined as the locus of points $P$ equidistant from a given point and a given line not containing the point, that is, such that $PF = PD$, where $F$ is the given point, called the *focus*, and $PD$ is the distance to the given line $l$, called the *directrix*. A line through the focus perpendicular to the directrix is called the *axis* (or *axis of symmetry*) and the point on the axis halfway between the directrix and the focus is called the *vertex*.

A parabola with axis parallel to one of the coordinate axes is said to be in *standard orientation*. If, in addition, the vertex of the parabola is at the origin, the parabola is said to be in one of four *standard positions*: opening right, opening left, opening up, and opening down.

## Graphs of Parabolas in Standard Position

Graphs of parabolas in standard position with their equations and characteristics are shown in Figs. 37-1 to 37-4.

| OPENING RIGHT | OPENING LEFT | OPENING UP | OPENING DOWN |
|---|---|---|---|
| Vertex: (0,0)<br>Focus: $F(p,0)$<br>Directrix: $x = -p$ | Vertex: (0,0)<br>Focus: $(-p,0)$<br>Directrix: $x = p$ | Vertex: (0,0)<br>Focus: $F(0,p)$<br>Directrix: $y = -p$ | Vertex: (0,0)<br>Focus: $F(0,-p)$<br>Directrix: $y = p$ |
| Equation:<br>$y^2 = 4px$ | Equation:<br>$y^2 = -4px$ | Equation:<br>$x^2 = 4py$ | Equation:<br>$x^2 = -4py$ |
| <br>Figure 37-1 | <br>Figure 37-2 | <br>Figure 37-3 | <br>Figure 37-4 |

## Parabolas in Standard Orientation

Replacing $x$ by $x - h$ has the effect of shifting the graph of an equation by $|h|$ units, to the right if $h$ is positive, to the left if $h$ is negative. Similarly, replacing $y$ by $y - k$ has the effect of shifting the graph by $|k|$ units, up if $k$ is positive and down if $k$ is negative. The equations and characteristics of parabolas in standard orientation, but not necessarily in standard position, are shown in the following table.

| OPENING RIGHT | OPENING LEFT | OPENING UP | OPENING DOWN |
|---|---|---|---|
| Equation:<br>$(y - k)^2 = 4p(x - h)$ | Equation:<br>$(y - k)^2 = -4p(x - h)$ | Equation:<br>$(x - h)^2 = 4p(y - k)$ | Equation:<br>$(x - h)^2 = -4p(y - k)$ |
| Vertex: $(h,k)$<br>Focus: $F(h + p,k)$<br>Directrix:<br>$x = h - p$ | Vertex: $(h,k)$<br>Focus: $F(h - p,k)$<br>Directrix:<br>$x = h + p$ | Vertex: $(h,k)$<br>Focus: $F(h,k + p)$<br>Directrix:<br>$y = k - p$ | Vertex: $(h,k)$<br>Focus: $F(h,k-p)$<br>Directrix:<br>$y = k + p$ |

## SOLVED PROBLEMS

**37.1.** Find the locus of points $P(x,y)$ such that the distance of $P$ from point $P_1(2,0)$ is twice the distance of $P$ from the origin.

Set $d(P_1,P) = 2d(O,P)$. Then $\sqrt{(x - 2)^2 + y^2} = 2\sqrt{x^2 + y^2}$. Simplifying yields:

$$(x - 2)^2 + y^2 = 4(x^2 + y^2)$$

$$x^2 - 4x + 4 + y^2 = 4x^2 + 4y^2$$

$$0 = 3x^2 + 3y^2 + 4x - 4$$

The locus is a circle with center on the $x$-axis.

**37.2.** Derive the formula $d = \dfrac{|Ax_1 + By_1 + C|}{\sqrt{A^2 + B^2}}$ for the perpendicular distance $d$ from a point $P_1(x_1,y_1)$ to a straight line $Ax + By + C = 0$.

Drop a perpendicular from $P_1$ to the line at point $L$. Then $d = \left|\overrightarrow{P_1 L}\right|$. Let $P(x,y)$ be an arbitrary point on the given straight line (see Fig. 37-5).

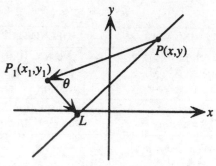

Figure 37-5

In the right triangle $PP_1 L$, $d = \left|\overrightarrow{P_1 L}\right| = \left|\overrightarrow{PP_1}\right| \cos\theta = \left|\overrightarrow{PP_1}\right| \dfrac{\overrightarrow{PP_1} \cdot \overrightarrow{P_1 L}}{\left|\overrightarrow{PP_1}\right|\left|\overrightarrow{P_1 L}\right|} = \dfrac{\overrightarrow{PP_1} \cdot \overrightarrow{P_1 L}}{\left|\overrightarrow{P_1 L}\right|}$.

Now $\overrightarrow{PP_1} = \langle x_1 - x, y_1 - y \rangle$. To find $\overrightarrow{P_1 L}$, note that the given line has slope $-\dfrac{A}{B}$, hence, any perpendicular line has slope $\dfrac{B}{A}$. Therefore, every vector perpendicular to the given line, including $\overrightarrow{P_1 L}$, can be written as $a\langle 1, B/A \rangle$ for some value of $a$. Therefore,

$$d = \frac{\overrightarrow{PP_1} \cdot \overrightarrow{P_1 L}}{\left|\overrightarrow{P_1 L}\right|} = \frac{\langle x_1 - x, y_1 - y \rangle \cdot a\langle 1, B/A \rangle}{|a\langle 1, B/A \rangle|} = \frac{a[(x_1 - x) + (B/A)(y_1 - y)]}{\sqrt{a^2(1 + B^2/A^2)}}$$

Since the signs of $a$ and $A$ are unspecified, and a distance must be a nonnegative quantity, take the absolute value of the right-hand side to ensure that $d$ is not calculated negative. Then

$$d = \left|\frac{a[(x_1 - x) + (B/A)(y_1 - y)]}{\sqrt{a^2(1 + B^2/A^2)}}\right| = \left|\frac{a[A(x_1 - x) + B(y_1 - y)]}{a\sqrt{A^2 + B^2}}\right| = \frac{|Ax_1 - Ax + By_1 - By|}{\sqrt{A^2 + B^2}}$$

Finally, since $(x,y)$ is on the line $Ax + By + C = 0$, it must satisfy the equation of the line, hence the quantity $-Ax - By$ can be replaced by $C$, and

$$d = \frac{|Ax_1 + By_1 + C|}{\sqrt{A^2 + B^2}}$$

**37.3.** Find the distance from (a) the point $(5, -3)$ to the line $3x + 7y - 6 = 0$; (b) the point $(5, 7)$ to the line $x = -4$.

(a) Use the formula $d = \dfrac{|Ax_1 + By_1 + C|}{\sqrt{A^2 + B^2}}$ with $x_1 = 5$ and $y_1 = -3$:

$$d = \frac{|3 \cdot 5 + 7(-3) - 6|}{\sqrt{3^2 + 7^2}} = \frac{12}{\sqrt{58}}$$

(b) Rewrite the equation of the line in standard form $1x + 0y + 4 = 0$, then use the formula $d = \dfrac{|Ax_1 + By_1 + C|}{\sqrt{A^2 + B^2}}$ with $x_1 = 5$ and $y_1 = 7$:

$$d = \frac{|1 \cdot 5 + 0 \cdot 7 + 4|}{\sqrt{1^2 + 0^2}} = 9$$

**37.4.** Show that the equation of a parabola with focus $F(p, 0)$ and directrix $x = -p$ can be written as $y^2 = 4px$.

The parabola is defined by the relation $PF = PD$. Let $P$ be an arbitrary point $(x,y)$ on the parabola. Then $PF$ is found from the distance-between-two-points formula to be $\sqrt{(x - p)^2 + (y - 0)^2}$. $PD$ is found from the formula for the distance from a point to a line to be $|x + p|$. Hence:

$$PF = PD$$
$$\sqrt{(x - p)^2 + (y - 0)^2} = |x + p|$$
$$(x - p)^2 + y^2 = (x + p)^2$$
$$x^2 - 2px + p^2 + y^2 = x^2 + 2px + p^2$$
$$y^2 = 4px$$

**37.5.** Show that the equation of a parabola with focus $F(0, -p)$ and directrix $y = p$ can be written as $x^2 = -4py$.

The parabola is defined by the relation $PF = PD$. Let $P$ be an arbitrary point $(x,y)$ on the parabola. Then $PF$ is found from the distance-between-two-points formula to be $\sqrt{(x - 0)^2 + (y + p)^2}$. $PD$ is found from the formula for the distance from a point to a line to be $|y - p|$. Hence:

$$PF = PD$$
$$\sqrt{(x - 0)^2 + (y + p)^2} = |y - p|$$
$$x^2 + (y + p)^2 = (y - p)^2$$
$$x^2 + y^2 + 2py + p^2 = y^2 - 2py + p^2$$
$$x^2 = -4py$$

**37.6.** For the parabola $y^2 = 12x$, find the focus, directrix, vertex, and axis, and sketch a graph.

The equation of the parabola is in the form $y^2 = 4px$ with $4p = 12$, thus $p = 3$. Hence the parabola is in standard position, with vertex $(0,0)$, opening right, and has focus at $(3,0)$, directrix the line $x = -3$, and axis the $x$-axis, $y = 0$. The graph is shown in Fig. 37-6.

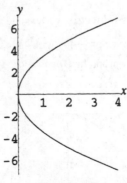

Figure 37-6

**37.7.** Show that $y^2 - 8x + 2y + 9 = 0$ is the equation of a parabola. Find the focus, directrix, vertex, and axis, and sketch a graph.

Complete the square on $y$ to obtain:

$$y^2 + 2y = 8x - 9$$
$$y^2 + 2y + 1 = 8x - 8$$
$$(y + 1)^2 = 8(x - 1)$$

Thus the equation is the equation of a parabola in the form $(y - k)^2 = 4p(x - h)$ with $p = 2$, $h = 1$, and $k = -1$. Hence the parabola is in standard orientation, with vertex $(1,-1)$, opening right, and thus has its focus at

$(h + p, k) = (2 + 1, -1) = (3, -1)$. Its directrix is the line $x = h - p = 1 - 2 = -1$, and its axis is the line $y = -1$. The graph is shown in Fig. 37-7.

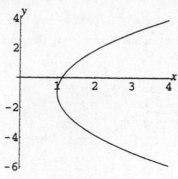

Figure 37-7

**37.8.** Find the equation of a parabola in standard position with focus $(5,0)$ and directrix $x = -5$.

Since the parabola is in standard position with focus on the positive $x$-axis, the focus is located at the point $(p, 0)$, hence, $p = 5$. The parabola is opening right, hence its equation must be of form $y^2 = 4px$. Substituting $p = 5$ yields $y^2 = 20x$.

**37.9.** Find the equation of a parabola in standard orientation with focus $(3, 4)$ and directrix the $y$-axis.

The equation can be found by direct substitution in the definition of the parabola $PF = PD$. Alternatively, note that the vertex is the point half the distance from the focus to the directrix, that is, the point $\left(\frac{3}{2}, 4\right)$. Since the focus is to the right of the directrix, the parabola opens to the right and has an equation of the form $(y - k)^2 = 4p(x - h)$, with $h = \frac{3}{2}$ and $k = 4$. The distance from the vertex at $\left(\frac{3}{2}, 4\right)$ to the focus at $(3, 4)$ is then also $\frac{3}{2}$, and this is the value of $p$. Substituting yields

$$(y - 4)^2 = 4\left(\frac{3}{2}\right)\left(x - \frac{3}{2}\right)$$
$$(y - 4)^2 = 6x - 9$$

**37.10.** For the parabola $x^2 = -2y$, find the focus, directrix, vertex, and axis, and sketch a graph.

The equation of the parabola is in the form $x^2 = -4py$ with $4p = 2$, thus $p = \frac{1}{2}$. Hence the parabola is in standard position, with vertex $(0, 0)$, opening down, and has focus at $\left(0, -\frac{1}{2}\right)$, directrix the line $y = \frac{1}{2}$, and axis the $y$-axis, $x = 0$. The graph is shown in Fig. 37-8.

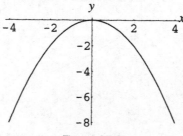

Figure 37-8

**37.11.** Show that $x^2 + 2x + 6y - 11 = 0$ is the equation of a parabola. Find the focus, directrix, vertex, and axis and sketch a graph.

Complete the square on $x$ to obtain:

$$x^2 + 2x = -6y + 11$$
$$x^2 + 2x + 1 = -6y + 12$$
$$(x + 1)^2 = -6(y - 2)$$

Thus the equation is the equation of a parabola in the form $(x - h)^2 = -4p(y - k)$ with $p = \frac{3}{2}$, $h = -1$, and $k = 2$. Hence the parabola is in standard orientation, with vertex $(-1, 2)$, opening down, and thus has its focus at $(h, k - p) = \left(-1, 2 - \frac{3}{2}\right) = \left(-1, \frac{1}{2}\right)$. The directrix is the line $y = k + p = 2 + \frac{3}{2} = \frac{7}{2}$, and its axis is the line $x = -1$. The graph is shown in Fig. 37-9.

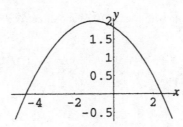

Figure 37-9

**37.12.** Find the equation of a parabola in standard position, opening down, with focus $(0, -4)$ and directrix the line $y = 4$.

Since the parabola is in standard position with focus on the negative $y$-axis, the focus is located at the point $(0, -p)$, hence $p = 4$. The parabola is opening down, hence its equation must be of form $x^2 = -4py$. Substituting $p = 4$ yields $x^2 = -16y$.

**37.13.** Find the equation of a parabola in standard orientation with focus $(3,4)$ and directrix the line $y = 6$.

The equation can be found by direct substitution in the definition of the parabola $PF = PD$. Alternatively, note that the vertex is the point half the distance from the focus to the directrix, that is, the point $(3,5)$. Since the focus is below the directrix, the parabola opens down and has an equation of the form $(x - h)^2 = -4p(y - k)$, with $h = 3$ and $k = 5$. The distance from the vertex at $(3,5)$ to the focus at $(3,6)$ is then 1, and this is the value of $p$. Substituting yields

$$(x - 3)^2 = -4(1)(y - 5)$$
$$(x - 3)^2 = -4y + 20$$

## SUPPLEMENTARY PROBLEMS

**37.14.** Find the locus of points $P(x,y)$ such that the distance from $P$ to the $y$-axis is 5.

*Ans.* $x = 5$ and $x = -5$, two straight lines parallel to the $y$-axis.

**37.15.** Find the locus of points $P(x,y)$ such that $P$ is equidistant from both axes.

*Ans.* $y = x$ and $y = -x$, two straight lines through the origin.

**37.16.** Find the locus of points $P(x,y)$ such that the distance of $P$ from $P_1(1,1)$ is one-half the distance of $P$ from $P_2(-2,-2)$.

*Ans.* $x^2 + y^2 - 4x - 4y = 0$, a circle passing through the origin.

**37.17.** Find the locus of points $P(x,y)$ equidistant from $(5,-1)$ and $(3,-8)$.

*Ans.* $4x + 14y + 47 = 0$, a straight line, the perpendicular bisector of the line segment joining the given points

**37.18.** Find the locus of points $P(x,y)$ equidistant from $(-5,3)$ and $x - y + 8 = 0$.

*Ans.* $x^2 + y^2 + 2xy + 4x + 4y + 4 = 0$, that is, $(x + y + 2)^2 = 0$, a straight line perpendicular to the given line at the given point.

**37.19.** Find the locus of points $P(x,y)$ such that the product of their distances from $(0,4)$ and $(0,-4)$ is 16.

  *Ans.* $x^4 + 2x^2y^2 + y^4 + 32x^2 - 32y^2 = 0$

**37.20.** Show that the equation of a parabola with focus $F(-p,0)$ and directrix $x = p$ can be written as $y^2 = -4px$.

**37.21.** Show that the equation of a parabola with focus $F(0,p)$ and directrix $y = -p$ can be written as $x^2 = 4py$.

**37.22.** Sketch graphs of the equations (a) $y^2 = -2x$; (b) $x^2 = 6y$.

  *Ans.* (a) Fig. 37-10; (b) Fig. 37-11

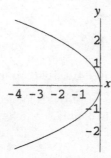

Figure 37-10

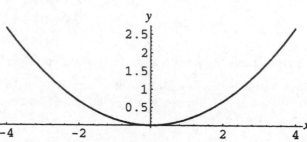

Figure 37-11

**37.23.** Find equations for parabolas in standard position (a) with focus at $(0,7)$ and directrix the line $y = -7$; (b) with focus at $\left(-\dfrac{5}{4},0\right)$ and directrix the line $x = \dfrac{5}{4}$.

  *Ans.* (a) $x^2 = 28y$; (b) $y^2 = -5x$

**37.24.** Find equations for parabolas in standard orientation (a) with focus at $(-2,3)$ and directrix the $y$-axis; (b) with focus at $(-2,3)$ and directrix the line $y = 1$.

  *Ans.* (a) $y^2 - 6y + 4x + 13 = 0$; (b) $x^2 + 4x - 4y + 12 = 0$

**37.25.** Sketch graphs of the equations (a) $y^2 - 2y - 3x - 2 = 0$; (b) $x^2 + 2x + 2y - 5 = 0$.

  *Ans.* (a) Fig. 37-12; (b) Fig. 37-13

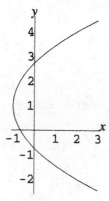

Figure 37-12

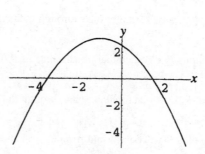

Figure 37-13

**37.26.** Use the definition of the parabola directly to find the equation of a parabola with focus $F(2,2)$ and directrix the line $x + y + 2 = 0$.

  *Ans.* $x^2 - 2xy + y^2 - 12x - 12y + 12 = 0$

# CHAPTER 38

# *Ellipses and Hyperbolas*

## Definition of Ellipse

The locus of points $P$ such that the sum of the distances from $P$ to two fixed points is a constant is called an *ellipse*. Thus, let $F_1$ and $F_2$ be the two points (called *foci*, the plural of *focus*), then the defining relation for the ellipse is $PF_1 + PF_2 = 2a$. The line through the foci is called the *focal axis* of the ellipse; the point on the focal axis halfway between the foci is called the *center*; the points where the ellipse crosses the focal axis are called the *vertices*. The line segment joining the two vertices is called the *major axis*, and the line segment through the center, perpendicular to the major axis, with both endpoints on the ellipse, is called the *minor axis*. (See Fig. 38-1.)

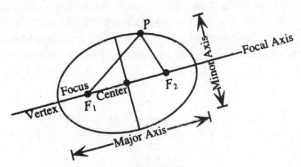

Figure 38-1

An ellipse with focal axis parallel to one of the coordinate axes is said to be in *standard orientation*. If, in addition, the center of the ellipse is at the origin, the ellipse is said to be in one of two *standard positions*: with foci on the $x$-axis or with foci on the $y$-axis.

## Graphs of Ellipses in Standard Position

Graphs of ellipses in standard position with their equations and characteristics are shown in the following table:

| FOCI ON $x$-AXIS | FOCI ON $y$-AXIS |
|---|---|
| Equation: $\dfrac{x^2}{a^2} + \dfrac{y^2}{b^2} = 1$ <br> where $b^2 = a^2 - c^2$ <br> Note: $a > b, a > c$ | Equation: $\dfrac{x^2}{b^2} + \dfrac{y^2}{a^2} = 1$ <br> where $b^2 = a^2 - c^2$ <br> Note: $a > b, a > c$ |
| Foci: $F_1(-c, 0), F_2(c, 0)$ <br> Vertices: $(-a, 0), (a, 0)$ <br> Center: $(0, 0)$ | Foci: $F_1(0, -c), F_2(0, c)$ <br> Vertices: $(0, -a), (0, a)$ <br> Center: $(0, 0)$ |

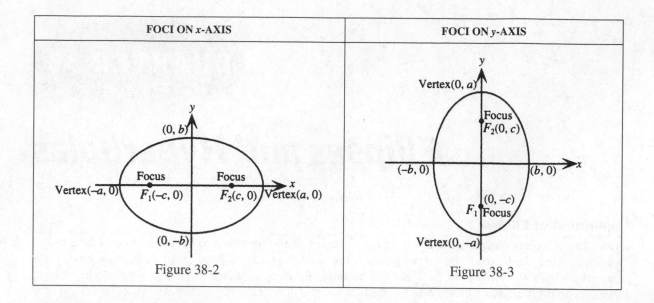

| FOCI ON *x*-AXIS | FOCI ON *y*-AXIS |
| --- | --- |
| Figure 38-2 | Figure 38-3 |

## Definition of Hyperbola

The locus of points $P$ such that the absolute value of the difference of the distances from $P$ to two fixed points is a constant is called a *hyperbola*. Thus, let $F_1$ and $F_2$ be the two points ( *foci*), then the defining relation for the hyperbola is $|PF_1 - PF_2| = 2a$. The line through the foci is called the *focal axis* of the hyperbola; the point on the focal axis halfway between the foci is called the *center*; the points where the hyperbola crosses the focal axis are called the *vertices*. The line segment joining the two vertices is called the *transverse axis*. (See Fig. 38-4.)

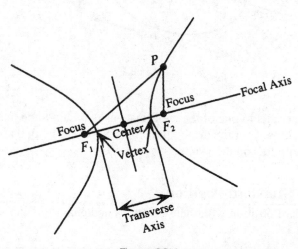

Figure 38-4

A hyperbola with focal axis parallel to one of the coordinate axes is said to be in *standard orientation*. If, in addition, the center of the hyperbola is at the origin, the hyperbola is said to be in one of two *standard positions*: with foci on the *x*-axis or with foci on the *y*-axis.

## Graphs of Hyperbolas in Standard Position

Graphs of hyperbolas in standard position with their equations and characteristics are shown in the following table:

| FOCI ON $x$-AXIS | FOCI ON $y$-AXIS |
|---|---|
| Foci: $F_1(-c, 0)$, $F_2(c, 0)$<br>Vertices: $(-a, 0)$, $(a, 0)$<br>Center: $(0, 0)$ | Foci: $F_1(0, -c)$, $F_2(0, c)$<br>Vertices: $(0, -a)$, $(0, a)$<br>Center: $(0, 0)$ |
| Equation: $\dfrac{x^2}{a^2} - \dfrac{y^2}{b^2} = 1$<br><br>where $b^2 = c^2 - a^2$<br>Note: $c > a$, $c > b$ | Equation: $\dfrac{y^2}{a^2} - \dfrac{x^2}{b^2} = 1$<br><br>where $b^2 = c^2 - a^2$<br>Note: $c > a$, $c > b$ |
| Asymptotes: $y = \pm\dfrac{b}{a}x$ | Asymptotes: $y = \pm\dfrac{a}{b}x$ |
| <br>Figure 38-5 | <br>Figure 38-6 |

## Definition of Eccentricity

A measure of the shape for an ellipse or hyperbola is the quantity $e = \dfrac{c}{a}$, called the *eccentricity*. For an ellipse, $0 < e < 1$, for a hyperbola $e > 1$.

## SOLVED PROBLEMS

**38.1.** Derive the equation of an ellipse in standard position with foci on the $x$-axis.

Let $P(x, y)$ be an arbitrary point on the ellipse. Given that the foci are $F_1(-c, 0)$ and $F_2(c, 0)$, then the definition of the ellipse $PF_1 + PF_2 = 2a$ yields:

$$\sqrt{(x + c)^2 + (y - 0)^2} + \sqrt{(x - c)^2 + (y - 0)^2} = 2a$$

Subtracting one of the square roots from both sides, squaring, and simplifying yields:

$$\sqrt{(x + c)^2 + (y - 0)^2} = 2a - \sqrt{(x - c)^2 + (y - 0)^2}$$

$$(x + c)^2 + y^2 = 4a^2 - 4a\sqrt{(x - c)^2 + (y - 0)^2} + (x - c)^2 + y^2$$

$$x^2 + 2xc + c^2 + y^2 = 4a^2 - 4a\sqrt{(x - c)^2 + (y - 0)^2} + x^2 - 2xc + c^2 + y^2$$

$$4xc - 4a^2 = -4a\sqrt{(x - c)^2 + (y - 0)^2}$$

$$xc - a^2 = -a\sqrt{(x - c)^2 + (y - 0)^2}$$

Now square both sides again and simplify:

$$x^2c^2 - 2xca^2 + a^4 = a^2[(x - c)^2 + y^2]$$

$$x^2c^2 - 2xca^2 + a^4 = a^2x^2 - 2a^2xc + a^2c^2 + a^2y^2$$

$$x^2c^2 - a^2x^2 - a^2y^2 = a^2c^2 - a^4$$

$$x^2(c^2 - a^2) - a^2y^2 = a^2(c^2 - a^2)$$

By the triangle inequality, the sum of two sides of a triangle is always greater than the third side. Hence (see Fig. 38-1)

$$PF_1 + PF_2 > F_1F_2$$

$$2a > 2c$$

$$a^2 > c^2$$

Thus the quantity $a^2 - c^2$ must be positive. Set $a^2 - c^2 = b^2$. Then $c^2 - a^2 = -b^2$ and the equation of the ellipse becomes:

$$-b^2x^2 - a^2y^2 = -a^2b^2$$

$$b^2x^2 + a^2y^2 = a^2b^2$$

This is generally written in standard form:

$$\frac{x^2}{a^2} + \frac{y^2}{b^2} = 1$$

Note that it follows from $a^2 - c^2 = b^2$ that $a > b$.

**38.2.** Analyze the equation $\dfrac{x^2}{a^2} + \dfrac{y^2}{b^2} = 1$ of an ellipse in standard position, foci on the $x$-axis.

Set $x = 0$, then $\dfrac{y^2}{b^2} = 1$; thus, $y = \pm b$. Hence $\pm b$ are the $y$-intercepts.

Set $y = 0$, then $\dfrac{x^2}{a^2} = 1$; thus $x = \pm a$. Hence $\pm a$ are the $x$-intercepts.

Substitute $-y$ for $y$: $\dfrac{x^2}{a^2} + \dfrac{(-y)^2}{b^2} = 1$; $\dfrac{x^2}{a^2} + \dfrac{y^2}{b^2} = 1$. Since the equation is unchanged, the graph has $x$-axis symmetry.

Substitute $-x$ for $x$: $\dfrac{(-x)^2}{a^2} + \dfrac{y^2}{b^2} = 1$; $\dfrac{x^2}{a^2} + \dfrac{y^2}{b^2} = 1$. Since the equation is unchanged, the graph has $y$-axis symmetry. It follows that the graph also has origin symmetry.

Note further that solving for $y$ in terms of $x$ yields $y = \pm\dfrac{b}{a}\sqrt{a^2 - x^2}$; hence $-a \le x \le a$ for $y$ to be real.

Similarly, $-b \le y \le b$ for $x$ to be real.

Summarizing, the graph is confined to the region between the intercepts $\pm a$ on the $x$-axis and $\pm b$ on the $y$-axis, and has all three symmetries. The graph of the ellipse is shown in Fig. 38-2.

**38.3.** Analyze the equation $\dfrac{x^2}{b^2} + \dfrac{y^2}{a^2} = 1$ of an ellipse in standard position, foci on the $y$-axis.

Set $x = 0$, then $\dfrac{y^2}{a^2} = 1$; thus $y = \pm a$. Hence $\pm a$ are the $y$-intercepts.

Set $y = 0$, then $\dfrac{x^2}{b^2} = 1$; thus $x = \pm b$. Hence $\pm b$ are the $x$-intercepts.

Substitute $-y$ for $y$: $\dfrac{x^2}{b^2} + \dfrac{(-y)^2}{a^2} = 1$; $\dfrac{x^2}{b^2} + \dfrac{y^2}{a^2} = 1$. Since the equation is unchanged, the graph has $x$-axis symmetry.

Substitute $-x$ for $x$: $\dfrac{(-x)^2}{b^2} + \dfrac{y^2}{a^2} = 1$; $\dfrac{x^2}{b^2} + \dfrac{y^2}{a^2} = 1$. Since the equation is unchanged, the graph has $y$-axis symmetry. It follows that the graph also has origin symmetry.

Note further that solving for $y$ in terms of $x$ yields $y = \pm\dfrac{a}{b}\sqrt{b^2 - x^2}$; hence $-b \le x \le b$ for $y$ to be real. Similarly $-a \le y \le a$ for $x$ to be real.

Summarizing, the graph is confined to the region between the intercepts $\pm b$ on the $x$-axis and $\pm a$ on the $y$-axis, and has all three symmetries. The graph of the ellipse is shown in Fig. 38-3.

**38.4.** Analyze and sketch graphs of the ellipses (a) $4x^2 + 9y^2 = 36$; (b) $4x^2 + y^2 = 36$.

(a) Written in standard form, the equation becomes

$$\frac{x^2}{9} + \frac{y^2}{4} = 1$$

Thus $a = 3, b = 2$.

Therefore $c = \sqrt{a^2 - b^2} = \sqrt{9 - 4} = \sqrt{5}$.

Hence the ellipse is in standard position with foci at $(\pm\sqrt{5}, 0)$ on the $x$-axis, $x$-intercepts $(\pm 3, 0)$, and $y$-intercepts $(0, \pm 2)$. The graph is shown in Fig. 38-7.

(b) Written in standard form, the equation becomes

$$\frac{x^2}{9} + \frac{y^2}{36} = 1$$

Thus $a = 6, b = 3$.

Therefore $c = \sqrt{a^2 - b^2} = \sqrt{36 - 9} = 3\sqrt{3}$.

Hence the ellipse is in standard position with foci at $(0, \pm 3\sqrt{3})$ on the $y$-axis, $x$-intercepts $(\pm 3, 0)$, and $y$-intercepts $(0, \pm 6)$. The graph is shown in Fig. 38-8.

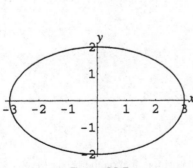

Figure 38-7

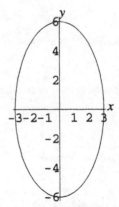

Figure 38-8

**38.5.** Derive the equation of a hyperbola in standard position with foci on the $x$-axis.

Let $P(x,y)$ be an arbitrary point on the hyperbola. Given that the foci are $F_1(-c,0)$ and $F_2(c,0)$, then the definition of the hyperbola $|PF_1 - PF_2| = 2a$; that is, $PF_1 - PF_2 = \pm 2a$ yields:

$$\sqrt{(x + c)^2 + (y - 0)^2} - \sqrt{(x - c)^2 + (y - 0)^2} = \pm 2a$$

Adding the second square root to both sides, squaring, and simplifying yields:

$$\sqrt{(x + c)^2 + (y - 0)^2} = \pm 2a + \sqrt{(x - c)^2 + (y - 0)^2}$$

$$(x + c)^2 + y^2 = 4a^2 \pm 4a\sqrt{(x - c)^2 + (y - 0)^2} + (x - c)^2 + y^2$$

$$x^2 + 2xc + c^2 + y^2 = 4a^2 \pm 4a\sqrt{(x - c)^2 + (y - 0)^2} + x^2 - 2xc + c^2 + y^2$$

$$4xc - 4a^2 = \pm 4a\sqrt{(x - c)^2 + (y - 0)^2}$$

$$xc - a^2 = \pm a\sqrt{(x - c)^2 + (y - 0)^2}$$

Now square both sides again and simplify:

$$x^2c^2 - 2xca^2 + a^4 = a^2[(x - c)^2 + y^2]$$

$$x^2c^2 - 2xca^2 + a^4 = a^2x^2 - 2a^2xc + a^2c^2 + a^2y^2$$

$$x^2c^2 - a^2x^2 - a^2y^2 = a^2c^2 - a^4$$

$$x^2(c^2 - a^2) - a^2y^2 = a^2(c^2 - a^2)$$

By the triangle inequality, the sum of two sides of a triangle is always greater than the third side. Hence (see Fig. 38-4)

$$PF_2 + F_1F_2 > PF_1$$

$$F_1F_2 > PF_1 - PF_2$$

$$2c > 2a$$

$$c > a$$

Thus the quantity $c^2 - a^2$ must be positive. Set $c^2 - a^2 = b^2$. Then the equation of the hyperbola becomes:

$$b^2x^2 - a^2y^2 = a^2b^2$$

This is generally written in standard form:

$$\frac{x^2}{a^2} - \frac{y^2}{b^2} = 1$$

Note that it follows from $c^2 - a^2 = b^2$ that $c > b$ and $c > a$.

**38.6.** Analyze the equation $\dfrac{x^2}{a^2} - \dfrac{y^2}{b^2} = 1$ of a hyperbola in standard position, foci on the *x*-axis.

Set $x = 0$, then $\dfrac{y^2}{b^2} = -1$; thus $y^2 = -b^2$. Hence there can be no *y*-intercepts.

Set $y = 0$, then $\dfrac{x^2}{a^2} = 1$; thus $x = \pm a$. Hence $\pm a$ are the *x*-intercepts.

Substitute $-y$ for $y$: $\dfrac{x^2}{a^2} - \dfrac{(-y)^2}{b^2} = 1$; $\dfrac{x^2}{a^2} - \dfrac{y^2}{b^2} = 1$. Since the equation is unchanged, the graph has *x*-axis symmetry.

Substitute $-x$ for $x$: $\dfrac{(-x)^2}{a^2} - \dfrac{y^2}{b^2} = 1$; $\dfrac{x^2}{a^2} - \dfrac{y^2}{b^2} = 1$. Since the equation is unchanged, the graph has *y*-axis symmetry. It follows that the graph also has origin symmetry.

Note further that solving for $y$ in terms of $x$ yields $y = \pm\frac{b}{a}\sqrt{x^2 - a^2}$; hence $x \geq a$ or $x \leq -a$ for $y$ to be real.

Solving for $x$ in terms of $y$ yields $x = \pm\frac{a}{b}\sqrt{y^2 + b^2}$; hence $y$ can take on any value.

It is left as an exercise to show that as $x$ becomes arbitrarily large, the distance between the graphs of $y = \pm\frac{b}{a}\sqrt{x^2 - b^2}$ and the lines $y = \pm\frac{b}{a}x$ becomes arbitrarily small, thus the lines are oblique asymptotes for the graph.

To draw the graph of the hyperbola, mark the intercepts $\pm a$ on the $x$-axis. Mark the points $\pm b$ on the $y$-axis. Draw vertical line segments through the points $x = \pm a$ and horizontal line segments through the points $y = \pm b$ to form the box shown in Fig. 38-5. Draw the diagonals of the box; these are the asymptotes of the hyperbola. Then sketch the hyperbola starting from the intercept $x = a$ and approaching the asymptote $y = bx/a$. The remainder of the hyperbola follows from the symmetry with respect to axes and origin, as shown in Fig. 38-5.

**38.7.** Analyze the equation $\dfrac{y^2}{a^2} - \dfrac{x^2}{b^2} = 1$ of a hyperbola in standard position, foci on the $y$-axis.

Set $x = 0$, then $\dfrac{y^2}{a^2} = 1$; thus $y = \pm a$. Hence $\pm a$ are the $y$-intercepts.

Set $y = 0$, then $-\dfrac{x^2}{b^2} = 1$; thus $x^2 = -b^2$. Hence there can be no $x$-intercepts.

Substitute $-y$ for $y$: $\dfrac{(-y)^2}{a^2} - \dfrac{x^2}{b^2} = 1$; $\dfrac{y^2}{a^2} - \dfrac{x^2}{b^2} = 1$. Since the equation is unchanged, the graph has $x$-axis symmetry.

Substitute $-x$ for $x$: $\dfrac{y^2}{a^2} - \dfrac{(-x)^2}{b^2} = 1$; $\dfrac{y^2}{a^2} - \dfrac{x^2}{b^2} = 1$. Since the equation is unchanged, the graph has $y$-axis symmetry. It follows that the graph also has origin symmetry.

Note further that solving for $y$ in terms of $x$ yields $y = \pm\frac{a}{b}\sqrt{b^2 + x^2}$; hence $x$ can take on any value.

Solving for $x$ in terms of $y$ yields $x = \pm\frac{b}{a}\sqrt{y^2 - a^2}$; hence $y \geq a$ or $y \leq -a$ for $x$ to be real.

It is left as an exercise to show that as $x$ becomes arbitrarily large, the distance between the graphs of $y = \pm\frac{a}{b}\sqrt{b^2 + x^2}$ and the lines $y = \pm\frac{a}{b}x$ becomes arbitrarily small, thus the lines are oblique asymptotes for the graph.

To draw the graph of the hyperbola, mark the intercepts $\pm a$ on the $y$-axis. Mark the points $\pm b$ on the $x$-axis. Draw vertical line segments through the points $x = \pm b$ and horizontal line segments through the points $y = \pm a$ to form the box shown in Fig. 38-6. Draw the diagonals of the box; these are the asymptotes of the hyperbola. Then sketch the hyperbola starting from the intercept $y = a$ and approaching the asymptote $y = ax/b$. The remainder of the hyperbola follows from the symmetry with respect to axes and origin, as shown in Fig. 38-6.

**38.8.** Analyze and sketch graphs of the hyperbolas (a) $4x^2 - 9y^2 = 36$; (b) $y^2 - 4x^2 = 36$.

(a) Written in standard form, the equation becomes

$$\frac{x^2}{9} - \frac{y^2}{4} = 1$$

(b) Written in standard form, the equation becomes

$$\frac{y^2}{36} - \frac{x^2}{9} = 1$$

Thus $a = 3$, $b = 2$.

Therefore $c = \sqrt{a^2 + b^2} = \sqrt{9 + 4} = \sqrt{13}$.
Hence the hyperbola is in standard position with foci at $(\pm\sqrt{13}, 0)$ on the $x$-axis, $x$-intercepts $(\pm 3, 0)$, and asymptotes $y = \pm\frac{2}{3}x$. The graph is shown in Fig. 38-9.

Thus $a = 6$, $b = 3$.

Therefore $c = \sqrt{a^2 + b^2} = \sqrt{36 + 9} = 3\sqrt{5}$.
Hence the hyperbola is in standard position with foci at $(0, \pm 3\sqrt{5})$ on the $y$-axis, $y$-intercepts $(0, \pm 6)$, and asymptotes $y = \pm 2x$. The graph is shown in Fig. 38-10.

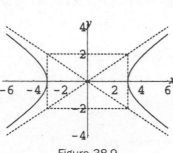

Figure 38-9

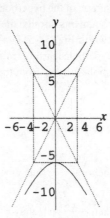

Figure 38-10

**38.9.** Show in a table the characteristics of hyperbolas and ellipses in standard orientation, with center at the point $(h,k)$.

Shifting the center of the curves from the origin to the point $(h, k)$ is reflected in the equations by replacing $x$ with $x - h$ and $y$ with $y - k$, respectively. Hence the shifted curves can be described as follows:

| Ellipse; equation | Ellipse; equation | Hyperbola; equation | Hyperbola; equation |
|---|---|---|---|
| $\dfrac{(x-h)^2}{a^2} + \dfrac{(y-k)^2}{b^2} = 1$ | $\dfrac{(x-h)^2}{b^2} + \dfrac{(y-k)^2}{a^2} = 1$ | $\dfrac{(x-h)^2}{a^2} - \dfrac{(y-k)^2}{b^2} = 1$ | $\dfrac{(y-k)^2}{a^2} - \dfrac{(x-h)^2}{b^2} = 1$ |
| Foci: $(h \pm c, k)$, <br> Vertices: $(h \pm a, k)$ <br> Endpoints of minor <br> axis: $(h, k \pm b)$ | Foci: $(h, k \pm c)$, <br> Vertices: $(h, k \pm a)$ <br> Endpoints of minor <br> axis: $(h \pm b, k)$ | Foci: $(h \pm c, k)$, <br> Vertices: $(h \pm a, k)$ <br> Asymptotes: <br> $(y - k) = \pm\dfrac{b}{a}(x - h)$ | Foci: $(h, k \pm c)$, <br> Vertices: $(h, k \pm a)$ <br> Asymptotes: <br> $(y - k) = \pm\dfrac{a}{b}(x - h)$ |

**38.10.** Analyze and sketch the graph of $9x^2 + 4y^2 - 18x + 8y = 23$.

Complete the square on $x$ and $y$.

$$9(x^2 - 2x) + 4(y^2 + 2y) = 23$$
$$9(x^2 - 2x + 1) + 4(y^2 + 2y + 1) = 23 + 9 \cdot 1 + 4 \cdot 1$$
$$9(x - 1)^2 + 4(y + 1)^2 = 36$$
$$\frac{(x - 1)^2}{4} + \frac{(y + 1)^2}{9} = 1$$

Comparing with the table in Problem 38.9, we find that this is the equation of an ellipse with center at $(1,-1)$.

Since $a > b$, $a^2 = 9$ and $b^2 = 4$, thus $a = 3$, $b = 2$, and $c = \sqrt{a^2 - b^2} = \sqrt{5}$; the focal axis is parallel to the $y$-axis. Foci: $(h, k \pm c) = (1,-1 \pm \sqrt{5})$. Vertices: $(h, k \pm a) = (1, -1 \pm 3)$, thus $(1,2)$ and $(1,-4)$. Endpoints of minor axes: $(h \pm b, k) = (1 \pm 2, -1)$, thus $(3,-1)$ and $(-1,-1)$. The graph is shown in Fig. 38-11.

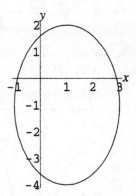

Figure 38-11

**38.11.** Analyze and sketch the graph of $9x^2 - 16y^2 - 36x + 32y = 124$.

Complete the square on $x$ and $y$.

$$9(x^2 - 4x) - 16(y^2 - 2y) = 124$$

$$9(x^2 - 4x + 4) - 16(y^2 - 2y + 1) = 124 + 9 \cdot 4 - 16 \cdot 1$$

$$9(x - 2)^2 - 16(y - 1)^2 = 144$$

$$\frac{(x - 2)^2}{16} - \frac{(y - 1)^2}{9} = 1$$

Comparing with the table in Problem 38.9, we find that this is the equation of a hyperbola with center at $(2,1)$. Since the coefficient of the square involving $x$ is positive, the focal axis is parallel to the $x$-axis. (*Note:* For a hyperbola there is no restriction that $a > b$.) Hence $a^2 = 16$ and $b^2 = 9$; thus, $a = 4$, $b = 3$, and $c = \sqrt{a^2 + b^2} = \sqrt{25} = 5$. Foci: $(h \pm c, k) = (2 \pm 5, 1)$, thus, $(7,1)$ and $(-3,1)$. Vertices: $(h \pm a, k) = (2 \pm 4, 1)$, thus, $(6,1)$ and $(-2,1)$. Asymptotes: $(y - k) = \pm\frac{b}{a}(x - h)$, thus, $(y - 1) = \pm\frac{3}{4}(x - 2)$. Draw vertical lines through the vertices and horizontal lines through the points $(h, k \pm b) = (2, 1 \pm 3)$, thus $(2,4)$ and $(2,-2)$. These form the box. Sketch in the asymptotes and the diagonals of the box, then draw the hyperbola from the vertices out toward the asymptotes. The graph is shown in Fig. 38-12.

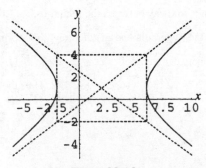

Figure 38-12

**38.12.** Analyze the quantity eccentricity $e = c/a$ for an ellipse and a hyperbola.

For an ellipse, $0 < c < a$, hence $0 < c/a = e < 1$. The eccentricity measures the shape of the ellipse as follows:

If $e$ is small, that is, close to 0, then $c$ is small compared to $a$; hence, $b = \sqrt{a^2 - c^2}$ is close to $a$. Then the minor and major axes of the ellipse are roughly equal in size and the ellipse resembles a circle (the word *eccentricity* means departure from the center).

If $e$ is large, that is, close to 1, then $c$ is roughly equal to $a$; hence $b = \sqrt{a^2 - c^2}$ is close to 0. Then the major axis of the ellipse is substantially larger than the minor axis, and the ellipse has an elongated shape.

For a hyperbola, $c > a$; hence $c/a = e > 1$. The eccentricity measures the shape of the hyperbola by constraining the slope of the asymptotes, as follows:

If $e$ is small, that is, close to 1, then $c$ is roughly equal to $a$; hence $b = \sqrt{c^2 - a^2}$ is close to 0. Then the asymptotes, having slopes $\pm b/a$ or $\pm a/b$, will seem close to the axes on which the vertices lie and the hyperbola will have a hairpin shape.

If $e$ is large, then $a$ is small compared to $c$; hence $b = \sqrt{c^2 - a^2}$ is close to $c$, and thus also large compared to $a$. Then the asymptotes will seem far from the axes on which the vertices lie and the hyperbola will seem wide.

**38.13.** Find the equation of an ellipse (a) in standard position with foci $(\pm 3, 0)$ and $y$-intercepts $(0, \pm 2)$; (b) in standard orientation with foci $(1,5)$ and $(1,7)$ and eccentricity $\frac{1}{2}$.

(a) The ellipse is in standard position with foci on the $x$-axis. Hence it has an equation of the form $\frac{x^2}{a^2} + \frac{y^2}{b^2} = 1$. From the position of the foci, $c = 3$; from the position of the $y$-intercepts, $b = 2$; hence $a = \sqrt{c^2 + b^2} = \sqrt{3^2 + 2^2} = \sqrt{13}$. Thus the equation of the ellipse is $\frac{x^2}{13} + \frac{y^2}{4} = 1$.

(b) The center of the ellipse is midway between the foci, thus, at $(1,6)$. Comparing with the table in Problem 38.9, the ellipse has an equation of the form $\frac{(x - h)^2}{b^2} + \frac{(y - k)^2}{a^2} = 1$, with $(h,k) = (1,6)$. The distance between the foci $= 2c = 2$, thus $c = 1$. Since $e = c/a = 1/2$, it follows that $a = 2$ and $b = \sqrt{a^2 - c^2} = \sqrt{2^2 - 1^2} = \sqrt{3}$. Thus the equation of the ellipse is $\frac{(x - 1)^2}{3} + \frac{(y - 6)^2}{4} = 1$.

**38.14.** Find the equation of a hyperbola (a) in standard position with foci $(\pm 3, 0)$ and $x$-intercepts $(\pm 2, 0)$; (b) in standard orientation with foci $(1,5)$ and $(1,7)$ and eccentricity 2.

(a) The hyperbola is in standard position with foci on the $x$-axis. Hence it has an equation of the form $\frac{x^2}{a^2} - \frac{y^2}{b^2} = 1$. From the position of the foci, $c = 3$, from the position of the vertices, $a = 2$; hence $b = \sqrt{c^2 - a^2} = \sqrt{3^2 - 2^2} = \sqrt{5}$. Thus the equation of the hyperbola is $\frac{x^2}{4} - \frac{y^2}{5} = 1$.

(b) The center of the hyperbola is midway between the foci, thus, at $(1,6)$. Comparing with the table in Problem 38.9, we find that the hyperbola has an equation of the form $\frac{(y - k)^2}{a^2} - \frac{(x - h)^2}{b^2} = 1$, with $(h,k) = (1,6)$. The distance between the foci $= 2c = 2$, thus $c = 1$. Since $e = c/a = 2$, it follows that $a = 1/2$ and $b = \sqrt{c^2 - a^2} = \sqrt{1^2 - (1/2)^2} = (\sqrt{3})/2$. Thus the equation of the hyperbola is $\frac{(y - 6)^2}{1/4} - \frac{(x - 1)^2}{3/4} = 1$.

## SUPPLEMENTARY PROBLEMS

**38.15.** Analyze and sketch graphs of the ellipses (a) $\frac{x^2}{9} + \frac{y^2}{5} = 1$; (b) $25x^2 + 16y^2 + 100x - 96y = 156$.

*Ans.* (a) Standard position, foci on $x$-axis at $(\pm 2,0)$, vertices $(\pm 3,0)$, endpoints of minor axis $(0,\pm\sqrt{5})$. See Fig. 38-13.

(b) Standard orientation, focal axis parallel to $y$-axis, center at $(-2,3)$, foci $(-2,0)$ and $(-2,6)$, vertices $(-2,-2)$ and $(-2,8)$, endpoints of minor axis $(2,3)$ and $(-6,3)$. See Fig. 38-14.

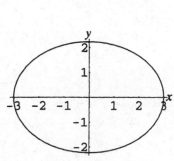

Figure 38-13

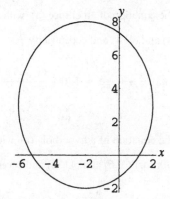

Figure 38-14

**38.16.** Analyze and sketch graphs of the hyperbolas (a) $\frac{x^2}{9} - \frac{y^2}{5} = 1$; (b) $x^2 - y^2 + 6x + 34 = 0$.

*Ans.* (a) Standard position, foci on $x$-axis at $(\pm\sqrt{14},0)$, vertices $(\pm 3,0)$, asymptotes $y = \pm\frac{\sqrt{5}}{3}x$. See Fig. 38-15.

(b) Standard orientation, focal axis parallel to $y$-axis, foci at $(-3,\pm 5\sqrt{2})$, vertices $(-3,\pm 5)$, asymptotes $y = \pm(x + 3)$. See Fig. 38-16.

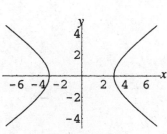

Figure 38-15

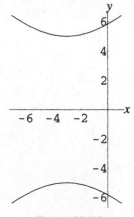

Figure 38-16

**38.17.** Show that as $x$ becomes arbitrarily large, the distance between the graphs of $y = \pm\frac{b}{a}\sqrt{x^2 - a^2}$ and the lines $y = \pm\frac{b}{a}x$ becomes arbitrarily small, thus the lines are oblique asymptotes for the graph.

**38.18.** Show that as $x$ becomes arbitrarily large, the distance between the graphs of $y = \pm\frac{a}{b}\sqrt{b^2 + x^2}$ and the lines $y = \pm\frac{a}{b}x$ becomes arbitrarily small, thus the lines are oblique asymptotes for the graph.

**38.19.** Find the eccentricity for each of the following:

(a) $\dfrac{x^2}{9} + \dfrac{y^2}{5} = 1$; (b) $25x^2 + 16y^2 + 100x - 96y = 156$; (c) $\dfrac{x^2}{9} - \dfrac{y^2}{5} = 1$; (d) $x^2 - y^2 + 6x + 34 = 0$.

*Ans.* (a) 2/3; (b) 3/5; (c) $\sqrt{14}/3$; (d) $\sqrt{2}$

**38.20.** Find the equation of an ellipse (a) with major vertices $(\pm4,0)$ and eccentricity $\frac{1}{4}$; (b) with minor vertices $(-3, 4)$ and $(1, 4)$ and eccentricity $\frac{4}{5}$.

*Ans.* (a) $\dfrac{x^2}{16} + \dfrac{y^2}{15} = 1$; (b) $\dfrac{(x + 1)^2}{4} + \dfrac{(y - 4)^2}{100/9} = 1$

**38.21.** Find the equation of a hyperbola (a) with vertices $(0, \pm12)$ and asymptotes $y = \pm3x$; (b) with foci $(3, 6)$ and $(11, 6)$ and eccentricity $\frac{4}{3}$.

*Ans.* (a) $\dfrac{y^2}{144} - \dfrac{x^2}{16} = 1$; (b) $\dfrac{(x - 7)^2}{9} - \dfrac{(y - 6)^2}{7} = 1$

**38.22.** Use the definition of an ellipse $PF_1 + PF_2 = 2a$ directly to find the equation of an ellipse with foci at $(0,0)$ and $(4,0)$ and major axis $2a = 6$.

*Ans.* $\dfrac{(x - 2)^2}{9} + \dfrac{y^2}{5} = 1$

# CHAPTER 39

# Rotation of Axes

## Rotation of Coordinate Systems

It is often convenient to analyze curves and equations in terms of a Cartesian coordinate system for which the axes have been rigidly rotated through a (normally acute) angle with respect to the standard Cartesian coordinate system.

## Transformation of Coordinates Under Rotation

Let $P$ be a point in the plane; then $P$ has coordinates $(x,y)$ in the standard Cartesian coordinate system (called the *old* system) and coordinates $(x', y')$ in the rotated system (called the *new* system). (See Fig. 39-1.) Then the coordinates in the old system can be expressed in terms of the coordinates in the new system by the *transformation equations*:

$$x = x' \cos\theta - y' \sin\theta$$

$$y = x' \sin\theta + y' \cos\theta$$

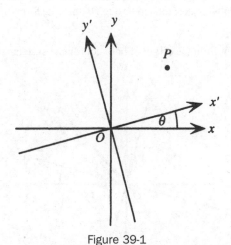

Figure 39-1

These equations can be applied to the coordinates of individual points; a frequent use is to transform equations of curves given in the old coordinate system into equations in the new system, where the form of the equation may be easier to analyze.

**EXAMPLE 39.1**  Analyze the effect on the equation $xy = 2$ of rotating the axes through a 45° angle.

If $\theta = 45°$, then $\cos\theta = \sin\theta = 1/\sqrt{2}$. Hence the transformation equations become

$$x = \frac{x' - y'}{\sqrt{2}}, \qquad y = \frac{x' + y'}{\sqrt{2}}$$

Performing these substitution in the original equation yields

$$xy = \left(\frac{x' - y'}{\sqrt{2}}\right)\left(\frac{x' + y'}{\sqrt{2}}\right) = 2$$

$$\frac{x'^2 - y'^2}{2} = 2$$

This can be written as

$$\frac{x'^2}{4} - \frac{y'^2}{4} = 1$$

which can be seen to be the equation of a hyperbola in standard position with foci on the $x'$-axis (that is, the *new x*-axis), rotated 45° with respect to the old.

## Analyzing Second-Degree Equations

In analyzing second-degree equations written in the standard form

$$Ax^2 + Bxy + Cy^2 + Dx + Ey + F = 0$$

it is useful to rotate axes. An angle $\theta$ can always be found such that rotating axes through this angle transforms the equation into the form $A'x'^2 + C'y'^2 + D'x' + E'y' + F = 0$. The angle $\theta$ is given by

1. If $A = C$, then $\theta = 45°$
2. Otherwise, $\theta$ is a solution of the equation $\tan 2\theta = \dfrac{B}{A - C}$.

### SOLVED PROBLEMS

**39.1.** Show that for any point $P$ that has coordinates $(x,y)$ in a standard Cartesian coordinate system and coordinates $(x',y')$ in a system with axes rotated through $\theta$, the transformation equations $x = x'\cos\theta - y'\sin\theta$, $y = x'\sin\theta + y'\cos\theta$ hold.

Consider the vector $\overrightarrow{OP}$ drawn from the origin of both coordinate systems to $P$ in Fig. 39-2.

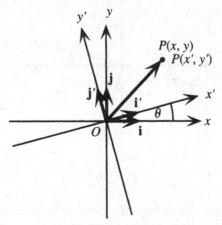

Figure 39-2

It is convenient to use the notation of Problem 27.13, in which **i** and **j** are defined, respectively, as the unit vectors in the positive $x$-and $y$-directions (in the old coordinate system). Then, in this coordinate system, $\overrightarrow{OP} = x\mathbf{i} + y\mathbf{j}$. Similarly, **i'** and **j'** are, respectively, the unit vectors in the positive $x'$- and $y'$-directions, and, in this coordinate system, $\overrightarrow{OP} = x'\mathbf{i'} + y'\mathbf{j'}$. Then $x\mathbf{i} + y\mathbf{j} = x'\mathbf{i'} + y'\mathbf{j'}$. If, now, the dot product of both sides of this identity is taken with vector **i,** it follows that:

$$(x\mathbf{i} + y\mathbf{j}) \cdot \mathbf{i} = (x'\mathbf{i'} + y'\mathbf{j'}) \cdot \mathbf{i}$$

$$x\mathbf{i} \cdot \mathbf{i} + y\mathbf{j} \cdot \mathbf{i} = x'\mathbf{i'} \cdot \mathbf{i} + y'\mathbf{j'} \cdot \mathbf{i}$$

In the latter identity, apply the theorem on the dot product:

Since the angle between $\mathbf{i}$ and $\mathbf{i}$ is $0°$, $\mathbf{i} \cdot \mathbf{i} = |\mathbf{i}| |\mathbf{i}| \cos 0° = 1 \cdot 1 \cdot 1 = 1$.
Since the angle between $\mathbf{i}$ and $\mathbf{j}$ is $90°$, $\mathbf{i} \cdot \mathbf{j} = |\mathbf{i}| |\mathbf{j}| \cos 90° = 1 \cdot 1 \cdot 0 = 0$.
Since the angle between $\mathbf{i}$ and $\mathbf{i}'$ is $\theta$, $\mathbf{i} \cdot \mathbf{i}' = |\mathbf{i}| |\mathbf{i}'| \cos \theta = 1 \cdot 1 \cdot \cos \theta = \cos \theta$.
Since the angle between $\mathbf{i}$ and $\mathbf{j}'$ is $\theta + 90°$, $\mathbf{i} \cdot \mathbf{j}' = |\mathbf{i}| |\mathbf{j}'| \cos(\theta + 90°) = 1 \cdot 1 \cdot (-\sin \theta) = -\sin \theta$.

Substituting yields:

$$x(1) + y(0) = x' \cos \theta - y' \sin \theta$$

$$x = x' \cos \theta - y' \sin \theta$$

The proof of the transformation equation for $y$, $y = x' \sin \theta + y' \cos \theta$, is left as an exercise.

**39.2.** Show that an angle $\theta$ can always be found such that rotating axes through this angle transforms the equation $Ax^2 + Bxy + Cy^2 + Dx + Ey + F = 0$ into the equation $A'x'^2 + C'y'^2 + D'x' + E'y' + F = 0$.

Rotating axes through an angle $\theta$ transforms the equation $Ax^2 + Bxy + Cy^2 + Dx + Ey + F = 0$ by making the substitutions $x = x' \cos \theta - y' \sin \theta$, $y = x' \sin \theta + y' \cos \theta$. Performing the substitutions yields:

$$A(x' \cos \theta - y' \sin \theta)^2 + B(x' \cos \theta - y' \sin \theta)(x' \sin \theta + y' \cos \theta) + C(x' \sin \theta + y' \cos \theta)^2$$
$$+ D(x' \cos \theta - y' \sin \theta) + E(x' \sin \theta + y' \cos \theta) + F = 0$$

Expanding and combining terms in $x'^2$, $y'^2$, $x'y'$, $x'$, and $y'$ yields:

$$x'^2(A\cos^2 \theta + B\cos \theta \sin \theta + C\sin^2 \theta) + x'y'[-2A\cos \theta \sin \theta + B(\cos^2 \theta - \sin^2 \theta) + 2C\sin \theta \cos \theta]$$
$$+ y'^2(A\sin^2 \theta - B\sin \theta \cos \theta + C\cos^2 \theta) + x'(D\cos \theta + E\sin \theta) + y'(-D\sin \theta + E\cos \theta) + F = 0$$

In order for the equation to have exactly the form $A'x'^2 + C'y'^2 + D'x' + E'y' + F = 0$, the coefficient of the $x'y'$ term must be zero, that is:

$$-2A\cos \theta \sin \theta + B(\cos^2 \theta - \sin^2 \theta) + 2C\sin \theta \cos \theta = 0$$

$$-A\sin 2\theta + B\cos 2\theta + C\sin 2\theta = 0$$

$$(A - C)\sin 2\theta = B\cos 2\theta$$

Thus if $A = C$, then $B\cos 2\theta = 0$, thus $2\theta = 90°$, or $\theta = 45°$. Otherwise, divide both sides by $(A - C)\cos 2\theta$ to obtain

$$\frac{\sin 2\theta}{\cos 2\theta} = \frac{B}{A - C}$$

$$\tan 2\theta = \frac{B}{A - C}$$

This equation will have an acute angle solution for $\theta$.

**39.3.** Find an appropriate angle through which to rotate axes and sketch a graph of the equation $3x^2 - 2\sqrt{3}xy + y^2 + 2x + 2\sqrt{3}y = 0$.

Here $A = 3$, $B = -2\sqrt{3}$, $C = 1$; hence set

$$\tan 2\theta = \frac{B}{A - C} = \frac{-2\sqrt{3}}{3 - 1} = -\sqrt{3}$$

The smallest solution of this equation is given by $2\theta = 120°$, that is, $\theta = 60°$. Since $\sin 60° = \sqrt{3}/2$ and $\cos 60° = \frac{1}{2}$, the transformation equations are

$$x = \frac{x' - y'\sqrt{3}}{2} \qquad y = \frac{x'\sqrt{3} + y'}{2}$$

Substituting these into the original equation followed by simplification yields:

$$3\left(\frac{x' - y'\sqrt{3}}{2}\right)^2 - 2\sqrt{3}\left(\frac{x' - y'\sqrt{3}}{2}\right)\left(\frac{x'\sqrt{3} + y'}{2}\right) + \left(\frac{x'\sqrt{3} + y'}{2}\right)^2 + 2\left(\frac{x' - y'\sqrt{3}}{2}\right)$$

$$+ 2\sqrt{3}\left(\frac{x'\sqrt{3} + y'}{2}\right) = 0$$

$$\frac{x'^2(3 - 6 + 3) + x'y'(-6\sqrt{3} + 4\sqrt{3} + 2\sqrt{3}) + y'^2(9 + 6 + 1)}{4} + \frac{x'(2 + 6) + y'(-2\sqrt{3} + 2\sqrt{3})}{2} = 0$$

$$4y'^2 + 4x' = 0$$

$$y'^2 = -x'$$

Thus, in the rotated system, the graph of the equation is a parabola, vertex at (0,0), opening left. The graph is shown in Fig. 39-3.

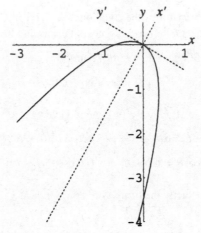

Figure 39-3

**39.4.** Solve the transformation equations for $x'$ and $y'$ in terms of $x$ and $y$ to find the *reverse transformation equations.*

Write the transformation equations in standard form for equations in $x'$ and $y'$:

$$x'\cos\theta - y'\sin\theta = x$$

$$x'\sin\theta + y'\cos\theta = y$$

Now apply Cramer's rule to obtain:

$$x' = \frac{\begin{vmatrix} x & -\sin\theta \\ y & \cos\theta \end{vmatrix}}{\begin{vmatrix} \cos\theta & -\sin\theta \\ \sin\theta & \cos\theta \end{vmatrix}} = x\cos\theta + y\sin\theta \qquad y' = \frac{\begin{vmatrix} \cos\theta & x \\ \sin\theta & y \end{vmatrix}}{\begin{vmatrix} \cos\theta & -\sin\theta \\ \sin\theta & \cos\theta \end{vmatrix}} = -x\sin\theta + y\cos\theta$$

**39.5.** Find appropriate transformation equations for rotation of axes and sketch a graph of the equation $2x^2 - 3xy - 2y^2 + 10 = 0$.

Here $A = 2$, $B = -3$, $C = -2$; hence set

$$\tan 2\theta = \frac{B}{A - C} = \frac{-3}{2 - (-2)} = -\frac{3}{4}$$

An exact solution of this equation is not possible, but it is also not necessary, since $\sin\theta$ and $\cos\theta$ can be found from the half-angle formulas. Assume the smallest solution of the equation, with $90° < 2\theta < 180°$, then, since $\sec 2\theta = -\sqrt{1 + \tan^2 2\theta} = -\sqrt{1 + \left(-\frac{3}{4}\right)^2} = -\frac{5}{4}$, $\cos 2\theta = \frac{1}{\sec 2\theta} = -\frac{4}{5}$.

Hence, since $45° < \theta < 90°$,

$$\sin\theta = \sqrt{\frac{1 - \cos 2\theta}{2}} = \sqrt{\frac{1 - (-4/5)}{2}} = \sqrt{\frac{9}{10}} = \frac{3}{\sqrt{10}}$$

$$\cos\theta = \sqrt{\frac{1 + \cos 2\theta}{2}} = \sqrt{\frac{1 + (-4/5)}{2}} = \sqrt{\frac{1}{10}} = \frac{1}{\sqrt{10}}$$

Thus the transformation equations to rotate axes in order to eliminate the $xy$ term are:

$$x = \frac{x' - 3y'}{\sqrt{10}} \qquad y = \frac{3x' + y'}{\sqrt{10}}$$

Substituting these into the original equation followed by simplification yields:

$$2\left(\frac{x' - 3y'}{\sqrt{10}}\right)^2 - 3\left(\frac{x' - 3y'}{\sqrt{10}}\right)\left(\frac{3x' + y'}{\sqrt{10}}\right) - 2\left(\frac{3x' + y'}{\sqrt{10}}\right)^2 + 10 = 0$$

$$\frac{x'^2(2 - 9 - 18) + x'y'(-12 + 24 - 12) + y'^2(18 + 9 - 2)}{10} + 10 = 0$$

$$\frac{-25x'^2 + 25y'^2}{10} + 10 = 0$$

$$\frac{x'^2}{4} - \frac{y'^2}{4} = 1$$

Thus, in the rotated system, the graph of the equation is a hyperbola in standard position, with focal axis on the $x'$ axis and asymptotes $y' = \pm x'$. To sketch, note that the axes have been rotated through an angle $\theta$ with $\tan\theta = 3$; hence the $x'$-axis has precisely slope 3 with respect to the old coordinate system. The graph, together with the asymptotes, is shown in Fig. 39-4.

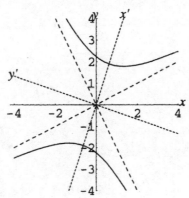

Figure 39-4

**39.6.** In the previous problem, (a) find the coordinates of the foci in the new system and in the old system and (b) find the equations of the asymptotes in the old system.

(a) From the equation of the hyperbola in the new system, $a = b = 2$; hence $c = \sqrt{a^2 + b^2} = 2\sqrt{2}$.

Thus the coordinates of the foci in the new system are $(x', y') = (\pm 2\sqrt{2}, 0)$. To transform these to the old system, use the transformation equations:

$$x = x'\cos\theta - y'\sin\theta = \pm 2\sqrt{2}\left(\frac{1}{\sqrt{10}}\right) - 0\left(\frac{3}{\sqrt{10}}\right) = \pm\frac{2}{\sqrt{5}}$$

$$y = x'\sin\theta + y'\cos\theta = \pm 2\sqrt{2}\left(\frac{3}{\sqrt{10}}\right) + 0\left(\frac{1}{\sqrt{10}}\right) = \pm\frac{6}{\sqrt{5}}$$

Hence the coordinates of the foci in the old system are $(x, y) = \left(\frac{2}{\sqrt{5}}, \frac{6}{\sqrt{5}}\right)$ and $(x, y) = \left(-\frac{2}{\sqrt{5}}, -\frac{6}{\sqrt{5}}\right)$.

(b) The equations of the asymptotes in the new system are $y' = \pm x'$. To transform these to the old system, use the *reverse* transformation equations:

$$x' = x\cos\theta + y\sin\theta = \frac{x + 3y}{\sqrt{10}} \qquad y' = -x\sin\theta + y\cos\theta = \frac{-3x + y}{\sqrt{10}}$$

Then $y' = x'$ becomes $\dfrac{-3x + y}{\sqrt{10}} = \dfrac{x + 3y}{\sqrt{10}}$, or, after simplification, $-2x = y$, and $y' = -x'$ becomes

$\dfrac{-3x + y}{\sqrt{10}} = -\left(\dfrac{x + 3y}{\sqrt{10}}\right)$, or, after simplification, $x = 2y$.

## SUPPLEMENTARY PROBLEMS

**39.7.** Complete Problem 39.1 by showing that a rotation through an angle $\theta$ transforms $y$ according to the transformation equation $y = x'\sin\theta + y'\cos\theta$.

**39.8.** Find an appropriate angle through which to rotate axes to eliminate the $xy$ term in the equation
$21x^2 - 10xy\sqrt{3} + 31y^2 = 144$.

*Ans.* 30°

**39.9.** Find the equation into which $21x^2 - 10xy\sqrt{3} + 31y^2 = 144$ is transformed by the rotation of the previous problems and sketch the graph.

*Ans.* Equation: $\dfrac{x'^2}{9} + \dfrac{y'^2}{4} = 1$. See Fig. 39-5.

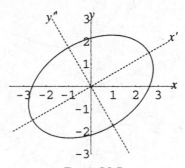

Figure 39-5

**39.10.** Show that the equation $x^2 + y^2 = r^2$ does not change (is *invariant*) under a rotation of the axes through any angle $\theta$.

**39.11.** Find the transformation equations to rotate axes through an appropriate angle to eliminate the $xy$ term in the equation $16x^2 + 24xy + 9y^2 + 60x - 80y + 100 = 0$.

*Ans.* $x = \dfrac{4x' - 3y'}{5}, \quad y = \dfrac{3x' + 4y'}{5}$

**39.12.** Find the equation into which $16x^2 + 24xy + 9y^2 + 60x - 80y + 100 = 0$ is transformed by the rotation of the previous problem and sketch the graph.

*Ans.* Equation: $x'^2 = 4(y' - 1)$. See Fig. 39-6.

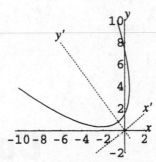

Figure 39-6

**39.13.** Show that in transforming the equation $Ax^2 + Bxy + Cy^2 + Dx + Ey + F = 0$ through any rotation of axes into an equation of form $A'x'^2 + B'x'y' + C'y'^2 + D'x' + E'y' + F = 0$, the quantity $A + C$ will equal the quantity $A' + C'$. (*Hint*: See Problem 39.2 for expressions for $A'$ and $C'$.)

**39.14.** Find the transformation equations to rotate axes through an appropriate angle to eliminate the $xy$ term in the equation $3x^2 + 8xy - 3y^2 - 4x\sqrt{5} + 8y\sqrt{5} = 0$.

*Ans.*   $x = \dfrac{2x' - y'}{\sqrt{5}}, y = \dfrac{x' + 2y'}{\sqrt{5}}.$

**39.15.** Find the equation into which $3x^2 + 8xy - 3y^2 - 4x\sqrt{5} + 8y\sqrt{5} = 0$ is transformed by the rotation of the previous problem and sketch the graph.

*Ans.*   Equation: $\dfrac{(y' - 2)^2}{4} - \dfrac{x'^2}{4} = 1$. See Fig. 39-7.

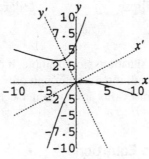

Figure 39-7

# CHAPTER 40

# Conic Sections

## Definition of Conic Sections

The curves that result from the intersection of a plane with a cone are called *conic sections*. Fig. 40-1 shows the four major possibilities: circle, ellipse, parabola, and hyperbola.

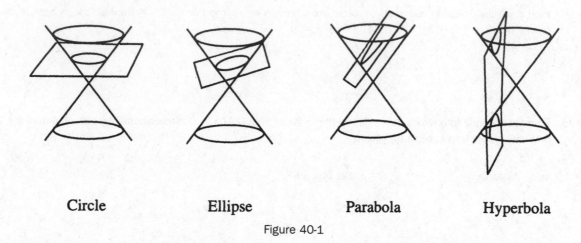

| Circle | Ellipse | Parabola | Hyperbola |

Figure 40-1

*Degenerate* cases arise from exceptional situations; for example, if the plane in the first figure that intersects the cone in a circle were to be lowered until it passes through only the vertex of the cone, the circle would "degenerate" into a point. Other degenerate cases are: two intersecting lines, two parallel lines, one line, or no graph at all.

## Classification of Second-Degree Equations

The graph of a second-degree equation in two variables $Ax^2 + Bxy + Cy^2 + Dx + Ey + F = 0$ is a conic section. Ignoring degenerate cases, the possibilities are as follows:

**A. If no $xy$ term is present ($B = 0$):**

1. If $A = C$ the graph is a circle. Otherwise, $A \neq C$; then:
2. If $AC = 0$ the graph is a parabola.
3. If $AC > 0$ the graph is an ellipse.
4. If $AC < 0$ the graph is a hyperbola.

**B. In General:**

1. If $B^2 - 4AC = 0$ the graph is a parabola.
2. If $B^2 - 4AC < 0$ the graph is an ellipse (or circle if $B = 0$, $A = C$).
3. If $B^2 - 4AC > 0$ the graph is a hyperbola.

The quantity $B^2 - 4AC$ is called the *discriminant* of the second-degree equation.

**EXAMPLE 40.1** Identify the curve with equation $x^2 + 3y^2 + 8x + 4y = 50$, assuming the graph exists. $B = 0$.

Since $A = 1$ and $C = 3$, $AC = 3 > 0$, thus the graph is an ellipse.

**EXAMPLE 40.2** Identify the curve with equation $x^2 + 8xy + 3y^2 + 4y = 50$, assuming the graph exists.

Since $A = 1$, $B = 8$, $C = 3$, $B^2 - 4AC = 8^2 - 4 \cdot 1 \cdot 3 = 52 > 0$, thus the graph is a hyperbola.

## SOLVED PROBLEMS

**40.1.** Derive the classification scheme for second degree equations with $B = 0$.

First note that any such equation has the form $Ax^2 + Cy^2 + Dx + Ey + F = 0$.

1. If $A = 0$, the square can be completed on $y$ to yield $C(y - k)^2 = -D(x - h)$; if $C = 0$, the square can be completed on $x$ to yield $A(x - h)^2 = -E(y - k)$. These are recognizable as equations of parabolas in standard orientation, corresponding to the case $AC = 0$.

2. Otherwise neither $A$ nor $C$ is zero. Then the square can be completed on both $x$ and $y$ to yield $A(x - h)^2 + C(y - k)^2 = G$. The following cases can be further distinguished.

3. $G = 0$. The equation represents a degenerate conic section, either a point or two straight lines.

4. $G \neq 0$. If $A$ and $C$ have opposite signs, the equation can be written as $\dfrac{(x - h)^2}{m^2} - \dfrac{(y - k)^2}{n^2} = \pm 1$, which is the equation of a hyperbola, corresponding to the case $AC < 0$. If $A$ and $C$ have both the same sign as $G$, the equation can be written as $\dfrac{(x - h)^2}{m^2} + \dfrac{(y - k)^2}{n^2} = 1$. Then if the denominators are equal, this is the equation of a circle; if not, it is the equation of an ellipse, with $AC > 0$. Finally, if $A$ and $C$ both have the opposite sign from $G$, the equation represents a degenerate conic section, consisting of no point at all.

**40.2.** Recall that a rotation of axes through any angle $\theta$ transforms a second-degree equation of the form $Ax^2 + Bxy + Cy^2 + Dx + Ey + F = 0$ into another second-degree equation in the form $A'x'^2 + B'x'y' + C'y'^2 + D'x' + E'y' + F = 0$. Show that, regardless of the value of $\theta$, $B^2 - 4AC = B'^2 - 4A'C'$.

In Problem 39.2, it was shown that rotating axes through an angle $\theta$ transforms the equation $Ax^2 + Bxy + Cy^2 + Dx + Ey + F = 0$ by making the substitutions $x = x'\cos\theta - y'\sin\theta$, $y = x'\sin\theta + y'\cos\theta$, yielding:

$$x'^2(A\cos^2\theta + B\cos\theta\sin\theta + C\sin^2\theta) + x'y'[-2A\cos\theta\sin\theta + B(\cos^2\theta - \sin^2\theta) + 2C\sin\theta\cos\theta]$$

$$+ y'^2(A\sin^2\theta - B\sin\theta\cos\theta + C\cos^2\theta) + x'(D\cos\theta + E\sin\theta) + y'(-D\sin\theta + E\cos\theta) + F = 0$$

Comparing this with the form $A'x'^2 + B'x'y' + C'y'^2 + D'x' + E'y' + F = 0$ shows that

$$A' = A\cos^2\theta + B\cos\theta\sin\theta + C\sin^2\theta$$

$$B' = -2A\cos\theta\sin\theta + B(\cos^2\theta - \sin^2\theta) + 2C\sin\theta\cos\theta$$

$$C' = A\sin^2\theta - B\sin\theta\cos\theta + C\cos^2\theta$$

Thus

$$B'^2 - 4A'C' = [-2A\cos\theta\sin\theta + B(\cos^2\theta - \sin^2\theta) + 2C\sin\theta\cos\theta]^2$$

$$-4(A\cos^2\theta + B\cos\theta\sin\theta + C\sin^2\theta)(A\sin^2\theta - B\sin\theta\cos\theta + C\cos^2\theta)$$

Expanding and collecting terms yields

$$B'^2 - 4A'C' = A^2(4\cos^2\theta\sin^2\theta - 4\cos^2\theta\sin^2\theta) + B^2(\cos^4\theta - 2\cos^2\theta\sin^2\theta + \sin^4\theta + 4\cos^2\theta\sin^2\theta)$$

$$+ C^2(4\cos^2\theta\sin^2\theta - 4\cos^2\theta\sin^2\theta) + AB(-4\cos^3\theta\sin\theta + 4\cos\theta\sin^3\theta + 4\cos^3\theta\sin\theta - 4\cos\theta\sin^3\theta)$$

$$+ AC(-8\sin^2\theta\cos^2\theta - 4\cos^4\theta - 4\sin^4\theta) + BC(4\cos^3\theta\sin\theta - 4\cos\theta\sin^3\theta - 4\cos^3\theta\sin\theta + 4\cos\theta\sin^3\theta)$$

The coefficients of $A^2$, $C^2$, $AB$, and $BC$ are seen to reduce to zero, and the right side reduces to:

$$B'^2 - 4A'C' = B^2(\cos^4\theta + 2\cos^2\theta\sin^2\theta + \sin^4\theta) - 4AC(\cos^4\theta + 2\cos^2\theta\sin^2\theta + \sin^4\theta)$$

$$= (B^2 - 4AC)(\cos^2\theta + \sin^2\theta)^2$$

$$= B^2 - 4AC$$

This equality is often stated as follows: The quantity $B^2 - 4AC$ is *invariant* under a rotation of axes through any angle.

**40.3.** Derive the general classification scheme for second-degree equations.

The general second-degree equation has the form $Ax^2 + Bxy + Cy^2 + Dx + Ey + F = 0$. If $B \neq 0$, then it was shown in the previous chapter (Problem 39.2) that there is an angle $\theta$ through which the axes can be rotated so that the equation takes the form $A'x'^2 + C'y'^2 + D'x' + E'y' + F = 0$. Then, this is the equation of

1. A parabola if $A'C' = 0$

2. An ellipse if $A'C' > 0$

3. A hyperbola if $A'C' < 0$

For the transformed equation, the discriminant becomes $B^2 - 4AC = -4A'C'$. Therefore, in case 1, the original discriminant $B^2 - 4AC = 0$; in case 2, $B^2 - 4AC < 0$; and in case 3, $B^2 - 4AC > 0$. Summarizing, the equation $Ax^2 + Bxy + Cy^2 + Dx + Ey + F = 0$ represents

1. A parabola if $B^2 - 4AC = 0$

2. An ellipse if $B^2 - 4AC < 0$ (or circle if $B = 0$, $A = C$)

3. A hyperbola if $B^2 - 4AC > 0$

Here degenerate cases are neglected and it is assumed that the equation has a graph.

**40.4.** Identify the following as the equations of a circle, an ellipse, a parabola, or a hyperbola:

(a) $3x^2 + 8x + 12y = 16$; (b) $3x^2 - 3y^2 + 8x + 12y = 16$;

(c) $3x^2 + 3y^2 + 8x + 12y = 16$; (d) $3x^2 + 4y^2 + 8x + 12y = 16$

(a) Here $B = 0$, $A = 3$, and $C = 0$. With $B = 0$, since $AC = 0$, this is the equation of a parabola.

(b) Here $B = 0$, $A = 3$, and $C = -3$. With $B = 0$, since $AC < 0$, this is the equation of a hyperbola.

(c) Here $B = 0$, and $A = C = 3$. Thus this is the equation of a circle.

(d) Here $B = 0$, $A = 3$, and $C = 4$. With $B = 0$, since $AC > 0$, this is the equation of an ellipse

**40.5.** Identify the following as the equations of a circle, an ellipse, a parabola, or a hyperbola:

(a) $3x^2 + 8xy + 12y = 16$; (b) $3x^2 + 8xy - 3y^2 + 8x + 12y = 16$;

(c) $3x^2 + 6xy + 3y^2 + 8x + 12y = 16$; (d) $3x^2 + 2xy + 3y^2 + 8x + 12y = 16$

(a) Here $A = 3$, $B = 8$, $C = 0$, so $B^2 - 4AC = 8^2 - 4 \cdot 3 \cdot 0 = 64 > 0$. Hence this is the equation of a hyperbola.

(b) Here $A = 3$, $B = 8$, $C = -3$, so $B^2 - 4AC = 8^2 - 4 \cdot 3(-3) = 100 > 0$. Hence this is the equation of a hyperbola.

(c) Here $A = 3$, $B = 6$, $C = 3$, so $B^2 - 4AC = 6^2 - 4 \cdot 3 \cdot 3 = 0$. Hence this is the equation of a parabola.

(d) Here $A = 3$, $B = 2$, $C = 3$, so $B^2 - 4AC = 2^2 - 4 \cdot 3 \cdot 3 = -32 < 0$. Hence this is the equation of an ellipse.

*Note*: Since $B \neq 0$ in all of these cases, none of these can be an equation of a circle.

**40.6.** It can be shown that, in general, for any ellipse and any hyperbola, there are two straight lines called *directrices*, perpendicular to the focal axis and at distance $a/e = a^2/c$ from the center, such that the equation of the curve can be derived from the relation $PF = e \cdot PD$, where $PF$ is the distance from a point on the curve to a focus and $PD$ is the perpendicular distance to the directrix. (See Figs. 40-2 and 40-3.)

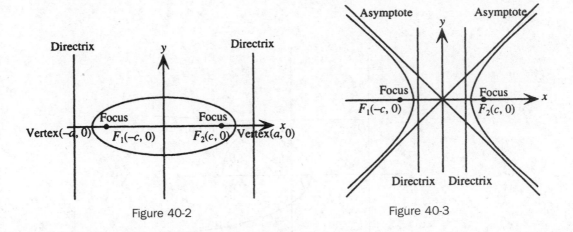

Figure 40-2          Figure 40-3

Find the directrices and verify the derivation of the equation from $PF = e \cdot PD$ for:

(a) the ellipse $\dfrac{x^2}{4} + y^2 = 1$; (b) the hyperbola $x^2 - y^2 = 1$

(a) Here $a = 2, b = 1, c = \sqrt{a^2 - b^2} = \sqrt{4 - 1} = \sqrt{3}$. Thus $e = c/a = \sqrt{3}/2$. The directrices then are the vertical lines $x = \pm a/e = \pm 2 \div \sqrt{3}/2 = \pm 4/\sqrt{3}$. The relation $PF = e \cdot PD$ then becomes

$$\sqrt{(x - \sqrt{3})^2 + y^2} = \frac{\sqrt{3}}{2}\left| x - \frac{4}{\sqrt{3}} \right| \qquad \text{(choosing the right-hand focus and directrix)}$$

Squaring both sides and simplifying yields:

$$x^2 - 2x\sqrt{3} + 3 + y^2 = \frac{3}{4}\left( x^2 - \frac{8}{\sqrt{3}}x + \frac{16}{3} \right)$$

$$x^2 - 2x\sqrt{3} + 3 + y^2 = \frac{3}{4}x^2 - 2x\sqrt{3} + 4$$

$$\frac{x^2}{4} + y^2 = 1$$

(b) Here $a = 1, b = 1, c = \sqrt{a^2 + b^2} = \sqrt{1 + 1} = \sqrt{2}$. Thus $e = c/a = \sqrt{2}$. The directrices then are the vertical lines $x = \pm a/e = \pm 1/\sqrt{2}$. The relation $PF = e \cdot PD$ then becomes

$$\sqrt{(x - \sqrt{2})^2 + y^2} = \sqrt{2}\left| x - \frac{1}{\sqrt{2}} \right| \qquad \text{(choosing the right-hand focus and directrix)}$$

Squaring both sides and simplifying yields:

$$x^2 - 2x\sqrt{2} + 2 + y^2 = 2\left( x^2 - \frac{2}{\sqrt{2}}x + \frac{1}{2} \right)$$

$$x^2 - 2x\sqrt{2} + 2 + y^2 = 2x^2 - 2x\sqrt{2} + 1$$

$$x^2 + y^2 + 2 = 2x^2 + 1$$

$$1 = x^2 - y^2$$

*Note*: If the eccentricity of a parabola is defined as 1, then the relation $PF = e \cdot PD$ can be viewed as describing all three noncircle conic sections: parabola, ellipse, and hyperbola.

**40.7.** Show that in polar coordinates the equation of a conic section with (one) focus at the pole and directrix, the line $r\cos\theta = -p$ can be written as

$$r = \frac{ep}{1 - e\cos\theta}$$

See Fig. 40-4.

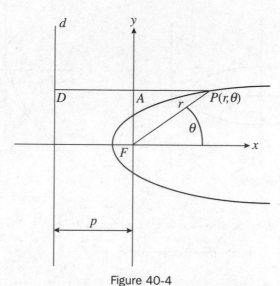

Figure 40-4

From the previous problem, a conic section can be defined by the relation $PF = e \cdot PD$. Here, the focus is at the origin, hence the distance from a point on the conic section to the focus $PF = r$. The distance from $P$ to the directrix is given by

$$PD = PA + AD = r\cos\theta + p.$$

Thus

$$PF = e \cdot PD$$
$$r = e(r\cos\theta + p)$$
$$r = re\cos\theta + ep$$
$$r - re\cos\theta = ep$$
$$r(1 - e\cos\theta) = ep$$
$$r = \frac{ep}{1 - e\cos\theta}$$

**40.8.** Identify each of the following as the equations of an ellipse, a hyperbola, or a parabola:

(a) $r = \dfrac{4}{1 - \cos\theta}$; (b) $r = \dfrac{4}{1 - 2\cos\theta}$; (c) $r = \dfrac{4}{2 - \cos\theta}$

(a) Comparing the given equation with $r = \dfrac{ep}{1 - e\cos\theta}$, $e = 1$ and $ep = p = 4$, hence this is the equation of a parabola.

(b) Comparing the given equation with $r = \dfrac{ep}{1 - e\cos\theta}$, $e = 2$ and $ep = 2p = 4$, $p = 2$, hence this is the equation of a hyperbola.

(c) To compare the given equation with $r = \dfrac{ep}{1 - e\cos\theta}$, rewrite it as $r = \dfrac{2}{1 - \frac{1}{2}\cos\theta}$. Then $e = \dfrac{1}{2}$ and $ep = \dfrac{1}{2}p = 2$, $p = 4$, hence this is the equation of an ellipse.

## SUPPLEMENTARY PROBLEMS

**40.9.** Identify the following as the equations of a circle, an ellipse, a parabola, or a hyperbola:

(a) $x^2 + 2y^2 - 2x + 3y = 50$; (b) $x^2 - 2x + 3y = 50$; (c) $x^2 - y^2 - 2x + 3y = 50$;

(d) $y^2 - 2x + 3y = x^2 + 50$; (e) $2x^2 + 2y^2 - 2x + 3y = 50$

*Ans.*  (a) ellipse; (b) parabola; (c) hyperbola; (d) hyperbola; (e) circle

**40.10.** Identify the following as the equations of a circle, an ellipse, a parabola, or a hyperbola:

(a) $x^2 + 2xy + y^2 - 2x + 3y = 50$; (b) $2xy + y^2 - 2x + 3y = 50$; (c) $x^2 + xy + y^2 - 2x + 3y = 50$;

(d) $x^2 + 4xy + y^2 - 2x + 3y = 50$; (e) $(x - y)^2 + (x + y)^2 - 2x = 50$

*Ans.*  (a) parabola; (b) hyperbola; (c) ellipse; (d) hyperbola; (e) circle

**40.11.** The following equations represent typical degenerate conic sections. By factoring or other algebraic techniques, identify the graphs:

(a) $x^2 + xy - 3x = 0$; (b) $x^2 - 2xy + y^2 = 81$; (c) $x^2 + 4xy + 4y^2 + 2x + 4y + 1 = 0$;

(d) $2x^2 + y^2 - 4y + 16 = 0$; (e) $2x^2 + 4x + y^2 - 4y + 6 = 0$

*Ans.*  (a) $x = 0$ or $x + y = 3$, two intersecting lines; (b) $x - y = \pm 9$, two parallel lines;

(c) $x + 2y + 1 = 0$, one line; (d) $2x^2 + (y - 2)^2 = -12$, no point;

(e) $2(x + 1)^2 + (y - 2)^2 = 0$, graph contains one point: $(-1,2)$

**40.12.** Find the directrices for the graphs of the following equations: (a) $\dfrac{x^2}{4} + \dfrac{y^2}{9} = 1$; (b) $\dfrac{y^2}{9} - \dfrac{x^2}{4} = 1$

*Ans.*  (a)  $y = \pm\dfrac{9}{\sqrt{5}}$; (b) $y = \pm\dfrac{9}{\sqrt{13}}$

**40.13.** Show that in polar coordinates the equation of a conic section with (one) focus at the pole and directrix the line $r\cos\theta = p$ can be written as

$$r = \frac{ep}{1 + e\cos\theta}$$

**40.14.** Identify each of the following as the equations of an ellipse, a hyperbola, or a parabola:

(a) $r = \dfrac{12}{3 - \cos\theta}$; (b) $r = \dfrac{12}{1 + 3\cos\theta}$; (c) $r = \dfrac{12}{1 + \cos\theta}$; (d) $r = \dfrac{12}{3 - 8\cos\theta}$

*Ans.*  (a) ellipse; (b) hyperbola; (c) parabola; (d) hyperbola

**40.15.** Show that in polar coordinates the equation of a conic section with (one) focus at the pole and directrix the line $r\sin\theta = p$ can be written as

$$r = \frac{ep}{1 + e\sin\theta}$$

**40.16.** Show that in polar coordinates the equation of a conic section with (one) focus at the pole and directrix the line $r\sin\theta = -p$ can be written as

$$r = \frac{ep}{1 - e\sin\theta}$$

# Sequences and Series

## Definition of Sequence

A sequence is a function with domain the natural numbers (*infinite sequence*) or some subset of the natural numbers from 1 up to some larger number (*finite sequence*). The notation $f(n) = a_n$ is used to denote range elements of the function: the $a_1, a_2, a_3, \ldots$ are called the first, second, third, etc. *terms* of the sequence, and $a_n$ is referred to as the $n$th term. The independent variable $n$ is referred to as the *index*. Unless otherwise specified, a sequence is assumed to be an infinite sequence.

**EXAMPLE 41.1**   Write the first four terms of the sequence specified by $a_n = 2n$.

$a_1 = 2 \cdot 1, a_2 = 2 \cdot 2, a_3 = 2 \cdot 3, a_4 = 2 \cdot 4$. The sequence would be written $2 \cdot 1, 2 \cdot 2, 2 \cdot 3, 2 \cdot 4, \ldots$ or $2, 4, 6, 8, \ldots$.

**EXAMPLE 41.2**   Write the first four terms of the sequence specified by $a_n = (-1)^n$.

$a_1 = (-1)^1, a_2 = (-1)^2, a_3 = (-1)^3, a_4 = (-1)^4$. The sequence would be written $(-1)^1, (-1)^2, (-1)^3, (-1)^4, \ldots$ or $-1, 1, -1, 1, \ldots$.

## Finding the *n*th Term of a Sequence

Given the first few terms of a sequence, a common exercise is to determine the $n$th term, that is, a formula which generates all the terms. In fact, such a formula is not uniquely determined, but in many cases a simple one can be developed.

**EXAMPLE 41.3**   Find a formula for the $n$th term of the sequence $1, 4, 9, 16, \ldots$.

Notice that the terms are all perfect squares, and the sequence could be written $1^2, 2^2, 3^2, 4^2, \ldots$.

Thus the $n$th term of the sequence can be given as $a_n = n^2$.

## Recursively Defined Sequence

A sequence is defined *recursively* by specifying the first term and defining later terms with respect to earlier terms.

**EXAMPLE 41.4**   Write the first four terms of the sequence defined by $a_1 = 3, a_n = a_{n-1} + 7, n > 1$.

For $n = 1, a_1 = 3$

For $n = 2, a_2 = a_{2-1} + 7 = a_1 + 7 = 3 + 7 = 10$

For $n = 3, a_3 = a_{3-1} + 7 = a_2 + 7 = 10 + 7 = 17$

For $n = 4, a_4 = a_{4-1} + 7 = a_3 + 7 = 17 + 7 = 24$

The sequence can be written $3, 10, 17, 24, \ldots$.

## Definition of Series

A series is the indicated sum of the terms of a sequence. Thus if $a_1, a_2, a_3, \ldots, a_m$ are the $m$ terms of a finite sequence, then associated with the sequence is the series given by $a_1 + a_2 + a_3 + \cdots + a_m$. Series are often written using the summation notation:

$$a_1 + a_2 + a_3 + \cdots + a_m = \sum_{k=1}^{m} a_k$$

Here $\Sigma$ is called the *summation symbol*, and $k$ is called the *index of summation* or just the index. The right-hand side of this definition is read, "the sum of the $a_k$, with $k$ going from 1 to $m$."

**EXAMPLE 41.5** Write in expanded form: $\sum_{k=1}^{5} \dfrac{1}{k^2}$,

Replace $k$, in turn, with the integers from 1 to 5 and add the results:

$$\sum_{k=1}^{5} \frac{1}{k^2} = \frac{1}{1^2} + \frac{1}{2^2} + \frac{1}{3^2} + \frac{1}{4^2} + \frac{1}{5^2} = 1 + \frac{1}{4} + \frac{1}{9} + \frac{1}{16} + \frac{1}{25}$$

## Infinite Series

The sum of all the terms of an infinite sequence is referred to as an *infinite series*, and is indicated by the symbol:

$$\sum_{k=1}^{\infty} a_k$$

Infinite series in general are discussed in calculus courses; a special case (infinite geometric series) is treated in Chapter 43.

## Factorial Symbol

A useful definition is the factorial symbol. For natural numbers $n$, $n!$ (pronounced $n$ factorial) is defined as the product of the natural numbers from 1 up to $n$. Then

$$1! = 1 \qquad 2! = 1 \cdot 2 = 2 \qquad 3! = 1 \cdot 2 \cdot 3 = 6 \qquad 4! = 1 \cdot 2 \cdot 3 \cdot 4 = 24$$

and so on. Separately, $0!$ is defined to equal 1.

## SOLVED PROBLEMS

**41.1.** Write the first four terms of the sequences specified by

(a) $a_n = 2n - 1$; (b) $b_n = 6 - 4n$; (c) $c_n = 2^n$; (d) $d_n = 3(-2)^n$

(a) $a_1 = 2 \cdot 1 - 1 = 1, a_2 = 2 \cdot 2 - 1 = 3, a_3 = 2 \cdot 3 - 1 = 5, a_4 = 2 \cdot 4 - 1 = 7$. The sequence would be written $1, 3, 5, 7, \ldots$.

(b) $b_1 = 6 - 4 \cdot 1 = 2, b_2 = 6 - 4 \cdot 2 = -2, b_3 = 6 - 4 \cdot 3 = -6, b_4 = 6 - 4 \cdot 4 = -10$. The sequence would be written $2, -2, -6, -10, \ldots$.

(c) $c_1 = 2^1 = 2, c_2 = 2^2 = 4, c_3 = 2^3 = 8, c_4 = 2^4 = 16$. The sequence would be written $2, 4, 8, 16, \ldots$.

(d) $d_1 = 3(-2)^1 = -6, d_2 = 3(-2)^2 = 12, d_3 = 3(-2)^3 = -24, d_4 = 3(-2)^4 = 48$. The sequence would be written $-6, 12, -24, 48, \ldots$.

**41.2.** Write the first four terms of the sequences specified by

(a) $a_n = \dfrac{1}{3n+1}$; (b) $b_n = \dfrac{n^2}{3n-2}$; (c) $c_n = \sin\dfrac{\pi n}{4}$; (d) $d_n = \dfrac{(-1)^n\sqrt{n}}{(n+1)(n+2)}$

(a) $a_1 = \dfrac{1}{3\cdot 1+1} = \dfrac{1}{4}, a_2 = \dfrac{1}{3\cdot 2+1} = \dfrac{1}{7}, a_3 = \dfrac{1}{3\cdot 3+1} = \dfrac{1}{10}, a_4 = \dfrac{1}{3\cdot 4+1} = \dfrac{1}{13}$

(b) $b_1 = \dfrac{1^2}{3\cdot 1-2} = 1, b_2 = \dfrac{2^2}{3\cdot 2-2} = 1, b_3 = \dfrac{3^2}{3\cdot 3-2} = \dfrac{9}{7}, b_4 = \dfrac{4^2}{3\cdot 4-2} = \dfrac{8}{5}$

(c) $c_1 = \sin\dfrac{\pi\cdot 1}{4} = \dfrac{1}{\sqrt{2}}, c_2 = \sin\dfrac{\pi\cdot 2}{4} = 1, c_3 = \sin\dfrac{\pi\cdot 3}{4} = \dfrac{1}{\sqrt{2}}, c_4 = \sin\dfrac{\pi\cdot 4}{4} = 0$

(d) $d_1 = \dfrac{(-1)^1\sqrt{1}}{(1+1)(1+2)} = -\dfrac{1}{6}, d_2 = \dfrac{(-1)^2\sqrt{2}}{(2+1)(2+2)} = \dfrac{\sqrt{2}}{12}, d_3 = \dfrac{(-1)^3\sqrt{3}}{(3+1)(3+2)} = -\dfrac{\sqrt{3}}{20},$

$d_4 = \dfrac{(-1)^4\sqrt{4}}{(4+1)(4+2)} = \dfrac{\sqrt{4}}{30} = \dfrac{1}{15}$

**41.3.** Write the tenth term of each of the sequences in the previous problem.

$$a_{10} = \frac{1}{3\cdot 10+1} = \frac{1}{31} \qquad\qquad b_{10} = \frac{10^2}{3\cdot 10-2} = \frac{25}{7}$$

$$c_{10} = \sin\frac{\pi\cdot 10}{4} = 1 \qquad\qquad d_{10} = \frac{(-1)^{10}\sqrt{10}}{(10+1)(10+2)} = \frac{\sqrt{10}}{132}$$

**41.4.** Write the first four terms of the following recursively defined sequences:

(a) $a_1 = 1, a_n = na_{n-1}, n>1$; (b) $a_1 = 1, a_n = a_{n-1}+2, n>1$; (c) $a_1 = 12, a_n = \dfrac{a_{n-1}}{4}, n>1$

(a) For $n=1, a_1 = 1$   (b) For $n=1, a_1 = 1$

For $n=2, a_2 = 2a_{2-1} = 2a_1 = 2\cdot 1 = 2$   For $n=2, a_2 = a_{2-1}+2 = a_1+2 = 1+2 = 3$

For $n=3, a_3 = 3a_{3-1} = 3a_2 = 3\cdot 2 = 6$   For $n=3, a_3 = a_{3-1}+2 = a_2+2 = 3+2 = 5$

For $n=4, a_4 = 4a_{4-1} = 4a_3 = 4\cdot 6 = 24$   For $n=4, a_4 = a_{4-1}+2 = a_3+2 = 5+2 = 7$

(c) For $n=1, a_1 = 12$

For $n=2, a_2 = \dfrac{a_{2-1}}{4} = \dfrac{a_1}{4} = \dfrac{12}{4} = 3$

For $n=3, a_3 = \dfrac{a_{3-1}}{4} = \dfrac{a_2}{4} = \dfrac{3}{4}$

For $n=4, a_4 = \dfrac{a_{4-1}}{4} = \dfrac{a_3}{4} = \dfrac{3/4}{4} = \dfrac{3}{16}$

**41.5.** The sequence defined by $a_1 = 1, a_2 = 1, a_n = a_{n-1}+a_{n-2}, n>2$, is called a Fibonacci sequence. Write the first six terms of this sequence.

For $n=1, a_1 = 1$   For $n=2, a_2 = 1$

For $n=3, a_3 = a_2+a_1 = 1+1 = 2$   For $n=4, a_4 = a_3+a_2 = 2+1 = 3$

For $n=5, a_5 = a_4+a_3 = 3+2 = 5$   For $n=6, a_6 = a_5+a_4 = 5+3 = 8$

The sequence would be written $1, 1, 2, 3, 5, 8, \ldots$.

**41.6.** Find a formula for the *n*th term of a sequence whose first four terms are given by:

(a) $2, 4, 6, 8, \ldots$; (b) $1, \dfrac{1}{3}, \dfrac{1}{5}, \dfrac{1}{7}, \ldots$; (c) $-1, 2, -4, 8, \ldots$; (d) $\dfrac{1}{2}, \dfrac{2}{5}, \dfrac{3}{10}, \dfrac{4}{17}, \ldots$

(a) Comparing the terms of the sequence with $n = 1, 2, 3, 4, \ldots$ shows that the individual terms are each 2 times the index *n* of the term. Thus a possible formula would be $a_n = 2n$.

(b) The sequence can be written as $\dfrac{1}{1}, \dfrac{1}{3}, \dfrac{1}{5}, \dfrac{1}{7}, \ldots$; thus, comparing the denominators with the previous sequence shows that each denominator is 1 less than $2, 4, 6, 8, \ldots$, hence, can be written as $2n - 1$. Thus a possible formula would be $a_n = \dfrac{1}{2n - 1}$.

(c) The absolute values of the terms of the sequence are powers of 2, that is, $2^0, 2^1, 2^2, 2^3$; comparing this to $n = 1, 2, 3, 4, \ldots$ suggests that the *n*th term has absolute value $2^{n-1}$. The fact that the signs of the terms alternate can be represented (in more than one way) by successive powers of $-1$, for example, $(-1)^1$, $(-1)^2, (-1)^3, (-1)^4, \ldots$. Thus a possible formula would be $a_n = (-1)^n 2^{n-1}$.

(d) A pattern for the denominators can be found by comparing to the sequence $1, 4, 9, 16, \ldots$ of Example 41.3; since each denominator is 1 more than the corresponding term of this sequence, they can be represented by $n^2 + 1$. The numerators are equal to the index of the terms; thus, a possible formula would be $a_n = \dfrac{n}{n^2 + 1}$.

**41.7.** Write each series in expanded form:

(a) $\displaystyle\sum_{k=1}^{4} (6k + 1)$; (b) $\displaystyle\sum_{j=1}^{5} \dfrac{j}{j^2 + 1}$; (c) $\displaystyle\sum_{j=3}^{20} (-1)^{j-1}(5j)$; (d) $\displaystyle\sum_{k=1}^{p} \dfrac{k^k}{k!}$

(a) Replace *k* in the expression $6k + 1$, in turn, with each natural number from 1 to 4, and place an addition symbol between the results:

$$\sum_{k=1}^{4} (6k + 1) = (6 \cdot 1 + 1) + (6 \cdot 2 + 1) + (6 \cdot 3 + 1) + (6 \cdot 4 + 1) = 7 + 13 + 19 + 25 = 64$$

(b) Note that the letter *j* is used for the index here; in general, any variable letter may be used, but the letters $i, j, k$ are the most common. Replace *j* in the expression after the summation symbol, in turn, with each natural number from 1 to 5, and place an addition symbol between the results.

$$\sum_{j=1}^{5} \dfrac{j}{j^2 + 1} = \dfrac{1}{1^2 + 1} + \dfrac{2}{2^2 + 1} + \dfrac{3}{3^2 + 1} + \dfrac{4}{4^2 + 1} + \dfrac{5}{5^2 + 1} = \dfrac{1}{2} + \dfrac{2}{5} + \dfrac{3}{10} + \dfrac{4}{17} + \dfrac{5}{26}$$

In this context, it is not always necessary to complete the arithmetic; if desired, the addition can be performed to yield $\dfrac{3597}{2210} \approx 1.6276$.

(c) Note that the index starts from 3; there is no requirement that a series must start from index 1. Replace *j* in the expression after the summation symbol, in turn, with each natural number from 3 to 20, and place an addition symbol between the results.

$$\sum_{j=3}^{20} (-1)^{j-1}(5j) = (-1)^{3-1}(5 \cdot 3) + (-1)^{4-1}(5 \cdot 4) + (-1)^{5-1}(5 \cdot 5) + \cdots + (-1)^{20-1}(5 \cdot 20)$$

$$= 15 - 20 + 25 - \cdots - 100$$

If there are an unwieldy number of terms, as in this case, not all terms are written out explicitly; the three dots (*ellipsis* ...) symbol is used.

(d) Here the variable on top of the summation symbol indicates that the number of terms is not explicitly stated. Write out the first few terms and the last term; use the ellipsis symbol.

$$\sum_{k=1}^{p} \dfrac{k^k}{k!} = \dfrac{1^1}{1!} + \dfrac{2^2}{2!} + \dfrac{3^3}{3!} + \cdots + \dfrac{p^p}{p!}$$

**41.8.** Write the following series in expanded form:

(a) $\displaystyle\sum_{k=1}^{3} \frac{x^{k+1}}{k}$; (b) $\displaystyle\sum_{k=1}^{5} (-1)^{k-1} x^k$; (c) $\displaystyle\sum_{k=0}^{4} \frac{(-1)^k x^k}{k!}$

(a) Replace $k$ in the expression after the summation symbol, in turn, with each natural number from 1 to 3, and place an addition symbol between the results.

$$\sum_{k=1}^{3} \frac{x^{k+1}}{k} = \frac{x^{1+1}}{1} + \frac{x^{2+1}}{2} + \frac{x^{3+1}}{3} = x^2 + \frac{x^3}{2} + \frac{x^4}{3}$$

(b) Replace $k$ in the expression after the summation symbol, in turn, with each natural number from 1 to 5, and place an addition symbol between the results.

$$\sum_{k=1}^{5} (-1)^{k-1} x^k = (-1)^{1-1} x^1 + (-1)^{2-1} x^2 + (-1)^{3-1} x^3 + (-1)^{4-1} x^4 + (-1)^{5-1} x^5$$
$$= x - x^2 + x^3 - x^4 + x^5$$

(c) Replace $k$ in the expression after the summation symbol, in turn, with each integer from 0 to 4, and place an addition symbol between the results.

$$\sum_{k=0}^{4} \frac{(-1)^k x^k}{k!} = \frac{(-1)^0 x^0}{0!} + \frac{(-1)^1 x^1}{1!} + \frac{(-1)^2 x^2}{2!} + \frac{(-1)^3 x^3}{3!} + \frac{(-1)^4 x^4}{4!}$$
$$= \frac{1}{0!} - \frac{x}{1!} + \frac{x^2}{2!} - \frac{x^3}{3!} + \frac{x^4}{4!} \text{ or } 1 - x + \frac{x^2}{2} - \frac{x^3}{6} + \frac{x^4}{24}$$

**41.9.** Write the following series in summation notation:

(a) $3 + 6 + 9 + 12 + 15$; (b) $\dfrac{1}{2} - \dfrac{1}{4} + \dfrac{1}{8} - \dfrac{1}{16}$; (c) $4 + \dfrac{4}{3} + \dfrac{4}{9} + \cdots + \dfrac{4}{729}$; (d) $\dfrac{x}{1} + \dfrac{x^2}{2} + \dfrac{x^3}{6} + \dfrac{x^4}{24}$

(a) Comparing the terms of the sequence with $k = 1, 2, 3, 4, \ldots$ shows that the individual terms are each 3 times the index $k$ of the term. Thus a possible formula for the terms would be $a_k = 3k$; there are 5 terms, hence the series can be written as $\displaystyle\sum_{k=1}^{5} 3k$.

(b) Comparing the terms of the sequence with $k = 1, 2, 3, 4, \ldots$ shows that the denominators are powers of 2: $2^1, 2^2, 2^3, \ldots$ The fact that the signs of the terms alternate can be represented by successive powers of $-1$, for example, $(-1)^0, (-1)^1, (-1)^2, \ldots$ Thus a possible formula for the terms would be $a_k = (-1)^{k-1}/2^k$; there are 4 terms, hence the series can be written as $\displaystyle\sum_{k=1}^{4} (-1)^{k-1}/2^k$.

(c) Comparing the terms of the sequence with $k = 1, 2, 3, 4, \ldots$ shows that the denominators are powers of 3: $3^0, 3^1, 3^2, \ldots$ Thus a possible formula for the terms would be $a_k = 4/3^{k-1}$. Since the last term has denominator $729 = 3^6$, setting $6 = k - 1$ yields $k = 7$; there are 7 terms, hence the series can be written as $\displaystyle\sum_{k=1}^{7} 4/3^{k-1}$.

(d) Comparing the terms of the sequence with $k = 1, 2, 3, 4, \ldots$ shows that the denominators are representable as factorials; a possible formula for the terms would be $a_k = x^k/k!$; there are 4 terms, hence the series can be written as $\displaystyle\sum_{k=1}^{4} x^k/k!$.

## SUPPLEMENTARY PROBLEMS

**41.10.** Write the first four terms of the following sequences: (a) $a_n = \dfrac{1}{10^n}$; (b) $a_n = \dfrac{3n}{n+5}$; (c) $a_n = 5 - 2n$;

(d) $a_n = n[1 - (-1)^n]$; (e) $a_1 = 5$, $a_n = 2a_{n-1} - 1$, $n > 1$; (f) $a_1 = 4$, $a_n = -a_{n-1}/5$, $n > 1$.

*Ans.* (a) $\dfrac{1}{10}$, $\dfrac{1}{100}$, $\dfrac{1}{1000}$, $\dfrac{1}{10,000}$; (b) $\dfrac{3}{6}$, $\dfrac{6}{7}$, $\dfrac{9}{8}$, $\dfrac{12}{9}$; (c) $3, 1, -1, -3$;

(d) $2, 0, 6, 0$; (e) $5, 9, 17, 33$; (f) $4, -\dfrac{4}{5}, \dfrac{4}{25}, -\dfrac{4}{125}$

**41.11.** The sequence recursively defined by $a_n = \dfrac{a_{n-1}^2 + x}{2a_{n-1}}$, with $a_1$ chosen arbitrarily, may be used to approximate

$\sqrt{x}$ to any desired degree of accuracy. Find the first four terms of the sequence $a_1 = 2$, $a_n = \dfrac{a_{n-1}^2 + 5}{2a_{n-1}}$, $n > 1$

and compare to the calculator approximation for $\sqrt{5}$.

*Ans.* $2, 2.25, 2.23611, 2.236068$, calculator: $\sqrt{5} \approx 2.236068$

**41.12.** Find a formula for the $n$th term of a sequence whose first four terms are given by:

(a) $4, 7, 10, 13, \ldots$; (b) $1, -3, 5, -7, \ldots$; (c) $\dfrac{6}{7}, -\dfrac{7}{9}, \dfrac{8}{11}, -\dfrac{9}{13}, \ldots$; (d) $\dfrac{x^2}{2!}, \dfrac{x^4}{4!}, \dfrac{x^6}{6!}, \dfrac{x^8}{8!}, \ldots$

*Ans.* (a) $a_n = 3n + 1$; (b) $a_n = (-1)^{n-1}(2n-1)$; (c) $a_n = (-1)^{n-1}\dfrac{n+5}{2n+5}$; (d) $a_n = \dfrac{x^{2n}}{(2n)!}$

**41.13.** Write in expanded form: (a) $\displaystyle\sum_{k=1}^{4} \dfrac{(-2)^k}{k+1}$; (b) $\displaystyle\sum_{k=3}^{6} \dfrac{x^k}{(k+1)!}$; (c) $\displaystyle\sum_{k=0}^{3} \dfrac{x^k}{(k+1)(k+3)}$

*Ans.* (a) $\dfrac{-2}{2} + \dfrac{4}{3} + \dfrac{-8}{4} + \dfrac{16}{5}$; (b) $\dfrac{x^3}{4!} + \dfrac{x^4}{5!} + \dfrac{x^5}{6!} + \dfrac{x^6}{7!}$; (c) $\dfrac{1}{1 \cdot 3} + \dfrac{x}{2 \cdot 4} + \dfrac{x^2}{3 \cdot 5} + \dfrac{x^3}{4 \cdot 6}$

**41.14.** Write the following in summation notation:

(a) $\dfrac{1}{3} + \dfrac{2}{5} + \dfrac{3}{7} + \dfrac{4}{9}$; (b) $x - 2x^2 + 3x^3 - 4x^4 + 5x^5 - 6x^6$; (c) $x - \dfrac{x^3}{3!} + \dfrac{x^5}{5!} - \dfrac{x^7}{7!}$

*Ans.* (a) $\displaystyle\sum_{k=1}^{4} \dfrac{k}{2k+1}$; (b) $\displaystyle\sum_{k=1}^{6} (-1)^{k-1} k x^k$; (c) $\displaystyle\sum_{k=1}^{4} \dfrac{(-1)^{k-1} x^{2k-1}}{(2k-1)!}$

CHAPTER 42

# The Principle of Mathematical Induction

## Sequences of Statements

Statements about the natural numbers can often be regarded as sequences of statements $P_n$.

**EXAMPLE 42.1** The statement, "The sum of the first $n$ natural numbers is equal to $\dfrac{n(n+1)}{2}$," can be written as $P_n: 1 + 2 + 3 + \cdots + n = \dfrac{n(n+1)}{2}$ or $P_n: \displaystyle\sum_{k=1}^{n} k = \dfrac{n(n+1)}{2}$. Then $P_1$ is the statement $1 = \dfrac{1(1+1)}{2}$, $P_2$ is the statement $1 + 2 = \dfrac{2(2+1)}{2}$, and so on.

**EXAMPLE 42.2** The statement, "For each natural number $n$, $n^2 - n + 41$ is a prime number," can be written as $P_n: n^2 - n + 41$ is a prime number. Then $P_1$ is the statement: $1^2 - 1 + 41$, or 41, is a prime number, $P_2$ is the statement: $2^2 - 2 + 41$, or 43, is a prime number, and so on.

## Principle of Mathematical Induction (PMI)

Given any statement about the natural numbers $P_n$, if the following conditions hold:

1. $P_1$ is true.
2. Whenever $P_k$ is true, $P_{k+1}$ is true.

Then $P_n$ is true for all $n$.

## Proof by Mathematical Induction

To apply the principle of mathematical induction to a sequence of statements $P_n$:

1. Write out the statements $P_1$, $P_k$, and $P_{k+1}$.
2. Show that $P_1$ is true.
3. Assume that $P_k$ is true. From this assumption (it is never necessary to prove $P_k$ explicitly), show that the truth of $P_{k+1}$ follows. This proof is often called the *induction step*.
4. Conclude that $P_n$ holds for all $n$.

## Failure of Proof by PMI

A sequence of statements may only be true for some values of $n$, or it may be true for no values of $n$. In these cases, the principle of mathematical induction will not apply, and the proof will fail.

**EXAMPLE 42.3** In the previous example, $P_1, P_2, P_3$, and so on, up to $P_{40}$ are all true. ($P_{40}$ is the statement: $40^2 - 40 + 41$, or 1601, is a prime number.) However, $P_{41}$, the statement: $41^2 - 41 + 41$, or $41^2$, is a prime number, is clearly false. Thus the sequence of statements $P_n$ is regarded as false in general, and certainly cannot be proved, although some individual statements are true.

## Extended Principle of Mathematical Induction

If there is some natural number $m$ such that all statements $P_n$ of a sequence are true for $n \geq m$, then the *extended* principle of mathematical induction may be used; if the following conditions hold:

1. $P_m$ is true.
2. Whenever $P_k$ is true, $P_{k+1}$ is true.

Then $P_n$ is true for all $n \geq m$.

**EXAMPLE 42.4** Let $P_n$ be the statement: $n! \geq 2^n$. $P_1$, $P_2$, and $P_3$ are false. (For example, $P_2$ is the false statement $2! \geq 2^2$ or $2 \geq 4$.) However, $P_4$ is the true statement $4! \geq 2^4$ or $24 \geq 16$, and the statement can be proved true for all $n \geq 4$ by the extended principle of mathematical induction.

## SOLVED PROBLEMS

**42.1.** Prove the statement $P_n$: $1 + 2 + 3 + \cdots + n = \dfrac{n(n+1)}{2}$ by mathematical induction.

Note that the left side can be thought of as the sum of the natural numbers up to and including $n$. Then

$P_1$ is the statement $1 = \dfrac{1(1+1)}{2}$.

$P_k$ is the statement $1 + 2 + 3 + \cdots + k = \dfrac{k(k+1)}{2}$.

$P_{k+1}$ is the statement $1 + 2 + 3 + \cdots + (k+1) = \dfrac{(k+1)[(k+1)+1]}{2}$.

Now $P_1$ is true, since $1 = \dfrac{1(1+1)}{2} = \dfrac{2}{2} = 1$ is true. Assume that $P_k$ is true for an arbitrary value of $k$.

To show that $P_{k+1}$ holds under this assumption, note that the left side can be thought of as the sum of the natural numbers up to and including $k+1$, thus the term on the left before the last is $k$, and $P_{k+1}$ can be rewritten as

$$P_{k+1}: 1 + 2 + 3 + \cdots + k + (k+1) = \frac{(k+1)(k+2)}{2}$$

Thus the left side of $P_{k+1}$ differs from the left side of $P_k$ only by the single additional term $(k+1)$. Hence, starting with $P_k$, which is assumed to be true, add $(k+1)$ to both sides:

$$1 + 2 + 3 + \cdots + k = \frac{k(k+1)}{2}$$

$$1 + 2 + 3 + \cdots + k + (k+1) = \frac{k(k+1)}{2} + (k+1)$$

Simplifying the right side yields:

$$\frac{k(k+1)}{2} + (k+1) = \frac{k(k+1)}{2} + \frac{2(k+1)}{2} = \frac{k(k+1) + 2(k+1)}{2} = \frac{(k+1)(k+2)}{2}$$

Thus, from the assumption that $P_k$ is true, it follows that

$$1 + 2 + 3 + \cdots + k + (k+1) = \frac{(k+1)(k+2)}{2}$$

holds. But this is precisely the statement $P_{k+1}$. Thus the truth of $P_{k+1}$ follows from the truth of $P_k$. Thus, by the principle of mathematical induction, $P_n$ holds for all $n$.

**42.2.** Prove the statement $P_n$: $1 + 3 + 5 + \cdots + (2n - 1) = n^2$ by mathematical induction.

Proceed as in the previous problem.

$P_1$ is the statement $1 = 1^2$.

$P_k$ is the statement $1 + 3 + 5 + \cdots + (2k - 1) = k^2$.

$P_{k+1}$ is the statement $1 + 3 + 5 + \cdots + [2(k + 1) - 1] = (k + 1)^2$, which can be rewritten as

$$1 + 3 + 5 + \cdots + (2k - 1) + (2k + 1) = (k + 1)^2$$

Now $P_1$ is obviously true. Assume the truth of $P_k$ and, comparing it to $P_{k+1}$, note that the left side of $P_{k+1}$ differs from the left side of $P_k$ only by the single additional term $(2k + 1)$. Hence, starting with $P_k$, add $(2k + 1)$ to both sides.

$$1 + 3 + 5 + \cdots + (2k - 1) = k^2$$
$$1 + 3 + 5 + \cdots + (2k - 1) + (2k + 1) = k^2 + (2k + 1)$$

The right side is immediately seen to be $k^2 + 2k + 1 = (k + 1)^2$, thus

$$1 + 3 + 5 + \cdots + (2k - 1) + (2k + 1) = (k + 1)^2$$

holds. But this is precisely the statement $P_{k+1}$. Thus the truth of $P_{k+1}$ follows from the truth of $P_k$. Thus, by the principle of mathematical induction, $P_n$ holds for all $n$.

**42.3.** Prove the statement $P_n$: $1 + 2 + 2^2 + \cdots + 2^{n-1} = 2^n - 1$ by mathematical induction.

Proceed as in the previous problems.

$P_1$ is the statement $1 = 2^1 - 1$,

$P_k$ is the statement $1 + 2 + 2^2 + \cdots + 2^{k-1} = 2^k - 1$.

$P_{k+1}$ is the statement $1 + 2 + 2^2 + \cdots + 2^{(k+1)-1} = 2^{k+1} - 1$, which can be rewritten as

$$1 + 2 + 2^2 + \cdots + 2^{k-1} + 2^k = 2^{k+1} - 1$$

Now $P_1$ is true, since $1 = 2^1 - 1 = 2 - 1 = 1$ is true. Assume the truth of $P_k$ and, comparing it to $P_{k+1}$, note that the left side of $P_{k+1}$ differs from the left side of $P_k$ only by the single additional term $2^k$. Hence, starting with $P_k$, add $2^k$ to both sides.

$$1 + 2 + 2^2 + \cdots + 2^{k-1} = 2^k - 1$$
$$1 + 2 + 2^2 + \cdots + 2^{k-1} + 2^k = 2^k + 2^k - 1$$

Simplifying the right side yields:

$$2^k + 2^k - 1 = 2 \cdot 2^k - 1 = 2^1 \cdot 2^k - 1 = 2^{k+1} - 1, \text{ thus}$$
$$1 + 2 + 2^2 + \cdots + 2^{k-1} + 2^k = 2^{k+1} - 1$$

holds. But this is precisely the statement $P_{k+1}$. Thus the truth of $P_{k+1}$ follows from the truth of $P_k$. Thus, by the principle of mathematical induction, $P_n$ holds for all $n$.

**42.4.** Prove the statement $P_n$: $\sum_{j=1}^{n} j^2 = \dfrac{n(n + 1)(2n + 1)}{6}$ by mathematical induction.

Proceed as in the previous problems.

$P_1$ is the statement $\sum_{j=1}^{1} j^2 = \dfrac{1(1 + 1)(2 \cdot 1 + 1)}{6}$.

$P_k$ is the statement $\sum_{j=1}^{k} j^2 = \dfrac{k(k + 1)(2k + 1)}{6}$ or $1^2 + 2^2 + \cdots + k^2 = \dfrac{k(k + 1)(2k + 1)}{6}$.

$P_{k+1}$ is the statement $\displaystyle\sum_{j=1}^{k+1} j^2 = \dfrac{(k+1)[(k+1)+1][2(k+1)+1]}{6}$, which can be rewritten as

$$1^2 + 2^2 + \cdots + k^2 + (k+1)^2 = \dfrac{(k+1)(k+2)(2k+3)}{6}$$

Now $P_1$ is true, since the left side is merely $1^2$ and the right side is $\dfrac{1 \cdot 2 \cdot 3}{6}$, that is, 1. Assume the truth of $P_k$ and, comparing it to $P_{k+1}$, note that the left side of $P_{k+1}$ differs from the left side of $P_k$ only by the single additional term $(k+1)^2$. Hence, starting with $P_k$, add $(k+1)^2$ to both sides.

$$1^2 + 2^2 + \cdots + k^2 = \dfrac{k(k+1)(2k+1)}{6}$$

$$1^2 + 2^2 + \cdots + k^2 + (k+1)^2 = \dfrac{k(k+1)(2k+1)}{6} + (k+1)^2$$

Simplifying the right side yields:

$$\dfrac{k(k+1)(2k+1)}{6} + (k+1)^2 = \dfrac{k(k+1)(2k+1)}{6} + \dfrac{6(k+1)^2}{6}$$

$$= \dfrac{(k+1)[k(2k+1) + 6(k+1)]}{6}$$

$$= \dfrac{(k+1)[2k^2 + 7k + 6]}{6}$$

$$= \dfrac{(k+1)(k+2)(2k+3)}{6}$$

Thus $1^2 + 2^2 + \cdots + k^2 + (k+1)^2 = \dfrac{(k+1)(k+2)(2k+3)}{6}$ holds. But this is precisely the statement $P_{k+1}$. Thus the truth of $P_{k+1}$ follows from the truth of $P_k$. Thus, by the principle of mathematical induction, $P_n$ holds for all $n$.

**42.5.** Prove the statement $P_n$: $n < 2^n$ for any positive integer $n$ by mathematical induction.

$P_1$ is the statement $1 < 2^1$.

$P_k$ is the statement $k < 2^k$.

$P_{k+1}$ is the statement $k + 1 < 2^{k+1}$.

Now $P_1$ is obviously true. Assume the truth of $P_k$, and, comparing it with $P_{k+1}$, note that the right side of $P_{k+1}$ is 2 times the right side of $P_k$. Hence, starting with $P_k$, multiply both sides by 2:

$$2^k > k$$
$$2 \cdot 2^k > 2k$$
$$2^{k+1} > 2k$$

But $2k = k + k \geq k + 1$. Hence

$$2^{k+1} > 2k \geq k + 1 \qquad \text{and} \qquad 2^{k+1} > k + 1.$$

But this is precisely the statement $P_{k+1}$. Thus the truth of $P_{k+1}$ follows from the truth of $P_k$. Thus, by the principle of mathematical induction, $P_n$ holds for all positive integers $n$.

**42.6.** Prove the statement $P_n$: $n! > 2^n$ is true for any integer $n \geq 4$ by the extended principle of mathematical induction.

$P_4$ is the statement $4! > 2^4$.

$P_k$ is the statement $k! > 2^k$, or $1 \cdot 2 \cdot 3 \cdot \cdots \cdot k > 2^k$.

$P_{k+1}$ is the statement $(k+1)! > 2^{k+1}$, or $1 \cdot 2 \cdot 3 \cdot \cdots \cdot k \cdot (k+1) > 2^{k+1}$.

Now $P_4$ is true, since $4! = 1 \cdot 2 \cdot 3 \cdot 4 = 24$ and $2^4 = 16$. Assume the truth of $P_k$, and, comparing it with $P_{k+1}$, note that the left side of $P_{k+1}$ is $k+1$ times the left side of $P_k$. Hence, starting with $P_k$, multiply both sides by $k+1$:

$$1 \cdot 2 \cdot 3 \cdot \cdots \cdot k > 2^k$$
$$1 \cdot 2 \cdot 3 \cdot \cdots \cdot k \cdot (k+1) > 2^k(k+1)$$

But, since $k > 1$, $k+1 > 2$, hence $2^k(k+1) > 2^k \cdot 2 = 2^k \cdot 2^1 = 2^{k+1}$. Therefore

$$1 \cdot 2 \cdot 3 \cdot \cdots \cdot k \cdot (k+1) > 2^{k+1}$$

holds. But this is precisely the statement $P_{k+1}$. Thus the truth of $P_{k+1}$ follows from the truth of $P_k$. Thus, by the extended principle of mathematical induction, $P_n$ holds for all $n \geq 4$.

**42.7.** Prove the statement $P_n$: $x - y$ is a factor of $x^n - y^n$ for any positive integer $n$ by mathematical induction.

$P_1$ is the statement: $x - y$ is a factor of $x^1 - y^1$.

$P_k$ is the statement: $x - y$ is a factor of $x^k - y^k$.

$P_{k+1}$ is the statement: $x - y$ is a factor of $x^{k+1} - y^{k+1}$.

Now $P_1$ is true, since any number is a factor of itself. Assume the truth of $P_k$; the statement can be rewritten as $x^k - y^k = (x - y)Q(x)$, where $Q(x)$ is some polynomial. Similarly $P_{k+1}$ can be rewritten as $x^{k+1} - y^{k+1} = (x - y)R(x)$, where $R(x)$ is some (other) polynomial. To show that $P_{k+1}$ holds under the assumption of the truth of $P_k$, note that

$$\begin{aligned}
x^{k+1} - y^{k+1} &= x^{k+1} - xy^k + xy^k - y^{k+1} \\
&= (x^{k+1} - xy^k) + (xy^k - y^{k+1}) \\
&= x(x^k - y^k) + y^k(x - y)
\end{aligned}$$

Since by assumption $x^k - y^k = (x - y)Q(x)$, then

$$\begin{aligned}
x^{k+1} - y^{k+1} &= x(x^k - y^k) + y^k(x - y) \\
&= x(x - y)Q(x) + y^k(x - y). \\
&= (x - y)[xQ(x) + y^k]
\end{aligned}$$

In other words, the required polynomial $R(x)$ is equal to $xQ(x) + y^k$ and $x - y$ is a factor of $x^{k+1} - y^{k+1}$. But this is precisely the statement $P_{k+1}$. Thus the truth of $P_{k+1}$ follows from the truth of $P_k$. Thus, by the principle of mathematical induction, $P_n$ holds for all positive integers $n$.

**42.8.** Use the principle of mathematical induction to prove DeMoivre's theorem: If $z = r(\cos\theta + i\sin\theta)$ is a complex number in trigonometric form, then for any positive integer $n$, $z^n = r^n(\cos n\theta + i\sin n\theta)$.

Here $P_n$ is the statement $z^n = r^n(\cos n\theta + i\sin n\theta)$. Then

$P_1$ is the statement $z^1 = r^1(\cos 1\theta + i\sin 1\theta)$.

$P_k$ is the statement $z^k = r^k(\cos k\theta + i\sin k\theta)$.

$P_{k+1}$ is the statement $z^{k+1} = r^{k+1}[\cos(k+1)\theta + i\sin(k+1)\theta]$.

Now $P_1$ is obviously true. Assume the truth of $P_k$, and, comparing it with $P_{k+1}$, note that the left side of $P_{k+1}$ is $z$ times the left side of $P_k$. Hence, starting with $P_k$, multiply both sides by $z$:

$$\begin{aligned}
z^k &= r^k(\cos k\theta + i\sin k\theta) \\
zz^k &= z(r^k(\cos k\theta + i\sin k\theta)) \\
z^{k+1} &= r(\cos\theta + i\sin\theta)r^k(\cos k\theta + i\sin k\theta) \\
z^{k+1} &= rr^k(\cos\theta + i\sin\theta)(\cos k\theta + i\sin k\theta) \\
z^{k+1} &= r^{k+1}(\cos\theta + i\sin\theta)(\cos k\theta + i\sin k\theta)
\end{aligned}$$

But by the rule for multiplying complex numbers in trigonometric form,

$$(\cos\theta + i\sin\theta)(\cos k\theta + i\sin k\theta) = \cos(\theta + k\theta) + i\sin(\theta + k\theta) = \cos(k + 1)\theta + i\sin(k + 1)\theta$$

Thus $z^{k+1} = r^{k+1}[\cos(k + 1)\theta + i\sin(k + 1)\theta]$ holds. But this is precisely the statement $P_{k+1}$. Thus the truth of $P_{k+1}$ follows from the truth of $P_k$. Thus, by the principle of mathematical induction, $P_n$, that is, DeMoivre's theorem, holds for all positive integers $n$.

## SUPPLEMENTARY PROBLEMS

**42.9.** Prove by mathematical induction: $2 + 4 + 6 + \cdots + 2n = n(n + 1)$.

**42.10.** Prove by mathematical induction: $3 + 7 + 11 + \cdots + (4n - 1) = n(2n + 1)$.

**42.11.** Prove by mathematical induction: $1 + 3 + 3^2 + \cdots + 3^{n-1} = \dfrac{3^n - 1}{2}$.

**42.12.** Prove by mathematical induction: $1^3 + 2^3 + 3^3 + \cdots + n^3 = \dfrac{n^2(n + 1)^2}{4}$.

**42.13.** Deduce from Problems 42.1 and 42.12 that $\displaystyle\sum_{k=1}^{n} k^3 = \left(\sum_{k=1}^{n} k\right)^2$.

**42.14.** Prove by mathematical induction: $\dfrac{1}{1 \cdot 3} + \dfrac{1}{3 \cdot 5} + \dfrac{1}{5 \cdot 7} + \cdots + \dfrac{1}{(2n - 1)(2n + 1)} = \dfrac{n}{2n + 1}$.

**42.15.** Prove $\dfrac{1}{\sqrt{1}} + \dfrac{1}{\sqrt{2}} + \dfrac{1}{\sqrt{3}} + \cdots + \dfrac{1}{\sqrt{n}} > \sqrt{n}, n \geq 2$, by the extended principle of mathematical induction.

**42.16.** Prove by mathematical induction: $x + y$ is a factor of $x^{2n-1} + y^{2n-1}$ for any positive integer $n$.

# CHAPTER 43

# Special Sequences and Series

## Definition of Arithmetic Sequence

A sequence of numbers $a_n$ is called an *arithmetic* sequence if successive terms differ by the same constant, called the *common difference*. Thus $a_n - a_{n-1} = d$ and $a_n = a_{n-1} + d$ for all terms of the sequence. It can be proved by mathematical induction that for any arithmetic sequence, $a_n = a_1 + (n - 1)d$.

## Definition of Arithmetic Series

An arithmetic series is the indicated sum of the terms of a finite arithmetic sequence. The notation $S_n$ is often used, thus, $S_n = \sum_{k=1}^{n} a_k$. For an arithmetic series,

$$S_n = \frac{n}{2}(a_1 + a_n) \qquad S_n = \frac{n}{2}[2a_1 + (n - 1)d]$$

**EXAMPLE 43.1**  Write the first 6 terms of the arithmetic sequence 4, 9, ....

Since the sequence is arithmetic, with $a_1 = 4$ and $a_2 = 9$, the common difference $d$ is given by $a_2 - a_1 = 9 - 4 = 5$. Thus, each term can be found from the previous term by adding 5, hence the first 6 terms are 4, 9, 14, 19, 24, 29.

**EXAMPLE 43.2**  Find the sum of the first 20 terms of the sequence of the previous example.

To find $S_{20}$, either of the formulas for an arithmetic series may be used. Since $a_1 = 4$, $n = 20$, and $d = 5$ are known, the second formula is more convenient:

$$S_n = \frac{n}{2}[2a_1 + (n - 1)d]$$

$$S_{20} = \frac{20}{2}[2 \cdot 4 + (20 - 1)5] = 1030$$

## Definition of Geometric Sequence

A sequence of numbers $a_n$ is called a *geometric sequence* if the quotient of successive terms is a constant, called the *common ratio*. Thus $a_n \div a_{n-1} = r$ or $a_n = ra_{n-1}$ for all terms of the sequence. It can be proved by mathematical induction that for any geometric sequence, $a_n = a_1 r^{n-1}$.

## Definition of Geometric Series

A geometric series is the indicated sum of the terms of a geometric sequence. For a geometric series with $r \neq 1$,

$$S_n = a_1 \frac{1 - r^n}{1 - r}$$

**EXAMPLE 43.3**  Write the first 6 terms of the geometric sequence 4, 6, ....

Since the sequence is geometric, with $a_1 = 4$ and $a_2 = 6$, the common ratio $r$ is given by $a_2 \div a_1 = 6 \div 4 = 3/2$. Thus, each term can be found from the previous term by multiplying by 3/2, hence the first 6 terms are 4, 6, 9, 27/2, 81/4, 243/8.

**EXAMPLE 43.4** Find the sum of the first 8 terms of the sequence of the previous example.

Use the sum formula with $a_1 = 4$, $n = 8$, and $r = 3/2$:

$$S_n = a_1 \frac{1 - r^n}{1 - r}$$

$$S_8 = 4 \frac{1 - (3/2)^8}{1 - (3/2)} = \frac{6305}{32}$$

## Infinite Geometric Series

It is not possible to add up all the terms of an infinite geometric sequence. In fact, if $|r| \geq 1$, the sum is not defined. However, it can be shown in calculus that if $|r| < 1$, then the sum of all the terms, denoted by $S_\infty$, is given by:

$$S_\infty = \frac{a_1}{1 - r}$$

**EXAMPLE 43.5** Find the sum of all the terms of the geometric sequence $6, 4, \ldots$.

Since the sequence is geometric, with $a_1 = 6$ and $a_2 = 4$, the common ratio $r$ is given by $a_2 \div a_1 = 4 \div 6 = 2/3$. Therefore $S_\infty = \dfrac{a_1}{1 - r} = \dfrac{6}{1 - 2/3} = 18$.

## Series Identities

The following identities can be proved by mathematical induction:

$$\sum_{k=1}^{n} a_k + \sum_{k=1}^{n} b_k = \sum_{k=1}^{n} (a_k + b_k) \qquad \sum_{k=1}^{n} a_k - \sum_{k=1}^{n} b_k = \sum_{k=1}^{n} (a_k - b_k) \qquad \sum_{k=1}^{n} c a_k = c \sum_{k=1}^{n} a_k$$

$$\sum_{k=1}^{n} c = cn \qquad \sum_{k=1}^{n} k = \frac{n(n + 1)}{2} \qquad \sum_{k=1}^{n} k^2 = \frac{n(n + 1)(2n + 1)}{6}$$

$$\sum_{k=1}^{n} k^3 = \frac{n^2(n + 1)^2}{4} \qquad \sum_{k=1}^{n} k^4 = \frac{n(n + 1)(2n + 1)(3n^2 + 3n - 1)}{30}$$

## SOLVED PROBLEMS

**43.1.** Identify the following sequences as arithmetic, geometric, or neither.

(a) $2, 4, 8, \ldots$; (b) $\frac{1}{2}, \frac{1}{3}, \frac{1}{4}, \ldots$; (c) $7, 5, 3, \ldots$; (d) $\frac{1}{4}, \frac{1}{8}, \frac{1}{16}, \ldots$

(a) Since $a_2 - a_1 = 4 - 2 = 2$ and $a_3 - a_2 = 8 - 4 = 4$, the sequence is not arithmetic. Since $a_2/a_1 = 4/2 = 2$ and $a_3/a_2 = 8/4 = 2$, the sequence is geometric with a common ratio of 2.

(b) Since $a_2 - a_1 = \frac{1}{3} - \frac{1}{2} = -\frac{1}{6}$ and $a_3 - a_2 = \frac{1}{4} - \frac{1}{3} = -\frac{1}{12}$, the sequence is not arithmetic. Since $a_2 \div a_1 = \frac{1}{3} \div \frac{1}{2} = \frac{2}{3}$ and $a_3 \div a_2 = \frac{1}{4} \div \frac{1}{3} = \frac{3}{4}$, the sequence is not geometric. Thus it is neither arithmetic nor geometric.

(c) Since $a_2 - a_1 = 5 - 7 = -2$ and $a_3 - a_2 = 3 - 5 = -2$, the sequence is arithmetic with a common difference of $-2$.

(d) Since $a_2 - a_1 = \frac{1}{8} - \frac{1}{4} = -\frac{1}{8}$ and $a_3 - a_2 = \frac{1}{16} - \frac{1}{8} = -\frac{1}{16}$, the sequence is not arithmetic. Since $a_2 \div a_1 = \frac{1}{8} \div \frac{1}{4} = \frac{1}{2}$ and $a_3 \div a_2 = \frac{1}{16} \div \frac{1}{8} = \frac{1}{2}$, the sequence is geometric with a common ratio of $\frac{1}{2}$.

**43.2.** Identify the following sequences as arithmetic, geometric, or neither.

(a) $3, \frac{15}{4}, \frac{9}{2}, \ldots$ ; (b) $\ln 1, \ln 2, \ln 3, \ldots$ ; (c) $x^{-1}, x^{-2}, x^{-3}, \ldots$ ; (d) $0.1, 0.11, 0.111, \ldots$

(a) Since $a_2 - a_1 = \frac{15}{4} - 3 = \frac{3}{4}$ and $a_3 - a_2 = \frac{9}{2} - \frac{15}{4} = \frac{3}{4}$, the sequence is arithmetic with a common difference of $\frac{3}{4}$.

(b) Since $a_2 - a_1 = \ln 2 - \ln 1 = \ln 2$ and $a_3 - a_2 = \ln 3 - \ln 2 = \ln \frac{3}{2}$, the sequence is not arithmetic. Since $a_2 \div a_1 = (\ln 2) \div (\ln 1)$ is not defined, the sequence is not geometric. Thus it is neither arithmetic nor geometric.

(c) Since $a_2 - a_1 = x^{-2} - x^{-1} = \dfrac{1 - x}{x^2}$ and $a_3 - a_2 = x^{-3} - x^{-2} = \dfrac{1 - x}{x^3}$, the sequence is not arithmetic except in the special case $x = 1$. Since $a_2 \div a_1 = x^{-2} \div x^{-1} = x^{-1}$ and $a_3 \div a_2 = x^{-3} \div x^{-2} = x^{-1}$, the sequence is geometric with a common ratio of $x^{-1}$ (except in the special case $x = 0$).

(d) Since $a_2 - a_1 = 0.11 - 0.1 = 0.01$ and $a_3 - a_2 = 0.111 - 0.11 = 0.001$, the sequence is not arithmetic. Since $a_2 \div a_1 = 0.11 \div 0.1 = 1.1$ and $a_3 \div a_2 = 0.111 \div 0.11 \approx 1.01$, the sequence is not geometric. Thus it is neither arithmetic nor geometric.

**43.3.** Prove that for an arithmetic sequence the $n$th term is given by $a_n = a_1 + (n - 1)d$.

An arithmetic sequence is defined by the relation $a_n = a_{n-1} + d$. Let $P_n$ be the statement that $a_n = a_1 + (n - 1)d$ and proceed by mathematical induction.

$P_1$ is the statement $a_1 = a_1 + (1 - 1)d$.

$P_k$ is the statement $a_k = a_1 + (k - 1)d$.

$P_{k+1}$ is the statement $a_{k+1} = a_1 + [(k + 1) - 1]d$, which can be rewritten as

$$a_{k+1} = a_1 + kd$$

Now $P_1$ is obviously true. Assume the truth of $P_k$ and note that by the definition of an arithmetic sequence, $a_{k+1} = a_k + d$. Therefore

$$a_{k+1} = a_k + d = a_1 + (k - 1)d + d = a_1 + kd.$$

But $a_{k+1} = a_1 + kd$ is precisely the statement $P_{k+1}$. Thus the truth of $P_{k+1}$ follows from the truth of $P_k$. Thus, by the principle of mathematical induction, $P_n$ holds for all $n$.

**43.4.** Given that the following sequences are arithmetic, find the common difference and write the next three terms and the $n$th term.

(a) $2, 5, \ldots$ ; (b) $9, \dfrac{17}{2}, \ldots$ ; (c) $\ln 1, \ln 2, \ldots$

(a) The common difference is $5 - 2 = 3$. Each term is found by adding 3 to the previous term, hence the next three terms are 8, 11, 14. The $n$th term is found from $a_n = a_1 + (n - 1)d$ with $a_1 = 2$ and $d = 3$; thus $a_n = 2 + (n - 1)3 = 3n - 1$.

(b) The common difference is $\frac{17}{2} - 9 = -\frac{1}{2}$. Each term is found by adding $-\frac{1}{2}$ to the previous term, hence the next three terms are $8, \frac{15}{2}, 7$. The $n$th term is found from $a_n = a_1 + (n - 1)d$ with $a_1 = 9$ and $d = -\frac{1}{2}$; thus
$$a_n = 9 + (n - 1)\left(-\tfrac{1}{2}\right) = \frac{19 - n}{2}.$$

(c) The common difference is $\ln 2 - \ln 1 = \ln 2$. Each term is found by adding $\ln 2$ to the previous term, hence the next three terms are given by $\ln 2 + \ln 2 = \ln 4$, $\ln 4 + \ln 2 = \ln 8$, and $\ln 8 + \ln 2 = \ln 16$. The $n$th term is found from $a_n = a_1 + (n - 1)d$ with $a_1 = \ln 1$ and $d = \ln 2$; thus
$$a_n = \ln 1 + (n - 1)\ln 2 = (n - 1)\ln 2 = \ln 2^{n-1}.$$

**43.5.** Given that the following sequences are geometric, find the common ratio and write the next three terms and the $n$th term.

(a) $5, 10, \ldots$; (b) $4, -2, \ldots$; (c) $0.03, 0.003, \ldots$

(a) The common ratio is $10 \div 5 = 2$. Each term is found by multiplying the previous term by 2, hence the next three terms are 20, 40, 80. The $n$th term is found from $a_n = a_1 r^{n-1}$ with $a_1 = 5$ and $r = 2$; thus $a_n = 5 \cdot 2^{n-1}$.

(b) The common ratio is $-2 \div 4 = -\frac{1}{2}$. Each term is found by multiplying the previous term by $-\frac{1}{2}$; hence the next three terms are $1, -\frac{1}{2}, \frac{1}{4}$. The $n$th term is found from $a_n = a_1 r^{n-1}$ with $a_1 = 4$ and $r = -\frac{1}{2}$; thus
$$a_n = 4\left(-\frac{1}{2}\right)^{n-1} = \frac{(-1)^{n-1}}{2^{n-3}}.$$

(c) The common ratio is $0.003 \div 0.03 = 0.1$. Each term is found by multiplying the previous term by 0.1; hence the next three terms are 0.0003, 0.00003, 0.000003. The $n$th term is found from
$$a_n = a_1 r^{n-1} \text{ with } a_1 = 0.03 \text{ and } r = 0.1, \text{ thus } a_n = 0.03(0.1)^{n-1} = 3 \times 10^{-2} \times 10^{1-n} = \frac{3}{10^{n+1}}.$$

**43.6.** Derive the formulas $S_n = \frac{n}{2}(a_1 + a_n)$ and $S_n = \frac{n}{2}[2a_1 + (n-1)d]$ for the value of an arithmetic series.

To derive the first formula, write out the terms of $S_n$;
$$S_n = a_1 + (a_1 + d) + (a_1 + 2d) + \cdots + [a_1 + (n-1)d]$$

Now write the terms in reverse order, noting that to begin with $a_n$, each term is found by *subtracting d,* the common difference, from the previous term.
$$S_n = a_n + (a_n - d) + (a_n - 2d) + \cdots + [a_1 - (n-1)d]$$

Adding these two identities, term by term, and noting that all terms involving $d$ add to zero, yields:
$$S_n + S_n = (a_1 + a_n) + (a_1 + a_n) + (a_1 + a_n) + \cdots + (a_1 + a_n)$$

Since there are $n$ identical terms on the right,
$$2S_n = n(a_1 + a_n)$$
$$S_n = \frac{n}{2}(a_1 + a_n)$$

For the second formula, substitute $a_n = a_1 + (n-1)d$ into the above to obtain
$$S_n = \frac{n}{2}[a_1 + a_1 + (n-1)d]$$
$$S_n = \frac{n}{2}[2a_1 + (n-1)d]$$

**43.7.** Find the sum of the first 10 terms of the arithmetic sequences given in Problem 43.4.

(a) Here $a_1 = 2$ and $a_n = 3n - 1$. For $n = 10$,
$$S_{10} = \frac{10}{2}[2 + (3 \cdot 10 - 1)] = 155$$

(b) Here $a_1 = 9$ and $a_n = \frac{19 - n}{2}$. For $n = 10$,
$$S_{10} = \frac{10}{2}\left[9 + \frac{19 - 10}{2}\right] = \frac{135}{2}$$

(c) Here $a_1 = \ln 1$ and $a_n = \ln 2^{n-1}$. For $n = 10$,
$$S_{10} = \frac{10}{2}[\ln 1 + \ln 2^{10-1}] = 5\ln 2^9 = 45\ln 2$$

**43.8.** Derive the formula $S_n = a_1 \dfrac{1 - r^n}{1 - r}$ for the value of a finite geometric series ($r \neq 1$).

Write out the terms of $S_n$.

$$S_n = a_1 + a_1 r + a_1 r^2 + \cdots + a_1 r^{n-1}$$

Multiply both sides by $r$ to obtain

$$rS_n = a_1 r + a_1 r^2 + a_1 r^3 + \cdots + a_1 r^n$$

Subtracting these two identities, term by term, yields:

$$S_n - rS_n = a_1 - a_1 r^n$$
$$S_n(1 - r) = a_1(1 - r^n)$$

Assuming $r \neq 1$, both sides may be divided by $1 - r$ to yield $S_n = a_1 \dfrac{1 - r^n}{1 - r}$. Note that if $r = 1$, then

$$S_n = a_1 + a_1 + a_1 + \cdots + a_1 = na_1$$

**43.9.** Find the sum of the first 7 terms of the geometric sequences given in Problem 43.5.

(a) Here $a_1 = 5$ and $r = 2$. For $n = 7$,

$$S_7 = 5 \left( \frac{1 - 2^7}{1 - 2} \right) = 635$$

(b) Here $a_1 = 4$ and $r = -\dfrac{1}{2}$. For $n = 7$,

$$S_7 = 4 \left[ \frac{1 - \left(-\frac{1}{2}\right)^7}{1 - \left(-\frac{1}{2}\right)} \right] = 4 \left( \frac{2^7 + 1}{2^7 + 2^6} \right) = \frac{129}{48}$$

(c) Here clearly $S_7 = 0.03333333$. Using the formula is more cumbersome, but yields the same result.

**43.10.** Give a plausibility argument to justify the formula $S_\infty = \dfrac{a_1}{1 - r}$ for the sum of all the terms of an infinite geometric sequence, $|r| < 1$.

First note that there are three possibilities: $r = 0$, $0 < r < 1$, and $-1 < r < 0$. For the first case, the formula is clearly valid, since all terms after the first are zero; hence $S_\infty = a_1 + 0 = \dfrac{a_1}{1 - 0}$. For $0 < r < 1$, consider the formula $S_n = a_1 \dfrac{1 - r^n}{1 - r}$ and let $n$ increase beyond all bounds. Since for real $n$, $r^n$ is then an exponential decay function, as $n \to \infty$, $r^n \to 0$. It seems plausible that this remains valid if $n$ is restricted to integer values. Thus, as $n \to \infty$, $S_n \to a_1 \dfrac{1}{1 - r}$ and $S_\infty = \dfrac{a_1}{1 - r}$. A similar, but more cumbersome, argument can be given if $-1 < r < 0$. A convincing proof is left for a calculus course.

**43.11.** Find the sum of all the terms of each geometric sequence given in Problem 43.5, or state that the sum is undefined.

(a) Since $r = 2$, the sum of all the terms is not defined. The sequence is said to *diverge*.

(b) Since $r = -\dfrac{1}{2}$ and $a_1 = 4$, $S_\infty = \dfrac{4}{1 - \left(-\frac{1}{2}\right)} = \dfrac{8}{3}$.

(c) Since $r = 0.1$ and $a_1 = 0.03$, $S_\infty = \dfrac{0.03}{1 - 0.1} = \dfrac{1}{30}$.

**43.12.** Use mathematical induction to show that $\sum_{j=1}^{n} a_j + \sum_{j=1}^{n} b_j = \sum_{j=1}^{n} (a_j + b_j)$ holds for all positive integers $n$.

Let $P_n$ be the above statement. Then

$P_1$ is the statement $\sum_{j=1}^{1} a_j + \sum_{j=1}^{1} b_j = \sum_{j=1}^{1} (a_j + b_j)$.

$P_k$ is the statement $\sum_{j=1}^{k} a_j + \sum_{j=1}^{k} b_j = \sum_{j=1}^{k} (a_j + b_j)$.

$P_{k+1}$ is the statement $\sum_{j=1}^{k+1} a_j + \sum_{j=1}^{k+1} b_j = \sum_{j=1}^{k+1} (a_j + b_j)$.

Now $P_1$ is true, since it reduces to $a_1 + b_1 = (a_1 + b_1)$. Assume the truth of $P_k$; then

$$a_1 + a_2 + \cdots + a_k + b_1 + b_2 + \cdots + b_k = (a_1 + b_1) + (a_2 + b_2) + \cdots + (a_k + b_k)$$

Add $a_{k+1} + b_{k+1}$ to both sides, then

$$a_1 + a_2 + \cdots + a_k + b_1 + b_2 + \cdots + b_k + a_{k+1} + b_{k+1} = (a_1 + b_1) + (a_2 + b_2) + \cdots + (a_k + b_k)$$
$$+ (a_{k+1} + b_{k+1})$$

Rearranging the terms on the left side yields

$$a_1 + a_2 + \cdots + a_k + a_{k+1} + b_1 + b_2 + \cdots + b_k + b_{k+1} = (a_1 + b_1) + (a_2 + b_2) + \cdots + (a_k + b_k)$$
$$+ (a_{k+1} + b_{k+1})$$

Writing this in the summation notation, it becomes

$$\sum_{j=1}^{k+1} a_j + \sum_{j=1}^{k+1} b_j = \sum_{j=1}^{k+1} (a_j + b_j)$$

But this is precisely the statement $P_{k+1}$. Thus the truth of $P_{k+1}$ follows from the truth of $P_k$. Thus, by the principle of mathematical induction, $P_n$ holds for all $n$.

**43.13.** Determine the seating capacity of a lecture hall if there are 32 rows of seats, with 18 seats in the first row, 21 seats in the second row, 24 seats in the third row, and so on.

The number of seats in each row forms an arithmetic sequence, with $a_1 = 18$, $d = 21 - 18 = 3$, and $n = 32$. Use the second formula for the value of an arithmetic series:

$$S_n = \frac{n}{2}[2a_1 + (n - 1)d]$$

$$S_{32} = \frac{32}{2}[2 \cdot 18 + (32 - 1)3] = 2064$$

**43.14.** A company buys a machine that is valued at \$87,500 and depreciates it at the rate of 30% per year. What is the value of the machine at the end of 5 years?

Note that depreciation of 30% of the value of the machine means that at the end of each year the value is 70% of what it was at the beginning. Thus the value at the end of each year is a constant multiple of the value at the end of the previous year. Hence the values form a geometric sequence, with $a_1 = (0.7)(87500)$ (the value at the *end* of the first year), $r = 0.7$ and $n = 5$. Thus

$$a_n = a_1 r^{n-1}$$
$$a_5 = (0.7)(87500)(0.7)^{5-1} \approx 14706$$

The value of the machine is \$14,706.

**43.15.** A ball is dropped from a height of 80 feet and bounces to three-fourths of its initial height. Assuming that this process continues indefinitely, find the total distance travelled by the ball before coming to rest.

Initially the ball travels 80 feet before hitting the ground. It then bounces up to a height of $\frac{3}{4}(80)$ and then back down this same distance. As this process repeats, the distance traveled can be written:

$$80 + 2\left(\frac{3}{4}\right)80 + \frac{3}{4}\left[2\left(\frac{3}{4}\right)80\right] + \cdots$$

Except for the first term, this may be regarded as an infinite geometric series with $a_1 = 2\left(\frac{3}{4}\right)80 = 120$ and $r = \frac{3}{4}$. Hence, if the process continues indefinitely, the entire distance travelled is given by

$$80 + S_\infty = 80 + \frac{120}{1 - \frac{3}{4}} = 560 \text{ feet}$$

## SUPPLEMENTARY PROBLEMS

**43.16.** Are the following sequences arithmetic, geometric, or neither?

(a) $\frac{3}{8}, \frac{3}{2}, 6, \ldots$; (b) $\frac{3}{8}, \frac{3}{4}, \frac{9}{8}, \ldots$; (c) $\frac{3}{4}, \frac{3}{5}, \frac{3}{6}, \ldots$; (d) $\frac{3}{4}, -\frac{3}{4}, \frac{3}{4}, -\frac{3}{4}, \ldots$; (e) $\frac{3}{4}, \frac{4}{5}, \frac{5}{6}, \ldots$

*Ans.* (a) geometric; (b) arithmetic; (c) neither; (d) geometric; (e) neither

**43.17.** For the following arithmetic sequences, state the common difference, and write the next three terms and the $n$th term: (a) $\frac{3}{5}, \frac{4}{5}, \ldots$; (b) $-8, -5, \ldots$; (c) $\pi, 3\pi, \ldots$

*Ans.* (a) $d = \frac{1}{5}$; $1, \frac{6}{5}, \frac{7}{5}$; $a_n = \dfrac{n+2}{5}$; (b) $d = 3$; $-2, 1, 4$; $a_n = 3n - 11$;

(c) $d = 2\pi$; $5\pi, 7\pi, 9\pi$; $a_n = (2n - 1)\pi$

**43.18.** Prove by mathematical induction: for a geometric sequence, the $n$th term is given by $a_n = a_1 r^{n-1}$.

**43.19.** For the following geometric sequences, state the common ratio, and write the next three terms and the $n$th term: (a) $\frac{3}{32}, \frac{3}{4}, \ldots$; (b) $-5, 5, -5, \ldots$; (c) $1, 1.05, \ldots$

*Ans.* (a) $r = 8$; $6, 48, 384$; $a_n = 3 \cdot 2^{3n-8}$; (b) $r = -1$; $5, -5, 5$; $a_n = 5(-1)^n$;

(c) $r = 1.05$; $(1.05)^2, (1.05)^3, (1.05)^4$; $a_n = (1.05)^{n-1}$

**43.20.** For the following geometric sequences, state the common ratio and find the sum of all terms, or state that the sum is undefined. (a) $4, \frac{1}{2}, \frac{1}{16}, \ldots$; (b) $\frac{1}{5}, -\frac{1}{5}, \frac{1}{5}, \ldots$; (c) $36, -12, 4, \ldots$; (d) $1, 0.95, \ldots$

*Ans.* (a) $r = \frac{1}{8}$, $S_\infty = \frac{32}{7}$; (b) $r = -1$, sum undefined; (c) $r = -\frac{1}{3}$, $S_\infty = 27$; (d) $r = 0.95$, $S_\infty = 20$

**43.21.** Use mathematical induction to show that $\displaystyle\sum_{k=1}^{n} a_k - \sum_{k=1}^{n} b_k = \sum_{k=1}^{n} (a_k - b_k)$ and $\displaystyle\sum_{k=1}^{n} c a_k = c \sum_{k=1}^{n} a_k$ hold for all integers $n$.

**43.22.** Suppose that \$0.01 were deposited into a bank account on the first day of June, \$0.02 on the second day, \$0.04 on the third day, and so on in a geometric sequence. (a) How much money would be deposited at this rate on June 30th? (b) How much money would be in the account after this last deposit?

*Ans.* (a) \$5,368,709.12; (b) \$10,737,418.23

# CHAPTER 44

# Binomial Theorem

## Binomial Expansions

Binomial expansions, that is, binomials or other two-term quantities raised to integer powers, are of frequent occurrence. If the general binomial expression is $a + b$, then the first few powers are given by:

$$(a + b)^0 = 1$$
$$(a + b)^1 = a + b$$
$$(a + b)^2 = a^2 + 2ab + b^2$$
$$(a + b)^3 = a^3 + 3a^2b + 3ab^2 + b^3$$

## Patterns in Binomial Expansions

Many patterns have been observed in the sequence of expansions of $(a + b)^n$. For example:

1. There are $n + 1$ terms in the expansion of $(a + b)^n$.
2. The exponent of $a$ starts in the first term as $n$, and decreases by 1 in each succeeding term down to 0 in the last term.
3. The exponent of $b$ starts in the first term as 0, and increases by 1 in each succeeding term up to $n$ in the last term.

## Binomial Theorem

The binomial theorem gives the expansion of $(a + b)^n$. In its most compact form, this is written as follows:

$$(a + b)^n = \sum_{r=0}^{n} \binom{n}{r} a^{n-r} b^r$$

The symbols $\binom{n}{r}$ are called the binomial coefficients, defined as: $\binom{n}{r} = \dfrac{n!}{r!(n-r)!}$.

**EXAMPLE 44.1**   Calculate the binomial coefficients $\binom{3}{r}$ and verify the expansion of $(a + b)^3$ above.

$$\binom{3}{0} = \frac{3!}{0!(3-0)!} = \frac{3!}{1 \cdot 3!} = 1 \qquad \binom{3}{1} = \frac{3!}{1!(3-1)!} = \frac{3!}{1!2!} = \frac{3 \cdot 2 \cdot 1}{1(2 \cdot 1)} = 3$$

$$\binom{3}{2} = \frac{3!}{2!(3-2)!} = \frac{3!}{2!1!} = \frac{3 \cdot 2 \cdot 1}{(2 \cdot 1)1} = 3 \qquad \binom{3}{3} = \frac{3!}{3!(3-3)!} = \frac{3!}{3!0!} = \frac{3!}{3! \cdot 1} = 1$$

Therefore

$$(a + b)^3 = \sum_{r=0}^{3} \binom{3}{r} a^{3-r} b^r = \binom{3}{0} a^{3-0} b^0 + \binom{3}{1} a^{3-1} b^1 + \binom{3}{2} a^{3-2} b^2 + \binom{3}{3} a^{3-3} b^3$$

$$= 1a^3 b^0 + 3a^2 b^1 + 3a^1 b^2 + 1a^0 b^3 = a^3 + 3a^2 b + 3ab^2 + b^3$$

## Properties of the Binomial Coefficients

The following are readily verifiable:

$$\binom{n}{0} = \binom{n}{n} = 1 \qquad \binom{n}{r} = \binom{n}{n-r} \qquad \binom{k}{r-1} + \binom{k}{r} = \binom{k+1}{r}$$

The binomial coefficients are also referred to as the combinatorial symbols. Then the designation $_nC_r$ is used, with

$$_nC_r = \binom{n}{r}$$

## Finding Particular Terms of a Binomial Expansion

In the binomial expansion of $(a + b)^n$, $r$, the index of the terms, starts at 0 in the first term and goes up to $n$ in the $n + 1$st term. Thus the index $r$ is equal to $j - 1$ in the $j$th term. If a particular term is desired, it is generally thought of as the $j + 1$st term; then $r$ is equal to $j$ and the value of the $j + 1$st term is given by

$$\binom{n}{j} a^{n-j} b^j$$

**EXAMPLE 44.2** Find the fifth term in the expansion of $(a + b)^{16}$.

Here $n = 16$ and $j + 1 = 5$, thus $j = 4$ and the term is given by

$$\binom{n}{j} a^{n-j} b^j = \binom{16}{4} a^{16-4} b^4 = \frac{16!}{4!(16-4)!} a^{16-4} b^4 = 1820 a^{12} b^4$$

## SOLVED PROBLEMS

**44.1.** Calculate the binomial coefficients:

(a) $\binom{4}{2}$; (b) $\binom{8}{5}$; (c) $\binom{12}{1}$; (d) $\binom{n}{n-1}$

(a) $\binom{4}{2} = \dfrac{4!}{2!(4-2)!} = \dfrac{4!}{2!2!} = \dfrac{4 \cdot 3 \cdot 2 \cdot 1}{2 \cdot 1 \cdot 2 \cdot 1} = 6$

(b) $\binom{8}{5} = \dfrac{8!}{5!(8-5)!} = \dfrac{8!}{5!3!} = \dfrac{8 \cdot 7 \cdot 6 \cdot 5!}{5!(3 \cdot 2 \cdot 1)} = \dfrac{8 \cdot 7 \cdot 6}{3 \cdot 2 \cdot 1} = 56$

(c) $\binom{12}{1} = \dfrac{12!}{1!(12-1)!} = \dfrac{12!}{1!11!} = \dfrac{12 \cdot 11!}{1 \cdot 11!} = 12$

(d) $\binom{n}{n-1} = \dfrac{n!}{(n-1)![n-(n-1)]!} = \dfrac{n(n-1)!}{(n-1)!1!} = n$

**44.2.** Show that $\binom{n}{n} = \binom{n}{0} = 1$.

$$\binom{n}{n} = \frac{n!}{n!(n-n)!} = \frac{n!}{n!0!} = \frac{n!}{n!(1)} = 1. \text{ Similarly } \binom{n}{0} = \frac{n!}{0!(n-0)!} = \frac{n!}{1(n!)} = 1$$

**44.3.** Show that $\binom{n}{r} = \dfrac{n(n-1) \cdot \cdots \cdot (r+1)}{(n-r)!} = \dfrac{n(n-1) \cdot \cdots \cdot (n-r+1)}{r!}$ for any integer $r < n$.

Note that $n! = n(n-1)(n-2) \cdot \cdots \cdot (r+1)r \cdot \cdots \cdot 1$. Hence

$$\binom{n}{r} = \frac{n!}{r!(n-r)!} = \frac{n(n-1)(n-2) \cdot \cdots \cdot (r+1)r \cdot \cdots \cdot 1}{r!(n-r)!} = \frac{n(n-1)(n-2) \cdot \cdots \cdot (r+1)r!}{r!(n-r)!}$$

$$= \frac{n(n-1) \cdot \cdots \cdot (r+1)}{(n-r)!}$$

Similarly, $n! = n(n - 1)(n - 2) \cdot \ldots \cdot (n - r + 1)(n - r) \cdot \ldots \cdot 1$. Hence

$$\binom{n}{r} = \frac{n!}{r!(n - r)!} = \frac{n(n - 1)(n - 2) \cdot \cdots \cdot (n - r + 1)(n - r) \cdot \cdots \cdot 1}{r!(n - r)!}$$

$$= \frac{n(n - 1)(n - 2) \cdot \cdots \cdot (n - r + 1)(n - r)!}{r!(n - r)!}$$

$$= \frac{n(n - 1) \cdot \cdots \cdot (n - r + 1)}{r!}$$

**44.4.** Use the results of the previous problems to write out the terms of $(a + b)^4$.

$$(a + b)^4 = \sum_{r=0}^{4} \binom{4}{r} a^{4-r} b^r$$

$$= \binom{4}{0} a^{4-0} b^0 + \binom{4}{1} a^{4-1} b^1 + \binom{4}{2} a^{4-2} b^2 + \binom{4}{3} a^{4-3} b^3 + \binom{4}{4} a^{4-4} b^4$$

$$= 1a^4 + \frac{4}{1} a^3 b + \frac{4 \cdot 3}{2 \cdot 1} a^2 b^2 + \frac{4 \cdot 3 \cdot 2}{3 \cdot 2 \cdot 1} a^1 b^3 + 1b^4$$

$$= a^4 + 4a^3 b + 6a^2 b^2 + 4ab^3 + b^4$$

**44.5.** Write the binomial expansion of $(3x - 5y)^4$.

Use the result of the previous problem with $a = 3x$ and $b = -5y$. Then

$$[(3x) + (-5y)]^4 = (3x)^4 + 4(3x)^3(-5y) + 6(3x)^2(-5y)^2 + 4(3x)(-5y)^3 + (-5y)^4$$

$$= 81x^4 - 540x^3 y + 1350x^2 y^2 - 1500xy^3 + 625y^4$$

**44.6.** Write the first three terms in the binomial expansion of $(a + b)^{20}$.

Since $(a + b)^{20} = \sum_{r=0}^{20} \binom{20}{r} a^{20-r} b^r$, the first three terms can be written as

$$\binom{20}{0} a^{20-0} b^0 + \binom{20}{1} a^{20-1} b^1 + \binom{20}{2} a^{20-2} b^2 = 1a^{20} + \frac{20}{1} a^{19} b + \frac{20 \cdot 19}{2 \cdot 1} a^{18} b^2$$

$$= a^{20} + 20a^{19} b + 190a^{18} b^2$$

**44.7.** Write the first three terms in the binomial expansion of $(2x^5 + 3t^2)^{12}$.

Take $a = 2x^5$ and $b = 3t^2$. Then $(2x^5 + 3t^2)^{12} = \sum_{r=0}^{12} \binom{12}{r}(2x^5)^{12-r}(3t^2)^r$. The first three terms of this can be written as

$$\binom{12}{0}(2x^5)^{12} + \binom{12}{1}(2x^5)^{11}(3t^2) + \binom{12}{2}(2x^5)^{10}(3t^2)^2$$

$$= 4096x^{60} + 12(2048x^{55})(3t^2) + \frac{12 \cdot 11}{2 \cdot 1}(1024x^{50})(9t^4) = 4096x^{60} + 73{,}728x^{55}t^2 + 608{,}256x^{50}t^4$$

**44.8.** Show that $\binom{n}{r} = \binom{n}{n - r}$.

Substitute $n - r$ for $r$ in the definition of $\binom{n}{r}$. Then

$$\binom{n}{n - r} = \frac{n!}{(n - r)![n - (n - r)]!} = \frac{n!}{(n - r)!r!} = \frac{n!}{r!(n - r)!} = \binom{n}{r}$$

**44.9.** Show that $\left( \begin{array}{c} k \\ r - 1 \end{array} \right) + \left( \begin{array}{c} k \\ r \end{array} \right) = \left( \begin{array}{c} k + 1 \\ r \end{array} \right)$.

Note first that $r! = r(r - 1)!$. Also, $(k + 1)! = (k + 1)k!$ and $(k - r + 1)! = (k - r + 1)(k - r)!$

Then

$$\left( \begin{array}{c} k \\ r - 1 \end{array} \right) + \left( \begin{array}{c} k \\ r \end{array} \right) = \frac{k!}{(r - 1)!(k - r + 1)!} + \frac{k!}{r!(k - r)!}$$

The LCD for the two fractional expressions on the right is $r!(k - r + 1)!$. Rewriting with this common denominator yields:

$$\frac{k!}{(r - 1)!(k - r + 1)!} + \frac{k!}{r!(k - r)!} = \frac{rk!}{r(r - 1)!(k - r + 1)!} + \frac{(k - r + 1)k!}{(k - r + 1)r!(k - r)!}$$

$$= \frac{rk!}{r!(k - r + 1)!} + \frac{(k - r + 1)k!}{r!(k - r + 1)!}$$

The two expressions on the right can be combined to yield:

$$\frac{rk!}{r!(k - r + 1)!} + \frac{(k - r + 1)k!}{r!(k - r + 1)!} = \frac{rk! + (k - r + 1)k!}{r!(k - r + 1)!} = \frac{(r + k - r + 1)k!}{r!(k - r + 1)!} = \frac{(k + 1)k!}{r!(k - r + 1)!}$$

The last expression is precisely

$$\frac{(k + 1)!}{r!(k + 1 - r)!} = \left( \begin{array}{c} k + 1 \\ r \end{array} \right)$$

**44.10.** Show that the binomial coefficients can be arranged in the form shown in Fig. 44-1.

$$1$$
$$1 \; 1$$
$$1 \; 2 \; 1$$
$$1 \; 3 \; 3 \; 1$$
$$1 \; 4 \; 6 \; 4 \; 1$$
$$\cdots\cdots\cdots\cdots$$

Figure 44-1

where each entry, except the 1's, is the sum of the two entries above it and to the right and left. (This triangular arrangement is often called Pascal's triangle.)

Clearly the first two rows represent $(a + b)^0 = 1$ and the coefficients of $(a + b)^1 = 1a + 1b$. For the other rows, note that the first and the last binomial coefficients in each are given by

$$\left( \begin{array}{c} n \\ 0 \end{array} \right) = 1 \qquad \text{and} \qquad \left( \begin{array}{c} n \\ n \end{array} \right) = 1$$

respectively. For all other coefficients, since, as proved in the previous problem,

$$\left( \begin{array}{c} k \\ r - 1 \end{array} \right) + \left( \begin{array}{c} k \\ r \end{array} \right) = \left( \begin{array}{c} k + 1 \\ r \end{array} \right)$$

each entry in the $k + 1$st row is the sum of the two entries in the $k$th row above it and to the right and left.

**44.11.** Use mathematical induction to prove the binomial theorem for positive integers $n$.

Let $P_n$ be the statement of the binomial theorem:

$$(a + b)^n = \sum_{r = 0}^{n} \left( \begin{array}{c} n \\ r \end{array} \right) a^{n - r} b^r$$

Then $P_1$ is the statement $(a + b)^1 = \sum_{r = 0}^{1} \left( \begin{array}{c} 1 \\ r \end{array} \right) a^{1 - r} b^r$.

$P_k$ is the statement $(a + b)^k = \sum_{r=0}^{k} \binom{k}{r} a^{k-r} b^r$.

$P_{k+1}$ is the statement $(a + b)^{k+1} = \sum_{r=0}^{k+1} \binom{k+1}{r} a^{k+1-r} b^r$.

Now $P_1$ is true, since the left side is $a + b$ and the right side is

$$\binom{1}{0} a^{1-0} b^0 + \binom{1}{1} a^{1-1} b^1 = 1a + 1b = a + b$$

Assume the truth of $P_k$, and, comparing it with $P_{k+1}$, note that the left side of $P_{k+1}$ is $a + b$ times the left side of $P_k$. Hence, starting with $P_k$, multiply both sides by $a + b$:

$$(a + b)(a + b)^k = (a + b) \sum_{r=0}^{k} \binom{k}{r} a^{k-r} b^r$$

$$(a + b)^{k+1} = a \sum_{r=0}^{k} \binom{k}{r} a^{k-r} b^r + b \sum_{r=0}^{k} \binom{k}{r} a^{k-r} b^r$$

$$= \sum_{r=0}^{k} \binom{k}{r} a^{k+1-r} b^r + \sum_{r=0}^{k} \binom{k}{r} a^{k-r} b^{r+1}$$

Writing out the terms of the sums yields:

$$\binom{k}{0} a^{k+1} + \binom{k}{1} a^k b + \binom{k}{2} a^{k-1} b^2 + \cdots + \binom{k}{k-1} a^2 b^{k-1} + \binom{k}{k} ab^k$$

$$+ \binom{k}{0} a^k b + \binom{k}{1} a^{k-1} b^2 + \cdots + \binom{k}{k-2} a^2 b^{k-1} + \binom{k}{k-1} ab^k + \binom{k}{k} b^{k+1}$$

Combining like terms, and noting that $\binom{k}{0} = 1 = \binom{k+1}{0}$ and $\binom{k}{k} = 1 = \binom{k+1}{k+1}$, yields

$$\binom{k+1}{0} a^{k+1} + \left( \binom{k}{1} + \binom{k}{0} \right) a^k b + \left( \binom{k}{2} + \binom{k}{1} \right) a^{k-1} b^2 + \cdots$$

$$+ \left( \binom{k}{k-1} + \binom{k}{k-2} \right) a^2 b^{k-1} + \left( \binom{k}{k} + \binom{k}{k-1} \right) ab^k + \binom{k+1}{k+1} b^{k+1}$$

$$= \binom{k+1}{0} a^{k+1} + \binom{k+1}{1} a^k b + \binom{k+1}{2} a^{k-1} b^2 + \cdots + \binom{k+1}{k-1} a^2 b^{k-1} + \binom{k+1}{k} ab^k + \binom{k+1}{k+1} b^{k+1}$$

Thus, writing the last expression in summation notation,

$$(a + b)^{k+1} = \sum_{r=0}^{k+1} \binom{k+1}{r} a^{k+1-r} b^r$$

But this is precisely the statement $P_{k+1}$. Thus the truth of $P_{k+1}$ follows from the truth of $P_k$. Thus, by the principle of mathematical induction, $P_n$, that is, the binomial theorem, holds for all positive integers $n$.

**44.12.** Write the eighth term in the expansion of $\left( \sqrt{x} + \dfrac{1}{\sqrt{x}} \right)^{13}$.

The $(j + 1)$st term in the expansion of $(a + b)^n$ is given by $\binom{n}{j} a^{n-j} b^j$. Here $n = 13$ and $j + 1 = 8$; hence $j = 7$. Thus the required term is

$$\binom{13}{7} (\sqrt{x})^{13-7} \left( \frac{1}{\sqrt{x}} \right)^7 = \frac{13!}{(13-7)!7!} \frac{(\sqrt{x})^6}{(\sqrt{x})^7} = \frac{13 \cdot 12 \cdot 11 \cdot 10 \cdot 9 \cdot 8}{6! \sqrt{x}} = \frac{1716}{\sqrt{x}}$$

**44.13.** Use the binomial theorem to approximate $(1.01)^{20}$ to three decimal places.

Expand $(1 + 0.01)^{20}$ to obtain:

$$\binom{20}{0}1^{20} + \binom{20}{1}1^{19}(0.01)^1 + \binom{20}{2}1^{18}(0.01)^2 + \binom{20}{3}1^{17}(0.01)^3 + \binom{20}{4}1^{16}(0.01)^4 + \cdots$$

$$= 1 + 20(0.01) + \frac{20 \cdot 19}{2 \cdot 1}(0.0001) + \frac{20 \cdot 19 \cdot 18}{3 \cdot 2 \cdot 1}(0.000001) + \frac{20 \cdot 19 \cdot 18 \cdot 17}{4 \cdot 3 \cdot 2 \cdot 1}(10^{-8}) + \cdots$$

$$= 1 + 0.2 + 0.019 + 0.00114 + 0.00004845 + \cdots$$

$$= 1.220188\ldots$$

where the neglected terms have no effect on the third decimal place. Thus $(1.01)^{20} \approx 1.220$ to three decimal places.

## SUPPLEMENTARY PROBLEMS

**44.14.** Calculate the binomial coefficients: (a) $\binom{15}{1}$; (b) $\binom{8}{6}$; (c) $\binom{12}{9}$; (d) $\binom{n}{n-2}$

    *Ans.* (a) 15; (b) 28; (c) 220; (d) $\dfrac{n(n-1)}{2}$

**44.15.** Write the binomial expansion of (a) $(a+b)^5$; (b) $(2x+y)^5$.

    *Ans.* (a) $a^5 + 5a^4b + 10a^3b^2 + 10a^2b^3 + 5ab^4 + b^5$; (b) $32x^5 + 80x^4y + 80x^3y^2 + 40x^2y^3 + 10xy^4 + y^5$

**44.16.** Write the binomial expansion of (a) $(4s - 3t)^3$; (b) $\left(2a - \dfrac{b}{5}\right)^5$.

    *Ans.* (a) $64s^3 - 144s^2t + 108st^2 - 27t^3$; (b) $32a^5 - 16a^4b + \dfrac{16}{5}a^3b^2 - \dfrac{8}{25}a^2b^3 + \dfrac{2}{125}ab^4 - \dfrac{b^5}{3125}$

**44.17.** Prove that $\binom{n}{0} + \binom{n}{1} + \cdots + \binom{n}{n-1} + \binom{n}{n} = 2^n$, that is, that the sum of the binomial coefficients for any power $n$ is equal to $2^n$. [*Hint*: Consider the binomial expansion of $(1 + 1)^n$.]

**44.18.** Find the middle term in the binomial expansion of (a) $\left(3x - \dfrac{y}{3}\right)^{14}$; (b) $(x^3 + 2y^3)^{10}$.

    *Ans.* (a) $-3432x^7y^7$; (b) $8064x^{15}y^{15}$

**44.19.** It is shown in calculus that if $|x| < 1$ and $\alpha$ is not a positive integer, then $(1 + x)^\alpha = \sum_{j=0}^{\infty} \binom{\alpha}{j}x^j$ with $\binom{\alpha}{j} = \dfrac{\alpha(\alpha - 1) \cdot \cdots \cdot (\alpha - j + 1)}{j!}$. Use this formula to write the first three terms of the binomial expansion of (a) $(1 + x)^{-2}$; (b) $(1 + x)^{1/2}$.

    *Ans.* (a) $1 - 2x + 3x^2$; (b) $1 + \dfrac{1}{2}x - \dfrac{1}{8}x^2$

# CHAPTER 45

# Limits, Continuity, Derivatives

## Informal Definition of Limit

If the values taken on by a function $f(x)$ can be made arbitrarily close to $L$ by taking input values $x$ arbitrarily close to $a$, then $L$ is called the limit of $f(x)$ as $x$ approaches $a$, written

$$\lim_{x \to a} f(x) = L$$

**EXAMPLE 45.1** $\lim_{x \to 4}(2x - 3) = 5$, since $2x - 3$ can be made arbitrarily close to 5 by taking values of $x$ arbitrarily close to 4, as suggested by the following table:

| $x$ | 3.5 | 3.9 | 3.99 | 3.999 | 4.5 | 4.1 | 4.01 | 4.001 |
|---------|-----|-----|------|-------|-----|-----|------|-------|
| $2x - 3$ | 4 | 4.8 | 4.98 | 4.998 | 6 | 5.2 | 5.02 | 5.002 |

## Formal Definition of Limit

$\lim_{x \to a} f(x) = L$ means that given any $\varepsilon > 0$, a number $\delta > 0$ can be found so that if $0 < |x - a| < \delta$, then $|f(x) - L| < \varepsilon$.

**EXAMPLE 45.2** In the previous example, given any $\varepsilon > 0$, take $0 < |x - 4| < \varepsilon/2$.

Then:

$$2|x - 4| < \varepsilon$$
$$|2x - 8| < \varepsilon$$
$$|(2x - 3) - 5| < \varepsilon$$

Thus $\lim_{x \to 4}(2x - 3) = 5$.

Note that the statement $\lim_{x \to a} f(x) = L$ says nothing about what happens at $a$. Possibly $f(a) = L$; however, possibly $f(a)$ is undefined, or defined but unequal to $L$.

## Properties of Limits

$$\lim_{x \to a} c = c \qquad \lim_{x \to a} x = a$$

If $\lim_{x \to a} f(x) = L$ and $\lim_{x \to a} g(x) = M$, then

$$\lim_{x \to a}[f(x) + g(x)] = L + M \qquad \lim_{x \to a}[f(x) - g(x)] = L - M$$

$$\lim_{x \to a}[f(x)g(x)] = LM \qquad \lim_{x \to a}[f(x)]^n = L^n$$

$$\lim_{x \to a}[f(x)/g(x)] = L/M \quad \text{provided} \quad M \neq 0$$

$$\lim_{x \to a} \sqrt[n]{f(x)} = \sqrt[n]{L} \text{ provided } n \text{ is an odd integer, or } n \text{ is an even integer and } L \text{ is positive.}$$

## Finding Limits Algebraically

As a result of these properties, many limits can be found algebraically.

**EXAMPLE 45.3**  Find $\lim_{x \to 4}(3x + 7)$.

$$\lim_{x \to 4}(3x + 7) = \lim_{x \to 4}3x + \lim_{x \to 4}7 = \lim_{x \to 4}3 \cdot \lim_{x \to 4}x + \lim_{x \to 4}7 = 3 \cdot 4 + 7 = 19$$

**EXAMPLE 45.4**  Find $\lim_{x \to 3}\dfrac{x^2 - 9}{x - 3}$

$$\lim_{x \to 3}\frac{x^2 - 9}{x - 3} = \lim_{x \to 3}\frac{(x - 3)(x + 3)}{x - 3} = \lim_{x \to 3}(x + 3) = \lim_{x \to 3}x + \lim_{x \to 3}3 = 3 + 3 = 6$$

There are, however, many situations where a limit does not exist.

**EXAMPLE 45.5**  Find $\lim_{x \to 0}\left(-\dfrac{1}{x}\right)$.

Consider the following table:

| $x$ | $-0.5$ | $-0.1$ | $-0.01$ | $-0.001$ | $0.5$ | $0.1$ | $0.01$ | $0.001$ |
|---|---|---|---|---|---|---|---|---|
| $-\dfrac{1}{x}$ | $2$ | $10$ | $100$ | $1000$ | $-2$ | $-10$ | $-100$ | $-1000$ |

The values are not approaching a limit; the limit does not exist.

## One-Sided Limits

1. If the values taken on by a function $f(x)$ can be made arbitrarily close to $L$ by taking input values $x$ arbitrarily close to (but greater than) $a$, then $L$ is called the limit of $f(x)$ as $x$ approaches $a$ from the right, written

$$\lim_{x \to a^+} f(x) = L$$

2. If the values taken on by a function $f(x)$ can be made arbitrarily close to $L$ by taking input values $x$ arbitrarily close to (but less than) $a$, then $L$ is called the limit of $f(x)$ as $x$ approaches $a$ from the left, written

$$\lim_{x \to a^-} f(x) = L$$

## Infinite Limits

If the values taken on by a function $f(x)$ can be made arbitrarily large and positive by taking input values $x$ arbitrarily close to $a$, then it is said that the limit of $f(x)$ as $x$ approaches $a$ is (positive) infinite, written $\lim_{x \to a} f(x) = \infty$. If the values taken on by a function $f(x)$ can be made arbitrarily large and negative by taking input values $x$ arbitrarily close to $a$, then it is said that the limit of $f(x)$ as $x$ approaches $a$ is negative infinite, written $\lim_{x \to a} f(x) = -\infty$.

**EXAMPLE 45.6**  $\lim_{x \to 3}\dfrac{1}{(x - 3)^2} = \infty$, since $\dfrac{1}{(x - 3)^2}$ can be made arbitrarily large by taking values of $x$ arbitrarily close to 3, as suggested by the following table:

| $x$ | $2.5$ | $2.9$ | $2.99$ | $2.999$ | $3.5$ | $3.1$ | $3.01$ | $3.001$ |
|---|---|---|---|---|---|---|---|---|
| $\dfrac{1}{(x - 3)^2}$ | $4$ | $100$ | $10{,}000$ | $1{,}000{,}000$ | $4$ | $100$ | $10{,}000$ | $1{,}000{,}000$ |

## One-Sided Infinite Limits

1. If the values taken on by a function $f(x)$ can be made arbitrarily large and positive by taking input values $x$ arbitrarily close to (but greater than) $a$, it is said that the limit of $f(x)$ as $x$ approaches $a$ from the right is (positive) infinite, written $\lim\limits_{x \to a^+} f(x) = \infty$. If the values taken on by a function $f(x)$ can be made arbitrarily large and negative by taking input values $x$ arbitrarily close to (but greater than) $a$, it is said that the limit of $f(x)$ as $x$ approaches $a$ from the right is negative infinite, written $\lim\limits_{x \to a^+} f(x) = -\infty$.

2. If the values taken on by a function $f(x)$ can be made arbitrarily large and positive by taking input values $x$ arbitrarily close to (but less than) $a$, it is said that the limit of $f(x)$ as $x$ approaches $a$ from the left is (positive) infinite, written $\lim\limits_{x \to a^-} f(x) = \infty$. If the values taken on by a function $f(x)$ can be made arbitrarily large and negative by taking input values $x$ arbitrarily close to (but less than) $a$, it is said that the limit of $f(x)$ as $x$ approaches $a$ from the left is negative infinite, written $\lim\limits_{x \to a^-} f(x) = -\infty$.

**EXAMPLE 45.7** Find $\lim\limits_{x \to 0^+}\left(-\dfrac{1}{x}\right)$ and $\lim\limits_{x \to 0^-}\left(-\dfrac{1}{x}\right)$.

From the table in Example 45.5, it appears that $\lim\limits_{x \to 0^+}\left(-\dfrac{1}{x}\right) = -\infty$ and $\lim\limits_{x \to 0^-}\left(-\dfrac{1}{x}\right) = \infty$.

## Limits at Infinity

1. If the values taken on by a function $f(x)$ can be made arbitrarily close to $L$ by taking input values $x$ arbitrarily large and positive, then $L$ is called the limit of $f(x)$ as $x$ approaches (positive) infinity, written

$$\lim_{x \to \infty} f(x) = L$$

2. If the values taken on by a function $f(x)$ can be made arbitrarily close to $L$ by taking input values $x$ arbitrarily large and negative, then $L$ is called the limit of $f(x)$ as $x$ approaches negative infinity, written

$$\lim_{x \to -\infty} f(x) = L$$

## Definition of Continuity

1. A function $f(x)$ is called *continuous* for a value $c$ if $\lim\limits_{x \to c} f(x) = f(c)$. This is usually referred to as continuity at a point $c$, or simply continuity at $c$.

2. A function $f(x)$ is called continuous on an open interval $(a,b)$ if it is continuous at every point on the interval.

3. A function $f(x)$ is called continuous on a closed interval $[a,b]$ if it is continuous at every point on the interval $(a,b)$ and also $\lim\limits_{x \to a^+} f(x) = f(a)$ and $\lim\limits_{x \to b^-} f(x) = f(b)$.

If a function is not continuous for a value $c$, it is called *discontinuous*, and $c$ is called a point of discontinuity.

**EXAMPLE 45.8** It can be shown that every polynomial function is continuous at every point in $R$ and that every rational function is continuous at every point in its domain. Thus if $f(x)$ is a polynomial function, the limit $\lim\limits_{x \to c} f(x)$ can always be calculated as $f(c)$. If $f(x) = p(x)/q(x)$ is a rational function, the limit $\lim\limits_{x \to c} f(x)$ can be calculated as $f(c) = p(c)/q(c)$ for any value $c$ as long as $f(c)$ is defined, that is, if $q(c) \neq 0$.

## Definition of Derivative

Given a function $f(x)$, the *derivative* of $f$, written $f'(x)$, is a function defined by the formula

$$f'(x) = \lim_{h \to 0} \frac{f(x + h) - f(x)}{h}$$

provided that the limit exists. If the limit exists for a value $a$ (also referred to as: at the point $a$), the function is called *differentiable* at $a$. The process of finding the derivative is called *differentiation*.

**EXAMPLE 45.9** Find the derivative of $f(x) = x^2$.

$$\lim_{h \to 0} \frac{f(x + h) - f(x)}{h} = \lim_{h \to 0} \frac{(x + h)^2 - x^2}{h}$$

$$= \lim_{h \to 0} \frac{x^2 + 2xh + h^2 - x^2}{h}$$

$$= \lim_{h \to 0} \frac{2xh + h^2}{h}$$

$$= \lim_{h \to 0} \frac{h(2x + h)}{h}$$

$$= \lim_{h \to 0} (2x + h)$$

$$= 2x$$

## Average and Instantaneous Rates of Change

In Chapter 9, the *average* rate of change of $f(x)$ over an interval from $x$ to $x + h$ was defined as

$$\frac{f(x + h) - f(x)}{h}$$

also referred to as the *difference quotient*.

The derivative of $f(x)$, $f'(x) = \lim_{h \to 0} \dfrac{f(x + h) - f(x)}{h}$, is also called the *instantaneous* rate of change of the function with respect to the variable $x$.

## Tangent Line

The tangent line to the graph of a function $f(x)$ at the point $(a, f(a))$ is the straight line through the point with slope $m$ equal to the derivative of the function at the point $a$,

$$m(a) = f'(a) = \lim_{h \to 0} \frac{f(a + h) - f(a)}{h}$$

## Average and Instantaneous Velocity

Given a function $s(t)$ that represents the position of an object at time $t$, the *average velocity* of the object on the interval $[a,b]$ is given by

$$\frac{\text{Change in position}}{\text{Change in time}} = \frac{s(b) - s(a)}{b - a}$$

The *instantaneous velocity* of the object at time $t$ is given by the derivative of $s(t)$:

$$v(t) = s'(t) = \lim_{h \to 0} \frac{s(t + h) - s(t)}{h}$$

**SOLVED PROBLEMS**

**45.1.** Use the formal definition of limit to show (a) $\lim_{x \to a} c = c$; (b) $\lim_{x \to a} x = a$.

(a) $\lim_{x \to a} c = c$ means that given any $\varepsilon > 0$ a number $\delta > 0$ can be found so that if $0 < |x - a| < \delta$, then $|c - c| < \varepsilon$. However, regardless of $x$ and $\delta$, $|c - c| = |0| = 0 < \varepsilon$ holds for any $\varepsilon > 0$. This proves the required result.

(b) $\lim_{x \to a} x = a$ means that given any $\varepsilon > 0$, a number $\delta > 0$ can be found so that if $0 < |x - a| < \delta$, then $|x - a| < \varepsilon$. Clearly, given $\varepsilon > 0$, choose $\delta = \varepsilon$, then $0 < |x - a| < \delta$ will guarantee $|x - a| < \varepsilon$. This proves the required result.

**45.2.** Use the formal definition of limit to show that if $\lim_{x \to a} f(x) = L$ and $\lim_{x \to a} g(x) = M$, then $\lim_{x \to a} [f(x) + g(x)] = L + M$.

$\lim_{x \to a} [f(x) + g(x)] = L + M$ means that given any $\varepsilon > 0$, a number $\delta > 0$ can be found so that if $0 < |x - a| < \delta$, then $|(f(x) + g(x)) - (L + M)| < \varepsilon$.

Note that $|(f(x) + g(x)) - (L + M)| = |(f(x) - L) + (g(x) - M)| \le |f(x) - L| + |g(x) - M|$. This inequality follows from the triangle inequality (Chapter 7).

Hence, since $\lim_{x \to a} f(x) = L$ and $\lim_{x \to a} g(x) = M$, given $\varepsilon > 0$, choose $\delta_1$ so that if $0 < |x - a| < \delta_1$, then $|f(x) - L| < \varepsilon/2$, and choose $\delta_2$ so that if $0 < |x - a| < \delta_2$ then $|g(x) - M| < \varepsilon/2$.

Therefore, choose $\delta$ to be the smaller of $\delta_1$ and $\delta_2$. Then, if $0 < |x - a| < \delta$,
$|(f(x) + g(x)) - (L + M)| \le |f(x) - L| + |g(x) - M| < \varepsilon/2 + \varepsilon/2 = \varepsilon$. This proves the required result.

**45.3.** Find $\lim_{x \to 5} (2 - 3x)$ (a) by examining a table of values near 5; (b) by using the formal definition of limit; (c) by using the continuity of polynomial functions.

(a) Form a table of values near 5:

| $x$ | 4.5 | 4.9 | 4.99 | 4.999 | 5.5 | 5.1 | 5.01 | 5.001 |
|---|---|---|---|---|---|---|---|---|
| $2 - 3x$ | $-11.5$ | $-12.7$ | $-12.97$ | $-12.997$ | $-14.5$ | $-13.3$ | $-13.03$ | $-13.003$ |

This suggests that $\lim_{x \to 5} (2 - 3x) = -13$.

(b) If $0 < |x - 5| < \delta_1$, then
$$0 < |-3(x - 5)| < |-3|\delta_1$$
$$0 < |15 - 3x| < 3\delta_1$$
$$0 < |(2 - 3x) - (-13)| < 3\delta_1$$

Therefore, given $\varepsilon > 0$, choose $\delta = \delta_1 = \varepsilon/3$. If $0 < |x - 5| < \delta$, then
$$|(2 - 3x) - (-13)| < 3(\varepsilon/3) = \varepsilon$$

Thus $\lim_{x \to 5} (2 - 3x) = -13$ as suggested by the table.

(c) Since $f(x) = 2 - 3x$ is a polynomial function, $\lim_{x \to 5} f(x) = f(5) = 2 - 3 \cdot 5 = -13$.

**45.4.** Find (a) $\lim_{x \to -2} (x^3 - 3x^2 + 2x + 8)$; (b) $\lim_{x \to 4} \dfrac{x - 1}{x^2 + 3}$.

(a) Since $f(x) = x^3 - 3x^2 + 2x + 8$ is a polynomial function,
$$\lim_{x \to -2} f(x) = f(-2) = (-2)^3 - 3(-2)^2 + 2(-2) + 8 = -16$$

(b) Since $f(x) = \dfrac{x-1}{x^2+3}$ is a rational function defined at $x = 4$,

$$\lim_{x \to 4} f(x) = f(4) = \frac{4-1}{4^2+3} = \frac{3}{19}$$

**45.5.** Find the following limits algebraically:

(a) $\displaystyle\lim_{x \to 5} \frac{x-5}{x^2-25}$; (b) $\displaystyle\lim_{x \to 16} \frac{\sqrt{x}-4}{x-16}$; (c) $\displaystyle\lim_{x \to 0} \frac{(3-x)^2-9}{x}$

(a) $\displaystyle\lim_{x \to 5} \frac{x-5}{x^2-25} = \lim_{x \to 5} \frac{x-5}{(x-5)(x+5)} = \lim_{x \to 5} \frac{1}{x+5} = \frac{1}{10}$

(b) $\displaystyle\lim_{x \to 16} \frac{\sqrt{x}-4}{x-16} = \lim_{x \to 16} \frac{\sqrt{x}-4}{\left(\sqrt{x}-4\right)\left(\sqrt{x}+4\right)} = \lim_{x \to 16} \frac{1}{\sqrt{x}+4} = \frac{\displaystyle\lim_{x \to 16} 1}{\sqrt{\displaystyle\lim_{x \to 16} x} + \displaystyle\lim_{x \to 16} 4} = \frac{1}{\sqrt{16}+4} = \frac{1}{8}$

(c) $\displaystyle\lim_{x \to 0} \frac{(3-x)^2-9}{x} = \lim_{x \to 0} \frac{9-6x+x^2-9}{x} = \lim_{x \to 0} \frac{-6x+x^2}{x} = \lim_{x \to 0} \frac{x(-6+x)}{x} = \lim_{x \to 0}(-6+x) = -6$

**45.6.** (a) Give formal definitions of $\displaystyle\lim_{x \to a^+} f(x) = L$ and $\displaystyle\lim_{x \to a^-} f(x) = M$. (b) Show that if $\displaystyle\lim_{x \to a^+} f(x) = L$ and $\displaystyle\lim_{x \to a^-} f(x) = L$, then $\displaystyle\lim_{x \to a} f(x) = L$.

(a) $\displaystyle\lim_{x \to a^+} f(x) = L$ means that given any $\varepsilon > 0$, a number $\delta > 0$ can be found so that if $0 < x - a < \delta$ then $|f(x) - L| < \varepsilon$. $\displaystyle\lim_{x \to a^-} f(x) = M$ means that given any $\varepsilon > 0$, a number $\delta > 0$ can be found so that if $0 < a - x < \delta$, then $|f(x) - M| < \varepsilon$.

(b) If $\displaystyle\lim_{x \to a^+} f(x) = L$ and $\displaystyle\lim_{x \to a^-} f(x) = L$, then given any $\varepsilon > 0$ a number $\delta_1 > 0$ can be found so that if $0 < x - a < \delta_1$, then $|f(x) - L| < \varepsilon$ and a number $\delta_2 > 0$ can be found so that if $0 < a - x < \delta_2$, then $|f(x) - L| < \varepsilon$. So given any $\varepsilon > 0$, choose $\delta$ to be the smaller of $\delta_1$ and $\delta_2$. Then if $0 < |x - a| < \delta$, both $0 < x - a < \delta_1$ and $0 < a - x < \delta_2$ will hold and $|f(x) - L| < \varepsilon$ as required.

**45.7.** Let $f(x) = \begin{cases} x-3 & \text{if } x < 2 \\ 6x & \text{if } x \geq 2 \end{cases}$. Find (a) $\displaystyle\lim_{x \to 2^+} f(x)$; (b) $\displaystyle\lim_{x \to 2^-} f(x)$; (c) $\displaystyle\lim_{x \to 2} f(x)$.

(a) $\displaystyle\lim_{x \to 2^+} f(x) = \lim_{x \to 2^+} 6x = 12$

(b) $\displaystyle\lim_{x \to 2^-} f(x) = \lim_{x \to 2^-}(x-3) = -1$

(c) Since $\displaystyle\lim_{x \to 2^+} f(x) \neq \lim_{x \to 2^-} f(x)$, $\displaystyle\lim_{x \to 2} f(x)$ does not exist.

**45.8.** Let $f(x) = \begin{cases} x^2-3 & \text{if } x < 2 \\ \frac{1}{2}x & \text{if } x \geq 2 \end{cases}$. Find (a) $\displaystyle\lim_{x \to 2^+} f(x)$; (b) $\displaystyle\lim_{x \to 2^-} f(x)$; (c) $\displaystyle\lim_{x \to 2} f(x)$.

(a) $\displaystyle\lim_{x \to 2^+} f(x) = \lim_{x \to 2^+} \frac{1}{2}x = 1$

(b) $\displaystyle\lim_{x \to 2^-} f(x) = \lim_{x \to 2^-}(x^2-3) = 2^2-3 = 1$

(c) Since $\displaystyle\lim_{x \to 2^+} f(x) = \lim_{x \to 2^-} f(x) = 1$, $\displaystyle\lim_{x \to 2} f(x) = 1$

**45.9.** Find (a) $\displaystyle\lim_{x \to 4} \sqrt{x}$; (b) $\displaystyle\lim_{x \to -4} \sqrt{x}$.

(a) $\displaystyle\lim_{x \to 4} \sqrt{x} = \sqrt{\lim_{x \to 4} x} = \sqrt{4} = 2$

(b) Since $\sqrt{x}$ is not a real number for any value of $x$ near $-4$, $\displaystyle\lim_{x \to -4} \sqrt{x}$ does not exist.

**45.10.** Let $f(x) = \dfrac{1}{x-2}$. Find (a) $\lim\limits_{x \to 2^+} f(x)$; (b) $\lim\limits_{x \to 2^-} f(x)$; (c) $\lim\limits_{x \to 2} f(x)$.

Consider the following table:

| $x$ | 1.5 | 1.9 | 1.99 | 1.999 | 2.5 | 2.1 | 2.01 | 2.001 |
|---|---|---|---|---|---|---|---|---|
| $\dfrac{1}{x-2}$ | $-2$ | $-10$ | $-100$ | $-1000$ | 2 | 10 | 100 | 1000 |

(a) From the table, it appears that $\lim\limits_{x \to 2^+} f(x)$ does not exist; however, it can be said that $\lim\limits_{x \to 2^+} f(x) = \infty$.

(b) From the table, it appears that $\lim\limits_{x \to 2^-} f(x)$ does not exist; however, it can be said that $\lim\limits_{x \to 2^-} f(x) = -\infty$.

(c) Since $\lim\limits_{x \to 2^+} f(x) \neq \lim\limits_{x \to 2^-} f(x)$, $\lim\limits_{x \to 2} f(x)$ does not exist.

**45.11.** Let $f(x) = \dfrac{x+3}{(x-2)^2}$. Find (a) $\lim\limits_{x \to 2^+} f(x)$; (b) $\lim\limits_{x \to 2^-} f(x)$; (c) $\lim\limits_{x \to 2} f(x)$.

Consider the following table:

| $x$ | 1.5 | 1.9 | 1.99 | 1.999 | 2.5 | 2.1 | 2.01 | 2.001 |
|---|---|---|---|---|---|---|---|---|
| $\dfrac{x+3}{(x-2)^2}$ | 18 | 490 | 49,900 | 4,999,000 | 22 | 510 | 50,100 | 5,001,000 |

(a) From the table, it appears that $\lim\limits_{x \to 2^+} f(x)$ does not exist; however, it can be said that $\lim\limits_{x \to 2^+} f(x) = \infty$.

(b) From the table, it appears that $\lim\limits_{x \to 2^-} f(x)$ does not exist; however, it can be said that $\lim\limits_{x \to 2^-} f(x) = \infty$.

(c) Since $\lim\limits_{x \to 2^+} f(x) = \lim\limits_{x \to 2^-} f(x) = \infty$, it can be said that $\lim\limits_{x \to 2} f(x) = \infty$.

**45.12.** It can be shown that $\lim\limits_{x \to \infty} \dfrac{1}{x^n} = 0$ and $\lim\limits_{x \to -\infty} \dfrac{1}{x^n} = 0$ for any positive value of $n$. Use these facts to find

(a) $\lim\limits_{x \to \infty} \dfrac{2x+2}{x+3}$; (b) $\lim\limits_{x \to -\infty} \dfrac{5}{x^2+6}$.

(a) $\lim\limits_{x \to \infty} \dfrac{2x+2}{x+3} = \lim\limits_{x \to \infty} \dfrac{2+2/x}{1+3/x} = \dfrac{\lim\limits_{x \to \infty} 2 + 2\lim\limits_{x \to \infty}(1/x)}{\lim\limits_{x \to \infty} 1 + 3\lim\limits_{x \to \infty}(1/x)} = \dfrac{2+2 \cdot 0}{1+3 \cdot 0} = 2$

(b) $\lim\limits_{x \to -\infty} \dfrac{5}{x^2+6} = \dfrac{\lim\limits_{x \to -\infty}(5/x^2)}{\lim\limits_{x \to -\infty} 1 + \lim\limits_{x \to -\infty}(6/x^2)} = \dfrac{5 \lim\limits_{x \to -\infty}(1/x^2)}{\lim\limits_{x \to -\infty} 1 + 6 \lim\limits_{x \to -\infty}(1/x^2)} = \dfrac{5 \cdot 0}{1+6 \cdot 0} = 0$

**45.13.** Discuss how a function may fail to be continuous for a particular value and give examples.

A function $f(x)$ is continuous for a value $a$ if $\lim\limits_{x \to a} f(x) = f(a)$. However:

1. $f(a)$ may be undefined. For example, consider $f(x) = \dfrac{1}{x^2}$ at $x = 0$, or $f(x) = \sqrt{x}$ at any negative value of $x$, or $f(x) = \dfrac{x^2-4}{x-2}$ at $x = 2$.

2. The limit $\lim\limits_{x \to a} f(x)$ may fail to exist. For example, consider $f(x) = \dfrac{1}{x}$ at $x = 0$ or $f(x) = \begin{cases} x^2 \text{ if } x < 2 \\ 3x \text{ if } x \geq 2 \end{cases}$ at $x = 2$.

3. The limit $\lim_{x \to a} f(x)$ may exist and $f(a)$ may be defined, but $\lim_{x \to a} f(x) \neq f(a)$. For example, consider

$$f(x) = \begin{cases} x^2 \text{ if } x \neq 2 \\ 3 \text{ if } x = 2 \end{cases} \text{ at } x = 2, \text{ where } \lim_{x \to 2} f(x) = 4, \text{ but } f(2) = 3.$$

**45.14.** Analyze the derivative of the function $f(x) = |x|$.

Find $\lim_{h \to 0} \dfrac{f(x + h) - f(x)}{h}$.

$$\lim_{h \to 0} \frac{f(x + h) - f(x)}{h} = \lim_{h \to 0} \frac{|x + h| - |x|}{h}$$

If $x > 0$, then for sufficiently small $h$, $x + h > 0$, hence

$$\lim_{h \to 0} \frac{f(x + h) - f(x)}{h} = \lim_{h \to 0} \frac{|x + h| - |x|}{h} = \lim_{h \to 0} \frac{x + h - x}{h} = \lim_{h \to 0} \frac{h}{h} = \lim_{h \to 0} 1 = 1$$

If $x < 0$, then for sufficiently small $h$, $x + h < 0$, hence

$$\lim_{h \to 0} \frac{f(x + h) - f(x)}{h} = \lim_{h \to 0} \frac{|x + h| - |x|}{h} = \lim_{h \to 0} \frac{-(x + h) - (-x)}{h} = \lim_{h \to 0} \frac{-h}{h} = \lim_{h \to 0} (-1) = -1$$

If $x = 0$, however,

$$\lim_{h \to 0} \frac{f(x + h) - f(x)}{h} = \lim_{h \to 0} \frac{|x + h| - |x|}{h} = \lim_{h \to 0} \frac{|h|}{h}$$

This limit does not exist, since

$$\lim_{h \to 0^+} \frac{|h|}{h} = \lim_{h \to 0^+} \frac{h}{h} = \lim_{h \to 0^+} 1 = 1 \text{ but}$$

$$\lim_{h \to 0^-} \frac{|h|}{h} = \lim_{h \to 0^-} \frac{-h}{h} = \lim_{h \to 0^-} (-1) = -1$$

Summarizing, if $f(x) = |x|$, then $f'(x) = \begin{cases} 1 \text{ if } x > 0 \\ -1 \text{ if } x < 0 \\ \text{undefined if } x = 0 \end{cases}$

**45.15.** Find the derivatives of the following functions:

(a) $f(x) = x^3$; (b) $f(x) = \sqrt{x}, x > 0$; (c) $f(x) = \dfrac{1}{x}, x \neq 0$.

(a) Find $\lim_{h \to 0} \dfrac{f(x + h) - f(x)}{h}$.

$$\lim_{h \to 0} \frac{f(x + h) - f(x)}{h} = \frac{(x + h)^3 - x^3}{h}$$

$$= \lim_{h \to 0} \frac{x^3 + 3x^2h + 3xh^2 + h^3 - x^3}{h}$$

$$= \lim_{h \to 0} \frac{3x^2h + 3xh^2 + h^3}{h}$$

$$= \lim_{h \to 0} \frac{h(3x^2 + 3xh + h^2)}{h}$$

$$= \lim_{h \to 0} (3x^2 + 3xh + h^2)$$

$$= 3x^2$$

(b) Find $\lim\limits_{h\to 0}\dfrac{f(x+h)-f(x)}{h}$.

$$\lim_{h\to 0}\frac{f(x+h)-f(x)}{h} = \lim_{h\to 0}\frac{\sqrt{x+h}-\sqrt{x}}{h}$$

$$= \lim_{h\to 0}\frac{\sqrt{x+h}-\sqrt{x}}{h}\frac{\sqrt{x+h}+\sqrt{x}}{\sqrt{x+h}+\sqrt{x}}$$

$$= \lim_{h\to 0}\frac{x+h-x}{h(\sqrt{x+h}+\sqrt{x})}$$

$$= \lim_{h\to 0}\frac{h}{h(\sqrt{x+h}+\sqrt{x})}$$

$$= \lim_{h\to 0}\frac{1}{\sqrt{x+h}+\sqrt{x}}$$

$$= \frac{1}{2\sqrt{x}}$$

This is defined as long as $x > 0$.

(c) Find $\lim\limits_{h\to 0}\dfrac{f(x+h)-f(x)}{h}$.

$$\lim_{h\to 0}\frac{f(x+h)-f(x)}{h} = \lim_{h\to 0}\frac{1/(x+h)-1/x}{h}$$

$$= \lim_{h\to 0}\frac{x-(x+h)}{hx(x+h)}$$

$$= \lim_{h\to 0}\frac{-h}{hx(x+h)}$$

$$= \lim_{h\to 0}\frac{-1}{x(x+h)}$$

$$= \frac{-1}{x^2}$$

**45.16.** Find the equation of the line tangent to the graph of

(a) $f(x) = x^3$, at (2,8); (b) $f(x) = \sqrt{x}$, at (4,2); (c) $f(x) = \frac{1}{x}$, at $(-1,-1)$

(a) Using the derivative found in the previous problem, we find that the slope of the tangent line at (2,8) is $f'(2) = 3\cdot 2^2 = 12$. Using the point-slope form of the equation of a line, we find that the equation of the line through (2,8) with slope 12 is

$$y - 8 = 12(x - 2)$$
$$y - 8 = 12x - 24$$
$$y = 12x - 16$$

(b) Using the derivative found in the previous problem, we find that the slope of the tangent line at (4,2) is $f'(4) = \dfrac{1}{2\sqrt{4}} = \dfrac{1}{4}$. Using the point-slope form of the equation of a line, we find that the equation of the line through (4,2) with slope $\frac{1}{4}$ is

$$y - 2 = \frac{1}{4}(x - 4)$$

$$y - 2 = \frac{1}{4}x - 1$$

$$y = \frac{1}{4}x + 1$$

(c) Using the derivative found in the previous problem, we find that the slope of the tangent line at $(-1, -1)$ is $f'(-1) = \dfrac{-1}{(-1)^2} = -1$. Using the point-slope form of the equation of a line, we find that the equation of the line through $(-1, -1)$ with slope $-1$ is

$$y - (-1) = (-1)[x - (-1)]$$
$$y + 1 = -x - 1$$
$$y = -x - 2$$

## SUPPLEMENTARY PROBLEMS

**45.17.** Find the following limits algebraically:

(a) $\lim_{x \to -3}(5x + 1)$; (b) $\lim_{x \to 2}(2x^2 - 8x + 7)$; (c) $\lim_{x \to -3}\dfrac{x^2 - 7}{x + 5}$

*Ans.*   (a) $-14$; (b) $-1$; (c) $1$

**45.18.** Find the following limits algebraically:

(a) $\lim_{x \to 2}\dfrac{x^2 - 4}{2x - 4}$; (b) $\lim_{x \to 100}\dfrac{\sqrt{x} - 10}{x - 100}$; (c) $\lim_{x \to 2}\dfrac{2x + 4}{x^2 - 4}$ (d) $\lim_{x \to 0}\dfrac{(x - 3)^3 + 27}{x}$

*Ans.*   (a) $2$; (b) $\frac{1}{20}$; (c) does not exist; (d) $27$

**45.19.** Use the formal definition of limit to prove:

If $\lim_{x \to a} f(x) = L$ and $\lim_{x \to a} g(x) = M$, then $\lim_{x \to a}[f(x) - g(x)] = L - M$.

**45.20.** Let $f(x) = \begin{cases} x^3 \text{ if } x < 3 \\ 3x \text{ if } x \geq 3 \end{cases}$. Find (a) $\lim_{x \to 3^+} f(x)$; (b) $\lim_{x \to 3^-} f(x)$; (c) $\lim_{x \to 3} f(x)$

*Ans.*   (a) $9$; (b) $27$; (c) does not exist

**45.21.** Let $f(x) = \begin{cases} x^3 - 18 & \text{if } x < 3 \\ 3x & \text{if } x \geq 3 \end{cases}$. Find (a) $\lim_{x \to 3^+} f(x)$; (b) $\lim_{x \to 3^-} f(x)$; (c) $\lim_{x \to 3} f(x)$

*Ans.*   (a) $9$; (b) $9$; (c) $9$

**45.22.** Find the following limits:

(a) $\lim_{x \to 4}\sqrt{2 - x}$; (b) $\lim_{x \to -4}\sqrt{5 - x}$

*Ans.*   (a) does not exist; (b) $3$

**45.22.** Let $f(x) = \dfrac{x + 1}{x - 3}$. Find (a) $\lim_{x \to 3^+} f(x)$; (b) $\lim_{x \to 3^-} f(x)$; (c) $\lim_{x \to 3} f(x)$.

*Ans.*   (a) $\infty$; (b) $-\infty$; (c) does not exist

**45.23.** Let $f(x) = \dfrac{1 - x}{(x - 3)^2}$. Find (a) $\lim_{x \to 3^+} f(x)$; (b) $\lim_{x \to 3^-} f(x)$; (c) $\lim_{x \to 3} f(x)$.

*Ans.*   (a) $-\infty$; (b) $-\infty$; (c) $-\infty$

**45.24.** Find the following limits:

(a) $\lim\limits_{x\to\infty}\dfrac{100x}{5x^2-1}$; (b) $\lim\limits_{x\to-\infty}\dfrac{100x^2}{5x^2-1}$

*Ans.*   (a) 0; (b) 20

**45.25.** Show that if $f(x) = mx + b$, where $m$ and $b$ are constants, then $f'(x) = m$.

**45.26.** Find the derivatives of the following functions:

(a) $f(x) = x^4$; (b) $f(x) = \sqrt{x-4}, x > 4$; (c) $f(x) = \dfrac{1}{x^2}$

*Ans.*   (a) $f'(x) = 4x^3$; (b) $f'(x) = \dfrac{1}{2\sqrt{x-4}}, x > 4$; (c) $f'(x) = \dfrac{-2}{x^3}$

**45.27.** Find the equation of the line tangent to the graph of

(a) $f(x) = x^4$, at $(1,1)$; (b) $f(x) = \sqrt{x-4}$, at $(5,1)$; (c) $f(x) = \dfrac{1}{x^2}$, at $\left(\dfrac{1}{2},4\right)$

*Ans.*   (a) $y = 4x - 3$; (b) $y = \dfrac{1}{2}x - \dfrac{3}{2}$; (c) $y = -16x + 12$

**45.28.** Show that if $f(x) = x^n$, where $n$ is any positive integer, then $f'(x) = nx^{n-1}$.

(Hint: Use the binomial theorem.)

# Index